New Insights into Cell and Membrane Transport Processes

NEW HORIZONS IN THERAPEUTICS

Smith Kline & French Laboratories Research Symposia Series

Series Editors: George Poste and Stanley T. Crooke
Smith Kline & French Laboratories, Philadelphia, Pennsylvania

---

DOPAMINE RECEPTOR AGONISTS
Edited by George Poste and Stanley T. Crooke

MECHANISMS OF RECEPTOR REGULATION
Edited by George Poste and Stanley T. Crooke

NEW INSIGHTS INTO CELL AND MEMBRANE TRANSPORT PROCESSES
Edited by George Poste and Stanley T. Crooke

New Insights into Cell and Membrane Transport Processes

Edited by

GEORGE POSTE and
STANLEY T. CROOKE

Smith Kline & French Laboratories
Philadelphia, Pennsylvania

PLENUM PRESS • NEW YORK AND LONDON

Library of Congress Cataloging in Publication Data

New insights into cell and membrane transport processes.

(New horizons in therapeutics)

Proceedings of the Third Smith, Kline & French Research Symposium on New Horizons in Therapeutics, held in Philadelphia in 1985.

Includes bibliographies and index.

1. Biological transport—Congresses. 2. Cell membranes—Congresses. I. Poste, George. II. Crooke, Stanley T. III. Smith, Kline, and French Research Symposium on New Horizons in Therapeutics (3rd: 1985: Philadelphia, Pa.) IV. Series. [DNLM: 1. Biological Transport—congresses. 2. Cell Membrane Permeability—congresses. QH 509 N532 1985]

QH509.N48 1986 574.87′5 86-8175

ISBN 0-306-32183-6

A Division of Plenum Publishing Corporation
233 Spring Street, New York, N.Y. 10013

Printed in the United States of America

Contributors

John P. Arcoleo, Division of Environmental Sciences, Department of Human Genetics, and Cancer Center/Institute of Cancer Research, Columbia University, New York, New York 10032

Paula Barrett, Departments of Cell Biology and Internal Medicine, Yale University School of Medicine, New Haven, Connecticut 06510

Michael J. Berridge, AFRC Unit, Department of Zoology, University of Cambridge, Cambridge CB2 3EJ, England

Günter Blobel, Laboratory of Cell Biology, The Rockefeller University, New York, New York 10021

Lewis Cantley, Department of Biochemistry and Molecular Biology, Harvard University, Cambridge, Massachusetts 02138

William A. Catterall, Department of Pharmacology, University of Washington, Seattle, Washington 98195

Kathleen E. Coll, Department of Biochemistry and Biophysics, University of Pennsylvania School of Medicine, Philadelphia, Pennsylvania 19104

John Cuppoletti, CURE, Veterans Administration Wadsworth Division, Los Angeles, California 90073; and Department of Medicine, UCLA School of Medicine, Los Angeles, California 90024

Leigh English, Department of Biochemistry and Molecular Biology, Harvard University, Cambridge, Massachusetts 02138

A. P. Fox, Department of Physiology, Yale University School of Medicine, New Haven, Connecticut 06510

P. Gengo, Department of Biochemical Pharmacology, School of Pharmacy, State University of New York at Buffalo, Buffalo, New York 14260

Hans J. Geuze, Laboratory of Cell Biology, Medical Faculty, University of Utrecht, Utrecht, The Netherlands

Reid Gilmore, Laboratory of Cell Biology, The Rockefeller University, 1230 York Avenue, New York, New York 10021

Robert D. Gunther, CURE, Veterans Administration Wadsworth Division, Los Angeles, California 90073; and Department of Medicine, UCLA School of Medicine, Los Angeles, California 90024

John W. Hanrahan, Department of Physiology, Yale Medical School, New Haven, Connecticut 06510

P. Harikumar, Roche Institute of Molecular Biology, Roche Research Center, Nutley, New Jersey 07110. *Present address:* B and FT Division, Fiply, B.A.R.C., Trombay, Bombay 400085, India

P. Hess, Department of Physiology, Yale University School of Medicine, New Haven, Connecticut 06510

Joseph F. Hoffman, Department of Physiology, Yale University School of Medicine, New Haven, Connecticut 06510

Colin R. Hopkins, Department of Medical Cell Biology, University of Liverpool, Liverpool L69 3BX, England

Gerard M. Housey, Division of Environmental Sciences, Department of Human Genetics, and Cancer Center/Institute of Cancer Research, Columbia University, New York, New York 10032

W.-L. Wendy Hsiao, Division of Environmental Sciences, Department of Human Genetics, and Cancer Center/Institute of Cancer Research, Columbia University, New York, New York 10032

Suresh K. Joseph, Department of Biochemistry and Biophysics, University of Pennsylvania School of Medicine, Philadelphia, Pennsylvania 19104

A. Joslyn, Department of Biochemical Pharmacology, School of Pharmacy, State University of New York at Buffalo, Buffalo, New York 14260

H. R. Kaback, Roche Institute of Molecular Biology, Roche Research Center, Nutley, New Jersey 07110

Jonathan Kaunitz, CURE, Veterans Administration Wadsworth Division, Los Angeles, California 90073; and Department of Medicine, UCLA School of Medicine, Los Angeles, California 90024

Itaru Kojima, Departments of Cell Biology and Internal Medicine, Yale University School of Medicine, New Haven, Connecticut 06510

Ron R. Kopito, Whitehead Institute for Biomedical Research, Cambridge, Massachusetts 02142

J. B. Lansman, Department of Physiology, Yale University School of Medicine, New Haven, Connecticut 06510

Simon A. Lewis, Department of Physiology, Yale Medical School, New Haven, Connecticut 06510

Rob M. Liskamp, Division of Environmental Sciences, Department of Human Genetics, and Cancer Center/Institute of Cancer Research, Columbia University, New York, New York 10032

Harvey F. Lodish, Department of Biology, Massachusetts Institute of Technology, and Whitehead Institute for Biomedical Research, Cambridge, Massachusetts 02139

Ira Mellman, Department of Cell Biology, Yale University of School of Medicine, New Haven, Connecticut 06510

John Mendlein, CURE, Veterans Administration, Wadsworth Division, Los Angeles, California 90073; and Department of Medicine, UCLA School of Medicine, Los Angeles, California 90024

Paul H. Naccache, Departments of Pathology and Physiology, University of Connecticut Health Center, Farmington, Connecticut 06032. *Present address:* Immunology and Rheumatism, Centre Hospitalier de l'Université Laual, Ste.Foy, Quebec G1V 4G2, Canada

B. Nilius, Department of Physiology, Yale University School of Medicine, New Haven, Connecticut 06510

M. C. Nowycky, Department of Physiology, Yale University School of Medicine, New Haven, Connecticut 06510

Catherine O'Brian, Division of Environmental Sciences, Department of Human Genetics, and Cancer Center/Institute of Cancer Research, Columbia University, New York, New York 10032

George K. Ojakian, Department of Anatomy and Cell Biology, State University of New York, Downstate Medical Center, Brooklyn, New York 11203

Lawrence G. Palmer, Department of Physiology, Cornell University Medical College, New York, New York 10021

Mark Prentki, Department of Biochemistry and Biophysics, University of Pennsylvania School of Medicine, Philadelphia, Pennsylvania 19104

Edwin C. Rabon, CURE, Veterans Administration, Wadsworth Division, Los Angeles, California 90073; and Department of Medicine, UCLA School of Medicine, Los Angeles, California 90024

D. Rampe, Department of Biochemical Pharmacology, School of Pharmacy, State University of New York at Buffalo, Buffalo, New York 14260

Howard Rasmussen, Departments of Cell Biology and Internal Medicine, Yale University School of Medicine, New Haven, Connecticut 06510

John P. Reeves, Roche Institute of Molecular Biology, Roche Research Center, Nutley, New Jersey 07110

George Sachs, CURE, Veterans Administration, Wadsworth Division, Los Angeles, California 90073; and Department of Medicine, UCLA School of Medicine, Los Angeles, California 90024

Alan L. Schwartz, Division of Pediatric Haematology/Oncology, Children's Hospital, Dana Farber Institute and Harvard Medical School, Boston, Massachusetts 02115

Ramadan I. Sha'afi, Departments of Pathology and Physiology, University of Connecticut Health Center, Farmington, Connecticut 06032

A. Skattebol, Department of Biochemical Pharmacology, School of Pharmacy, State University of New York at Buffalo, Buffalo, New York 14260

Jan W. Slot, Laboratory of Cell Biology, Medical Faculty, University of Utrecht, Utrecht, The Netherlands

Ger J. Strous, Laboratory of Cell Biology, Medical Faculty, University of Utrecht, Utrecht, The Netherlands

Andrew P. Thomas, Department of Biochemistry and Biophysics, University of Pennsylvania School of Medicine, Philadelphia, Pennsylvania 19104

D. J. Triggle, Department of Biochemical Pharmacology, School of Pharmacy, State University of New York at Buffalo, Buffalo, New York 14260

Richard W. Tsien, Department of Physiology, Yale University School of Medicine, New Haven, Connecticut 06510

Simon Van Heyningen, Department of Biochemistry, University of Edinburgh, Edinburgh EH8 9XD, Scotland

Arthur Verhoeven, Department of Biochemistry and Biophysics, University of Pennsylvania School of Medicine, Philadelphia, Pennsylvania 19104

Bjorn Wallmark, AB Hassle, Molndal, Sweden

Peter Walter, Laboratory of Cell Biology, The Rockefeller University, New York, New York 10021

I. Bernard Weinstein, Division of Environmental Sciences, Department of Human Genetics, and Cancer Center/Institute of Cancer Research, Columbia University, New York, New York 10032

Benjamin White, Department of Biochemistry and Molecular Biology, Harvard University, Cambridge, Massachusetts 02138

John R. Williamson, Department of Biochemistry and Biophysics, University of Pennsylvania School of Medicine, Philadelphia, Pennsylvania 19104

Jos E. Zijderhand-Bleekemolen, Laboratory of Cell Biology, Medical Faculty, University of Utrecht, Utrecht, The Netherlands

Contents

I. Na^+, K^+, and H^+ TRANSPORT

II. PROTON PUMPS AND ELECTROCHEMICAL GRADIENTS

Chapter 3

The Erythrocyre Anion-Exchange Protein: Primary Structure Deduced from the cDNA Sequence and a Model for Its Arrangement within the Plasma Membrane

Ron R. Kopito and Harvey F. Lodish

Chapter 4

The Lysosomal Proton Pump

P. Harikumar and John P. Reeves

Chapter 5

Ion Pumps, Ion Pathways, Ion Sites

George Sachs, John Cuppoletti, Robert D. Gunther, Jonathan Kaunitz, John Mendlein, Edwin C. Rabon, and Bjorn Wallmark

III. CALCIUM CHANNELS AND THE CALCIUM PUMP

Chapter 6

Shifts between Modes of Calcium Channel Gating as a Basis for Pharmacologic Modulation of Calcium Influx in Cardiac, Neuronal, and Smooth-Muscle-Derived Cells

A. P. Fox, P. Hess, J. B. Lansman, B. Nilius, M. C. Nowycky, and R. W. Tsien

Chapter 7

Chemical Pharmacology of Ca^{2+} Channel Ligands

D. J. Triggle, A. Skattelbol, D. Rampe, A. Joslyn, and P. Gengo

Chapter 8

Information Flow in the Calcium Messenger System

Howard Rasmussen, Itaru Kojima, and Paula Barrett

Chapter 9

Neutrophil Activation, Polyphosphoinositide Hydrolysis, and the Guanine Nucleotide Regulatory Proteins

Paul H. Naccache and Ramadan I. Sha'afi

IV. INOSITOL LIPIDS AND THE CALCIUM-MEDIATED CELLULAR RESPONSES

Chapter 10

Agonist-Dependent Phosphoinositide Metabolism: A Bifurcating Signal Pathway

Michael J. Berridge

Chapter 11

Hormone-Induced Inositol Lipid Breakdown and Calcium-Mediated Cellular Responses in Liver

John R. Williamson, Suresh K. Joseph, Kathleen E. Coll, Andrew P. Thomas, Arthur Verhoeven, and Marc Prentki

Chapter 12

Comparison of the Na^+ Pump and the Ouabain-Resistant K^+ Transport System with Other Metal Ion Transport ATPases

Leigh English, Benjamin White, and Lewis Cantley

Chapter 13

Current Concepts of Tumor Promotion by Phorbol Esters and Related Compounds

Catherine A. O'Brian, Rob M. Liskamp, John P. Arcoleo, W.-L. Wendy Hsiao, Gerard M. Housey, and I. Bernard Weinstein

V. CELL POLARITY AND MEMBRANE TRANSPORT PROCESSES

Chapter 17

The Epithelial Sodium Channel

Lawrence G. Palmer

VI. ENDOCYTOSIS AS A CELL TRANSPORT PATHWAY

Chapter 18

Uptake and Intracellular Processing of Cell Surface Receptors: Current Concepts and Prospects

Colin R. Hopkins

Chapter 21

Transport of Protein Toxins across Cell Membranes

Simon Van Heyningen

I

Na^+, K^+, AND H^+ TRANSPORT

1

Molecular Properties of Voltage-Sensitive Sodium Channels

WILLIAM A. CATTERALL

1. Introduction

Electrical excitability is one of the most important and characteristic properties of neurons. Most vertebrate cells, including neurons, maintain large ionic gradients across their surface membranes such that the intracellular fluid contains a high concentration of potassium ions and low concentrations of sodium ions and calcium ions relative to the extracellular fluid. These large ion gradients are maintained by the action of energy-dependent ion pumps specific for Na^+ and K^+, or for Ca^{2+}. In addition, essentially all vertebrate cells maintain an internally negative membrane potential of the order of -60 mV, since their surface membranes are specifically permeable to K^+, and this allows K^+ to leak out of cells faster than Na^+ and Ca^{2+} can leak in. Nerve cells are electrically excitable because of the presence, in their surface membranes, of voltage-sensitive ion channels that are selective for Na^+, K^+, or Ca^{2+}. One class of Na^+ channels and many classes of Ca^{2+} and K^+ channels have been described in neurons. The channels open and close as a function of membrane voltage, allowing rapid movement of the appropriate ions down their concentration gradient so that ionic current passes into or out of the cell, depolarizing or hyperpolarizing the membrane.

2. Physiological Properties of Sodium Channels

The ionic mechanisms underlying electrical excitability have been defined by the voltage-clamp method (Hodgkin and Huxley, 1952). In this

WILLIAM A. CATTERALL • Department of Pharmacology, University of Washington, Seattle, Washington 98195.

approach, the voltage across the excitable membrane is controlled by means of a feedback amplifier circuit, and the ionic currents moving across the membrane in response to step changes in the membrane potential imposed by the experimenter are measured. The voltage-clamp technique has been used to show that the initial rapid depolarization during an action potential in nerve axons results from rapid voltage-dependent increases in membrane permeability to sodium ions. Many different lines of evidence indicate that a selective transmembrane sodium channel is responsible for the rapid increase in sodium permeability during the action potential.

Selective ion permeation is mediated by a hydrophilic pore containing a sodium-selective ion coordination site designated the ion selectivity filter (Hille, 1972). Ion conductance through the sodium channel is regulated or "gated" by two experimentally separable processes: (1) activation, which controls the rate and voltage dependence of the opening of the sodium channel after depolarization, and (2) inactivation, which controls the rate and voltage dependence of the subsequent closing of the sodium channel during a maintained depolarization (Hodgkin and Huxley, 1952). Estimates derived from analysis of membrane current noise or recordings of individual sodium channel currents of the rate of sodium movement through an activated sodium channel range from 8 to 25 pS, corresponding to more than 10^7 ions per second per channel at physiological temperature and Na^+ concentration (Conti *et al.*, 1976; Sigworth and Neher, 1980). These rates approach those for diffusion through free solution and imply that the residence time of Na^+ ions in the channel is very short and that their interactions with the channel are weak.

Voltage-clamp analysis has elucidated the three essential functional properties of sodium channels: voltage-dependent activation, voltage-dependent inactivation, and selective ion transport. However, an understanding of the molecular basis of neuronal excitability requires the determination of the density and distribution of voltage-sensitive ion channels in neurons, identification of the membrane macromolecules that comprise the ionic channels, solubilization and purification of these channel components, and correlation of their structural features with the known functional properties of sodium channels.

3. *Neurotoxins as Molecular Probes of Sodium Channels*

Neurotoxins that bind with high affinity and specificity to voltage-sensitive sodium channels and modify their properties have provided the essential tools for the identification and purification of sodium channels.

Table I. Neurotoxin Receptor Sites on the Sodium Channel

Site	Neurotoxins	Physiological effect
1	Tetrodotoxin	Inhibit ion transport
2	Veratridine Batrachotoxin Grayanotoxin Aconitine	Cause persistent activation
3	North African α scorpion toxins Sea anemone toxins	Slow activation
4	American β scorpion toxins	Enhance activation
5	*Ptychodiscus brevis* toxins	Repetitive firing

Several different groups of neurotoxins that act at five different neurotoxin receptor sites on the sodium channel have been useful in these studies (Table I; reviewed by Catterall, 1980, 1984).

Neurotoxin receptor site 1 binds the water-soluble heterocyclic guanidines tetrodotoxin and saxitoxin. These toxins inhibit sodium channel ion transport by binding to a common receptor site that is thought to be located near the extracellular opening of the ion-conducting pore of the sodium channel (Ritchie and Rogart, 1977).

Neurotoxin receptor site 2 binds several lipid-soluble toxins including grayanotoxin and the alkaloids veratridine, aconitine, and batrachotoxin. The competitive interactions of these four toxins at neurotoxin receptor site 2 have been confirmed by direct measurements of specific binding of [^{3}H]-labeled batrachotoxinin A 20α-benzoate to sodium channels. These toxins cause persistent activation of sodium channels at the resting membrane potential by blocking sodium channel inactivation and shifting the voltage dependence of the channel activation to more negative membrane potentials. Therefore, neurotoxin receptor site 2 is likely to be localized on a region of the sodium channel involved in voltage-dependent activation and inactivation.

Neurotoxin receptor site 3 binds polypeptide toxins purified from North African scorpion venoms or sea anemone nematocysts. These toxins slow or block sodium channel inactivation. They also enhance the persistent activation of sodium channels by the lipid-soluble toxins acting at neurotoxin receptor site 2. The affinity for binding of [^{125}I]-labeled derivatives of the polypeptide toxins to neurotoxin receptor site 3 is reduced by depolarization. The voltage dependence of scorpion toxin binding is correlated with the voltage dependence of sodium channel activation. These data indicate that neurotoxin receptor site 3 is located on the part of the sodium channel that undergoes a conformational change during voltage-dependent channel activation, leading to reduced affinity for scorpion toxin. Therefore, scorpion toxin and sea anemone toxin bind to voltage-sensing or gating structures of sodium channels.

Neurotoxin receptor site 4 binds a new class of scorpion toxins that has also proved valuable in studies of sodium channels. The venom of the American scorpion *Centruroides sculpturatus* and pure toxins from several American scorpions modify sodium channel activation rather than inactivation. These toxins bind to a new receptor site on the sodium channel (Jover *et al.*, 1980) and have been designated β scorpion toxins.

These neurotoxins provide specific high-affinity probes for distinct regions of the sodium channel structure. They have been used to detect and localize sodium channels in neuronal cells as well as to identify and purify the protein components of sodium channels that bind these toxins and to analyze their structural and functional properties.

4. *Identification of Protein Components of Sodium Channels in Neurons*

Measurements of the distribution and density of sodium channels indicate that, with the exception of the very small amount of specialized membrane at the node of Ranvier, sodium channels are a minor component of excitable membranes. These results emphasize the need for highly specific probes to identify the macromolecules that comprise the sodium channel. The neurotoxins that bind to sodium channels with high affinity and specificity have provided the tools needed in such experiments. Direct chemical identification of sodium channel components *in situ* was first achieved by specific covalent labeling of neurotoxin receptor site 3 with a photoreactive azidonitrobenzoyl derivative of the α scorpion toxin from *Leiurus quinquestriatus*. The photoreactive toxin derivative is allowed to bind specifically to sodium channels in the dark. Irradiation with ultraviolet light then chemically activates the arylazide group, which covalently reacts with the scorpion toxin receptor site on the sodium channel. Analysis of covalently labeled synaptosomes by polyacrylamide gel electrophoresis under denaturing conditions in sodium dodecyl sulfate (SDS) to separate synaptosomal proteins by size reveals specific covalent labeling of two polypeptides that were subsequently designated the α and β_1 subunits of the sodium channel (Beneski and Catterall, 1980). These proteins, as assessed by polyacrylamide gel electrophoresis in SDS, have molecular sizes of 270,000 and 36,000 daltons, respectively.

The β scorpion toxins derived from American scorpion venoms have been used to label neurotoxin receptor site 4 on the sodium channel (Barhanin *et al.*, 1983). Toxin from *Tityus serrulatus* was covalently attached to its receptor site by cross linking with disuccinimidyl suberate. A single polypeptide of 270,000 daltons was labeled in rat brain synaptosmes. In contrast, photoreactive derivatives of the β scorpion toxin from *Centruroides suffusus* label both the α and β_1 subunits (Darbon *et al.*, 1983). Thus, the receptor site for the β scorpion toxins is located on or near the α and β_1 subunits of the sodium channel, as previously found for the α scorpion toxins acting at neurotoxin receptor site 3.

5. *Molecular Size of the Sodium Channel*

The first indications of the molecular size of the neuronal sodium channel *in situ* were derived from radiation inactivation studies (Levinson and Ellory, 1973). In these experiments, membrane preparations from pig brain were irradiated with X rays, and the decrease in the number of functional tetrodotoxin-binding sites was measured as a function of radiation dose. The size of the membrane target can be determined from these data, since larger targets are more likely to be hit and are therefore inactivated at a lower radiation dose. Applying target theory, Levinson and Ellory concluded that a structure of 230,000 daltons was required for toxin binding. These experiments have recently been repeated by Barhanin *et al.* (1983), who compared the target size of the sodium channel assessed by either tetrodotoxin binding or by *Tityus serrulatus* toxin binding. In each case, the target size was approximately 270,000 daltons, in reasonable agreement with the earlier work. This size estimate might correspond to the molecular size of the entire sodium channel or to that of a protein subunit that is essential for binding these neurotoxins.

The molecular size of the intact sodium channel protein has been measured by hydrodynamic studies of the detergent-solubilized channed. The saxitoxin- and tetrodoxin-binding component of sodium channels can be solubilized with retention of high affinity and specificity of toxin binding by treatment with nonionic detergents (Benzer and Raftery, 1973; Henderson and Wang, 1973). In contrast to the ease of solubilization of the sodium channel with retention of saxitoxin- and tetrodotoxin-binding activity at neurotoxin receptor site 1, both neurotoxin receptor site 2 and neurotoxin receptor site 3 (Catterall *et al.*, 1979) lose their high-affinity neurotoxin binding activity on solubilization. The molecular size of the solubilized sodium channel from rat brain has been estimated by hydrodynamic studies to be 601,000 daltons (Hartshorne *et al.*, 1980). Since the detergent–channel complex contains 0.9 g of Triton X-100 and phosphatidylcholine per gram of protein, the size of the sodium channel protein solubilized from rat brain is 316,000 daltons. This represents the size of the entire sodium channel as solubilized in detergents and corresponds to a complex of three nonidentical protein subunits as described below. As estimated by this method, the sodium channel from rat skeletal muscle has a similar size (Barchi and Murphy, 1981).

6. *Protein Subunits of the Purified Channel*

The ability to solubilize the sodium channel in a well-defined monomeric form with retention of binding activity for saxitoxin and tetrodotoxin has allowed purification by a sequence of conventional protein separation procedures. The current purification schemes involve anion-exchange chromatography on DEAE-Sephadex, adsorption chromatography on hydroxylapatite gel, affinity chromatography on wheat germ agglutinin covalently attached to Sepharose 4B, and velocity sedimentation through sucrose gradients (Agnew *et al.*, 1980; Hartshorne and Catterall, 1981, 1984; Barchi, 1983). The purified sodium channel preparation from rat brain binds 0.9 mole of saxitoxin per mole of sodium channel of 316,000 daltons. If the sodium channel binds only one saxitoxin molecule, these data indicate that at least 90% of the protein in the purified preparation must be associated with the sodium channel.

The protein subunits of sodium channel preparations for these different tissues have been analyzed by denaturation in sodium dodecyl sulfate (SDS) and electrophoresis in polyacrylamide gels. The sodium channel from rat brain consists of three polypeptide subunits as illustrated in Fig. 1: α of 260,000 daltons, β_1 of 36,000 daltons, and β_2 of 33,000 daltons, as estimated by several techniques (Hartshorne *et al.*, 1982; Messner and Catterall, 1985). The α and β_2 subunits are attached by disulfide bonds, whereas the β_1 subunit is noncovalently associated with the complex. These subunits are present in a 1:1:1 molar ration in the purified protein.

Sodium channels are specifically adsorbed and eluted from columns of wheat germ agglutinin attached to Sepharose, indicating that they are glycoproteins containing complex carbohydrate chains with N-acetylglucosamine and sialic acid (Hartshorne and Catterall, 1981; Cohen and Barchi, 1981). Each of the subunits of the sodium channel from rat brain binds [^{125}I]-labeled wheat germ agglutinin and thus contains complex carbohydrate chains (Messner and Catterall, 1985). Removal of this carbohydrate by treatment of the purified subunits with neuraminidase plus endoglycosidase F or by chemical deglycosylation gives carbohydrate-free polypeptides of 220,000 daltons, 23,000 daltons, and 21,000 daltons

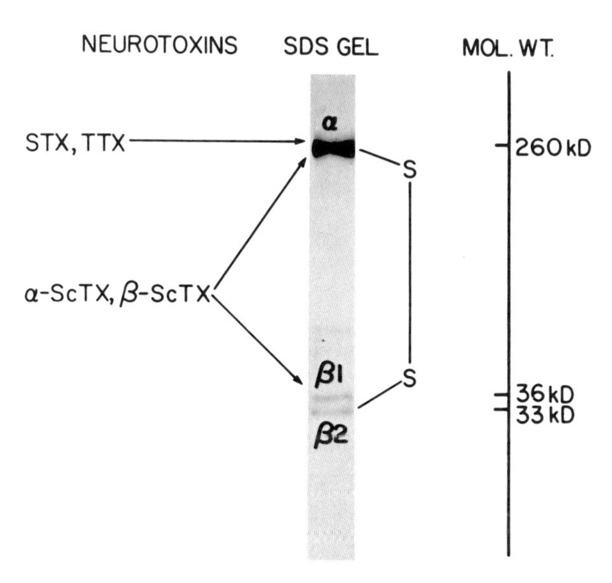

Figure 1. Properties of the subunits of the sodium channel from rat brain. Sodium channels purified from rat brain as described by Hartshorne and Catterall (1984) were denatured, and their component subunits analyzed by SDS gel electrophoresis. The apparent molecular weights of the α, β_1, and β_2 subunits are indicated along with the disulfide linkage of α and β_2 and the sites of covalent labeling with neurotoxin derivatives.

for the α, β_1, and β_2 subunits, respectively (Table II). Thus, the native α subunit is approximately 15% carbohydrate by weight, whereas the two β subunits are more than 30% carbohydrate.

Sodium channels purified from eel electroplax (Agnew *et al.*, 1980), rat brain (Beneski and Catterall, 1980; Hartshorne and Catterall, 1981, 1984), rat skeletal muscle (Casadei *et al.*, 1984), and chicken heart (Lombet and Lazdunski, 1984) all contain a large glycoprotein of 260,000 dal-

Table II. Subunit Structure of the Sodium Channel from Rat Brain

Component	Native molecular weight	Molecular weight after deglycosylation	Stoichiometry
Native complex	320,000		
α subunit	260,000	220,000	1.0
β_1 subunit	36,000	23,000	1.0
β_2 subunit	33,000	21,000	1.0

Table III. Subunit Composition of Sodium Channel

Tissue	Species	Subunit molecular weight	Reference
Brain	Rat	260,000	Hartshorne *et al.* (1982)
		39,000	
		37,000	
Skeletal	Rat	260,000	Barchi (1983)
		45,000	
		38,000	
Electroplax	Eel	260,000	Miller *et al.* (1983)
Heart	Chick	260,000	Lombet and Lazdunski (1984)

tons. For eel electroplax and chicken heart, this is the only polypeptide present in the purified preparations (Table III). Additional smaller polypeptides of 45,000 daltons and 38,000 daltons are present in purified preparations of the sodium channel from rat skeletal muscle. The finding that purified sodium channels from electroplax and chicken heart are competent to bind tetrodotoxin and saxitoxin and contain only a single polypeptide indicates that neurotoxin receptor site 1 must reside on the subunit of 260,000 daltons (Fig. 1). Covalent labeling of this polypeptide in purified preparations of electroplax sodium channel provides further support for this conclusion (Lombet *et al.*, 1983). It is likely that this large glycoprotein, which is common to sodium channels is these four tissues, is the principle functional component of the channel.

7. *Reconstitution of Sodium Channel Function from Purified Components*

The purified sodium channel preparations bind [^{3}H]-saxitoxin and tetrodotoxin with the same affinity as the native sodium channel and therefore contain neurotoxin receptor site 1 of the sodium channel in an active form. The purified channel also contains the α and β_1 subunits that were identified as components of neurotoxin receptor site 3 by photoaffinity labeling with scorpion toxin, although after solubilization binding activity for scorpion toxin is lost. However, purified channels do not have binding activity for neurotoxins at receptor site 2 and cannot transport sodium in the detergent-solubilized state. Reconstitution of these sodium channel functions from purified components is the only rigorous proof that the proteins identified and purified on the basis of their neurotoxin-binding activity are indeed sufficient to form a functional voltage-sensitive ion channel. In addition, successful reconstitution will provide a valuable experimental preparation for biochemical analysis of the structure and function of sodium channels.

Sodium channel ion transport was first successfully reconstituted from sodium channels substantially purified from rat brain and skeletal muscle and eel electroplax (Weigele and Barchi, 1982; Talvenheimo *et al.*, 1982). We have now applied these methods to essentially homogeneous preparations of sodium channels from rat brain (Tamkun *et al.*, 1984). Purified sodium channels in Triton X-100 solution are supplemented with phosphatidylcholine dispersed in Triton X-100, and the detergent is removed by adsorption to polystyrene beads. As the detergent is removed, phosphatidylcholine vesicles with a mean diameter of 1800 Å containing an average of 0.75 to 2 sodium channels per vesicle are formed. The functional activities of the sodium channel can then be assessed in neurotoxin-binding and ion-flux experiments.

The time course of $^{22}Na^+$ influx into phosphatidylcholine vesicles containing purified sodium channels from rat brain is illustrated in Fig. 2. The vesicle preparation was incubated for 2 min with veratridine to activate sodium channels and then diluted into medium containing $^{22}Na^+$. Influx into vesicles under control conditions was slow (Fig. 3). Incubation with veratridine increased the initial rate of influx ten- to 15-fold. When tetrodotoxin was present in both the intravesicular and extravesicular phases, the veratridine-dependent increase in initial rate of $^{22}Na^+$ influx was nearly completely blocked (Fig. 3). Half-maximal activation was observed with 28 μM veratridine and half-maximal inhibition with 14 nM tetrodotoxin, in close agreement with the corresponding values for the action of these toxins on native sodium channels. These results show that the purified sodium channel regains the ability to mediate neurotoxin-stimulated ion flux after incorporation into phosphatidylcholine vesicles. Evidently, the purified channel retains neurotoxin receptor site 2 and the ion-conducting pore of the sodium channel.

The above results show that at least some of the sodium channels in our most highly purified preparations can mediate selective neurotoxin-activated ion transport after incorporation into phospholipid vesicles. My colleagues and I have attempted to estimate how many of the reconstituted sodium channels contribute to our ion-flux measurements. First, the ion transport rates in purified and reconstituted sodium channel preparations were compared to those of veratridine-activated sodium channels in neuroblastoma cells and synaptosomes (Tamkun *et al.*, 1984). This comparison shows that the transport rate measured in ions per minute per saxitoxin receptor site is 33 to 70% of that in native membranes, suggesting

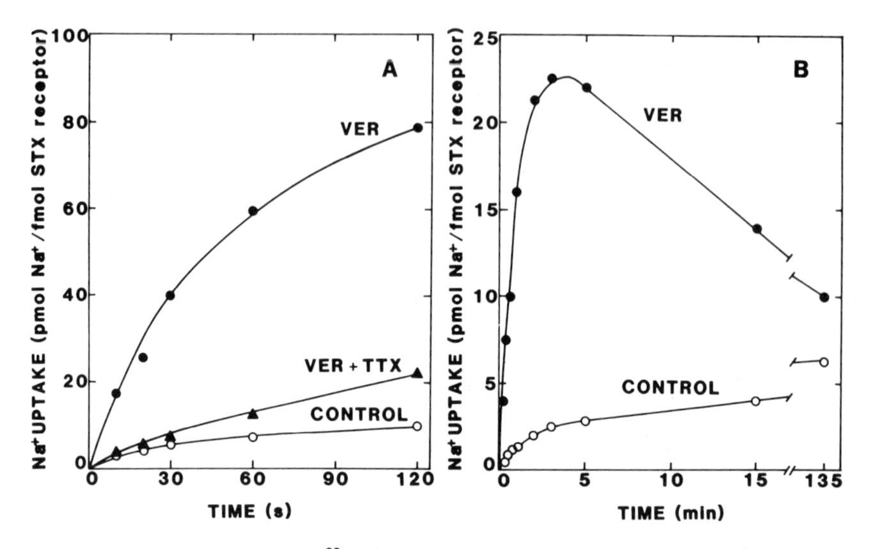

Figure 2. Neurotoxin-stimulated $^{22}Na^+$ influx mediated by the purified sodium channel. Sodium channels were purified from rat brain and incorporated into phosphatidylcholine vesicles as described by Tamkun *et al.* (1984). At $t = 0$, the vesicles containing Na_2SO_4 medium were diluted tenfold into medium containing $^{22}Na^+$ and no additions (○), 100 μM veratridine (●), or 100 μM veratridine plus 1 μM tetrodotoxin (▲). Uptake of $^{22}Na^+$ was measured by rapid filtration and γ counting.

that at least 33 to 70% of the reconstituted sodium channels are active. Second, we have compared the proportion of vesicles that contain sodium channels to the proportion of vesicles whose internal volume is accessible to veratridine-activated sodium channels. If sodium channels are distributed among vesicles according to a Poisson distribution, this comparison leads to a range of 30 to 70% for the fraction of active channels, depending on whether active vesicles containing more than one channel are assumed to have one active channel or all active channels. Both of these estimates indicate that a minimum of 30% of the reconstitued sodium channels are active. Since the sodium channel preparation is 90% pure, and no single contaminant comprises as much as 2% of the protein, we conclude that the purified complex of α, β_1, and β_2 is sufficient to mediate selective neurotoxin-activated ion flux.

Although sodium channels reconstituted into phosphatidylcholine vesicles can transport sodium, these channels do not bind α scorpion toxin at neurotoxin receptor site 3 (Tamkun *et al.*, 1984). In contrast, if purified sodium channels are incorporated into vesicles composed of a mixture of phosphatidylcholine and brain lipids, scorpion toxin binding is recovered.

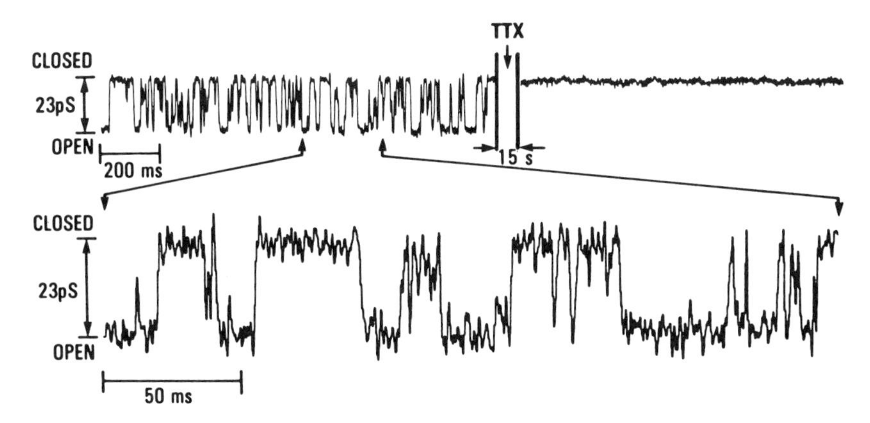

Figure 3. Single sodium channel currents in planar phospholipid bilayers. Sodium channels were purified and reconstituted as described by Feller *et al* (1985). These reconstituted vesicles were incubated with 1 μM batrachotoxin to block channel inactivation and allowed to fuse to preformed planar phospholipid bilayers as described by Hartshorne *et al.* (1985). The recording illustrates the opening and closing of a single purified and reconstituted sodium channel with a conductance of 23 pS. Where indicated, 1 μM tetrodotoxin was added to block the channel.

The toxin-binding reaction is of high affinity (K_D = 57 nM), and a mean of 0.76 ± 0.08 mole of scorpion toxin is bound per mole of purified sodium channel. In order to determine which phospholipids in the brain lipid fraction are required for restoration of scorpion toxin binding, purified sodium channels were reconstituted into vesicles composed of phosphatidylcholine and different individual brain lipids. The other two major lipids of brain plasma membranes, phosphatidylethanolamine and phosphatidylserine, enhanced high-affinity binding of scorpion toxin to the purified sodium channel in reconstituted phosphatidylcholine vesicles to an extent similar to that of brain lipid when added at levels comparable to those in synaptic plasma membranes. Optimum restoration of toxin binding was obtained with 35 to 50% phosphatidylethanolamine and 20 to 40% phosphatidylserine (Feller *et al.*, 1985). Phosphatidylethanolamine restored toxin binding more effectively at all ratios tested, and further addition of phosphatidylserine to vesicles having an optimum ration of phosphatidylethanolamine to phosphatidycholine did not improve scorpion toxin binding. Minor brain lipids including sphingomyelin, gangliosides, and cholesterol did not increase scorpion toxin binding when added at levels similar to those in synaptic plasma membranes. Thus, incorporation of purified sodium channels into phospholipid vesicles with an appropriate ration of phosphatidylethanolamine and phosphatidylcholine

Table IV. Voltage Dependence of Scorpion Toxin Binding to Purified and Reconstituted Sodium Channels

$[Na^+]_{out}/[Na^+]_{in}$	Voltage (mV)	$[^{125}I]$-LqTx bound $[(mol/mol) \times 10^{-3}]$
0.07	−70	3.31
0.15	−50	2.27
0.32	−30	1.32
1.0	0	0.87

is sufficient to satisfy the lipid requirement for high-affinity binding of scorpion toxin.

Binding of α scorpion toxins to sodium channels in intact neuronal membranes is voltage dependent, with K_D values in the range of 1 to 2 nM at resting membrane potentials of −40 to −50 mV. If a membrane potential of −60 mV is generated by diluting reconstituted vesicles containing 135 mM Na^+ into Na^+-free choline medium to give a tenfold outward sodium gradient, scorpion toxin binding is increased threefold at 0.5 to 2 min after dilution but then decays back to the value in the absence of a Na^+ gradient over 30 to 60 min (Feller *et al.*, 1985). Measurements of membrane potential show that it follows a similar time course, indicating that the increase in scorpion toxin binding caused by the Na^+ gradient results from the potential generated. More stable membrane potentials can be generated by dilution into a sucrose-substituted Na^+-free medium.

Table IV illustrates the results obtained when the concentration of Na^+ within the vesicles is varied during reconstitution and the resulting vesicles are diluted to a constant final concentration of 13.5 mM Na^+ outside. Binding of scorpion toxin is progressively reduced as the membrane potential changes from −60 to 0 mV, consistent with an increase in K_D on depolarization. At −60 mV, the K_D is 1.9 nM, in close agreement with values measured in intact synaptosomes or neuroblastoma cells. These results show that the purified sodium channel mediates high-affinity voltage-dependent binding of scorpion toxin when incorporated into phospholipid vesicles of appropriate composition. Since the voltage dependence of scorpion toxin binding is closely correlated with the voltage dependence of activation of sodium channels (Catterall, 1980), these data indicate that the purified channels retain voltage-dependent gating as well as selective ion transport. Thus, the purified sodium channel preparation from rat brain consisting of a stoichiometric complex of the α, β_1, and

Table V. Functional Properties of Purified and Native Sodium Channels

Property	Purified	Native
V_{50} for channel opening (in the presence of batrachotoxin)	−91 mV	−93 mV
Apparent gating charge	3.8	4–6
$K_{0.5}$ (BTX)	2 μM	0.5 μM
$K_{0.5}$ (veratridine)	30 mM	14 μM
$K_{0.5}$ (TTX)	8 μM	7–16 nM
Voltage-dependent ScTx binding	Yes	Yes
Conductance	25 pS	30 pS
Ion selectivity	Na > K > Rb	Na > K > Rb

β_2 subunits is sufficient to mediate most of the functions of the sodium channel that can be measured biochemically. These include neurotoxin binding and action at neurotoxin receptor sites 1 through 3 and selective neurotoxin-activated ion flux. However, in excitable membranes, sodium channels are normally activated and inactivated by changes in membrane potential. Electrical recording of sodium conductance mediated by the purified sodium channel on the millisecond time scale is required to demonstrate the full functional integrity of the sodium channel.

In order to record sodium currents mediated by single purified sodium channels, phospholipid bilayers composed of 33% phosphatidylcholine and 67% phosphatidylethanolamine were prepared by a modification of the method of Mueller and Rudin (Hartshorne *et al.*, 1985). Vesicles of phosphatidylcholine/phosphatidylethanolamine (65:35) containing purified sodium channels were treated with batrachotoxin to block inactivation of sodium channels and then incubated in the solution bathing one side of the preformed planar bilayer to allow fusion of vesicles. Electrical currents across the bilayer were recorded until single-channel currents appeared, indicating incorporation of functional sodium channels. A typical record of sodium conductance fluctuations is illustrated in Fig. 3. The single-channel conductance value for the purified sodium channel recorded in this figure was 23 pS compared to an average value of 25.2 pS for all channels studied (Hartshorne *et al.*, 1985). The single-channel conductance events are completely blocked by tetrodotoxin (Fig. 3) within a few seconds after addition, confirming that the conductance fluctuations recorded are mediated by sodium channels.

Further analysis of these single-channel currents showed that they have all the properties of sodium currents in intact membranes (Table V). They are blocked by tetrodotoxin with a K_D of 8 nM at −50 mV, and the block is voltage dependent. The fraction of time a purified sodium channel spends in the open state is voltage dependent. Half-maximal activation is achieved at -91 mV, in close agreement with native sodium channels modified by batrachotoxin, and the slope of the voltage dependence is similar. The purified channels are highly selective for Na^+ with $P_{Na}:P_K:P_{Rb} = 1:0.14:0.05$. Thus, the purified sodium channel retains essentially all of the functional properties of native sodium channels when returned to an appropriate phospholipid environment. The results show directly that this single protein is sufficient to confer electrical excitability on a phospholipid bilayer membrane *in vitro* and presumably also *in vivo*.

8. The Amino Acid Sequence of the Sodium Channel as Inferred from cDNA Clones

Purification and characterization of the sodium channel have made possible the application of molecular genetic techniques to further elucidate channel structuure. With the background information described above, Noda *et al*., (1984) pursued a two-pronged attack on the isolation of cDNA clones encoding the electroplax sodium channel. A cDNA library was prepared from electroplax mRNA in the plasmid vector pUC8, which synthesized polypeptides encoded by the inserted segments of cDNA under the direction of the *lac Z* gene regulatory sequences. Clones synthesizing portions of the sodium channel polypeptide were recognized with a polyclonal antiserum against the pure channel. In parallel, segments of the pure sodium channel were subjected to microsequence analysis, and several short segments of amino acid sequence were deduced. An oligonucleotide probe was prepared from one of these and used to screen the pUC8 plasmid clones that appeared positive in the immunologic screen. One clone was found positive in both screening assays. The nucleotide sequence of this clone confirmed that it encoded a portion of the sodium channel polypeptide. This clone was then used to screen a cDNA library containing longer inserts in the plasmid vector pBR322 by DNA hybridization. Several overlapping clones ranging in size from 500 to 5000 base pairs were isolated and subjected to nucleotide sequence analysis to give the complete sequence of the cDNA encoding the sodium channel protein.

The amino acid sequence deduced from these data leads directly to two major conclusions about the sodium channel protein: it contains fourfold internal sequence homology, and it does not have a hydrophobic leader sequence at the N terminal. The polypeptide of 208 kD (1820 amino acid residues) consists of four homologous domains of approximately 300 amino acid residues, which are connected and flanked by shorter stretches of nonhomologous residues. The high degree of sequence homology among these four domains (approximately 50% for each) is strong evidence that all four arose from a common ancestor and that they adopt similar secondary structures in pseudosymmetric orientations in the protein. On depolarization, sodium channels pass through multiple closed states before activation. A requirement that each domain undergo conformational change in order to achieve activation would provide a structural basis for the sigmoid time course of channel activation.

In addition to these two conclusions, which derive rather clearly from the amino acid sequence, analysis of the predicted secondary structure and hydropathy of the segments of the polypeptide led Noda *et al.*, to propose a detailed topological model of the folding of the channel polypeptide across the membrane and the general location of the transmembrane pore and gating charge. In this model, each of the homologous domains contains six segments. Four of these are membrane-spanning α helices. Two are intracellularly located, as are the amino and carboxyl termini and the bulk of the protein mass. In this model, the walls of the transmembrane pore are formed by segments from each of the four domains, which surround it in square array (as viewed perpendicular to the plane of the membrane). This proposal is reminiscent of the hexagonal array of ideal of identical subunits forming the gap junction channel and the pentagonal array of homologous subunits that form the ion channel of the nicotinic acetylcholine receptor. The sodium channel may accomplish with homologous domains within a single polypeptide what the other channels accomplish with homologous or identical subunits.

In addition to the four homologous domains, the sequence contains a 200-residue stretch in the center of the polypeptide between domains 2 and 3 that has four equally spaced clusters of negatively charged residues. It is proposed that each of these negatively charged clusters acts in concert with a segment of positively charged residues present in each homologous domain to provide the gating charges for channel activation and that the negatively charged segment mediates channel inactivation.

The α subunit of the sodium channel from rat brain resembles this polypeptide from electroplax in size and chemical properties. It is likely, therefore, that it adopts a similar transmembrane orientation. In support of this suggestion, the α subunit is phosphorylated at four sites by cAMP-dependent protein kinase from cytoplasm of synaptosomes (Costa and Catterall, 1984) and is glycosylated and covalently labeled by neurotoxins on its extracellular face, indicating a transmembrane location. These features plus information from the biochemical and molecular genetic experiments described above are incorporated into the speculative model of the structure of the rat brain sodium channel illustrated in Fig. 4. Most

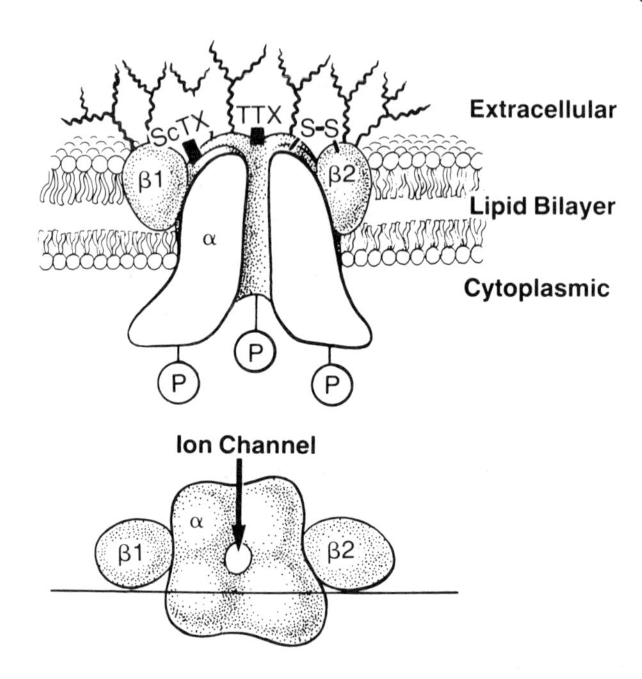

Figure 4. Structural model of the rat brain sodium channel. A hypothetical model of a sodium channel from rat brain is constructed based on the evidence from biochemical and molecular genetic experiments described in the text.

of the mass of the α subunit is intracellular, whereas most of the mass of the β subunits is extracellular. The ion channel is formed among the four homologous domains identified by Noda *et al.* (1984).

These hypotheses regarding the topology of the sodium channel and the location of its gating charge must be considered speculative at this time. Similar hypotheses for the nicotinic acetylcholine receptor are currently being revised. However, they provide a provocative template for the next phase of work on the molecular properties of the sodium channel, which will no doubt consist of locating the functional components of the channel within its primary, secondary, and tertiary structures. Recent experience with other proteins points to three major avenues of approach: site-directed mutagenesis, antipeptide antibodies, and covalent labeling. All three will probably be pursued concurrently in various laboratories. Development of a functional map of the sodium channel will be a long-term undertaking to which numerous laboratories will undoubtedly contribute. The availability of the amino acid sequence of the polypeptides now allows this process to begin. It is likely that it will provide important new insights into the molecular basis of electrical excitability.

References

Agnew, W. S., Moore, A. C., Levinson, S. R., and Raftery, M. A., 1980, Identification of a large molecular weight peptide associated with a tetrodotoxin binding protein from the electroplax of *Electrophorus electricus, Biochem. Biophys. Res. Commun.* **92:**860–866.

Barchi, R. L., 1983, Protein components of the purified sodium channel from rat skeletal muscle sarcolemma, *J. Neurochem.* **40:**1377–1385.

Barchi, R. L., and Murphy, L. E., 1981, Estimate of the molecular weight of the sarcolemmal sodium channel using H_2O–D_2O centrifugation, *J. Neurochem.* **36:**2097–2100.

Barhanin, J., Schmid, A., Lombet, A., Wheeler, K. P., and Lazdunski, M., 1983, Molecular size of different neurotoxin receptors on the voltage-sensitive Na^+ channel, *J. Biol. Chem.* **258:**700–702.

Beneski, D. A., and Catterall, W. A., 1980, Covalent labeling of protein components of the sodium channel with a photoactivable derivative of scorpion toxin, *Proc. Natl. Acad. Sci. U.S.A.* **77:**639–643.

Benzer, T. I., and Raftery, M. A., 1973, Solubilization and partial characterization of the tetrodotoxin binding component from nerve axons, *Biochem. Biophys. Res. Commun.* **51:**939–944.

Casadei, J. M., Gordon, R. D., Lampson, L. A., Schotland, D. L., and Barchi, R. L., 1984, Monoclonal antibodies against the voltage-sensitive Na^+ channel from mammalian skeletal muscle, *Proc. Natl. Acad. Sci. U.S.A.* **81:**6227–6231.

Catterall, W. A., 1980, Neurotoxins that act on voltage-sensitive sodium channels in excitable membranes, *Annu. Rev. Pharmacol. Toxicol.* **20:**15–43.

Catterall, W. A., 1984, The molecular basis of neuronal excitability, *Science* **223:**653–661.

Catterall, W. A., Morrow, C. S., and Hartshorne, R. P., 1979, Neurotoxin binding to receptor sites associated with voltage-sensitive sodium channels in intact, lysed, and detergent-solubilized brain membranes, *J. Biol. Chem.* **254:**11379–11387.

Cohen, S. A., and Barchi, R. L., 1981, Glycoprotein characteristics of the sodium channel saxitoxin-binding component from mammalian sarcolemma, *Biochem. Biophys. Acta* **645:**253–261.

Conti, F., Hille, B., Neumcke, B., Nonner, W., and Stampfli, R., 1976, Conductance of the sodium channel in myelinated nerve fibres with modified sodium inactivation, *J. Physiol.* (*Lond.*) **262:**729–742.

Costa, M. R. C., and Catterall, W. A., 1984, Cyclic AMP-dependent phosphorylation of the α subunit of the sodium channel in synaptic nerve ending particles, *J. Biol. Chem.* **259:**8210–8218.

Darbon, H. Jover, E., Couraud, F., and Rochat, H., 1983, Photoaffinity labeling of α- and β-scorpion toxin receptors associated with rat brain sodium channel, *Biochem. Biophys. Res. Commun.* **115:**415–422.

Feller, D., Talvenheimo, J. A., and Catterall, W. A., 1985, The sodium channel from rat brain: Reconstitution of voltage-dependent scorpion toxin binding in vesicles of defined lipid composition, *J. Biol. Chem.* **260:**11542–11547.

Hartshorne, R. P., and Catterall, W. A., 1981, Purification of the saxitoxin receptor of the sodium channel from rat brain, *Proc. Natl. Acad. Sci. U.S.A.* **78:**4620–4624.

Hartshorne, R. P., and Catterall, W. A., 1984, The sodium channel from rat brain. Purification and subunit composition, *J. Biol. Chem.* **259:**1667–1675.

Hartshorne, R. P., Coppersmith, J., and Catterall, W. A., 1980, Size characteristics of the solubilized saxitoxin receptor of the voltage sensitive sodium channel from rat brain, *J. Biol. Chem.***255:**10572–10575.

Hartshorne, R. P., Messner, D. J., Coppersmith, J. C., and Catterall, W. A., 1982, The saxitoxin receptor of the sodium channel from rat brain. Evidence for two nonidentical β subunits, *J. Biol. Chem.***257:**13888–13891.

Hartshorne, R. P., Keller, B. U., Talvenheimo, J. A., Catterall, W. A. and Montal, M., 1985, Functional reconstitution of the purified brain sodium channel in planar lipid bilayers, *Proc. Natl. Acad. Sci. U.S.A.* **82:**240–244.

Henderson, R., and Wang, J. H., 1972, Solubilization of a specific tetrodotoxin-binding component from garfish olfactory nerve membrane, *Biochemistry* **11:**4565–4569.

Hille, B., 1972, The permeability of the sodium channel to metal cations in myelinated nerve, *J. Gen. Physiol.* **59:**637–658.

Hodgkin, A. L., and Huxley, A. F., 1952, A quantitive description of membrane current and its application to conduction and excitation in nerve, *J. Physiol.* (*Lond.*) **117:**500–544.

Jover, E., Covraud, F., and Rochat, H., 1980, Two types of scorpion neurotoxins characterized by their binding to two separate receptor sites on rat brain synaptosomes, *Biochem. Biophys. Res. Comm.* **95:**1607–1614.

Levinson, S. R., and Ellory, J. C., 1973, Molecular size of the tetrodotoxin binding site estimated by irradiation inactivation, *Nature* (*New Biol.*) **245:**122–123.

Lombet, A., and Lazdunski, M., 1984, Characterization, solubilization, affinity labeling and purification of the cardiac Na^+ channel using *Tityus* toxin, *Eur. J. Biochem.* **141:**651–660.

Lombet, A., Norman, R. I., and Lazdunski, M., 1983, Affinity labeling of the tetrodotoxin-binding component of the Na^+ channel, *Biochem. Biophys. Res. Commun.* **114:**126–130.

Messner, D. J., and Catterall, W. A., 1985, The sodium channel from rat brain. Separation and characterization of subunits, *J. Biol. Chem.* **260:**10597–10604.

Noda, M., Shimizu, S., Tanabe, T., Takai, T., Kayano, T., Ikeda, T., Takahashi, H., Nakayama, H., Kanaoka, Y., Minamino, N., Kangawa, K., Mutsuo, H., Raftery, M. A., Hirose, T., Inayama, S., Hayashida, H., Miyata, T., and Numa, S., 1984, Primary structure of *Electrophorus electricus* sodium channel deduced from cDNA sequence, *Nature* **312:**121–127.

Ritchie, J. M., and Rogart, R. B., 1977, The binding of saxitoxin and tetrodotoxin to excitable tissue, *Rev. Physiol. Biochem. Pharmacol.* **79:**1–51.

Sigworth, F. J., and Neher, E., 1980, Single Na^+ channel currents observed in cultured rat muscle cells, *Nature* **287:**447–449.

Talvenheimo, J. A., Tamkun, M. M., and Catterall, W. A., 1982, Reconstitution of neurotoxin-stimulated sodium transport by the voltage-sensitive sodium channel purified from rat brain, *J. Biol. Chem.* **257:**11868–11871.

Tamkun, M. M., Talvenheimo, J. A., and Catterall, W. A., 1984, The sodium channel from rat brain. Reconstitution of neurotoxin-activated ion flux and scorpion toxins binding from purified components, *J. Biol. Chem.* **259:**1676–1688.

Weigele, J. B., and Barchi, R. L., 1982, Functional reconstitution of the purified sodium channel protein from rat sarcolemma, *Proc. Natl. Acad. Sci. U.S.A.* **79:**3651–3655.

II

PROTON PUMPS AND ELECTROCHEMICAL GRADIENTS

2

Studies of a Biological Energy Transducer

The lac Permease of Escherichia coli

H. R. KABACK

1. Introduction

In much the same way that the double helix model of DNA has provided the backbone for molecular genetics, the chemiosmotic hypothesis formulated and refined by Peter Mitchell during the 1960s (Mitchell, 1961, 1963, 1966) is now the conceptual framework for a wide variety of bioenergetic phenomena. In its most general form (Fig. 1), the chemiosmotic concept postulates that the immediate driving force for many processes in energy-coupling membranes is a H^+ electrochemical gradient ($\Delta\bar{\mu}_{H^+}$) composed of electrical and chemical parameters according to the following relationship:

$$\Delta\bar{\mu}_{H^+}/F = \Delta\psi - 2.3RT/F\,\Delta pH$$

where $\Delta\psi$ represents the electrical potential across the membrane and the ΔpH is the chemical difference in H^+ concentration across the membrane (R is the gas constant, T is absolute temperature, F is the Faraday constant; $2.3RT/F$ is equal to 58.8 at room temperature) (Mitchell, 1961).

Accordingly, the basic energy-yielding processes of the cell—respiration or the absorption of light—generate $\Delta\bar{\mu}_{H^+}$, and the energy stored therein is used to drive a number of seemingly unrelated phenomena such

H. R. KABACK • Roche Institute of Molecular Biology, Roche Research Center, Nutley, New Jersey 07110.

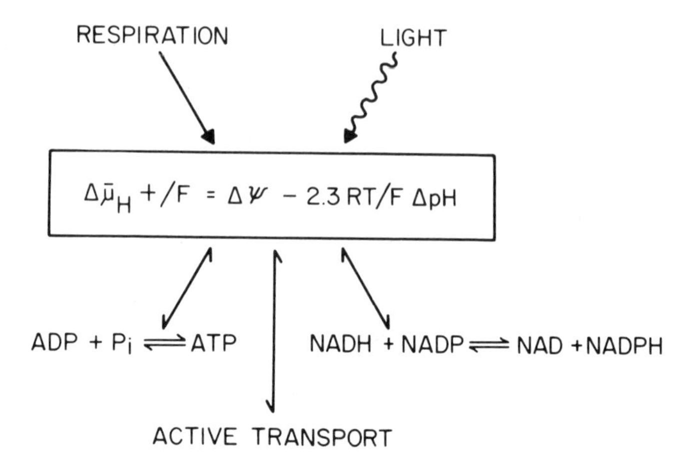

Figure 1. Generalized chemiosmotic hypothesis. Respiration or absorption of light leads to the generation of a H^+ electrochemical gradient ($\Delta\mu_{H^+}$) that provides the immediate driving force for oxidative and photophosphorylation, active transport, and transhydrogenation.

as the formation of ATP from ADP and inorganic phosphate (oxidative and photophosphorylation), active transport, and transhydrogenation of NADP by NADH. More recently, it has become apparent that $\Delta\bar{\mu}_{H^+}$ or one of its components is involved in a diversity of other phenomena, which include bacterial motility, transfer of genetic information, and sensitivity and resistance to certain antibiotics. Importantly, many of the processes driven by $\Delta\bar{\mu}_{H^+}$ are reversible. Thus, hydrolysis of ATP by the H^+ ATPase leads to the generation of $\Delta\bar{\mu}_{H^+}$. Similarly, transport of solutes down a concentration gradient (i.e., the reverse of active transport) can also lead to the generation of $\Delta\bar{\mu}_{H^+}$. Therefore, the "common currency of energy exchange," particularly in the bacterial cell, is not ATP but $\Delta\bar{\mu}_{H^+}$.

Developments in our understanding of bacterial active transport exemplify the type of advances being made in this area. Although concentration of solutes at the expense of metabolic energy (i.e., active transport) has been a recognized phenomenon for some time, insight into the molecules involved has begun to occur only recently. During the 1970s, a wealth of experimental findings, which are not reviewed here, provided strong support for the chemiosmotic nature of active transport in bacterial systems, particularly *Escherichia coli* (Kaback, 1976, 1984). With this aspect of the problem resolved, focus has shifted to a molecular level,

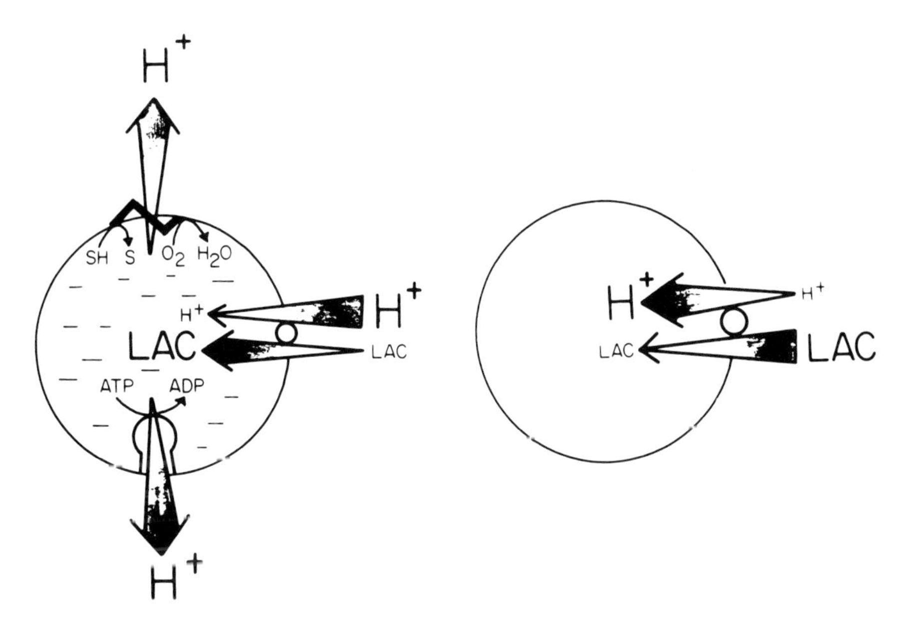

Figure 2. Lactose transport in *E. coli*. *Left*: Uphill lactose transport in response to $\Delta\mu_{H^+}$ (interior negative and alkaline) generated either by respiration or ATP hydrolysis. *Right*: Uphill H^+ transport in response to an inwardly directed lactose gradient. ⁓, membrane-bound respiratory chain with alternating proton and electron carrier; ⌒, the H^+ ATPase; O, the *lac* permease.

with emphasis on the β-galactoside transport system of *E. coli* as a model (cf. Kaback, 1976, 1983, 1984 for recent reviews).*

β-Galactoside transport in *E. coli* is catalyzed by the *lac* permease, which is encoded by the second structural gene in the *lac* operon, the *lac y* gene. This intrinsic membrane protein translocates substrate with H^+ in a symport (cotransport) reaction. In the presence of $\Delta\bar{\mu}_{H^+}$, hydrogen ion moves down its electrochemical gradient and drives the uphill translocation of β-galactosides (Fig. 2). Conversely, under nonenergized conditions, downhill movement of substrate along a concentration gradient drives uphill transport of H^+, leading to generation of $\Delta\bar{\mu}_{H^+}$, the polarity of which is determined by the direction of the substrate concentration gradient. Thus, the *lac* permease is representative of a large class of substrate-specific polypeptides that utilize the energy released from the downhill movement of a cation in response to a transmembrane electro-

* For brevity, the literature cited here is limited to a large extent to those papers that appear in print after Kaback, 1983, 1984; Overath and Wright, 1983.

chemical gradient to drive the uphill accumulation of solute. Therefore, studies with this transport system have immediate relevance to a wide range of biological systems, from bacteria to eucaryotes and their intracellular organelles and to various epithelia.

By use of a strain of *E. coli* with multiple copies of the *lac y* gene, a highly specific photoaffinity label for the permease, and reconstitution of transport activity in artificial membranes (i.e., proteoliposomes), the *lac* permease has been purified to homogeneity and demonstrated to be the product of the *lac y* gene. Proteoliposomes reconstituted with the purified permease exhibit all of the translocation reactions typical of the β-galactoside transport system in intact cells and right-side-out (RSO) membrane vesicles. Moreover, the *lac y* gene has been sequenced, and a secondary structure model for the permease has been proposed. Its structure and function are being studied in diverse ways.

2. *Morphology and Permeability of Proteoliposomes Reconstituted with lac Permease*

Proteoliposomes formed by octylglucoside dilution followed by one cycle of freeze–thaw/sonication are unilamellar and exhibit no internal structure. Furthermore, about 80% of the proteoliposomes fall between 50 and 150 nm when the cross-sectional diameters of the vesicles are measured statistically (Kaback, 1984).

Low-magnification electron microscopy of platinum/carbon replicas of freeze-fractured proteoliposomes containing purified permease confirms the unilamellar nature of the preparations. Higher magnification reveals that both the convex and concave fracture faces exhibit a uniform distribution of particles that are about 70 Å in diameter. Importantly, both fracture faces are completely devoid of particles when liposomes are formed from *E. coli* phospholipids in the absence of protein. Since particles, but no pits, are observed on both faces of the membrane, it seems likely that the permease has equal affinity for the phospholipids in each leaflet of the bilayer. Given the mass of the permease (46,504), a particle size of 70 Å suggests that the particles may contain one to two polypeptides, depending on the degree to which the shadowing increases the observed particle diameter and on other considerations (Costello *et al.*, 1984).

Proteoliposomes prepared in this manner are ideally suited for studies of H^+/solute symport. Morphologically, the preparation consists of a population of unilamellar, closed, unit-membrane-bound sacs that are relatively uniform in diameter and have no internal structure, characteristics

Table I. Comparison of Turnover Numbers for the *lac* Carrier Protein: ML 308–225 Membrane Vesicles versus Proteoliposomes Reconstituted with Purified Carrier

Reaction[a]	Turnover numbers (sec^{-1})	
	Membrane vesicles[b]	Proteoliposomes
$\Delta\psi$-driven influx ($\Delta\psi$ = 100 mV)	16 (K_m = 0.2 mM)	16–21 (K_m = 0.5 mM)
Counterflow	16–39 (K_m = 0.45 mM)	28 (K_m = 0.6 mM)
Facilitated diffusion	8–15 ($K_m \simeq$ 20 mM)	8–9 ($K_m \simeq$ 3.1 mM)
Efflux	8 (K_m = 2.1 mM)	6–9 (K_m = 2.5 mM)

[a] All reactions were carried out at pH 7.5 and 25°C.
[b] Determination of the amount of *lac* carrier protein in ML 308–225 membrane vesicles is based on photolabeling experiments with [^{3}H]-NPG, which indicate that the carrier represents about 0.5% of the membrane protein.

that are entirely consistent with the efflux and exchange kinetics observed for Rb^+ and lactose (Garcia *et al.*, 1983). The proteoliposomes are also passively impermeable to many ions, a property that is highly advantageous. Thus, certain aspects of H^+/lactose symport that are difficult to document with RSO vesicles (e.g., the stimulation of efflux by ionophores) are readily elucidated with the reconstituted system. Generally, proteoliposomes reconstituted with *lac* permease exhibit all of the phenomena described in RSO membrane vesicles, but the results are significantly more clear-cut and provide firmer support for certain ideas concerning reaction mechanisms (Garcia *et al.*, 1983; Viitanen *et al.*, 1983).

3. A Single Polypeptide is Required for Lactose Transport

Since evidence has been presented suggesting that active lactose transport may require more than a single gene product (Hong, 1977; Plate and Suit, 1981), careful kinetic experiments have been performed on proteoliposomes reconstituted with *lac* permease purified to a single polypeptide species (Viitanen, 1984). Turnover numbers were calculated for the permease operating in various modes and compared to those calculated from V_{max} values obtained with RSO vesicles in which the amount of permease was estimated by photoaffinity labeling (Table I). Both the turnover numbers of the permease and the apparent K_ms for lactose are virtually identical in proteoliposomes and membrane vesicles with respect to various modes of translocation. In addition, more detailed studies demonstrate that the reconstituted system exhibits kinetic properties analogous to those observed in RSO vesicles (e.g., the primary effect of $\Delta\bar{\mu}_{H^+}$ is to lower the apparent K_m for lactose, and the initial velocity of lactose influx at relatively low concentrations varies linearly with the square of $\Delta\bar{\mu}_{H^+}$). Finally, it should be noted that C. A. Plate (personal communication) has recently found that deletion of the *eup* gene (one of the loci reported to have a pleiotropic effect on transport activity with no effect on $\Delta\bar{\mu}_{H^+}$) causes stimulation of transport. Thus, the product of the *eup* gene does not play a direct role in transport but apparently acts in a regulatory capacity.

Taken as a whole, the results provide strong support for the contention that a single polypeptide species, the product of the *lac y* gene, is responsible for each of the transport reactions catalyzed by the β-galactoside transport system.

4. *Reconstitution of Active Transport in Proteoliposomes Containing lac Permease and Cytochrome o Oxidase*

Since purified *lac* permease catalyzes active transport in proteoliposomes when a $\Delta\Psi$ (interior negative) or a ΔpH (interior alkaline) is imposed artifically and the *o*-type cytochrome oxidase purified from *E. coli* generates $\Delta\bar{\mu}_{H^+}$ when reconstituted into proteoliposomes (Matsushita *et al.*, 1983, 1984), it follows that reconstituted preparations containing both proteins should exhibit respiration-driven lactose accumulation. In other words, by incorporating both of these proteins into the same proteoliposomes, it might be possible to reconstitute one of the basic physiological functions of the bacterial cell membrane, respiration-driven transport.

Purified cytochrome *o* oxidase contains four polypeptides (M_rs $\simeq$ 66, 35, 22, and 17 kD) and two *b*-type cytochromes (b_{558} and b_{563}). The preparation catalyzes oxidation of ubiquinol-1 (Q_1H_2) and other electron donors with specific activities 20- to 30-fold higher than crude membranes. Turnover of the oxidase in proteoliposomes generates a $\Delta\psi$ (interior negative) as a result of vectorial electron flow from the outer to the inner surface of the membrane. However, the pH gradient (interior negative) that is also generated results from scalar (i.e., nonvectorial) reactions that consume and release protons at the inner and/or outer surface of the membrane, respectively. That is, *o*-type cytochrome oxidase from *E. coli* does not appear to catalyze vectorial H^+ translocation (Matsushita *et al.*, 1984).

Proteoliposomes reconstituted simultaneously with purified oxidase and *lac* permease generate $\Delta\bar{\mu}_{H^+}$ (interior negative and alkaline) with

Q_1H_2 as electron donor, the magnitude of which is dependent on the concentration of the oxidase in the proteoliposomes. In the presence of Q_1H_2, the proteoliposomes accumulate lactose against a concentration gradient. The phenomenon is completely abolished by the addition of cyanide, which blocks the oxidase, or by the addition of the ionophores valinomycin and nigericin, which dissipate $\Delta\bar{\mu}_{H^+}$. Since transport in the absence of electron donors or in the presence of Q_1H_2 plus cyanide or valinomycin and nigericin represents equilibration with the medium, it is apparent that the steady-state level of lactose accumulation observed during oxidase turnover represents a concentration gradient of at least 10- to 20-fold. Moreover, by comparing lactose transport induced by Q_1H_2 to that induced by artifically imposed diffusion potentials and quantifying the magnitude of the $\Delta\bar{\mu}_{H^+}$ generated under each condition, it is clear that the lactose transport activity observed is commensurate with the magnitude of the bulk phase $\Delta\bar{\mu}_{H^+}$.

5. *Secondary Structure of lac Permease*

The permease is a very hydrophobic polypeptide of molecular weight 46,504, containing 417 amino acid residues of known sequence. Circular dichroic measurements on purified permease solubilized in octylglucoside or reconstituted into proteoliposomes demonstrate that 85 ± 5% of the amino acid residues are arranged in helical secondary structures. These findings lead to a systematic examination of primary structure as determined from the DNA sequence of the *lac y* gene. When the hydrophilicity and hydrophobicity (i.e., hydropathy) of the protein are evaluated along the amino acid sequence, it is apparent that the permease contains a number of relatively long hydrophobic segments punctuated by short hydrophilic regions. In light of the circular dichroism data, this finding suggests strongly that most, if not all, of these segments are α-helical. Furthermore, since the segments are markedly hydrophobic, it seems likely that they are embedded in the lipid bilayer. About 12 of the longest hydrophobic domains exhibit a mean length of 24 ± 4 amino acid residues, and they comprise approximately 70% of the length of the polypeptide. Additionally, 18 regions of the permease should contain reverse turns (180° reversals) (Fig. 3. cf. arrows). Fifteen of the putative turns (83%) fall within hydrophilic regions between the hydrophobic domains postulated to traverse the bilayer. The three exceptions are reverse turns that fall within two long hydrophobic segments included by amino acids 73–98 and 289–336. Based on these considerations, it is proposed that *lac*

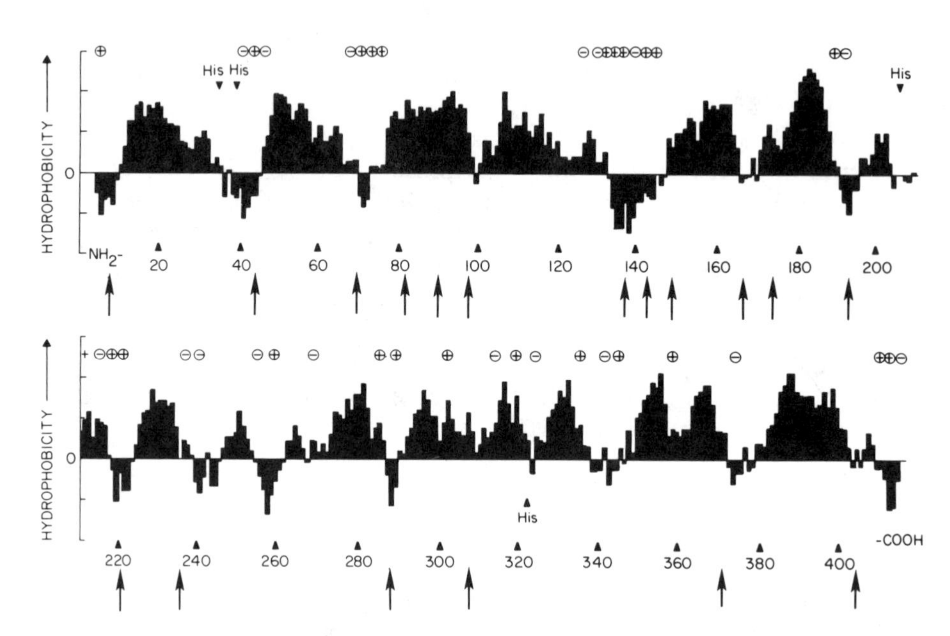

Figure 3. Hydropathic profile of the *lac* permease as determined from the DNA sequence of *lac y* (Foster *et al.*, 1983).

permease consists of 12 helical segments that traverse the membrane in a zig-zag fashion, as suggested for bacteriorhodopsin (Fig. 4).

Experiments utilizing proteolytic enzymes and site-directed polyclonal antibodies (i.e., antibodies directed against synthetic peptides corresponding in sequence to specified regions of the permease) provide preliminary support for the model. Thus, photoaffinity-labeled permease in RSO or inside-out (ISO) membrane vesicles is accessible to chymotrypsin, trypsin, or papain, demonstrating that the permease extends through the bilayer. Moreover, antibodies directed against peptides corresponding to the C terminus (Seckler *et al.*, 1983; Carrasco *et al.*, 1983, 1984a) and hydrophilic segments 5 and 7 (Fig., 4) bind preferentially to ISO vesicles relative to RSO vesicles, indicating that each of these portions of the permease is present on the same side of the membrane, the cytoplasmic surface. On the other hand, antibodies directed against a number of other hydrophilic segments do not bind to vesicles of either orientation, although the antibodies react with the permease after immunoblotting. Presumably, these portions of the protein are either buried

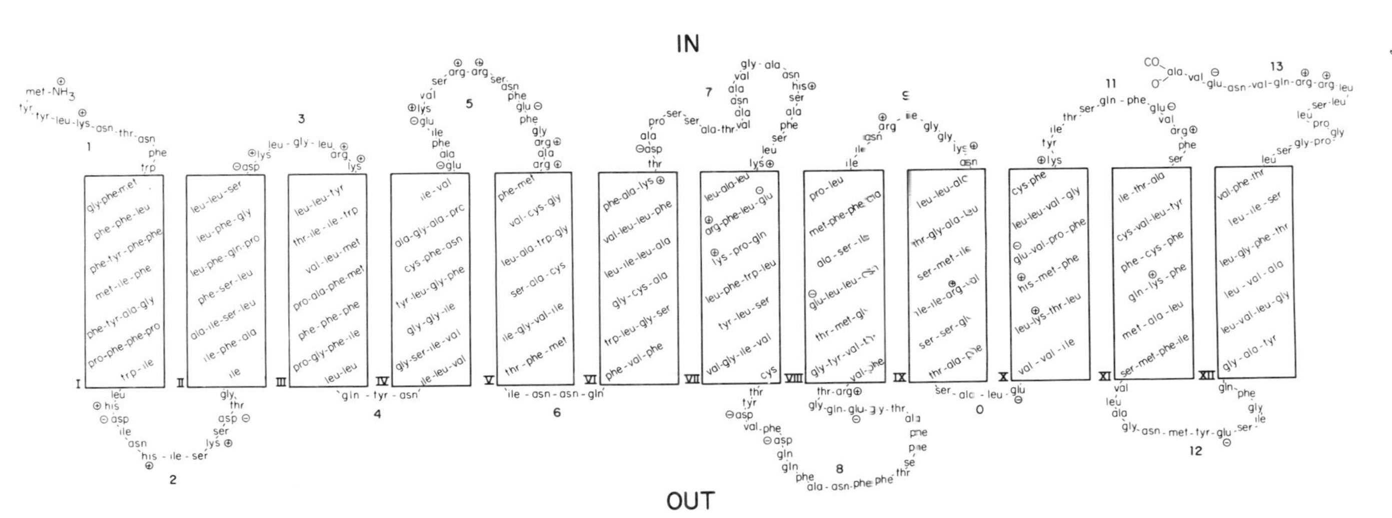

Figure 4. Secondary structure of *lac* permease as predicted from the hydropathic profile. Hydrophobic segments are shown in boxes as transmembrane α-helical domains connected by hydrophilic segments. As discussed in the text, the C terminus and hydrophilic segments 5 and 7 (with the N terminus as hydrophilic segment 1) are on the cytoplasmic surface of the bacterial membrane.

in the membrane or are inaccessible within the tertiary structure of the native polypeptide.

6. *Immunologic Reagents as Structure/Function Probes*

Monoclonal antibodies (Mabs) directed against purified *lac* permease have been prepared and shown to be highly specified. The effects of the Mabs on lactose transport were studied on RSO membrane vesicles and in proteoliposomes reconstituted with purified permease (Carrasco *et al.*, 1984b). Out of more than 60 Mabs tested, only one, designated 4B1, inhibits transport. Furthermore, the nature of the inhibition is remarkable in that 4B1 inhibits only those transport reactions that involve net H^+ translocation (i.e., active transport, carrier-mediated reflux and efflux under nonenergized conditions, and lactose-induced H^+ influx) with little or no effect on equilibrium exchange, the generation of $\Delta\bar{\mu}_{H^+}$, or the ability of the permease to bind ligand. Clearly, therefore, 4B1 alters the relationship between lactose and H^+ translocation at the level of the permease.

In studies of entrance counterflow with external [^{14}C]-lactose at saturating and subsaturating concentrations, it is apparent that 4B1 mimics the effects of deuterium oxide (Viitanen *et al.*, 1983), and the results suggest that the Mab either inhibits the rate of deprotonation or alters the equilibrium between protonated and deprotonated forms of the permease. Monovalent Fab fragments prepared from 4B1 inhibit transport in a manner that is similar qualitatively to that of the intact IgG. However, 4B1 IgG is approximately twice as effective as the Fab fragments on a molar basis, suggesting that the intact IgG binds bivalently, whereas the Fab fragments bind 1:1. Direct support for this conclusion is provided by binding experiments with radiolabeled 4B1 and 4B1 Fab fragments (Herzlinger *et al.*, 1984).

Radioiodinated 4B1 and 5F7 (another Mab obtained from the same fusion that does not inhibit transport) bind to distinct, nonoverlapping epitopes in the permease (Herzlinger *et al.*, 1984). It is evident from immunofluorescence microscopy and radioionated IgGs and Fab fragments that both Mabs bind to spheroplasts and RSO membrane vesicles but only to a small extent to ISO vesicles. Therefore, the 4B1 and 5F7 epitopes are on the external (i.e., periplasmic) surface of the membrane as opposed to C terminus and hydrophilic segments 5 and 7, which are on the cytoplasmic surface of the membrane (see above). In RSO vesicles, [^{125}I]-4B1 binds with a stoichiometry of 1 mole of antibody per 2 moles of permease, whereas [^{125}I]-4B1 Fab fragments bind 1:1. Importantly, intact

4B1 and its Fab fragments bind to proteoliposomes reconstituted with purified *lac* permease with a stoichiometry very similar to that observed in RSO membrane vesicles. Therefore, with respect to the 4B1 epitope, the orientation of the permease in the reconstituted system appears to be similar to that in the bacterial cytoplasmic membrane.

In contradistinction to these results, experiments with site-directed polyclonal antibodies directed against the C terminus of the permease indicate that the reconstituted system may not be entirely representative of the native membrane (Carrasco *et al.*, 1984a; Seckler and Wright, 1984). [^{125}I]-Anti-C-terminal Fab fragments, which bind to ISO vesicles relatively exclusively, bind to proteoliposomes containing the permease, indicating that a significant percentage of the C terminus is on the outside of the membrane in the reconstituted system (i.e., on the wrong side of the membrane). In addition, treatment of reconstituted proteoliposomes with carboxypeptidase extensively degrades the C terminus while having no effect on the 4B1 epitope or on transport activity. Finally, antibody to hydrophilic segment 7, which also binds to ISO vesicles, does not bind to proteoliposomes, indicating that this portion of the molecule—like the 4B1 epitope—maintains the proper orientation after reconstitution. Therefore, although the more obvious possibility that reconstituted permease molecules are scrambled has been considered, the data are more consistent with the notion that a portion of the permease molecules undergoes intramolecular dislocation of the C terminus during reconstitution with no effect on catalytic activity. In this context, it is also significant that 4B1 causes enhanced binding of Mab 4A10R in the reconstituted system. The Mab 4A10R is directed against a cytoplasmically disposed epitope that is partially related to the C terminus of the permease. In the native membrane, the epitopes for 4A10R and 4B1 are on opposite sides of the membrane (Herzlinger *et al.*, 1985). Since it is unlikely that this effect of 4B1 is intermolecular, a more reasonable interpretation is that these conformationally coupled epitopes, which are normally on opposite sides of the native membrane, are present on the external surface of the proteoliposomes within the same permease molecules.

Right-side-out vesicles from *E. coli* ML 308-22, a mutant "uncoupled" for β-galactoside/H^+ symport (Wong *et al.*, 1970), are specifically defective in the ability to catalyze accumulation of methyl-1-thio-β-D-galactopyranoside (TMG) in the presence of $\Delta\bar{\mu}_{H^+}$ (Herzlinger *et al.*, 1985). Furthermore, the rate of carrier-mediated efflux under nonenergized conditions is slow and unaffected by ambient pH from pH 5.5 to 7.5, and TMG-induced H^+ influx is only about 15% of that observed in vesicles containing wild-type *lac* permease. Alternatively, ML 308-22 vesicles bind *p*-nitrophenyl-α-D-galactopyranoside and Mab 4B1 to the same extent that wild type vesicles do and catalyze facilitated diffusion and equilibrium exchange as well as wild-type vesicles. When entrance counterflow is studied with external substrate at saturating and subsaturating concentrations, it is apparent that the mutation, like Mab 4B1, also simulates the effects of deuterium oxide. That is, the mutation has no effect on the rate or extent of counterflow when external substrate is saturating but stimulates tthe efficiency of counterflow when external substrate is below the apparent K_m. Moreover, although replacement of protium with deuterium stimulates counterflow in wild-type vesicles when external substrate is limiting, the isotope has no effect on the mutant vesicles under the same conditions. It is suggested that the mutation in ML 308-22 results in a *lac* permease with a higher pK_a, thereby either limiting the rate of deprotonation or altering the equilibrium between protonated and deprotonated forms of the permease.

Although Mab 4B1 binds identically to wild-type and mutant RSO vesicles, Mab 4A10R binds to ISO vesicles from the mutant only 30% as well as it binds to the same preparation from the wild type. Moreover, antibodies against hydrophilic domains 5 and 7 bind threefold better and one-fifth as well, respectively, to ISO vesicles from the mutant type relative to ISO vesicles from the wild type. Clearly, therefore, these immunologic probes are able to discriminate between wild-type and "uncoupled" permease molecules. The results also suggest that mutation causes a significant alteration in the conformation of the permease.

7. Subunit Structure

Although the permease is monomeric when solubilized in dodecylmaltoside (Wright *et al.*, 1983), use of radiation inactivation analysis indicates that the situation may be more complex when the permease is in the membrane (Goldkorn *et al.*, 1983). In these experiments, vesicles containing the permease are frozen rapidly in liquid N_2 before and after energization and subjected to a high-intensity electron beam for various periods of time. After irradiation, the samples are extracted with octylglucoside, reconstituted into proteoliposomes, and tested for activity. Under all conditions, the decrease in activity exhibits pseudo-first-order kinetics as a function of radiation dosage, allowing straightforward application of target theory for determination of functional molecular size.

When permease activity solubilized from nonenergized vesicles is assayed under these conditions, the results obtained yield a functional molecular mass of 45–50 kD, a value similar to the molecular mass of the permease as determined by other means. Moreover, similar values are obtained when the octylglucoside extract is irradiated, and target volumes observed for the D-lactate dehydrogenase (D-LDH) and the H^+ ATPase complex in the same vesicles are in reasonable agreement with the known molecular weights of these enzymes.

Strikingly, when the same procedures are carried out with vesicles that are energized with appropriate electron donors prior to freezing and irradiation, a functional molecular mass of 85–100 kD is obtained for the permease with no change in the target size of D-LDH. In contrast, when the vesicles are energized in the presence of a potent protonophore that collapses $\Delta\bar{\mu}_{H^+}$, the target size of the permease returns to 45–50 kD. In addition, genetic studies indicating that certain *lac y* mutations may be dominant (Mieschendahl, 1981) are also consistent with the idea that oligomer formation may be important for *lac* permease function.

8. Oligonucleotide-Directed Site-Specific Mutagenesis

Based on substrate protection against N-ethylmaleimide (NEM) inactivation, Fox and Kennedy (1965) postulated that there is an essential sulfhydryl group in the *lac* permease located at or near the active site, and Cys^{148} has been shown to be the critical residue (Beyreuther *et al.*, 1981). Although chemical modification of specific amino acid residues in a protein provides important information, there are obvious drawbacks to this approach. Recently, site-directed mutagenesis has been utilized to introduce single amino acid changes into certain proteins (Zoller and Smith, 1983), and this strategy has been used to evaluate the role of Cys^{148} in the *lac* permease (Trumble *et al.*, 1984).

By cloning the *lac y* gene into single-stranded M13 phage DNA and utilizing a synthetic deoxyoligonucleotide primer 21 bases in length (which is complementary to the *lac y* template with the exception of a single mismatch), Cys^{148} in the permease is converted into a glycine residue. Cells bearing the mutated *lac y* gene exhibit initial rates of lactose transport that are about fourfold lower than cells bearing the wild-type gene on a recombinant plasmid. Furthermore, transport activity is less sensitive to inactivation by NEM, and, strikingly, galactosyl-l-thio-β-D-galactopyranoside affords no protection whatsoever against inactivation. The findings suggest that although Cys^{148} is essential for substrate protection against sulfhydryl inactivation, it is not obligatory for lactose : H^+ symport and that another sulfhydryl group elsewhere within the *lac* premease may be required for full activity.

More recently, oligonucleotide-directed site-specific mutagenesis has been utilized to introduce other alterations into the *lac* permease (Kaback,

1984) (see Fig. 4). Experiments include the following. (1) Gln^{60} has been replaced with a glutamic acid residue, thereby introducing a negative charge into the second putative α-helix of the permease. Surprisingly, the replacement has no effect on transport activity, although preliminary experiments indicate that the altered permease is less stable to heat. (2) His^{35} and His^{39} have been simultaneously replaced with arginine residues, and in another instance, His^{322} has been changed to an arginine residue. The $His^{35}His^{39}$ mutation has no effect on transport. However, the replacement of His^{322} with arginine results in a dramatic loss of activity. Currently, the role of His^{205}, the fourth histidine residue in the polypeptide, is being evaluated. (3) A segment of the polypeptide from Met^{372} to Pro^{405} has been deleted, thereby excising the last transmembrane α-helical segment from the permease. Cells harboring this altered *lac y* gene express extremely low levels of permease. Thus, this form of the permease is either not inserted into the membrane or is inserted and then proteolyzed. In any event, the effects of these and other mutations on transport activity and on the disposition of various epitopes in the membranes are currently being examined.

References

Beyreuther, K., Beiseler, B., Ehring, R., and Müller-Hill, B., 1981, Identification of internal residues of lactose permease of *Escherichia coli* by radiolabel of peptide mixtures, in: *Methods in Protein Sequence Analysis* (M. Elzina, ed.), Humana Press, Clifton, NJ, p. 139.

Carrasco, N., Herzlinger, D., DeChiara, S., Danho, W., Gabriel, T. F., and Kaback, H. R., 1983, Topology of the *lac* protein in the membrane of *Escherichia coli, Biophys. J.* **45:**83a.

Carrasco, N., Herzlinger, D., Mitchell, R., DeChiara, S., Danho, W., Gabriel, T. F., and Kaback, H. R., 1984a, Intramolecular dislocation of the COOH-terminus of the *lac* carrier protein in reconstituted proteoliposomes, *Proc. Natl. Acad. Sci. U.S.A.* **81:**4672–4676.

Carrasco, N., Viitanen, P., Herzlinger, D., and Kaback, H. R., 1984b, Monoclonal antibodies against the *lac* carrier protein from *Escherichia coli*, I. Functional studies, *Biochemistry* **23:**3681–3687.

Costello, M. J., Viitanen, P., Carrasco, N., Foster, D. L., and Kaback, H. R., 1984, Morphology of proteoliposomes reconstituted with purified *lac* carrier protein from *Escherichia coli, J. Biol. Chem.* **259:**15579–15586.

Foster, D. L., Boublik, M., and Kaback, H. R., 1983, Structure of the *lac* carrier protein of *Escherichia coli, J. Biol. Chem.* **258:**31.

Fox, C. F., and Kennedy, E. P., 1965, Specific labeling and partial purification of the M protein, a component of the β-galactoside transport system of *Escherichia coli, Proc. Natl. Acad. Sci. U.S.A.* **54:**891–899.

Garcia, M. L., Viitanen, P., Foster, D. L., and Kaback, H. R., 1983, Mechanism of lactose translocation in proteoliposomes reconstituted with *lac* carrier protein purified from

Escherichia coli. I. Effect of pH and imposed membrane potential on efflux, exchange and counterflow, *Biochemistry* **22:**2524–2531.

Goldkorn, T., Rimon, G., and Kaback, H. R., 1983, Topology of the *lac* carrier protein in the membrane of *Escherichia coli, Proc. Natl. Acad. Sci. U.S.A.* **80:**3322–3326.

Herzlinger, D., Viitanen, P., Carrasco, N., and Kaback, H. R., 1984, Monoclonal antibodies against the *lac* carrier protein from *Escherichia coli*. II. Binding studies with membrane vesicles and proteoliposomes reconstituted with purified *lac* carrier protein, *Biochemistry* **23:**3688–3693.

Herzlinger, D., Carrasco, N., and Kaback, H. R., 1985, Functional and immunochemical characterization of a mutant of *Escherichia coli* energy uncoupled for lactose transport, *Biochemistry* **24:**221–229.

Hong, J.-S., 1977, An *ecf* mutation in *Escherichia coli* pleiotropically affecting energy coupling in active transport but not generation or maintenance of membrane potential, *J. Biol. Chem.* **252:**8582–8588.

Kaback, H. R., 1976, Molecular biology and energetics of membrane transport, *J. Cell, Physiol.* **89:**575–594

Kaback, H. R., 1983, The *lac* carrier protein in *Escherichia coli:* From membrane to molecule, *J. Membr. Biol,* **76:**95

Kaback, H. R., 1986, Active transport in *Escherichia coli:* From membrane to molecule, in: *Physiology of Membrane Disorders* (T. E. Andreoli, J. F. Hoffman, D. D. Fanestil, and S. G. Schultz, eds.), Plenum Press, New York, pp. 387–407.

Matsushita, K., Patel, L., Gennis, R. B., and Kaback, H. R., 1983, Reconstitution of active transport in proteoliposomes containing cytochome *o* oxidase and *lac* carrier protein purified from *Escherichia coli, Proc. Natl. Acad. Sci. U.S.A.* **80:**4889–4893.

Matsushita, K., Patel, L., and Kaback, H. R., 1984, Cytochrome *o* oxidase from *Escherichia coli*. Characterization of the enzyme and mechanism of electrochemical proton gradient generation, *Biochemistry* **23:**4703–4714.

Mieschendahl, M., Büchel, D., Bocklage, H., and Müller-Hill, B., 1981, Mutations in the *lac y* gene of *Escherichia coli* define functional organization of lactose permease, *Proc. Natl. Acad. Sci. U.S.A.* **78:**7652–7656.

Mitchell, P., 1961, Coupling of phosphorylation to electron hydrogen transfer by a chemiosmotic type of mechanism, *Nature* **191:**144–148–169.

Mitchell, P., 1963, Molecule, group, and electron translocation through natural membranes, *Biochem. Soc. Symp.* **22:**142–148.

Mitchell, P., 1966, *Chemiosmotic Coupling and Energy Transduction,* Glynn Research Ltd., Bodmin, UK.

Overath, P., and Wright, J. K., 1983, Lactose permease: A carrier on the move, *Trends Biochem. Sci.* **8:**404–408.

Plate, C. A., and Suit, J. L., 1981, The *eup* genetic locus of *Escherichia coli* and its role in H^+/solute symport, *J. Biol. Chem.* **256:**12974–12980.

Seckler, R., and Wright, J. K., 1984, Sidedness of native membrane vesicles of *Escherichia coli* and orientation of the reconstituted lactose:H^+ carrier, *Eur. J. Biochem.* **142:**269–279.

Seckler, R., Wright, J. K., and Overath, P., 1983, Peptide-specific antibody locates the COOH terminus of the lactose carrier of *Escherichia coli* on the cytoplasmic side of the plasma membrane, *J. Biol. Chem.* **258:**10817–10820.

Trumble, W. R., Viitanen, P. V., Sarkar, H. K., Poonian, M. S., and Kaback, H. R., 1984, Site-directed mutagenesis of Cys_{148} in the *lac* carrier protein of *Escherichia coli, Biochem. Biophys. Res. Commun.* **119:**860–867.

Viitanen, P., Garcia, M. L., Foster, D. L., Kaczorowski, G. J., and Kaback, H. R., 1983,

Mechanism of lactose translocation in proteoliposomes reconstituted with *lac* carrier protein purified from *Escherichia coli*. 2. Deuterium solvent isotope effects, *Biochemistry* **22**:2531–2536.

Viitanen, P., Garcia, M. L., and Kaback, H. R., 1984, Purified, reconstituted *lac* carrier protein from *Escherichia coli* is fully functional, *Proc. Natl. Acad. Sci. U.S.A.* **81**:1629–1633.

Wong, P. T. S., Kashket, E. R., and Wilson, T. H., 1970, Energy coupling in the lactose transport system of *Escherichia coli, Proc. Natl. Acad. Sci. U.S.A.* **65**:63–69.

Wright, J. K., Weigel, U., Lustig, A., Bocklage, H., Mieschendahl, M., Müller-Hill, B., and Overath, P., 1983, Does the lactose carrier of *Escherichia coli* function as a monomer? *FEBS Lett.* **162**:11–15.

Zoller, M. J., and Smith, M., 1983, Oligonucleotide-directed mutagenesis of DNA fragments cloned into M13 vectors, *Methods Enzymol.* **100**:468–500.

3

The Erythrocyte Anion-Exchange Protein

Primary Structure Deduced from the cDNA Sequence and a Model for Its Arrangement within the Plasma Membrane

RON R. KOPITO and HARVEY F. LODISH

1. Overview

A full-length cDNA encoding the mouse erythrocyte anion-exchange protein band 3 has been isolated and sequenced. Homology between the amino acid sequence deduced from this cDNA and that of published fragments of human band 3 confirms its identity. A model of the topology of band 3 within the plasma membrane is proposed that is based on published biochemical data and the deduced amino acid sequence. Twelve hydrophobic and amphipathic regions in the anion-exchange domain are proposed to span the membrane as α-helices, resulting in both C and N termini in the interior of the cell. The possibility is considered that these transmembrane helices are organized to form two hydrophilic channels per band 3 monomer, which undergo conformational changes during the anion-exchange cycle.

2. Introduction

The exchange of bicarbonate and chloride across the plasma membrane of the erythrocyte, mediated by the major integral membrane pro-

---

RON R. KOPITO • Whitehead Institute for Biomedical Research, Cambridge, Massachusetts 02142. *HARVEY F. LODISH* • Department of Biology, Massachusetts Institute of Technology, and Whitehead Institute for Biomedical Research, Cambridge, Massachusetts 02142.

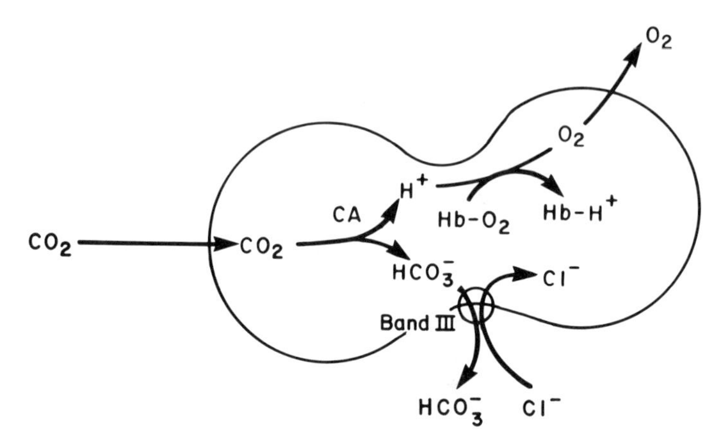

Figure 1. The role of band 3 in CO_2 transport. Abbreviations: Hb, hemoglobin; CA, carbonic anhydrase.

tein band 3 (Fairbanks *et al.*, 1971), is a key step in the overall transport of carbon dioxide from respiring tissues to the lungs (Fig. 1). Within the capillaries, CO_2 readily diffuses into erythrocytes, where it is hydrated by carbonic anhydrase to form bicarbonate and a proton. The direction of this equilibrium depends both on the P_{CO_2} and on the efficient removal of the reaction products. The liberated proton binds to hemoglobin, thereby facilitating the release of oxygen (the Bohr effect), and the bicarbonate anion is exchanged for extracellular Cl^- by band 3. The one million molecules of band 3 on the surface of each erythrocyte (Steck, 1974) permit the rapid equilibration of the intracellularly formed bicarbonate with the extracellular aqueous space. Indeed, about 80% of the total carbon dioxide carried in the blood is in the form of bicarbonate anion dissolved in the water phases of erythrocytes and plasma (Wieth *et al.*, 1982a). All of the reactions shown in Fig. 1 are reversed in the pulmonary circulation by the high P_{O_2} and low P_{CO_2}. The anion-exchange activity of band 3 has been extensively studied (reviewed in Wieth *et al.*, 1982a; Steck, 1974; Macara and Cantley, 1982; Cabantchik *et al.*, 1978; Knauf, 1979). It behaves as a sequential process in which the transmembrane movement of anions is reversible, tightly coupled, and electrically silent.

In addition to its role as an anion exchanger, band 3 mediates another essential function in the erythrocyte. It binds tightly to ankyrin (Bennett and Stenbuck, 1979, 1980), through which it links the plasma membrane to the meshlike network of spectrin and actin that constitutes the erythrocyte submembrane cytoskeleton (reviewed in Bennett, 1982).

Extensive studies of band 3 structure and topology have led to the identification of two distinct structural domains into which its anion-exchange activity and cytoskeletal interactions are segregated. Digestion of erythrocyte ghosts with trypsin (Steck *et al.*, 1976) yields a ~41-kD negatively charged, hydrophilic amino terminal fragment (TR41) that contains the high-affinity binding site for ankyrin (Bennett and Stenbuck, 1980) as well as, in human band 3, binding sites for hemoglobin (Shaklai *et al.*, 1977; Walder *et al.*, 1984) and several glycolytic enzymes (Strapazon and Steck, 1976). The C-terminal 52-kD fragment (TR452) remains tightly associated with the lipid bilayer and possesses the anion-exchange activity of the native protein. Studies on the topology of this domain suggest that it traverses the membrane at least seven times (Jennings *et al.*, 1984; Jennings and Nicknish, 1984) (see Fig. 5). Further proteolysis of TR52 with chymotrypsin results in its cleavage into 17-kD (CH17) and 35-kD (CH35) subfragments (Steck *et al.*, 1976). Interestingly, this proteolysis still results in no loss of anion-exchange activity (Grinstein *et al.*, 1978). Digestion of intact erythrocytes with papain, however, which removes a 5-kD fragment from the N terminus of CH35 and a few residues from the C terminus of CH17 (Jennings *et al.*, 1984), dramatically alters the anion-exchange process (Jennings and Adams, 1981).

Investigations of the membrane-associated anion-exchange domain of band 3 (TR52) using proteolytic mapping, inhibitor binding, and covalent modification have permitted the identification of functional groups and domains of the protein that may be involved in the exchange process. Several models of anion exchange have been postulated that attempt to integrate the kinetic and biophysical data with the results of structural studies. Unfortunately, the intimate association with the lipid bilayer of the anion-exchange domain of band 3 has hampered efforts to determine the sequence of all but two short fragments of this domain in the human protein. The molecular mechanisms and the structural organization of this important protein have, therefore, remained enigmatic.

We have isolated and sequenced a full-length cDNA clone encoding the band 3 message from mouse (Kopito and Lodish, 1985). Previous studies have established that human and mouse band 3 are very similar in both size and biochemical properties (Braell and Lodish, 1981). The amino acid sequence of band 3, deduced from the nucleotide sequence, allows us to design a testable model for protein-mediated transmembrane anion exchange. In this chapter we discuss features of the amino acid sequence of the band 3 polypeptide deduced from the cDNA and propose a novel model for its transmembrane orientation and the possible spatial arrangement of its membrane-spanning regions that incorporates data from the large number of biochemical and physiological studies on the human protein.

3. *Materials and Methods*

3.1. *Isolation of cDNA Clones and DNA Sequencing*

Details of the construction of the cDNA libraries, antibody screening, and DNA sequence have been described elsewhere (Kopito and Lodish, 1985). Briefly, a cDNA library from anemic mouse spleen RNA, in the expression vector λ-gt11 (Young and Davis, 1983), was probed with a polyclonal antibody to mouse erythrocyte band 3 (Braell and Lodish, 1981), and several partial-length cDNA clones were isolated. These were used to isolate the full-length cDNA from a size-selected λ-gt11 library. The complete nucleotide sequence (Maxam and Gilbert, 1980; Sanger *et al.*, 1977) was obtained from the sequence of overlapping cDNA clones.

3.2. *Computer-Assisted DNA Sequence Analysis*

The sequence was compiled with the assistance of the DB programs of Staden (Staden, 1982). Subsequent analysis of the sequence was performed using the ALIGN, RELATE, SEARCH, and DOTMATRIX programs (Dayhoff *et al.*, 1983) and the data base of Dayhoff *et al.* (1983).

3.3. *Cell Culture*

Friend murine erythroleukemia cells (MEL) (Friend *et al.*, 1978) were maintained in culture as previously described (Patel and Lodish, 1984). Induction was achieved by incubating the cells in media containing 5% BSA, 1.8% DMSO, and 1.8 mM Imferron as previously described (Patel and Lodish, 1984).

3.4. *RNA Blot Hybridization*

RNA was prepared by disruption of the tissue in 5 M guanidinium isothiocyanate followed by centrifugation through a gradient of CsCl essentially as previously described (Chirgwin *et al.*, 1979). Total RNA or poly(A)-enriched RNA (Aviv and Leder, 1972) was separated on a 1.5% agarose gel containing 6% formaldehyde (Maniatis *et al.*, 1982). The RNA was transferred to a nylon filter and hybridized sequentially with nick-translated probes from mouse band 3 cDNA and rat α-tubulin cDNA (Lemishka *et al.*, 1981). Hybridization was performed in 5× SSC and 50% formamide at 42°C as described (Maniatis *et al.*, 1982).

4. Results

4.1. Mouse and Human Band 3 Share Sequence Homology

Figure 2 shows the complete sequence of 4257 nucleotides compiled from overlapping cDNA clones. The size of the full-length cDNA agrees well with that of the mRNA (see Fig. 7). This sequence contains a single open reading frame (2919 bp) flanked by 127-bp and 1214-bp untranslated regions at the 5′ and the 3′ ends, respectively. The open reading frame extends for 929 codons from the ATG at nucleotide 1 to TGA at base 2919. There are two in-frame ATG codons at the beginning of the open reading frame (residues 1 and 4). We have assigned the first Met codon as the initiator because it is the first in-frame ATG downstream of the (in-phase) stop codon at base −63. The sequences flanking this ATG, but not the other, are homologous to the highly conserved sequence that flanks functional initiation sites in eucaryotic mRNAs: AXX AUG G (Kozak, 1981). (An out-of-frame ATG at base −50 defines a reading frame of only 41 codons and is probably not used.)

The molecular weight of the predicted polypeptide is 103,000, in good agreement with estimates of 90,000 to 100,000 obtained from SDS-polyacrylamide electrophoresis of the native human (Fairbanks *et al.*, 1971) and mouse proteins (Braell, 1981). The predicted C-terminal amino acid of the mouse sequence is valine, which is also the C-terminal residue of the human protein (Drickamer, 1976). Indeed, of the C-terminal 12 residues predicted from the mouse cDNA sequence, 11 are identical with those determined by carboxypeptidase-Y digestion of human band 3 (R. Reithmeier, personal communication). Parts of the amino acid sequence deduced from the cDNA sequence are highly homologous to that of the several known fragments of human erythrocyte band 3 (Fig. 3), proving unequivocally that the clones we have isolated encode the mouse erythrocyte anion-exchange protein.

The most striking homology is between residues 456 and 492 of mouse band 3 and the 38-amino-acid fragment from the human protein designated H3 (Mawby and Findlay, 1983) in Fig. 3. There is only a single conservative (Leu → Val) substitution between the two sequences; there are no insertions or deletions. Excellent alignment is also observed between the deduced sequence of mouse band 3 and the sequence of a peptic fragment

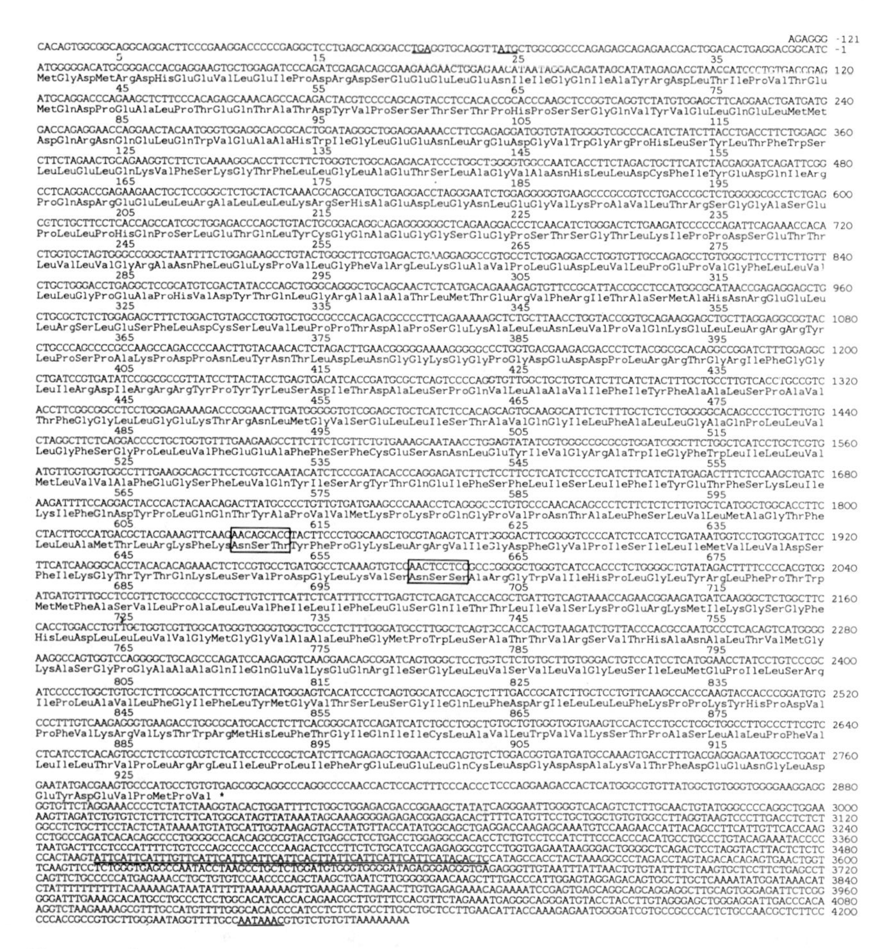

Figure 2. Complete nucleotide sequence of band 3 cDNA showing deduced amino acid sequence. Nucleotide numbers are indicated in the right-hand margin, and the sequence is written with the 5′ end of the cDNA at the top left corner. The in-frame nonsense codon at base −63 and an out-of-frame ATG at −50 are underlined. The two potential glycosylation sites at amino acids 611 and 660 are designated by boxes. Also underlined are the tandem repeat at nucleotide positions 3490–3550 and the polyadenylation signal at base 4232. Other relevant residues and structure are referred to by number in the text.

(P5) of the human protein designated H4 (Brock *et al.*, 1983) (Fig. 3); only five of the 72 residues are different, and alignment of the sequences required no insertions or deletions. The sequence of the amino-terminal 201 residues of human band 3 (Kaul *et al.*, 1983), designated H1, exhibits slightly lower homology with the sequence deduced from the mouse band

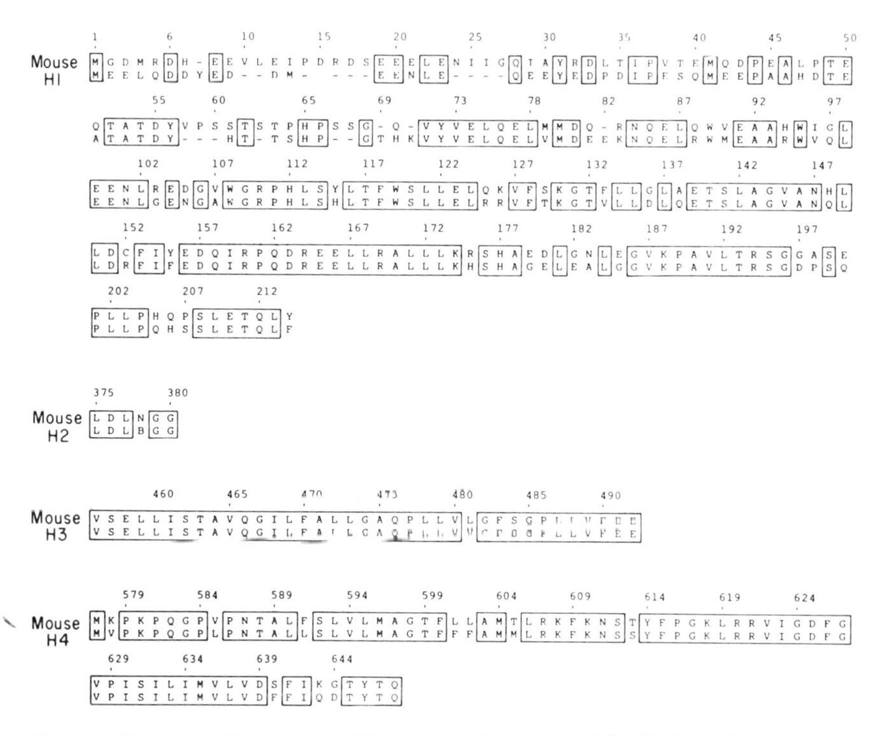

Figure 3. Alignment of sequences of fragments of human band 3 with the deduced sequence of the mouse protein. Shown are the alignments for an N-terminal fragment, H1 (Kaul *et al.*, 1983), a short peptide from the N terminus of a 14,000-dalton chymotryptic fragment, H2 (Mawby and Findlay, 1983), and two fragments from the membrane-associated domain, H3 (Mawby and Findlay, 1983) and H4 (Brock *et al.*, 1983) (see Fig. 5 for their locations).

3 clone. Computer-assisted alignment, using the mutation data matrix (Dayhoff *et al.*, 1983), which emphasizes evolutionary relationships in scoring amino acid homologies, suggests that mouse band 3 is slightly longer than its human homologue. As shown in Fig. 3, the homology is lowest at the extreme N terminus of the protein and seems to occur in clusters distributed throughout the segment.

4.2. *Band 3 Has Three Structural Domains*

A hydropathy plot (Kyte and Doolittle, 1982) (Fig. 4) reveals that mouse band 3 can be roughly divided into three domains, shown schematically in Fig. 5. The N-terminal 420 residues constitute the hydrophilic cytoplasmic domain with a decidely negative net charge (65 acidic and

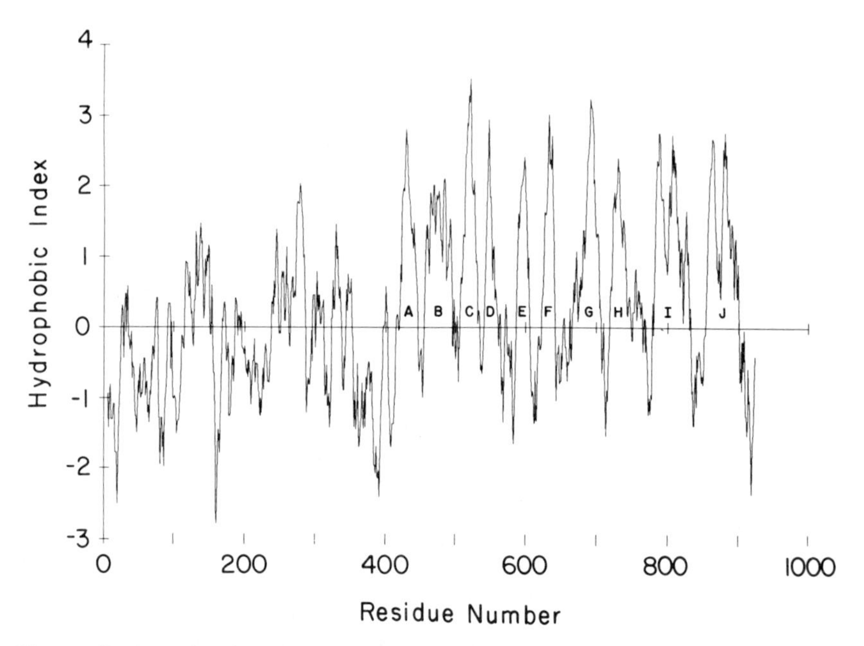

Figure 4. Hydropathy plot of the deduced amino acid sequence of murine band 3. Hydrophobicities were averaged over windows of 11 amino acids. The residue number is plotted on the horizontal scale. The ten hydrophobic regions are designated A–J.

37 basic residues). The central region consists of 450 amino acids of mixed polar and apolar composition, which are intimately associated with plasma membrane lipid. In this domain, groups of hydrophobic residues that correspond to the peaks A–J in Fig. 4 are interspersed with polar, predominantly basic, residues (Fig. 5). The third domain of band 3 encompasses the extreme 32 C-terminal residues. Eleven of these are either glutamate or aspartate, making it unlikely that the C-terminus is buried within the lipid bilayer.

4.2.1. The Cytoplasmic Domain

The binding of hemoglobin (Shaklai *et al.*, 1977; Walder *et al.*, 1984) and glycolytic enzymes (Strapazon and Steck, 1976) to the cytoplasmic domain of human band 3 is mediated primarily by the first 11 residues of the band 3 polypeptide (Walder *et al.*, 1984). The absence of detectable glyceraldehyde-3-phosphate dehydrogenase from mouse erythrocyte ghosts (V. Patel, personal communication) is consistent with the lack of homology between mouse and human sequences in this N-terminal un-

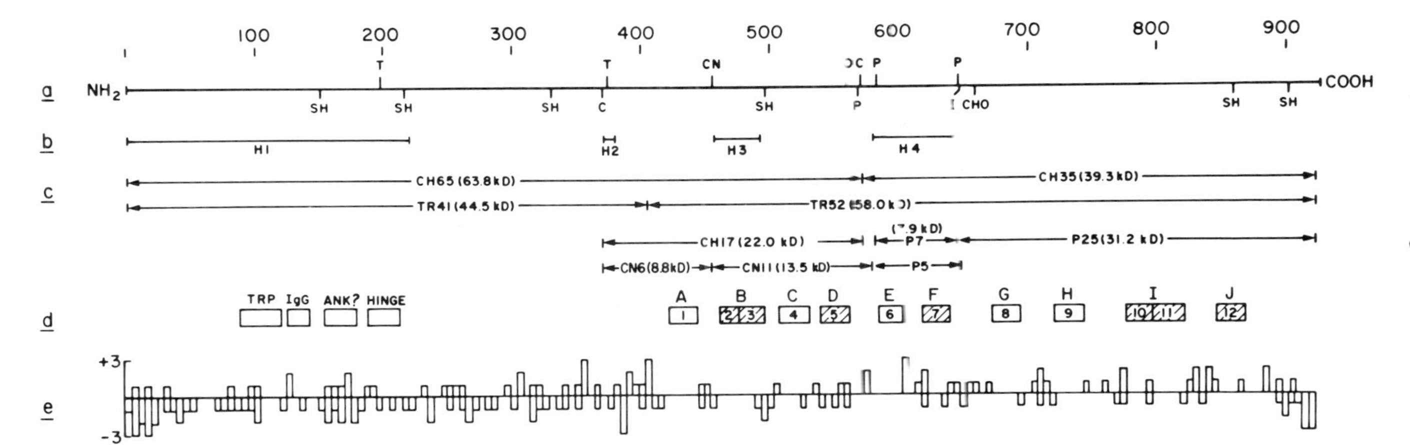

Figure 5. Domain map of band 3. a. The scale is the complete amino acid sequence of mouse band 3 deduced from the nucleotide sequence. Probable sites are indicated as follows: for proteolytic cleavage by trypsin (T) at position 194 (R. Reithmeyer, personal communication) and 381 (Mawby and Findlay, 1983; Jennings and Nicknish, 1984); chymotrypsin (C) (Steck *et al.*, 1976); cyanogen bromide (CN) (Mawby and Findlay, 1983; Jennings and Nicknish, 1984); and papain (P) (Jennings *et al.*, 1984). Cysteine residues (SH), H_2DIDS-binding sites (D), and the known site for exofacial radioiodination (I) (Brock *et al.*, 1983) are shown. The proposed site for attachment of the N-linked oligosaccharide is designated (CHO). b. Alignment of the fragments from human band 3 is shown. H1, N-terminal fragment; H2, short peptide from the N terminus of a 14,000-dalton chymotryptic fragment; H3 and H4, two fragments from the membrane-associated domain. c. The primary fragments arising from digestion of human band 3 with extracellular chymotrypsin, CH65 and CH35 (Grinstein *et al.*, 1978; Drickamer, 1976); intracellular trypsin, TR41 and TR52 (Steck *et al.*, 1976); extracellular papain, P7 and P25 (Jennings *et al.*, 1984); extracellular pepsin, P5 (Brock *et al.*, 1983); and cyanogen bromide cleavage of CH17, CN6, and CN11 (Mawby and Findlay, 1983; Jennings and Nicknish, 1984). The numbers in parentheses are the actual molecular weights of the fragments calculated from their amino acid composition. These values do not take into account the contribution of the single N-linked oligosaccharide. d. Structural features within the N-terminal cytoplasmic domain are tryptophan-rich (TRP) (Low *et al.*, 1984), antibody-binding (IgG) (Low *et al.*, 1984; England *et al.*, 1980), ankyrin-binding (ANK), and HINGE. These features are indicated and are discussed in the text. The locations of the hydrophobic peaks from the hydropathy plot (Fig. 4) are indicated byy the letters A–J, and the corresponding presumed membrane-spanning regions are designated by the numbers 1–12. Hatched boxes indicate that the region is likely to be amphipathic (see text for discussion). e. Charge plot. The location of charged residues in the mouse band 3 sequence is indicated by the bar graph, with the positions corresponding to the sequence numbers at the top of the figure. The basic residues (Arg and Lys) are assigned a value of +1, and the acidic residues (Glu and Asp) are assigned the value of −1. The bars represent the total values above and below the horizontal axis over five residues.

decapeptide. Other features of the cytoplasmic domain of band 3 are more conserved. Human erythrocyte band 3 possesses six cysteine residues, five of which are reactive with N-ethylmaleimide (Rao, 1979). Mouse band 3, as predicted from the cDNA sequence, also has six cysteines, shown in Fig. 5. Three of these are located within the cytoplasmic domain (CH65), which is consistent with the number determined in studies on human band 3.

Based on studies on pH-dependent reversible conformational equilibrium, Low *et al.*, (1984) proposed a domain structure for the cytoplasmic domain of band 3. They suggested that this domain is a homodimer with a "regulated hinge" rich in proline residues somewhere in the middle of CH65. These prolines apparently disrupt the α-helical conformation of the polypeptide in this region, rendering it susceptible to proteolysis. It is noteworthy that these prolines (at positions 161, 188, 201, and 204) are all conserved between human and mouse band 3 (cf. Fig. 3); Fig. 5 shows them to be clustered around a known trypsin cleavage site at position 194 (R. Reithmeier, personal communication). (Interestingly, two other proline-rich regions in the predicted mouse sequence are found at the known primary sites of intracellular and extracellular proteolysis, between residues 361–368 and 567–586, respectively; Fig. 5.) Between the proposed "hinge" and a region rich in tryptophan (residues 89–119), also highly conserved, is the purported antigenic region of band 3 (Low *et al.*, 1984). Monospecific polyclonal antibodies against total human band 3 recognize determinants located within a 20-kD fragment of the cytoplasmic domain (England *et al.*, 1980); this site is sufficiently distant from the N terminus that the binding of antibody does not interfere with the binding of hemoglobin (Low *et al.*, 1984). The region between residues 125 and 138 is only 57% homologous between human and mouse, whereas the residues flanking it on either side are 90–95% homologous. This observation is consistent with the lack of crossreactivity of antihuman band 3 antisera with the mouse protein and *vice versa* (Braell, 1981).

The ankyrin-binding site on band 3 has not been identified; Low *et al.*, (1984) tentatively place it between the "hinge" region and the "IgG" region, since antibodies against band 3 compete with labeled ankyrin for binding (Low *et al.*, 1984). It is likely that the high-affinity ankyrin-binding site is highly conserved among different species and perhaps among different nonerythroid ankyrin-binding proteins. In Fig. 5 we have tentatively positioned this binding site.

4.2.2. The Membrane-Associated Domain

Figure 4 indicates that the C-terminal ~500 amino acids are arranged in ten long hydrophobic stretches, labeled A–J. Regions of 20–24 amino

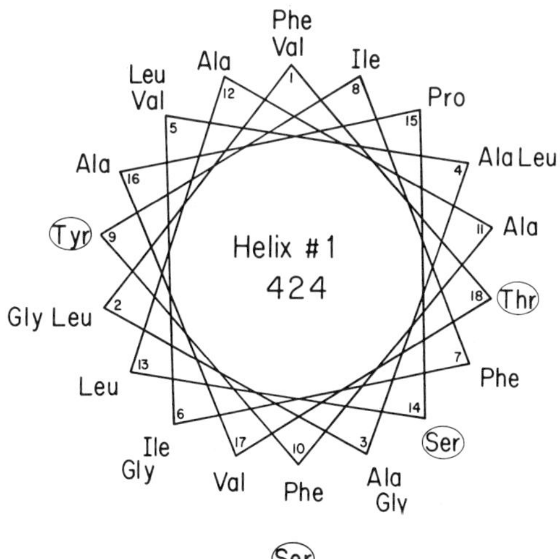

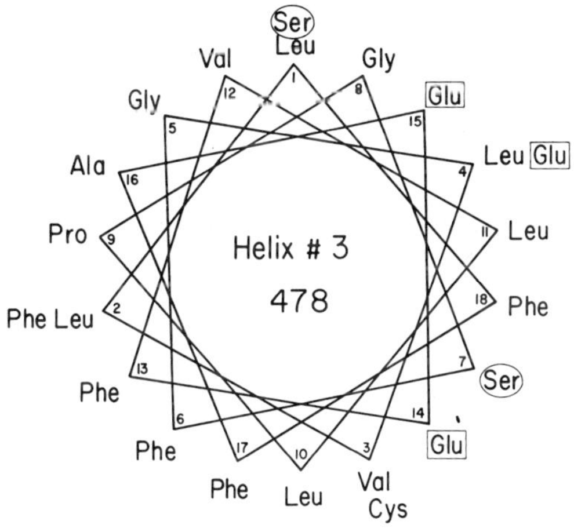

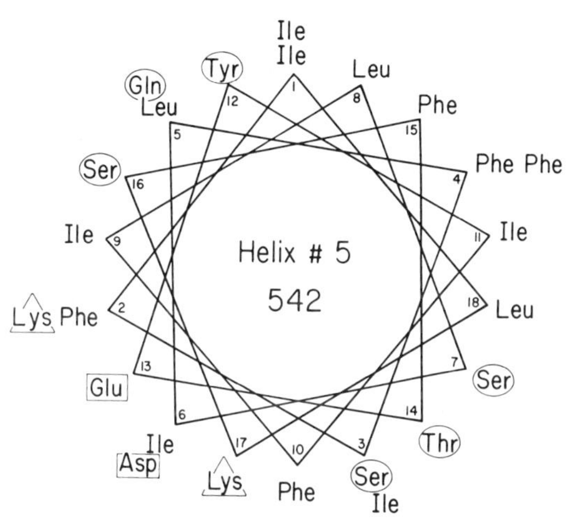

Figure 6. Two dimensional "wheel diagrams" of (a) helix 1 and (b) helices 3 and 5. Residues with polar, uncharged side chains are circled. Squares and triangles denote amino acids with acidic or basic side chains, respectively.

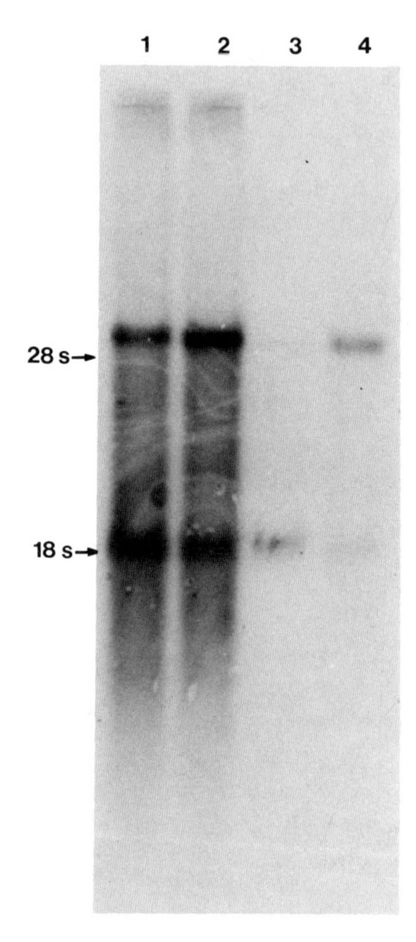

Figure 7. RNA blot analysis of RNA from anemic mouse spleen and Friend erythroleukemia (MEL) cells. Lanes 1 and 2 contain total and poly(A+) RNA, respectively, from anemic mouse spleen. Lanes 3 and 4 contain poly(A+) RNA from uninduced and 3-day induced MEL cells, respectively. The upper band results from hybridization of the blot with a mouse band 3 cDNA probe, whereas the lower band in each lane was obtained by reprobing the filters with a rat α-tubulin cDNA probe. The locations of the 28 S and 18 S ribosomal bands are indicated as relative size markers.

acids with hydropathy averages exceeding 1.5 are usually tightly associated with lipid and, in the case of several proteins, span the membrane in an α-helical configuration (Kyte and Doolittle, 1982; Eisenberg, 1984).

Examination of the band 3 sequence for potential membrane-spanning regions suggests that the regions corresponding to seven of the peaks in Fig. 4 span the phospholipid bilayer once (A, C, D, E, F, G, H) and that two (B, I) and possibly a third (J) span twice. The putative membrane-spanning regions corresponding to these peaks have been numbered 1–12 as designated in Fig. 5. We analyzed the sequences of these potential transmembrane segments by constructing helical "wheel diagrams" (Schiffer and Edmundson, 1967), assuming 3.6 amino acids per turn. These diagrams, examples of which are shown in Fig. 6, have been used to predict amphipathic structures in other proteins (Eisenberg, 1984). One important observation from this analysis is that all of the potential membrane-spanning helices in band 3 (with the possible exception of helices 1 and 8) exhibit some degree of amphipathicity. In these helices, the amino acids bearing charged or polar side groups are all oriented toward a single face (Fig. 6b).

4.3. Band 3 Expression Is Regulated during Differentiation of MEL Cells

Figure 7 shows that pB33 hybridizes to a single species of RNA from anemic spleen cells that migrate in a denaturing agarose gel at 4.3kb. Friend erythroleukemia cells (MEL) (Friend *et al.*, 1978) are induced to undergo terminal erythroid differentiation by treatment with DMSO or other agents (Marks and Rifkind, 1978). Differentiation is accompanied by the accumulation of erythrocyte-specific mRNAs and the selective reduction in expression of nonerythroid genes (Eisen *et al.*, 1977). Furthermore, induction of these cells results in the accumulation of several erythrocyte proteins including globin (Eisen *et al.*, 1977), band 3, and ankyrin (Patel and Lodish, 1984; Chirgwin *et al.*, 1979). Figure 7 shows that induction of MEL cells results in a significant increase in the level of band 3 mRNA and a concomitant loss of message for α-tubulin. The time course of the increase of band 3 RNA parallels the appearance of immunoprecipitable band 3 protein (Patel and Lodish, 1984). No additional bands were observable after overexposure (1 week).

5. Discussion

5.1. Transmembrane Orientation of the Band 3 Polypeptide

Figure 8 is a model for the proposed transmembrane orientation of the putative membrane-spanning regions of band 3. Digestion of intact human and mouse erythrocytes with chymotrypsin cleaves band 3 into a carboxyl-terminal glycosylated peptide, CH35, and an N-terminal nonglycosylated CH65 (Fig. 5). Sequence data on the N terminus of human CH35 (Brock *et al.*, 1983) position this site between the putative membrane-spanning segments 5 and 6. It thus follows that this segment must be extracytoplasmic. As deduced by Brock *et al.* (1983) from their sequence, the region between segments 7 and 8 is extracellular, whereas the residues between 6 and 7 are exposed to the cytoplasm. Of the two potential sites (Asn–X–Ser/Thr) for attachment of the single N-linked oligosaccharide of band 3, the one at residue 611 corresponds to the site

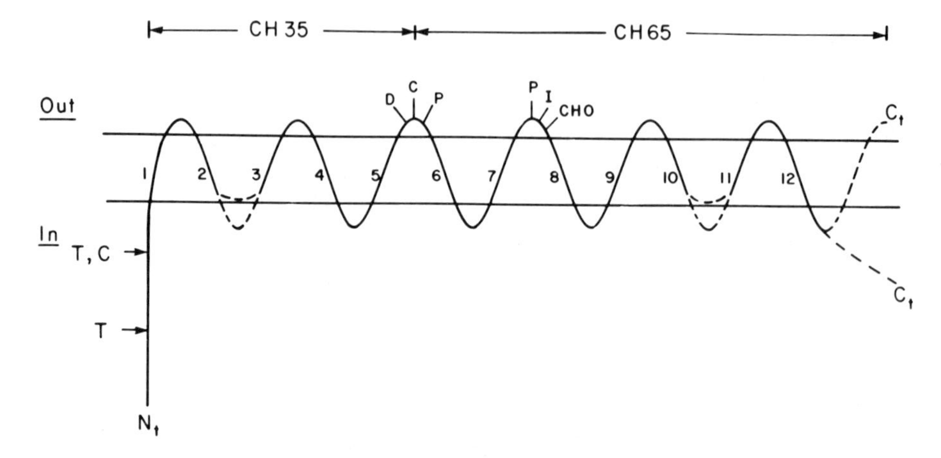

Figure 8. A model for the proposed transmembrane orientation of band 3. T, trypsin; C, chymotrypsin; CN, cyanogen bromide; P, papain; SH, cysteine residues; D, binding sites for H_2DIDS; I, exofacial radioiodination; CHO, proposed site for attachment of N-linked oligosaccharide; TRP, tryptophan-rich; IgG, antibody-binding; ANK, HINGE, ankyrin-binding.

in the human protein shown by Brock *et al.* (1983) not to contain carbohydrate. Therefore, the site at Asn^{660} must be the attachment site and thus extracellular. This analysis, taken together with the finding that the human Tyr^{646} is a substrate for extracellular radioiodination (Brock *et al.*, 1983), supports the conclusion that the region between helices 7 and 8 is extracellular.

The region encompassing segments 1 to 5 corresponds to part of the 17- to 22-kD fragment, CH17, that is generated by chymotryptic digestion of ghosts (Fig. 5). It must span the membrane an odd number of times, since it results from both an intra- and an extracellular proteolytic cleavage (Steck *et al.*, 1976). Indeed, studies with surface-specific radiolabeling reagents demonstrate that this fragment traverses the bilayer a minimum of three times (Jennings and Nicknish, 1984).

Assuming that peaks A, C, and D of the hydrophobicity plot correspond to three membrane-spanning segments, 1, 4, and 5, then region B (residues 453 to 491) either does not span the membrane or must do so twice. Hydrophobicity peak B is quite broad, encompassing 38 moderately apolar residues. These residues can be arranged in a structure composed of two antiparallel amphipathic helices with Pro^{477} forming a turn within the membrane or just at the interface between the bilayer and the cytoplasm. Analysis of the helical amphiphilicity (Fig. 6b) suggests that both the helices formed by this region of the protein would have all their polar residues projecting from a single face of each helix. Thus, we conclude tentatively that CH17 possesses five membrane-spanning α-helices. This conclusion is further supported by the identification of the three lysine residues in CH17 that are accessible to extracellular impermeant labeling reagents (see below).

The human counterpart to murine-spanning segments 6 and 7, a pepsin fragment (P5), has been sequenced by Brock *et al.*, (1983), who concluded that segment 7 forms an amphipathic helix through the lipid bilayer. The observation that papain cleavage of intact erythrocytes inhibits anion exchange and the substantial sequence homology between the mouse and human proteins in this region (Fig. 3) support its importance in the anion-exchange process.

Virtually nothing is known about the transmembrane orientation of the C terminus of band 3 corresponding to hydrophobic peaks G–J. Our model proposes four or five membrane crossings. The region corresponding to the 43 residues of peak I in Fig. 4 resembles membrane spans 2 and 3 in its ability to form an α-helical hairpin in which Arg^{800} is situated just at the cytoplasmic surface and the C-terminal 24 amino acids return to the extracellular surface. The proline residues at positions 796 and 802 may be involved in forming the hairpin structure. These two spanning segments, along with segment 12, could form moderately amphipathic helices. Following an extremely polar stretch between residues 832 and 850, there is a short region of 18 mixed hydrophobic and polar residues, which forms the N-terminal part of peak J in Fig. 4. This is the longest stretch of apolar amino acids between position 850 and the C terminus of the protein. We propose that this segment spans the membrane once, as an amphipathic α-helix (segment 12; see Fig. 5). This would position the C terminus of the protein inside the cell. Alternatively, the protein may traverse the membrane once more, leaving the C terminus outside the cell (shown by the dotted lines in Fig. 8). The exact disposition of the hydrophilic segments is currently under investigation with antibodies generated against synthetic peptides.

5.2. *Location of the H_2DIDS-Binding Lysines in Band 3*

Three lysines in human CH17 can be reductively methylated in intact cells by a membrane-impermeant reagent (Jennings and Nicknish, 1984) and thus must be exofacial. Two of these are within the C-terminal CNBr fragment of human CH17; the third has been mapped to the N-terminal CNBr fragment of CH17 (Jennings and Nicknish, 1984). There are only three lysine residues in murine CH17 (at positions 449, 558, and 561).

Given the high degree of homology between the human and mouse proteins in at least part of CH17 (H3), we assume that the positions of these lysine residues in human and mouse CH17 are the same.

One of the two lysines in the C-terminal CNBr fragment of human CH17 (CN11) is the one that binds extracellular 4,4′-diisthiocyano-2,2′-dihydrostilbene disulfonate (H_2DIDS), a bifunctional reagent that is a potent inhibitor of anion transport and a covalent cross linker of CH17 and CH35 (Jennings and Pasow, 1979). Segment 5 forms the C terminus of mouse CH17. The helical wheel diagram shown in Fig. 6b shows that segment 5 could form a membrane-spanning structure in which both Lys^{558} and Lys^{561} of mouse CH17 project together near the extracytoplasmic surface of the membrane from the same side of an amphipathic α-helix as the two acidic residues Glu^{554} and Asp^{565}. We therefore propose that the H_2DIDS-binding residue in CH17 is either Lys ^{558} or Lys^{561}.

The remaining lysine in mouse CH17 is located near the N terminus of the fragment (Lys^{449}). Since this residue is exofacial in human erythrocytes, it is likely that Lys^{449} in the mouse sequence must also face the extracytoplasmic surface, confirming that region A forms membrane-spanning domain 1 shown in Figs. 5 and 6 (see above discussion).

5.3. *The Structure and Mechanism of the Anion-Exchange Site*

Analysis of the predicted sequence of murine band 3 reveals that the protein possesses 12 potential membrane-spanning regions. Of these, five are quite hydrophobic and contain no residues with charged side chains. The seven remaining membrane-spanning regions contain hydrophobic, polar, and charged residues. We propose that the membrane-spanning regions of band 3 cross the plasma membrane in amphipathic helical structures that confine the polar and charged side chains to a single face of the helix. These helices, oriented with their axes transverse to the plane of the membrane, could cluster to form within the bilayer one or two hydrophilic regions in which the charged residues involved in the anion-exchange process are allowed to pair and achieve a stable structure. Such clustering of amphipathic helices has been proposed for other transport proteins that span the membrane multiple times (Eisenberg, 1984; Finer-Moore and Stroud, 1984). Significantly, the lysine residues that we propose bind to stilbene disulfonate inhibitors of anion exchange are associated with these amphipathic regions. The cross-linking of CH17 and CH35 by H_2DIDS suggests that these regions must be no more than 20 Å (the length of the H_2DIDS molecule) away from each other despite the fact that they are separated by at least 80 amino acids. Identification of the H_2DIDS-binding lysine residue in P25 will be an important contribution to understanding the spatial arrangement of the membrane-spanning regions of band 3.

The pH titration of chloride exchange in resealed ghosts (Wieth and Brahm, 1982) and treatment of ghosts with phenylglyoxal (Wieth *et al.*, 1982b) suggest the involvement in anion exchange of an arginine residue probably located near the H_2DIDS-binding Lys in CH35 (Wieth *et al.*, 1982a). This observation is consistent with our sequence of band 3; however, without additional experimental data, it is impossible to identify this residue, since any of the prospective H_2DIDS-binding lysines are in close proximity to arginines. Inhibition of anion exchange at acidic pH suggests the involvement in this process of a glutamate or aspartate residue (Wieth and Brahm, 1982). However, identification of these functional groups must await a better understanding of the three-dimensional arrangement of the protein in the bilayer and of the oligomeric interaction of band 3 protomers (reviewed in Jennings, 1984).

Band-3-mediated anion-exchange activity has been described in terms of a sequential (ping-pong) mechanism involving the oscillation of the protein between two conformational states in which the anion-binding site of each functional unit can exist at only one face of the membrane at any given time. An elegant model (Macara and Cantley, 1982) proposes that the pairing of an acidic and a basic residue within a positively charged cavity in the membrane forms an "anionic gate." Transport is initiated by the displacement of this charge pair by an exogenous anion and the formation of a new gate in an alternate conformational state. This hypothesis can account for the observed phenomena of substrate inhibition and "recruitment" of anion-binding sites to a single side of the membrane (Knauf, 1979).

All models of band-3-mediated anion exchange that have been proposed thus far have focused on the biochemistry and kinetics of this process and reflect the limitations imposed by the paucity of detailed structural information. In particular, these models have assumed that the anions are translocated through a single hydrophilic "channel" formed in the lipid bilayer by mono- or oligomeric band 3.

In Fig. 9 we present a hypothetical structure for band 3 in which the hydrophilic faces of the 12 amphipathic helices of a single band 3 monomer are oriented to form the interiors of two separate channels. The protein could alternate between two conformational states as a consequence of the binding of an anion to a site at one face of the membrane. The affinity of the anion-binding sites at the inner and outer faces of the membrane would be dependent on the conformation.

This model may help to explain the tight coupling of inward and outward translocation steps in the exchange process and the apparent

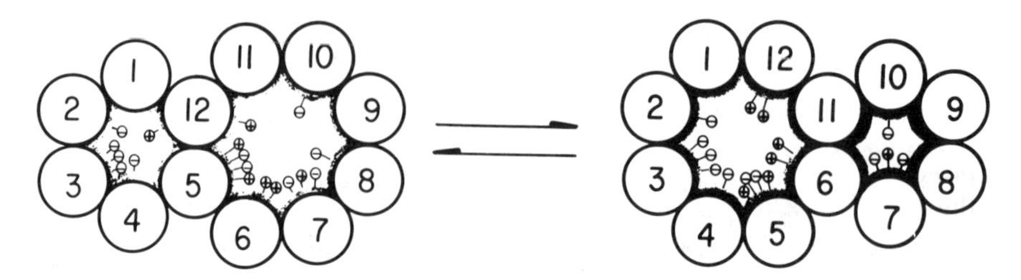

Figure 9. A model for the arrangement of the membrane-spanning helices of band 3. The shaded area designates the polar faces of the amphipathic helices.

involvement of both CH17 (helices 1–5) and CH35 (helices 6–12). Treatment of intact erythrocytes with papain, which cleaves the N terminus of CH35 (Jennings and Pasow, 1979), specifically inhibits the outward translocation step of anion exchange while actually accelerating the inward process (Jennings and Adams, 1981). The structure of bacteriorhodopsin (Henderson and Unwin, 1975; Khorna *et al.*, 1979) suggests that seven membrane spans appears to be sufficient to form a channel for the vectorial translocation of a single ion. It is intriguing that CH17, embedded in lipid bilayers, appears to behave as a DIDS-sensitive, voltage-dependent chloride channel (Galvez *et al.*, 1984).

Our model is presented as an alternative to previously described models that, implicitly and explicitly, envision a single passageway through which anions can pass across the membrane. It remains to be determined whether our model can account for all of the kinetic and biophysical data. Most importantly, however, several features of the structure proposed in Fig. 9 are testable. The loops on both sides of the membrane, which connect the membrane-spanning regions, are rich in basic amino acids. At least Lys^{558} and Lys^{561} and probably others may form part of the anion-binding domain. It will be interesting to mutate these and other lysine residues individually; such mutations could alter the activity of the protein and/or cause it to accumulate in one of two conformational states. Such studies are in progress and could contribute to our understanding of the molecular biology of anion transport.

ACKNOWLEDGMENTS. We thank V. Patel for advice on the MEL cell cultures and G. Natsoulis and M. Kreitman for assistance with DNA sequencing. We are grateful to S. Alper, R. Doolittle, M. Jennings, J. Kyte, A. Garcia, and R. Reithmeir for helpful discussions. This work was supported by grant #PO1-HL-27375 from the NIH. R.R.K. is supported by

NIH postdoctoral grant #F32 HL06680. We also thank Miriam Boucher for help with the manuscript.

References

Aviv, H., and Leder, P., 1972, Purification of biologically active globin messenger RNA by chromatography on oligothymidylic acid–cellulose, *Proc. Natl. Acad. Sci. U.S.A.* **69:**1408–1412.

Bennett, V., 1982, The molecular basis for membrane–cytoskeletal association in human erythrocytes, *J. Cell Biochem.* **18:**49–65.

Bennett, V., and Stenbuck, P. J., 1979, The membrane attachment protein for spectrin is associated with band 3 in human erythrocyte membranes, *Nature* **280:**468–473.

Bennett, V., and Stenbuck, P. J., 1980, Association between ankyrin and the cytoplasmic domain of band 3 isolated from the human erythrocyte membrane, *J. Biol. Chem.* **255:**6424–6432.

Braell, W. A., 1981, *Synthesis and Assembly of the Erythrocyte Anion Transport Protein*, Ph.D. Dissertation, Massachusetts Institute of Technology, Cambridge, MA.

Braell, W. A., and Lodish, H. F., 1981, Biosynthesis of the erythrocyte anion transport protein, *J. Biol. Chem.* **256:**11337–11344.

Brock, C. J., Tanner, M. J. A., and Kempf, C., 1983, The human erythrocyte anion-transport protein. Partial amino acid sequence, conformation and a possible molecular mechanism for anion exchange, *Biochem. J.* **213:**577–586.

Cabantchik, Z. I., Knauf, P. A., and Rothstein, A., 1978, The anion transport system of the red blood cell. The role of membrane protein evaluated by the use of "probes," *Biochem. Biophys. Acta* **515:**239–302.

Chirgwin, J. M., Przybyla, A. E., McDonald, R. J., and Rutter, W. J., 1979, Isolation of biologically active ribonucleic acid from sources enriched in ribonuclease, *Biochemistry* **18:**5294–5299.

Dayhoff, M. O., Barker, W. C., and Hunt, T. L., 1983, Establishing homologies in protein sequences, *Methods Enzymol.* **91:**524–545.

Drickamer, L. K., 1976, Fragmentation of the 95,00-dalton transmembrane polypeptide in human erythrocyte membranes, *J. Biol. Chem.* **251:**5115–5123.

Eisen, H., Bach, R., and Emery, R., 1977, Induction of spectrin in erythroleukemic cells transformed by Friend virus, *Proc. Natl. Acad. U.S.A.* **74:**3898–3902.

Eisenberg, D., 1984, Three-dimensional structure of membrane and surface proteins, *Annu. Rev. Biochem.* **53:**595–673.

England, B. J., Gunn, R. B., and Steck, T. L., 1980, An immunological study of band 3, the anion transport protein of the human red blood cell membrane, *Biochim Biophys. Acta* **623:**171–182.

Fairbanks, G., Steck, T. L., and Wallach, D. F. H., 1971, Electrophoretic analysis of the major polypeptides of the human erythrocyte membrane, *Biochemistry* **10:**2606–2617.

Finer-Moore, J., and Stroud, R. M., 1984, Amphipathic analysis and possible formation of the ion channel in an acetylcholine receptor, *Proc. Natl. Acad. Sci. U.S.A.* **81:**155–159.

Friend, C., Scher, W., Holland, J. G., and Sato, T., 1978, Hemoglobin synthesis in murine virus-induced leukemic cells *in vitro.*: Stimulation of erythroid differentiation by dimethyl sulfoxide, *Proc. Natl. Acad. Sci. U.S.A.* **68:**378–382.

Galvez, L. M., Jennings, M. L., and Tosteson, M. T., 1984, Incorporation of the DIDS-binding protein from the anion transport protein into bilayers, *Fed. Proc.* **43**:315.
Grinstein, S., Ship, S., and Rothstein, A., 1978, Anion transport in relation to proteolytic dissection of band 3 protein, *Biochim. Biophys. Acta* **507**:294–304.
Henderson, R., and Unwin, P. N. T., 1975, Three-dimensional model of purple membrane obtained by electron microscopy, *Nature* **257**:28–32.
Jennings, M. L., 1984, Oligomeric structure and the anion transport function of human erythrocyte band 3 protein, *J. Membr. Biol.* **80**:105–117.
Jennings, M. L., and Adams, M. F., 1981, Modification by papain of the structure and function of band 3, the erythrocyte anion transport protein, *Biochemistry* **20**:7118–7123.
Jennings, M. L., and Nicknish, J. S., 1984, Erythrocyte band 3 protein: Evidence for multiple membrane-crossing segments in the 1700-dalton chymotrypsin fragments, *Biochemistry* **23**:6432–6436.
Jennings, M. L., and Pasow, H., 1979, Anion transport across the erythrocyte membrane, *in situ* proteolysis of band 3 protein, and crosslinking of proteolytic fragments by 4,4′-diisothiocyano dihydrostilbene-2,2′-disulfonate, *Biochim. Biophys. Acta* **554**:498–519.
Jennings, M. L., Lackey, M. A., and Denney, D. H., 1984, Peptides of human erythrocyte band 3 protein. Protein produced by extracellular papain cleavage, *J. Biol. Chem.* **259**:4652–4660.
Kaul, R. K., Murthy, P. S. N., Reddy, A. G., Steck, T. L., and Kohler, H., 1983, Amino acid sequence of the N alpha-terminal 201 residues of human erythrocyte membrane band 3, *J. Biol. Chem.* **258**:7981–7990.
Khorna, H. G., Gerber, G. G., Herlihy, W. C., Gray, C. P., Anderess, R. J., Nihei, K., and Biemarin, K., 1979, Amino acid sequence of bacteriorhodopsin, *Proc. Natl. Acad. Sci. U.S.A.* **77**:5046–5050.
Knauf, P. A., 1979, Erythrocyte anion exchange and the band 3 protein: Transport kinetics and molecular structure, *Curr. Top. Membr. Transport* **912**:249.
Kopito, R. R., and Lodish, H. F., 1985, Primary structure and transmembrane orientation of the murine anion transport protein, *Nature* **316**:234–238.
Kozak, M., 1981, Possible role of flanking nucleotides in recognition of the AUG initiator codon by eukaryotic ribosomes, *Nucl. Acids Res.* **9**:5223–5262.
Kyte, J., and Doolittle, R. F., 1982, A simple method for displaying the hydropathic character of a protein, *J. Mol. Biol.* **157**:105–132.
Lemishka, I. R., Farmer, S., Rocaniello, V. R., and Sharp, P. A., 1981, Nucleotide sequence and evolution of a mammalian alpha-tubulin messenger RNA, *J. Mol. Biol.* **151**:101–120.
Low, P. S., Westfall, M. A., Allen, D. P., and Appell, K. C., 1984, Characterization of the reversible conformational equilibrium of the cytoplasmic domain of erythrocyte membrane band 3, *J. Biol. Chem.* **259**:13070–13076.
Macara, I. G., and Cantley, L. C., 1982, The structure and function of band 3, in: *Cell Membranes, Methods and Reviews* (E. Elson, W. Frazier, and L. Glaser, eds.), Plenum Press, New York, pp. 41–87.
Maniatis, T., Fritsch, E. F., and Sambrook, J., 1982, *Molecular Cloning: A Laboratory Manual*, Cold Spring Harbor Laboratory, New York.
Marks, P. A., and Rifkind, R. A., 1978, Erythroleukemic differentiation, *Annu. Rev. Biochem.* **47**:419–448.
Mawby, W. J., and Findlay, J. B. C., 1983, Characterization and partial sequence of di-iodosulphophenyl isothiocyanate-binding peptide from human erythrocyte anion-transport protein, *Biochem. J.* **205**:465–475.
Maxam, A., and Gilbert, W., 1980, Sequencing and end-labeled DNA with base-specific chemical cleavages, *Methods Enzymol.* **65**:499–560.

Patel, V. P., and Lodish, H. F., 1984, Loss of adhesion of murine erythroleukemia cells to fibronectin during erythroid differentiation, *Science* **224**:996–998.

Rao, A., 1979, Disposition of the band 3 polypeptide in the human erythrocyte membrane. The reactive sulfhydryl groups, *J. Biol. Chem.* **254**:3503–3511.

Sabban, E. L., Sabatini, D. D., Marchesi, J. T., and Adesnik, M., 1980, Biosynthesis of erythrocyte membrane protein band 3 in DMSO-induced Friend erythroleukemia cells, *J. Cell Physiol.* **104**:261–268.

Sanger, F., Nicklen, S., and Coulson, A. R., 1977, DNA sequencing with chain-terminating inhibitors, *Proc. Natl. Acad. Sci.* **74**:5463–5467.

Schiffer, M., and Edmundson, A. B., 1967, Use of helical wheels to represent the structures of proteins and to identify segments with helical potential, *Biophys. J.* **7**:121–135.

Shaklai, N., Yguerabide, J., and Ranney, H. M., 1977, Classification and localization of hemoglobin binding sites on the red blood cell membrane, *Biochemistry* **16**:5593–5597.

Staden, R., 1982, Automation of the computer handling of gel reading data produced by the shotgun method of DNA sequencing, *Nucl. Acids Res.* **10**:4731–4751.

Steck, T. L., 1974, The organization of proteins in the human red blood cell membrane. A review, *J. Cell Biol.* **62**:1–19.

Steck, T. L., Ramos, B., and Strapazon, E., 1976, Proteolytic dissection of band 3, the predominant transmembrane polypeptide of the human erythrocyte membrane, *Biochemistry* **15**:1153–1161.

Strapazon, E., and Steck, T. L., 1976, Binding of rabbit muscle aldoslase to band 3, the predominant polypeptide of the human erythrocyte membrane, *Biochemistry* **15**:1421–1424.

Walder, J. A., Chatterjee, R., Steck, T. L., Low, P. S., Musso, G. F., Kaiser, E. T., Rogers, P. H., and Arnone, A., 1984, The interaction of hemoglobin with the cytoplasmic domain of band 3 of the human erythrocyte membrane, *J. Biol. Chem.* **259**:10238–10246.

Wieth, J. O., and Brahm, J., 1985, Cellular anion transport, in: *The Kidney: Physiology and Pathophysiology* (G. Giebisch, and D. W. Seldin, eds.), Raven Press, New York, pp. 48–89.

Wieth, J. O., Anderson, O. S., Brahm, J., Bjerrum, P. J., and Borders, C. L., 1982a, Chloride–bicarbonate exchange in red blood cells: Physiology of transport and chemical modification of binding sites, *Phil. Trans. R. Soc. Lond.* **B299**:383–399.

Wieth, J. O., Bierrum, P. J., and Borders, C. L., Jr., 1982b, Irreversible inactivation of red cell chloride-exchange with phenylglycoxal, an arginine-specific reagent, *J. Gen. Physiol.* **79**:283–312.

Young, R. A., and Davis, R. W., 1983, Efficient isolation of genes by using antibody probes, *Proc. Natl. Acad. Sci. U.S.A.* **80**:1194–1198.

4

The Lysosomal Proton Pump

P. HARIKUMAR and JOHN P. REEVES

1. Introduction

Acid assists digestion. That this is true at the subcellular as well as the organismal level was recognized more than 90 years ago by Metchnikoff (1893), who observed that blue litmus turned red when engulfed by protozoans. Today it is universally recognized that lysosomes, the intracellular digestive organelles, show an internal pH approximately 2 units below that of their environment. This not only provides favorable conditions for the activity of lysosomal enzymes, which have acidic pH optima, but also assists directly in the digestive process by denaturing ingested proteins to make them more susceptible to enzymatic proteolysis. In other organelles, internal acidification serves a variety of more subtle purposes. The dissociation of ligands from their receptors in endosomes, the accumulation of catecholamines in storage granules, and the processing and sorting of newly synthesized proteins in the Golgi are processes that are regulated in part by intraorganellar acidity. The transmembrane pH gradient in these organelles is generated and maintained by an ATP-dependent proton pump residing in the organellar membrane. This chapter summarizes the evidence for, and the characteristics of, the proton pump found in lysosomes.

2. ATP-Dependent Acidification of Lysosomes

Early measurements of the internal pH of isolated lysosomes were conducted by measuring the accumulation of weak bases such as meth-

P. HARIKUMAR and JOHN P. REEVES • Roche Institute of Molecular Biology, Roche Research Center, Nutley, New Jersey 07110. *Present address of P.H.:* B and FT Division, Fiply, B.A.R.C., Trombay, Bombay 400085, India.

ylamine (Goldman and Rottenberg, 1973; Reijngoud and Tager, 1973; Henning, 1975). These studies revealed that the intralysosomal pH was 1.0–1.7 units lower than the external pH. It was suggested that this distribution of protons reflected a Donnan equilibrium across the lysosomal membrane (reviewed in Reijngoud and Tager, 1977; Reeves, 1984). Other investigators, however, accumulated evidence that suggested an involvement of ATP in intralysosomal acidification. Mego and his co-workers (Mego *et al.*, 1972; Mego, 1975) found that the presence of Mg-ATP stimulated the proteolysis of previously ingested [^{125}I]labeled bovine serum albumin by isolated lysosomes; this effect of ATP was blocked by ionophores that dissipated pH gradients, suggesting that ATP had stimulated the internal acidification of the lysosomes. Schneider (1979, 1981) and Dell'Antone (1979) reported that in rat liver lysosomes, the accumulation of weak bases (a measure of the intralysosomal acidity) was stimulated by the presence of Mg-ATP; the effects of ATP were blocked in an appropriate manner by proton-conducting ionophores. Reeves and Reames (1981) demonstrated that Mg-ATP exerted a time-dependent effect on the intralysosomal hydrolysis of amino acid methyl esters, suggesting that ATP had brought about a decrease in the intralysosomal pH. Again, the effects were blocked by proton-conducting ionophores.

The most convincing evidence in support of a lysosomal proton pump was provided by the elegant experiments of Ohkuma and his co-workers. These investigators coupled a pH-sensitive fluorescent dye, fluorescein isothiocyanate, to an indigestible dextran polymer (FITC-dextran) and allowed the polymer to be ingested by macrophages so that it eventually accumulated within the lysosomal compartment (Ohkuma and Poole, 1978). Fluorescence measurements of the cells indicated that the FITC-dextran was in an environment of pH 4.7–4.8. Lowering ATP levels in the cells by treatment with a combination of azide and 2-deoxyglucose led to a gradual increase in the lysosomal pH, which the authors suggested reflected an interference in the energy supply for a lysosomal proton pump. Direct evidence in support of this interpretation was later provided by Ohkuma *et al.* (1982), who labeled rat liver lysosomes with FITC-dextran. After isolation of the labeled lysosomes, it was shown that the addition of Mg-ATP to the medium produced a rapid fall in the internal pH (pH_i) of 0.3–0.5 units. This decline in pH_i was blocked by protonophores and by low concentrations of the sulfhydryl agent N-ethylmaleimide (NEM).

These results provide compelling evidence that an ATP-dependent proton pump is primarily responsible for the internal acidification of lysosomes.

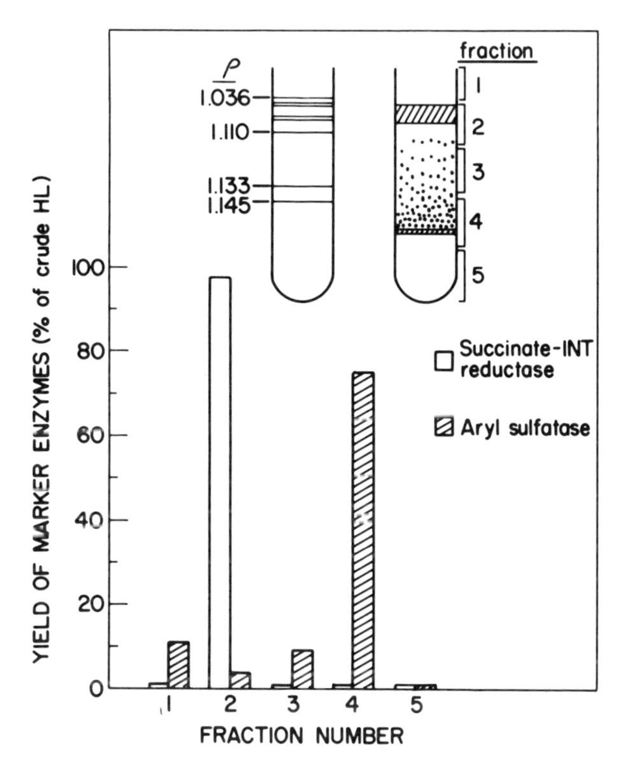

Figure 1. Separation of mitochondrial and lysosomal markers on a Percoll gradient. Lysosomes were prepared from rat kidney cortex as described by Harikumar and Reeves (1983). (From Harikumar and Reeves, 1983, with permission.)

3. Characteristics of the Lysosomal Proton Pump

In this section, the properties of the lysosomal proton pump are illustrated by data obtained in the authors' laboratory. However, it should be emphasized that many investigators, including those working with intracellular organelles other than lysosomes, have contributed to the current consensus regarding the characteristics of the proton pump.

We have described a simple procedure, based on earlier studies by Maunsbach (1969, 1974), for preparing highly purified lysosomes from rat kidney cortex using differential and Percoll density gradient centrifugation (Harikumar and Reeves, 1983). As illustrated in Fig. 1, the kidney lysosomes are considerably more dense than mitochondria, and the two can readily be separated on a Percoll gradient. The lysosomes obtained by

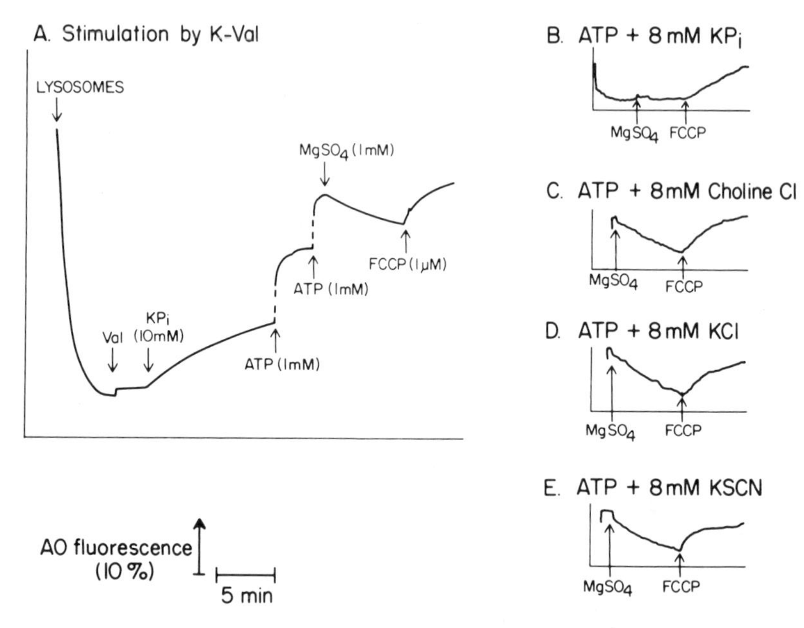

Figure 2. Acridine orange uptake by rabbit kidney lysosomes. Lysosomes were prepared from rabbit kidney using the method by Harikumar and Reeves (1983). A 20-μl aliquot of the lysosome suspension (approximately 200 μg protein) was placed in a cuvette containing 2 ml of 300 mM sucrose–50 mM MOPS/tris, pH 7.0, and 0.5 μM acridine orange. Fluorescence was monitored at 546 nm with excitation at 493 nm. The valinomycin concentration, when present (trace A), was 1 μM, and other additions were made as indicated.

this technique are purified approximately 40-fold over the crude homogenate with respect to aryl sulfatase and acid phosphatase activities and show practically no contamination with mitochondria, basolateral membranes, or endoplasmic reticulum. The lysosome preparations appear quite homogeneous when examined by electron microscopy and consist of spherical membrane-bounded electron-dense particles with diameters ranging from 0.5 to 1.4 μm. Mitochondrial profiles were seen with a frequency of approximately 1/160 lysosomes.

In a sucrose medium buffered at pH 7.0, the internal pH of these lysosomes is approximately 5.8, as determined by measuring the distribution of [^{14}C]-methylamine in the absence of ATP. This is also reflected in the accumulation of a fluorescent weak base, acridine orange (AO), a process that results in the quenching of the dye's fluorescence (Fig. 2A; see also Dell'Antone, 1979; Moriyama *et al.*, 1982). Thus, the AO fluorescence signal provides a qualitative measure of the intralysosomal pH such that a further decrease in fluorescence represents acidification of the lysosome. As shown in Fig. 2A, when the K^+ ionophore valinomycin was added to rabbit kidney lysosomes, a gradual increase in fluorescence occurred on the addition to external potassium phosphate (KP_i); this is best explained as an increase in the lysosomal pH as a result of the gradual loss of protons in exchange for the valinomycin-mediated entry of K^+. A much more rapid increase in fluorescence occurs if a K–H exchange ionophore such as nigericin is used in place of valinomycin (data not shown). As shown in Fig. 2B, little or no change in the AO fluorescence signal occurs if KP_i is added in the absence of an ionophore. The results indicate that the lysosomal membrane is relatively impermeable both to protons and to K^+. Similar results have been presented by Reinjgoud *et al.* (1976), who used methylamine accumulation as a measure of the intralysosomal pH.

When ATP is added to a suspension of lysosomes, there is an abrupt increase in the AO fluorescence signal (Fig. 2A); this, however, has nothing to do with changes in the lysosomal pH, since it is observed even in the absence of lysosomes (i.e., with AO and ATP alone). In lysosomes treated with K-valinomycin (Fig. 3A), the addition of Mg^{2+} in the presence of ATP activates the proton pump and brings about a gradual decline in the fluorescence signal, indicating intralysosomal acidification. This Mg-ATP-dependent change in AO fluorescence is reversed by the subsequent addition of a protonophore such as FCCP (Fig. 2A) and is completely blocked in FCCP is added prior to Mg-ATP (not shown). When Mg^{2+} is added to lysosomes incubated under similar conditions but without the prior addition of valinomycin, little or no additional acidification occurs (see trace B, Fig. 2). This indicates that net acidification of the lysosome requires an exchange of actively transported protons for internal cations (e.g., K^+ in the presence of valinomycin) or the coentry of external permeable anions. This interpretation is supported by the data in traces C, D, and E of Fig. 3, which indicate that Mg-ATP-dependent acidification occurs in the absence of valinomycin when external permeable anions such as Cl^- are present. The fact that external phosphate does not support net acidification by the lysosomal proton pump appears to rule out Schneider's suggestion (1983) that proton pumping involves an electroneutral H–P_i cotransport process.

3.1. *Electrogenicity*

One of the major questions concerning the lysosomal proton pump is whether or not proton translocation is directly coupled to the movement of other ions, i.e., whether pumping activity is an electroneutral or an

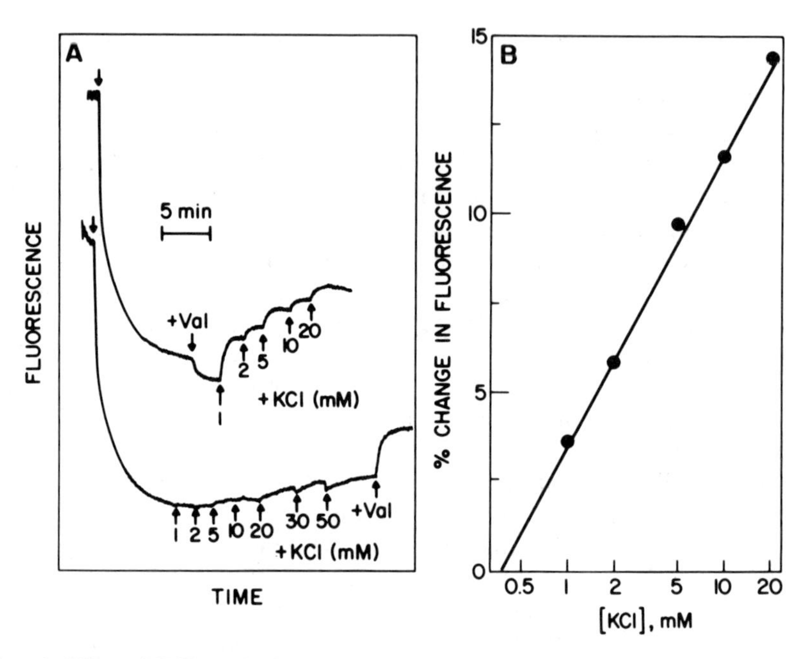

Figure 3. Effect of KCl on the fluorescence of a potential-sensitive dye in the presence of lysosomes. A: Kidney cortex lysosomes were added at the arrow to a cuvette containing 0.5 μM diS-C_3-(5) in 300 mM sucrose, 50 mM MOPS/tris, pH 7.4. Fluorescence was monitored at 670 nm with excitation at 622 nm. The initial decrease in fluorescence corresponds to approximately 65% of the total fluorescence. Upper trace: Effect of KCl in the presence of 0.5 μM valinomycin. Lower trace: Effect of KCl in the absence of valinomycin. B: The data in the upper trace are replotted versus the logarithm of the external K^+ concentration. (From Harikumar and Reeves, 1983, with permission.)

electrogenic process. The results presented above are most easily interpreted in terms of an electrogenic mechanism. Thus, net acidification occurs when permeable ions are present to short-circuit the positive (inside) membrane potential established by the proton pump; in the absence of permeable ions, the positive potential quickly generates an electrical force that opposes net pumping of protons into the lysosomes. Experiments using the voltage-sensitive carbocyanine dye diS-C_3-(5) provide firm evidence supporting an electrogenic mechanism (Harikumar and Reeves, 1983; see also Ohkuma *et al.*, 1983). This dye is positively charged and membrane permeable; thus, it accumulates in compartments that are electrically negative with respect to their surroundings, a process that (as in the case of acridine orange) results in the quenching of the dye's fluorescence.

As shown in Fig. 3, when lysosomes are added to a cuvette containing 0.5 μM diS-C_3-(5) in a sucrose medium (pH 7.0), 60–70% of the dye's fluorescence is quenched in a time-dependent manner. Most of this quenching reflects the accumulation of the dye by the lysosomes, but it is uncertain how much of this can be attributed to a negative membrane potential as opposed to the binding of the dye to lysosomal constituents. Once the steady state of dye accumulation has been attained, the fluorescence signal responds to imposed changes in the membrane potential in an appropriate manner. As shown in Fig. 3A, when the vesicles are treated with the K^+ ionophore valinomycin, the addition of external K^+ produces a dose-dependent increase in fluorescence. Indeed, the magnitude of the fluorescence change is a logarithmic function of the external K^+ concentration (Fig. 3A), suggesting that lysosomes behave like K^+ electrodes in the presence of valinomycin and that the diS-C_3-(5) signal provides an index of the membrane potential (Harikumar and Reeves, 1983).

The traces in Fig. 4 show that addition of Mg-ATP to the lysosomes in the presence of diS-C_3-(5) produces a rapid increase in fluorescence, suggesting that $\Delta\Psi$ becomes more positive inside. This fluorescence increase is reversed by protonophores (Fig. 4) and is completely prevented if the protonophores are added prior to Mg-ATP (not shown). The effect of Mg-ATP is also blocked by valinomycin in the presence of 10 mM K^+ (data not shown). Traces B and C in Fig. 4 show that the effect of Mg-ATP is still evident when the lysosomal pH gradient is dissipated by the addition of NH_4^+ or nigericin; thus, the fluorescence change reflects the activity of the proton pump *per se* rather than secondary changes caused by increased lysosomal acidity. (It should be noted that K^+-nigericin and NH_4^+ themselves produce an increase in fluorescence; this presumably reflects the decrease in the proton diffusion potential when the lysosomal pH gradient is collapsed by these agents.)

The measurements with diS-C_3-(5) have been confirmed by measuring the distribution of permeant ions such as $^{86}Rb^+$ (in the presence of valinomycin) and [^{14}C]-SCN^-. With $^{86}Rb^+$, the presence of Mg-ATP was shown to cause a shift in $\Delta\Psi$ from -90 mV to -50 mV; the results with [^{14}C]-SCN were similar except that the calculated values for $\Delta\Psi$ were each considerably more positive than for $^{86}Rb^+$ (an observation that may reflect binding of SCN^- to lysosomal constituents).

The effect of Mg-ATP on the diS-C_3-(5) signal is markedly reduced by the presence of permeant anions (e.g., $C1^-$, No_3^-, or SCN^-) in the external medium; P_i, however, did not attenuate the response of the dye to Mg-ATP. These results are consistent with the data presented above

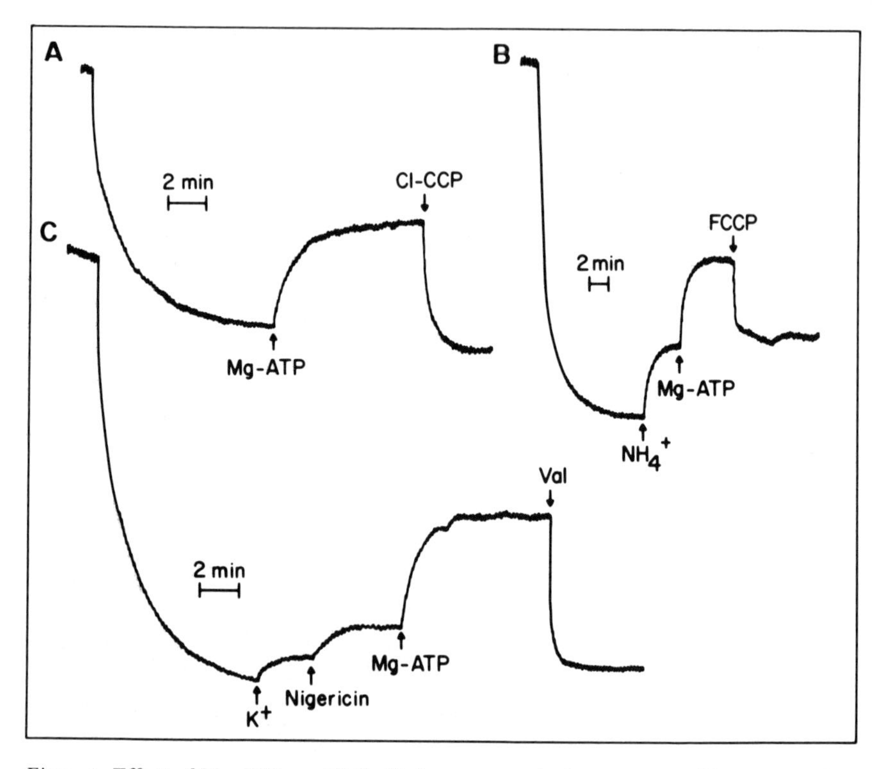

Figure 4. Effect of Mg-ATP on diS-C_3-(5) fluorescence in the presence of lysosomes. Conditions are as described in the legend to Fig. 3. Final concentrations of the various additions are as follows: ATP, 1 mM; $MgSO_4$, 1 mM; $(NH_4)_2SO_4$, 5 mM; K_2SO_4, 5 mM; carbonyl cyanide *m*-chlorophenylhydrazone (Cl-CCP), 1 μM; carbonyl cyanide *p*-trifluoromethoxyphenylhydrazone (FCCP), 1 μM; nigericin, μg/ml; and valinomycin, 1 μM. (From Harikumar and Reeves, 1983, with permission.)

(Fig. 2) showing that Cl^- and SCN^- stimulated ATP-dependent acidification of lysosomes, although P_i was ineffective.

3.2. Inhibitors

One difficulty in working with lysosomes is that the preparations are likely to be contaminated to a greater or lesser degree with mitochondria or submitochondrial particles. The latter are particularly troublesome because they exhibit an inverted topology with respect to the intact organelles and therefore show inwardly directed proton-pumping activity. Because of the high activity of the mitochondrial F_0F_1 ATPase, even a small contamination of the lysosomal preparations by submitochondrial particles could give misleading results. Fortunately, the organellar proton pump and the F_0F_1 ATPase can be distinguished by their differing specificities toward inhibitors. Thus, efrapeptin and oligomycin strongly inhibit the mitochondrial ATPase at low concentrations but have no effect on the lysosomal proton pump. Moreover, the lysosomal system is exceedingly sensitive to the sulfhydryl agent NEM (complete inhibition at concentrations of 50 μM or less), whereas the mitochondrial ATPase is not affected at these concentrations. The inhibitor profile for proton pumps in other intracellular organelles such as endosomes, chromaffin granules, coated vesicles, and Golgi membranes is similar to that of the lysosomal system (Rudnick, 1985).

3.3. Nucleotide Specificity

The order of effectiveness of various nucleotide substrates for the lysosomal proton pump is ATP > GTP ~ ITP > UTP ~ CTP (Reeves and Reames, 1981; Ohkuma *et al.*, 1982, 1983; Harikumar and Reeves, 1983). This order is observed for both net acidification and membrane potential measurements. In contrast, the proton pumps in other intracellular organelles such as endosomes, Golgi membranes, and endoplasmic reticulum membranes cannot utilize GTP in place of ATP (Merion *et al.*, 1983; Glickman *et al.*, 1983; Rees-Jones and Al-Awquati, 1984). This suggests that the lysosomal proton pump may be different from that of other organelles. Alternatively, it is possible that the lysosomes contain enzymes that can convert GTP or other nucleotides to ATP; nucleoside diphosphokinase is an example of such an enzyme, although it can only generate ATP from other nucleotides if ADP is present to accept the terminal phosphate. In any event, the observation that certain mutant cell lines exhibit a defect in endosomal acidification but show a normal lysosomal proton pump lends support to the idea that the two systems are different (Merion *et al.*, 1983).

4. Lysosome Vesicles

Most studies of the lysosomal proton pump have been carried out using intact lysosomes. However, Schneider (1983) reported that lysosomal membranes prepared by a freeze–thawing procedure exhibited ATP-dependent acidification as measured by methylamine accumulation. Moriyama *et al.* (1984a) prepared vesicles from purified lysosomes by osmotic shock and observed ATP-dependent changes in AO uptake and

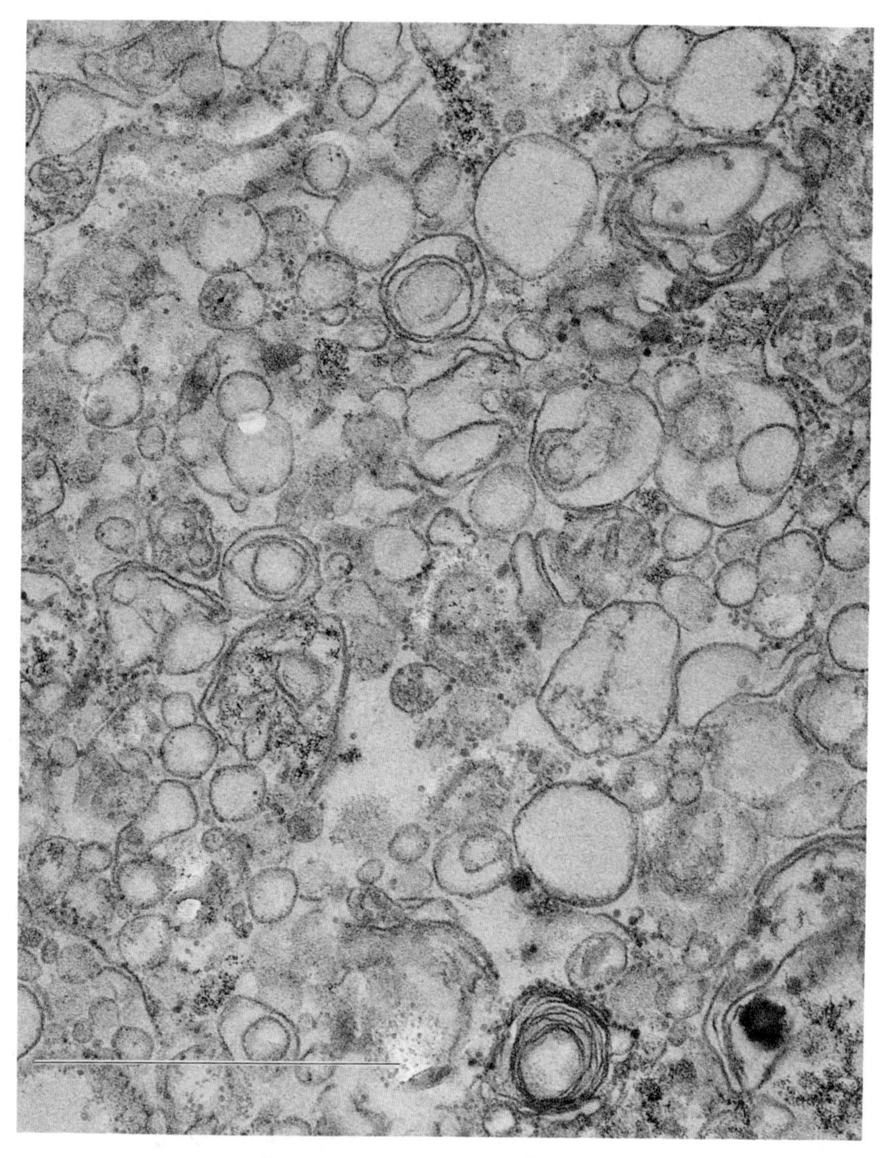

Figure 5. Electron micrograph of membrane vesicles produced by osmotic lysis of rat kidney lysosomes (bar = 1 μm). Vesicles were prepared by the following method. Intact lysosomes were prepared from rat kidney cortex as described by Harikumar and Reeves (1983); with some vesicle preparations, crude heavy lysosomes (HL), as described by Harikumar and Reeves (1983), were used as the starting material. The lysosomes were incubated in 100 mM KP_i (pH 7.0) containing 0.2 mM dithiothreitol and 10 mM L-glutamate dimethyl ester for 6 min at 37°C. The suspension was then centrifuged at 4400 × g_{max} for 10 min; the supernatant

diS-C_3-(5) fluorescence that were similar to those seen with the intact organelles. We have recently prepared membrane vesicles from kidney cortex lysosomes using glutamate dimethyl ester (10 mM) to bring about osmotic lysis of the lysosomes. As first shown by Goldman and Kaplan (1973), amino acid methyl esters penetrate the lysosome by passive diffusion, and the methyl groups are removed by the hydrolases therein. The resulting tree amino acid does not readily diffuse back out of the lysosome, so it accumulates to concentrations that eventually bring about osmotic disruption of the organelle. Lysosomal membrane vesicles prepared by this procedure are shown in Fig. 5. They have the typical appearance of vesicular membrane preparations and are devoid of the electron-dense internal constituents characteristic of the intact organelle. These vesicles show less than 1% of the aryl sulfatase activity found in the original lysosomes; the remainder is found in the supernatant after treatment with glutamate dimethyl ester, indicating extensive lysosomal disruption.

As shown in Fig. 6A, when vesicles were prepared in 100 mM KP_i (pH 7.0) and placed in a sucrose medium, the addition of the K–H exchange ionophore nigericin brings about a marked acidification of the vesicles, as reflected in AO accumulation. This demonstrates that the vesicles are osmotically intact and capable of generating and maintaining transmembrane ion gradients under appropriate conditions. When Mg-ATP is added to the vesicles in the presence of 100 mM choline chloride, a gradual acidification is observed that is reversed by the protonophore FCCP (Fig. 6, trace B). Oligomycin did not inhibit the ATP-dependent changes in AO fluorescence (trace C), but NEM (100 μM) completely blocked the response (trace F); these data indicate that the observed fluorescence changes originated from lysosomal vesicles rather than from contaminating submitochondrial particles. Also shown in Fig. 6 are traces indicating that vanadate does not inhibit proton-pumping activity (trace D), although dicyclohexylcarbodiimide (DCCD) does (trace E); and that ADP cannot substitute for ATP in stimulating acidification (trace G).

The vesicles also exhibit a Mg-dependent ATPase activity. Although this activity is similar in certain respects to proton-pumping activity (e.g., DCCD inhibits, whereas oligomycin and efrapeptin do not), the ATPase

---

was centrifuged again at 4400 × g_{max} for 10 min. The final supernatant was layered on a cushion of 34.5% sucrose and centrifuged at 35,000 rpm (218,000 × g_{max}) for 90 min in a Beckman SW40 rotor. The vesicles at the sucrose interface were collected by aspiration, diluted two- to threefold with KP_i, and centrifuged at 45,000 rpm in a Beckman Ti60 rotor for 90 min. The light brown pellet was resuspended in 100 mM KP_i, pH 7.0. We thank Mr. Frank Jenkins and Dr. Miloslav Boublik for producing the electron micrograph.

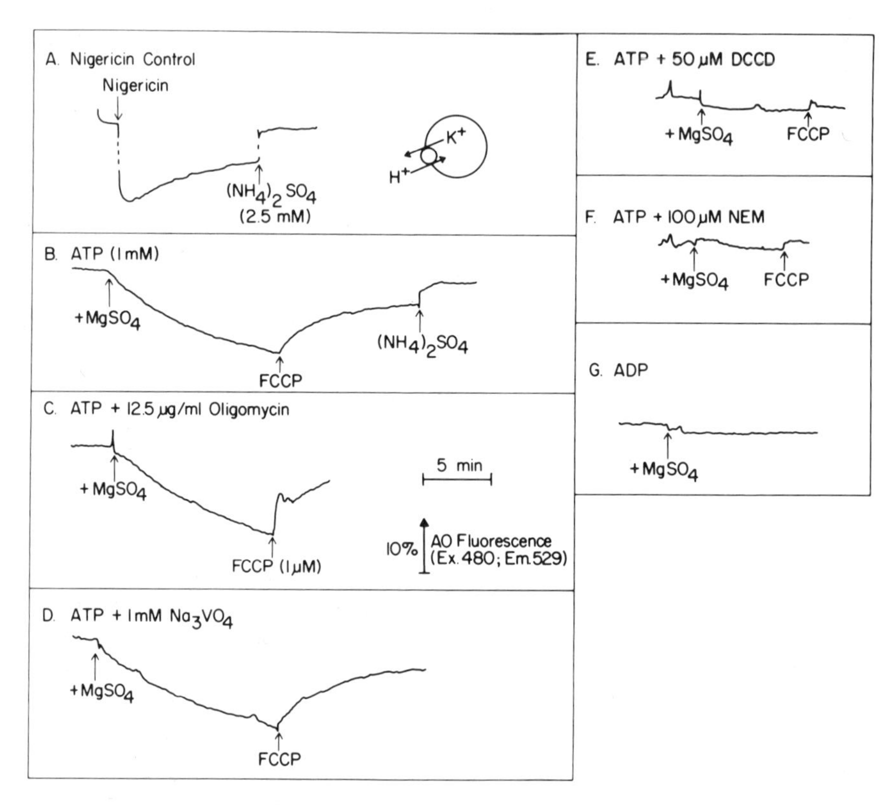

Figure 6. Acridine orange uptake by lysosomal membrane vesicles. Conditions were as described in Fig. 2 except that for traces B–G, the medium contained 100 mM choline in addition to the 300 mM sucrose. Vesicles were prepared as described in the legend to Fig. 5. The vesicles did not exhibit the initial decline in AO fluorescence shown by the intact lysosomes (cf. Fig. 2), indicating that there was no pH gradient across the vesicle membrane in the absence of Mg-ATP.

is quite insensitive to NEM. Thus, 100 μM NEM completely blocks ATP-dependent acidification but has little effect on ATPase activity, whereas 1 mM NEM inhibits ATPase activity by only 20% (Harikumar and Reeves, 1984; Moriyama *et al.*, 1984a). A similar observation for the chromaffin granule ATPase was reported by Flatmark *et al.* (1982). More recent results suggest that the chromaffin granule H-translocating ATPase is in fact sensitive to low concentrations of NEM once contaminating ATPases have been purified away (G. E. Dean, T. J. Nelson, and G. Rudnick, personal communication). Thus, it seems likely that the lysosomal vesicles contain other ATPases in addition to the proton pump.

Recently, Moriyama *et al.* (1984b) reported that the lysosomal ATPase could be solubilized with lysophosphatidylcholine and incorporated into proteoliposomes by detergent-dilution procedures. The reconstituted system showed ATP-dependent changes in the fluorescence of pH- and potential-sensitive dyes, suggesting that proton-pumping activity had survived this procedure. These results provide hope that the protein(s) corresponding to the lysosomal proton pump can be identified and their activity studied in the purified state. The purification of this ATPase and the development of antibodies directed against it will undoubtedly provide important new insights into the molecular mechanisms of proton pumps, the sorting processes involved in membrane and receptor recycling, and the role of pH gradients in regulating these processes.

REFERENCES

Dell'Antone, P., 1979, Evidence for an ATP-driven "proton pump" in rat liver lysosomes by basic dyes uptake, *Biochem. Biophys. Res. Commun.* **86:**180–189.

Flatmark, T., Gronberg, M., Husebye, E., Jr., and Berge, S. V., 1982, Inhibition by N-ethylmaleimide of the Mg-ATP-driven proton pump of the chromaffin granules, *FEBS Lett.* **149:**71–74

Glickman, J., Croen, K., Kelly, S., and Al-Awqati, Q., 1983, Golgi membranes contain an electrogenic H^+ pump in parallel to a chloride conductance, *J. Cell Biol.* **97:**1303–1308.

Goldman, R., and Kaplan, A., 1973, Rupture of rat liver lysosomes mediated by L-amino acid esters, *Biochim. Biophys. Acta* 318:205–216.

Goldman, R., and Rottenberg, H., 1973, Ion distribution in lysosomal suspensions, *FEBS Lett.* **33:**233–238.

Harikumar, P., and Reeves, J. P., 1983, The lysosomal proton pump is electrogenic, *J. Biol. Chem.* **258:**10403–10410.

Harikumar, P., and Reeves, J. P., 1984, The lysosomal proton pump, *Biochem. Soc. Trans.* **12:**906–908.

Henning, R., 1975, pH gradient across the lysosomal membrane generated by selective cation permeability and Donnan equilibrium, *Biochim. Biophys. Acta* **401:**307–316.

Maunsbach, A., 1969, Functions of lysosomes in kidney cells, in: *Lysosomes in Biology and Pathology*, Vol. 1 (J. T. Dingle and H. R. Fell, eds.), North-Holland, Amsterdam, pp. 115–153.

Maunsbach, A. B., 1974, Isolation of kidney lysosomes, *Methods Enzymol.* **31A:**330–339.

Mego, J. L., 1975, Further evidence for a proton pump in mouse kidney phagolysosomes: Effect of nigericin and 2,4-dinitrophenol on the stimulation of intralysosomal proteolysis by ATP, *Biochem. Biophys. Res. Commun.* **67:**571–575.

Mego, J. L., Farb, R. M., and Barnes, J., 1972, An adenosine triphosphate-dependent stabilization of proteolytic activity in heterolysosomes. Evidence for a proton pump, *Biochem. J.* **128:**763–769.

Merion, M., Schlesinger, P,. Brooks, R. M., Moehring, J. M., Moehring, T. J., and Sly, W. S., 1983, Defective acidification of endosomes in Chinese hamster ovary cell mutants "cross-resistant" to toxins and viruses, *Proc. Natl. Acad. Sci. U.S.A.* **80:**5315–5319.

Metchnikoff, E., 1983, *Lectures on the Comparative Pathology of Inflammation*, Kegan, Paul, Trench, Trübner and Co., London.

Moriyama, Y., Takano, T., and Ohkuma, S., 1982, Acridine orange as a fluorescent probe for lysosomal proton pump, *J. Biochem.* **92:**1333–1336.

Moriyama, Y., Takano, T., and Ohkuma, S., 1984a, Proton translocating ATPase in lysosomal membrane ghosts. Evidence that alkaline Mg^{2+}-ATPase acts as a proton pump, *J. Biochem.* **95:**995–1007.

Moriyama, Y., Takano, T., and Ohkuma, S., 1984b, Solubilization and reconstitution of a lysosomal H^+-pump, *J. Biochem.* **96:**927–930.

Ohkuma, S., and Poole, B., 1978, Fluorescence probe measurement of the intralysosomal pH in living cells and the perturbation of pH by various agents, *Proc. Natl. Acad. Sci. U.S.A.* **75:**3327–3331.

Ohkuma, S., Moriyama, Y., and Takano, T., 1982, Identification and characterization of a proton pump on lysosomes by fluorescein isothiocyanate dextran fluorescence, *Proc. Natl. Acad. Sci. U.S.A.* **79:**2758–2762.

Ohkuma, S., Moriyama, Y., and Takano, T., 1983, Electrogenic nature of lysosomal proton pump as revealed with a cyanine dye, *J. Biochem.* **94:**1935–1943.

Rees-Jones, R., and Al-Awqati, Q., 1984, Proton-translocating adenosine-triphosphate in rough and smooth microsomes from rat liver, *Biochemistry* **23:**2236–2240.

Reeves, J. P., 1984, The mechanism of lysosomal acidification, in: *Lysosomes in Biology and Pathology*, Vol. 7 (J. T. Dingle, R. T., Dean, and W. Sly, eds.), Elsevier, Amsterdam, pp. 175–199.

Reeves, J. P., and Reames, T., 1981, ATP stimulates amino acid accumulation by lysosomes incubated with amino acid methyl esters. Evidence for a lysosomal proton pump, *J. Biol. Chem.* **256:**6047–6053.

Reijngoud, D.-J., and Tager, J. M., 1973, Measurement of intralysosomal pH, *Biochim. Biophys. Acta.* **297:**174–178.

Reijngoud, D.-J., and Tager, J. M., 1977, The permeability properties of the lysosomal membrane, *Biochim. Biophys. Acta* **472:**419–449.

Reijngoud, D.-J., Oud, P. S., and Tager, J. M., 1976, Effect of ionophores on intralysosomal pH, *Biochim. Biophys. Acta* **448:**303–313.

Rudnick, G., 1986, Acidification of intracellular organelles: Mechanism and function, in: *Physiology of Membrane Disorders*, ed. 2 (T. E. Andreoli, D. D. Fanestil, J. F. Hoffman, and S. Schultz, eds.), Plenum Press, New York, 409–422.

Schneider, D. L., 1979, The acidification of rat liver lysosomes *in vitro*: A role for the membranous ATPase as a proton pump, *Biochem. Biophys. Res. Commun.* **87:**559–565.

Schneider, D. L., 1981, ATP-dependent acidification of intact and disrupted lysosomes. Evidence for an ATP-driven proton pump, *J. Biol. Chem.* **256:**3858–3864.

Schneider, D. L., 1983, ATP-dependent acidification of membrane vesicles isolated from purified rat liver lysosomes. Acidification activity requires phosphate, *J. Biol. Chem.* **258:**1833–1838.

5

Ion Pumps, Ion Pathways, Ion Sites

GEORGE SACHS, JOHN CUPPOLETTI, ROBERT D. GUNTHER, JONATHAN KAUNITZ, JOHN MENDLEIN, EDWIN C. RABON, and BJORN WALLMARK

1. Introduction

There has been a rapidly growing list of ATP-fueled ion pumps. Most of these are H^+-transporting ATPases ranging from the mitochondrial F_1F_0 ATPase through the H^+ pump of intracellular organelles to the gastric H^+–K^+ ATPase (Racker, 1976; Cidon and Nelson, 1983; Sachs *et al.*, 1976). In addition to the Ca^{2+} ATPase of sarcoplasmic reticulum and plasma membrane (Schatzman, 1982), the role of the Ca^{2+} pump of rough endoplasmic reticulum in fueling a facultative Ca^{2+} store has been recognized (Berridge, 1984), and its properties are currently of interest.

These ATPases can be classified in various ways. A biochemical definition is most frequently based on whether a phosphorylated intermediate is present, as demonstrated for the gastric H^+–K^+ ATPase or the Ca^{2+} ATPase (Heilmann *et al.*, 1984), or absent, as seen in lysosomal or renal medullary ATPase. It has been the practice to focus physiologically on the electrogenicity of a particular pump. Electrogenic pumps can be of various types, as was originally stated clearly by Mitchell (1966). A uniport pump can be considered a simple battery with an internal resistance, and function in the ion-transport sense requires a conductance in parallel. An antiport pump, if electrogenic, requires both a continuing supply of the two or more transported ions to appropriately located sites

GEORGE SACHS, JOHN CUPPOLETTI, ROBERT D. GUNTHER, JONATHAN KAUNITZ, JOHN MENDLEIN, and EDWIN C. RABON • CURE, Veterans Administration, Wadsworth Division, Los Angeles, CA 90073; and Department of Medicine, UCLA School of Medicine, Los Angeles, CA 90024. *BJORN WALLMARK* • AB Hassle, Molndal, Sweden.

on the pump and also a conductance in parallel to the pump to satisfy electroneutrality. A symport pump may also be electrogenic, and its rate will be determined by the supply of the pumped ions and the potential developed. On the other hand, an antiport or symport pump may be electroneutral when the pump mechanism requires exact balance of the charges flowing through it but again requires adequate access of the transported ions to the ion-binding sites.

There are various ways of regulating of ion pumps and their function. A major mechanism must involve alteration of adjunct ion pathways. In the case of an electrogenic pump, for example, the rate of the pump will be set by the electrochemical gradient facing the pump. For a univalent ion the equation is $\Delta\bar{\mu} = -[RT \ln [I']/[I''] - F \ln \text{PD}]$. In the absence of any conductance in the membrane other than the pump, the chemical term becomes zero, and the gradient will be expressed entirely as a potential. With infinite conductance, on the other hand, the potential term will be zero, and the chemical gradient will predominate. The conductance of the pump itself is also relevant. If the pump has a high conductance, then a small variation in membrane voltage will result in a large change in current. With a low pump conductance, variation in voltage will result in little change of current. Hence, the pump acts as a constant-voltage generator in the former and as a constant-current generator in the latter. Explicitly, the F_1F_0 ATPase generates constant voltage, and the Na^+–K^+ ATPase constant current.

An electrogenic pump will invariably generate a potential difference for variation between these extremes. The magnitude of this on the addition of ATP will depend on the value of the shunt conductance, but it will initially be rapid and will then decay as the concentration gradient rises. The actual potential difference will be comprised of a diffusional term, the Goldman equation relating to the permeant species, the pumped ion and its coion, and a pump term. In the case of a proton pump and a Cl^- conductance, the equations are (Heinz, 1981):

$$A_{\text{diff}} = (P_{\text{H}}[\text{H}]'' + P_{\text{Cl}}[\text{Cl}]')/(P_{\text{H}}[\text{H}]' + P_{\text{Cl}}[\text{Cl}]'')$$

and

$$j = J/(P_{\text{H}}[\text{H}]' + P_{\text{Cl}}[\text{Cl}]'')$$

therefore,

$$\exp(-ZF\Delta\psi/RT) = a_{\text{diff}} + j^2/2 + j\,(a_{\text{diff}} + j^2/4)^{1/2}$$

An electroneutral pump can also generate a potential difference depending on the diffusional terms and conductances of the permeant species.

Thus, modification of conductance can significantly alter the pump rate of an electrogenic pump of any type, but especially one of the uniport class. In the case of electrogenic symport or antiport or electroneutral antiport, modification of ionic pathways may alter the supply of ions to ligand sites on the pump. In this review we discuss the ion pathways associated with four pumps, each illustrating a different mechanism: the H^+ pump of lysosomes; the H^+ pump of renal medulla; the gastric H^+–K^+ ATPase; and the Ca^{2+} ATPase of rough endoplasmic reticulum. We then discuss briefly the pharmacology of the gastric ATPase.

2. The Lysosomal H^+ Pump

After some years of disagreement, it seems likely that lysosomes and perhaps also zymogen granules maintain proton gradients by both a Donnan distribution of negative charge across a proton-permeable membrane and a H^+-translocating ATPase (Reeves and Reames, 1981). The characteristics of this proton pump can be defined in several ways. Figure 1 shows that acidification of the intralysosomal space occurs with an impermeant cation and the addition of ATP in the presence of Cl^-. The presence of a proton gradient is established by reversal of the H^+ gradient following the addition to the medium of nigericin and K^+. In this granule, Cl^- could be acting as a simple conductance shunting a uniport pump, or it could be involved in a pump-coupled symport mechanism.

Figure 2 shows that Cl^- is acting to provide a shunt conductance for a uniport pump. Tritosomes are loaded with either K^+ or TMA^+ in the presence of sulfate, and hence gradients of K^+ can be formed by diluting the vesicles into either K^+ or TMA^+. In the case of an inward gradient of K^+, there is no proton transport even in the presence of valinomycin. This is interpreted as resulting from the presence of a large inwardly directed positive potential. With K^+ equal and the addition of valinomycin, there is significant proton transport, but less than in the presence of Cl^-. When internal K^+ is greater than external K^+, transport occurs in the absence of valinomycin, suggesting the presence of a significant K^+ conductance, but the proton transport rate and magnitude are much enhanced by the addition of valinomycin. Under these conditions, therefore, with a K^+ gradient inducing an internal negative potential, the transport rate is brought to the rate in the presence of inward Cl^- gradients by the addition of valinomycin. Thus, Cl^- is not crucial for proton trans-

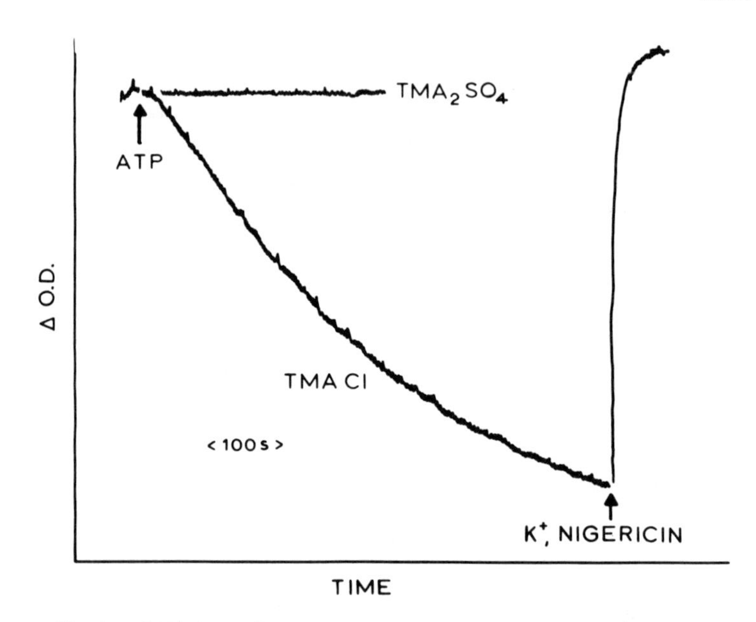

Figure 1. Uptake of H^+ into tritosomes occurs on the addition of Mg^{2+}-ATP to a particle suspension in the presence of Cl^- but not when SO_4^{2-} replaces Cl^-. The uptake of acridine orange monitors the ΔpH, as in Figs. 2 and 3.

port by lysosomes provided a conductance and potential are present across the membrane.

If a Cl^- conductance is indeed present, it should be possible to generate pH gradients across the tritosome membrane by applying a Cl^- gradient in the presence of the protonophore TCS, and this indeed occurs. In the absence of Cl^-, the ATPase should be able to generate an inward positive potential. Lysosomes take up the lipophilic cation diSC$_3$(5) in the absence of any permeant ion. This is probably because of the presence of a Donnan potential gradient. The addition of ATP results in the release of the cation along with the development of an inward positive potential. The presence of a pump inhibitor such as diethylstilbestrol or NEM prevents the ATP effect. These data allow the unique interpretation of the lysosomal ATPase as an electrogenic uniport pump (see Fig. 12). The regulation of the activity of this pump will thus depend on the concentration and gradient of the conducting ions, particularly Cl^-. Perhaps in secondary lysosomes additional Cl^--conductive elements or pump units are recruited.

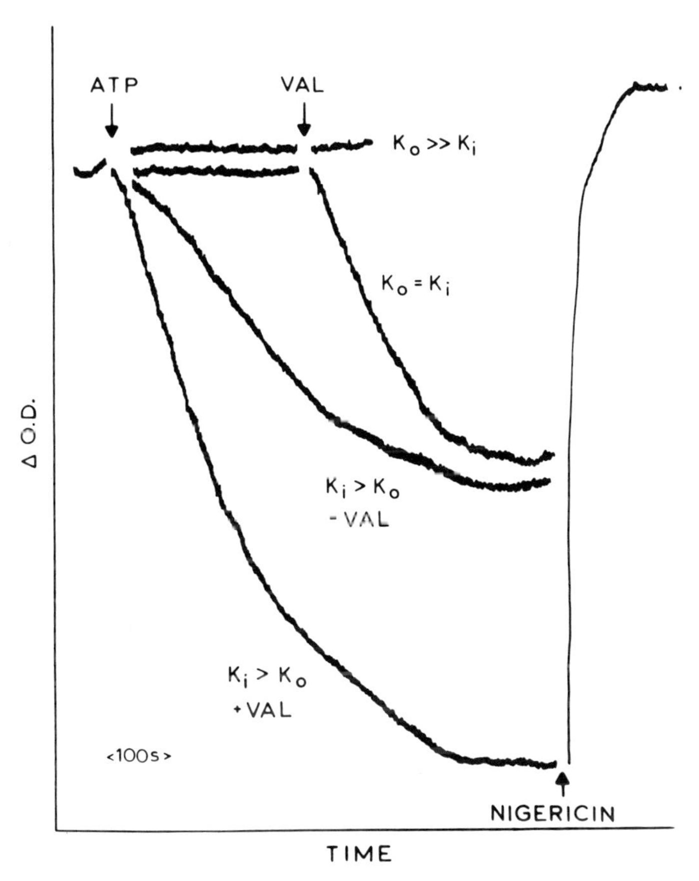

Figure 2. In the absence of Cl^-, valinomycin in the presence of K^+ is able to support ATP-dependent lysosomal acidification. The upper curve is in the presence of an inward K^+ gradient, and hence the ensuing potential does not allow electrogenic H^+ pumping. The second curve is in the absence of a K^+ gradient, and the lower two curves are in the presence of an outward K^+ gradient without and with valinomycin.

3. *The H^+ Pump of Renal Medulla*

The collecting duct of mammalian kidney is able to generate a luminal pH of about 4, giving a pH gradient of greater than 3 units. A microsomal fraction isolated from renal medulla has been shown to contain an NEM-sensitive, ATP-dependent H^+ pump (Gluck and Al-Awqati, 1984), and it has been concluded that the H^+ pump is electrogenic, as it is for clathrin-

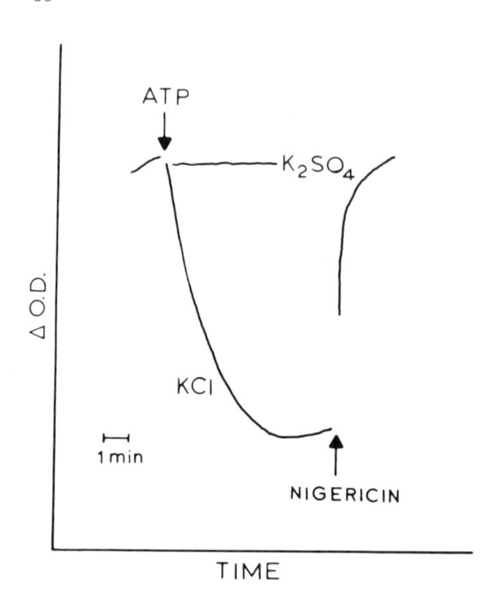

Figure 3. Acidification of renal microsomes occurs in the presence of Cl^- (or Br^-) only. Valinomycin-K^+ does not support transport in the absence of Cl^-.

coated vesicles. However, as described in this section, it appears that the H^+ pump is not of the simple uniport type but is rather an electrogenic symport.

Transport of H^+ by the renal proton pump occurs on addition of ATP in the presence of Cl^-. This transport is also NEM sensitive and vanadate insensitive. There is no detectable vesicular acidification in the presence of K^+ sulfate irrespective of the gradient of K^+ (Fig. 3). The addition of valinomycin in the absence of any K^+ gradient, with an impermeant ion, has only a small effect on proton transport, in contrast to the lysosomal proton pump. This argues against the Cl^- requirements reflecting a requirement for a simple shunt conductance for the proton pump. Searching for a Cl^- conductance by applying an internal negative potential by an inward Cl^- gradient, modifying the membrane conductance for H^+ by the addition of TCS, and then determining whether acidification occurs yields largely negative results. However, TCS does have an effect on the proton gradient developed by the ATPase: TCS blocks the proton gradient and dissipates it when it is formed by the addition of ATP and, moreover, increases the dissipation rate of the ATP-generated gradient. This is evidence that there is an electrogenic pump and some level of Cl^- conductance in the vesicles that transport H^+ in response to the addition of ATP. Perhaps the conductance is small enough that the gradient method for detection of conductance was too insensitive to detect it. Alternatively, the majority of vesicles in the preparation may have been leaky or may not have contained the H^+ pump.

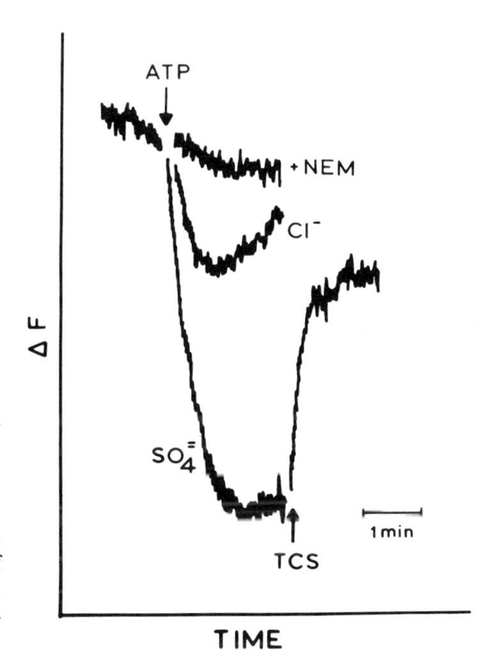

Figure 4. Renal vesicles develop a positive interior potential in the presence of ATP. The lower curve is in the absence of Cl^-, and uptake of the lipophilic anion diBAC(4)5 is reversed by the electrogenic protonophore TCS. The middle curve is the response in the presence of Cl^- showing a transient response, whereas the upper curve shows the inhibitory effect of NEM.

The H^+ pump is electrogenic as shown by ATP-dependent uptake of the lipophilic anion diBAC(4)5. In the absence of permeant ions there is uptake of this dye on addition of ATP that is dissipated by the addition of TCS (Fig. 4). In the presence of NEM, the pump inhibitor, no dye signal is found, and in the presence of Cl^- the potential is suppressed but not absent. It is, in fact, transient, as would be expected of an electrogenic pump that is allowed to develop a gradient of the pumped ion. Thus, this H^+ pump is electrogenic, as is the lysosomal pump, but it shows an absolute anion (Cl^-, Br^-) dependence.

This Cl^- dependence can be investigated further. When vesicles are K^+ loaded in the presence of K^+ sulfate and the Cl^- concentration is varied, there is an apparent saturation of transport rate as a function of external Cl^- concentration (data not shown). At saturation of Cl^-, there is no effect of addition of valinomycin with an outward K^+ gradient. This would result in a negative interior potential and would be expected to accelerate a uniport pump. In fact, if external K^+ is now varied, the surprising result obtained is that the pump rate increases with decreasing internal negativity. This implies that a secondary phenomenon is crucial for transport activity such as the activation of a Cl^- conductance or the moving of Cl^- to a pump site.

One interpretation of the data is that this renal H^+ transport mechanism is an electrogenic symport pump, not a uniport pump. Approximately similar data have been published for clathrin-coated vesicles, and a similar Cl^- dependence has been shown for synaptosomal vesicles (Cidon *et al.*, 1983). Regulation of this type of pump can therefore be achieved by a modification of the Cl^- site. In the case of the renal medulla and synaptosomes, fusion with the plasma membrane occurs. Here the pump membrane will face a positive potential, and with cytosolic Cl^- at about 15–20 mM, the activation of transport will occur. This interpretation emphasizes the point that if under no circumstances valinomycin-K^+ substitutes for Cl^-, a simple uniport mechanism is excluded. A model is presented in Fig. 12.

4. Gastric H^+–K^+ ATPase

The properties of the gastric ATPase vesicles differ depending on whether the vesicles are isolated from resting or from stimulated gastric mucosa. This section describes vesicles isolated from resting and stimulated mucosa in separate subsections.

4.1. "Resting" Vesicles

This ATPase falls into a different class from those previously discussed. It is responsible for a proton gradient 1000 times greater than the pump from renal medulla, and it is vanadate sensitive and relatively NEM sensitive. However, it forms a phosphorylated intermediate and appears to consist of a single peptide of 94 kD.

The transport catalyzed by the enzyme is shown in Fig. 5. Placing resting vesicles in KCl and then adding ATP results in little H^+ transport. However, if valinomycin is added, H^+ transport occurs. The upper right of the figure shows the effect of preincubation of the vesicles in KCl for 24 hr prior to the addition of ATP. There is now a very rapid uptake of H^+ sensitive to nigericin. Uptake also occurs if the vesicles are preincubated in K^+-sulfate solutions, but longer preincubation is necessary. The combination of these data shows that K^+ is required on the luminal face of this ATPase for H^+ transport to occur. That an H^+-for-K^+ exchange is occurring is shown at the bottom of the figure, where Rb^+ efflux is shown occurring with the addition of ATP to vesicles equilibrated in $^{86}Rb^+$ solutions. This efflux of Rb^+ is insensitive to TCS; thus, it is not coupled to a potential generated by an electrogenic proton pump but is directly transported by the ATPase (Schackmann *et al.*, 1977). This can

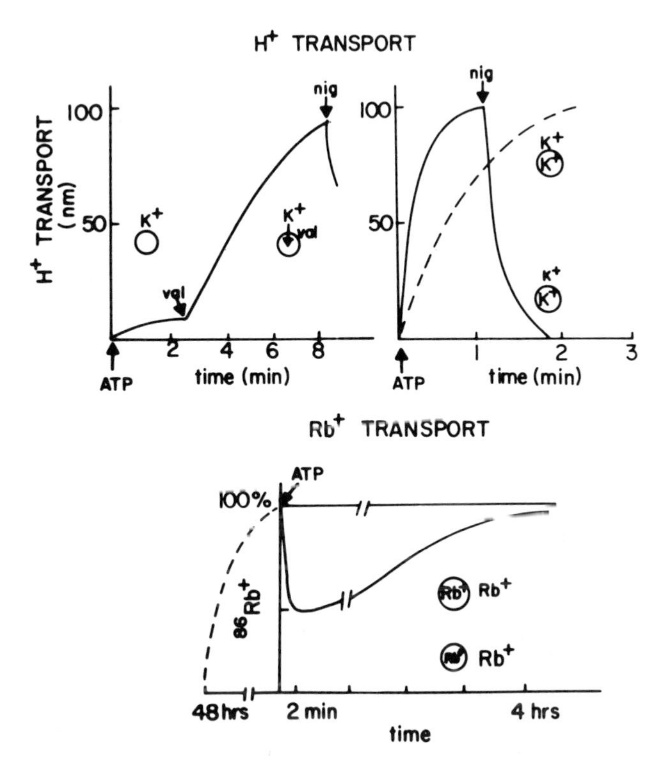

Figure 5. H^+ and Rb^+ transport by gastric vesicles from unstimulated hog mucosa. The upper left figure shows that when added to KCl, ATP-dependent H^+ uptake requires the addition of valinomycin to allow K^+ access to vesicle lumen. The upper right sector of the figure shows that preincubation in the presence of KCl abolishes the valinomycin requirement and results in more rapid H^+ transport than above. The lower figure shows that Rb^+ (as a K^+ substitute) actively effluxes from the vesicles at isotopic equilibrium on addition of ATP, showing direct coupling between H^+ uptake and cation efflux through the gastric H^+–K^+ ATPase.

also be shown directly by measuring ATPase activity in vesicles that have been preloaded with K^+ sulfate. In KCl solutions, the ATPase activity is stimulated by K^+, as shown in the right-hand panel of Fig. 6. The vesicles were ion tight, and K^+ alone stimulated the enzyme activity about twofold. The addition of valinomycin stimulated further, but maximal stimulation was obtained in the presence of the electroneutral K^+:H^+ exchange ionophore nigericin. This result shows that both transport and enzyme activity are internal K^+ limited. In the K^+-sulfate-loaded vesicles, ATPase activity is transient, falling to control levels as K^+ is depleted from the vesicles. A model consistent with these data is

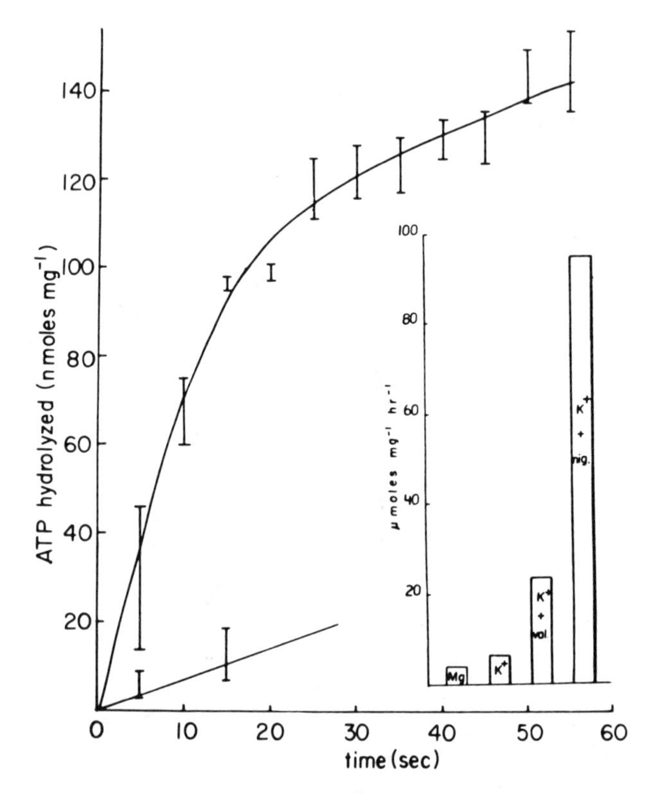

Figure 6. ATPase activity in the presence of KCl or K_2SO_4. The insert shows K^+ activation of ATPase activity in the presence of incubated KCl gradients in intact, resting gastric vesicles. K^+ alone only doubles activity; valinomycin increases activity fourfold; whereas nigericin stimulates 30-fold. The preloading of vesicles with K_2SO_4 allows transient activation of the K^+ ATPase because of K^+ efflux from the vesicles dependent on the H^+–K^+ ATPase (large figure).

that the vesicles lack KCl transport capability and that until K^+ enters the vesicles, no enzyme activity is possible.

Ion flux measurements can show which pathway (K^+ or Cl^-) is restricted. Thus, net RbCl flux is slow, $Cl^-:Cl^-$ exchange is equivalently slow, but $Rb^+:Rb^+$ exchange is fast, and its rate is commensurate with proton transport and ATPase rates. Thus, the rate of KCl entry is Cl^- limited, and the finding of a $Rb^+:Rb^+$ exchange independent of Cl^- but sensitive to vanadate, an inhibitor of the ATPase, suggests that the Rb^+ pathway is linked to the catalytic peptide of the ATPase (Rabon *et al.*, 1985).

Although it can be concluded from the above that the pump is an

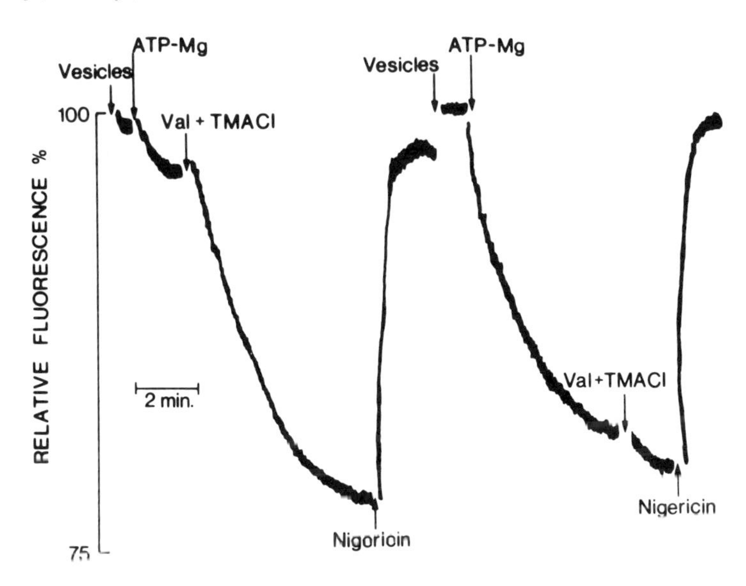

Figure 7. The difference in KCl dependence of transport in vesicles isolated from resting and stimulated mucosa. The left shows transport by vesicles isolated from resting tissue when added to KCl, showing the valinomycin requirement for H^+ transport following ATP addition. On the right, with vesicles isolated from stimulated tissue, no valinomycin is required because of the rapid unaided entry of KCl.

antiport type, its electrogenicity has not been established. An experiment using DOCC, a lipophilic cation, shows that a potential, interior negative, occurs only in the presence of TCS, an electrogenic protonophore; TCS also does not inhibit proton transport, which argues against the electrogenicity of the pump (data not shown). A lipophilic anion such as SCN^- or ANS^- should be taken up if an electrogenic pump potential develops in the H^+-pumping direction. This does not occur, but a potential does develop in the presence of valinomycin, an electrogenic K^+ ionophore. This shows that with K^+ export by the pump, a K^+ gradient develops, and with valinomycin turning the vesicle into a K^+ electrode, the diffusion potential of K^+ in the presence of valinomycin becomes detectable. These data generate the model of Fig. 12, an electroneutral exchange pump and a KCl pathway lacking in the vesicles isolated from resting gastric mucosa.

4.2. "Stimulated" Vesicles

If the above membranes represented the gastric pump, they would be ineffective in the secretion of HCl, since K^+ would not be supplied

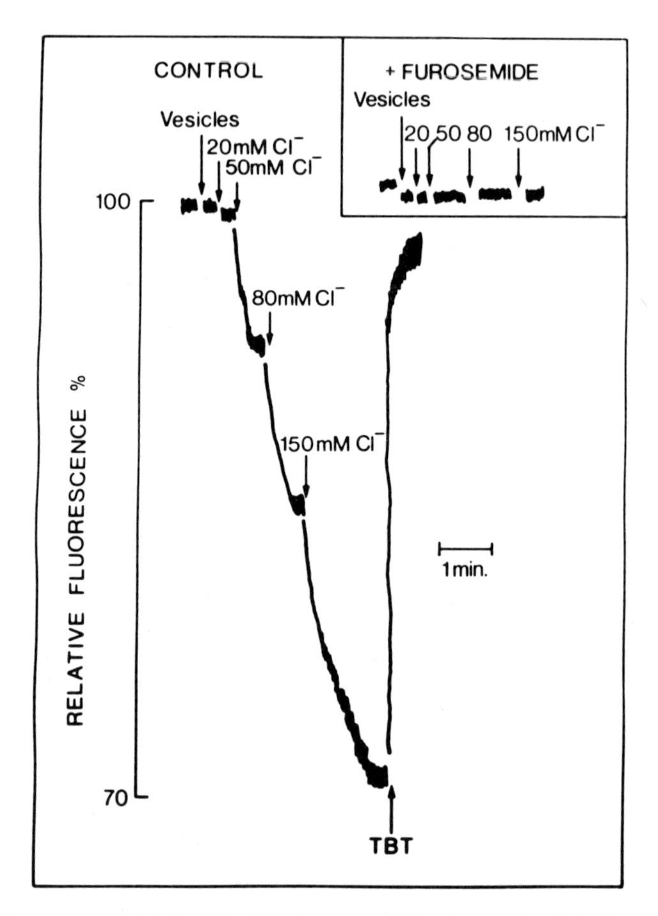

Figure 8. The measurement of a Cl^- diffusion potential by the uptake of the lipophilic cation $diSC_3(5)$ in response to applied inward Cl^- gradients in vesicles isolated from stimulated rabbit mucosa.

to the luminal face of the vesicles. Regulation of this proton pump must therefore be by the activation or insertion of a KCl pathway in parallel to the ATPase (Wolosin and Forte, 1983). Figure 7 (Cuppoletti and Sachs, 1984) shows that this is the case. When vesicles are isolated from stimulated mucosa, H^+ uptake into the vesicles occurs when ATP is added to vesicles placed in KCl medium with no valinomycin or preincubation. Indeed, valinomycin has little or no effect. Thus, a KCl pathway is present in the stimulated membranes.

If the K^+ conductance is measured by dilution of the K^+-sulfate-loaded vesicles into K^+-free medium, and the uptake of $diSC_3(5)$ is monitored, a K^+ conductance can be shown to be present, since the addition of external K^+ but not TMA^+ reverses the uptake of this lipophilic cation. Peculiarly, similar data are found for resting vesicles, consonant with the presence discussed above of Rb^+:Rb^+ exchange. However, it should be remembered that this technique measures only the K^+ conductance relative to sulfate. In contrast, when Cl^- conductance is measured by increasing the external Cl^- concentrations in TMA-sulfate medium, there is a Cl^- concentration gradient-dependent uptake of $diSC_3(5)$ dissipated by the electroneutral OH^-:Cl^- exchanger tributyl tin (Fig. 8). This phenomenon is not observed in resting vesicles, showing that the addition of a Cl^- conductance to the vesicles isolated from stimulated mucosa is a major change. Given the Cl^- conductance, it can be predicted that even if the pump is electroneutral, TCS will inhibit acid secretion, in contrast to the resting membranes. These data allow the model of Fig. 12, in which the secretion of HCl depends on the conductive entry of KCl followed by electroneutral H^+-for-K^+ exchange by the ATPase. Thus, in this system, regulation of the pump activity occurs not by alteration of the pump but by alteration of associated ion pathways.

5. *The Ca^{2+} Pump of Rough Endoplasmic Reticulum*

It has taken three decades since the demonstration of the phospholipid pathways mediating cholinergic and peptidergic responses of many cell types (Hokin and Hokin, 1952) to begin to understand the significance of these findings. The release of inositol triphosphate is central to the release of Ca^{2+} from intracellular stores in the rough endoplasmic reticulum (RER) (Berridge, 1984). Again, a pump is found in an enclosed intracellular vesicle, and uptake and release of Ca^{2+} must depend on the pump and parallel ion pathways.

There is ATP-dependent uptake of Ca^{2+} into RER isolated from rat liver. The suspension medium has significant effects. The largest and most rapid uptake is obtained in the presence of KCl, and replacement of either Cl^- or K^+ reduces uptake. In the presence of TCS and the absence of any ions, uptake dependent on ATP is significantly stimulated. This suggests that the provision of only a conductance satisfies the ionic requirements of the pump and, hence, that this ATPase is a uniport electrogenic system in spite of being of the phosphorylating type. If so, then a conductance for either Cl^- or K^+ must be present in the vesicles. The presence of a K^+ conductance is demonstrated by diluting K^+-loaded vesicles into tracer Rb^+ and measuring uptake of the tracer. The Rb^+ uptake following dilution is blocked by TEA, a blocker of K^+ conductance; hence, conductance for this cation is the physiological means of providing the counterion for Ca^{2+} uptake.

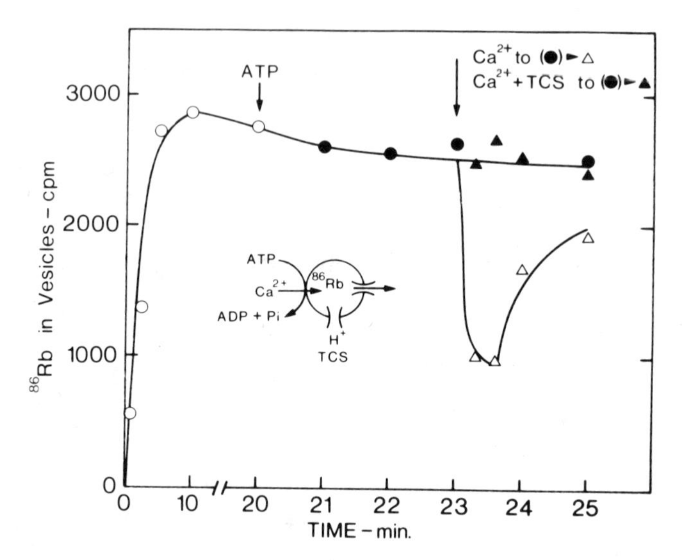

Figure 9. The uptake of $^{86}Rb^+$ in RER vesicles from rat liver showing a rapid approach to equilibrium in the presence of Cl^- and no effect of ATP on the distribution until Ca^{2+} is added. The transient $^{86}Rb^+$ efflux was blocked by TCS, showing that the $^{86}Rb^+$ distribution was measuring a pump-induced vesicle potential.

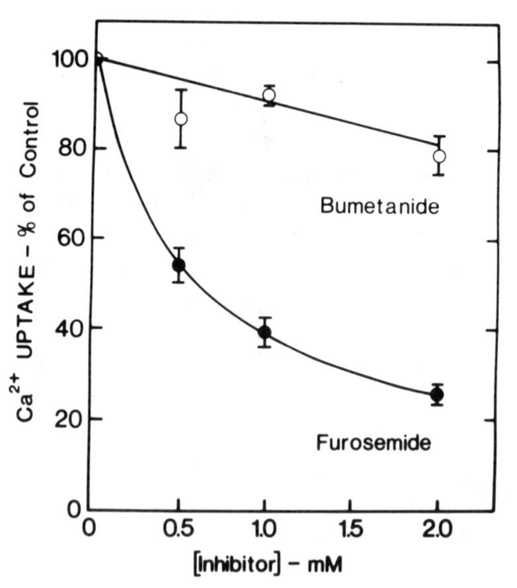

Figure 10. The inhibition of ATP-driven Ca^{2+} uptake into RER in the presence of KCl by furosemide but not bumetamide, showing the requirement for coupled KCl influx.

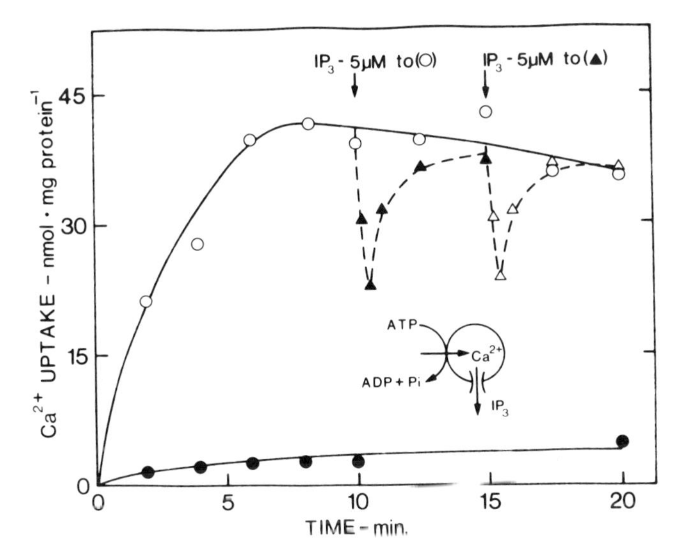

Figure 11. The release of Ca^{2+} from RER following ATP-dependent Ca^{2+} uptake on addition of IP_3.

If the pump is electrogenic, a potential should develop, and Fig. 9 demonstrates that this occurs. Here, tracer Rb^+ was allowed to accumulate in the RER. The addition of ATP had no effect on the distribution of Rb^+, but the subsequent addition of Ca^{2+} resulted in Rb^+ efflux. This efflux was conductance mediated, as shown by the finding that TCS prevented Rb^+ release by providing an alternate conductive pathway, in contrast to the data with gastric vesicles. The transient nature of the potential is to be expected, since a Ca^{2+} gradient develops over the time of this tracer experiment. Thus, we have an electrogenic pump and a K^+ conductance; K^+ must be provided inside the vesicles for the system to be able to accumulate large quantities of Ca^{2+}. This could occur by means of a coupled KCl pathway. To demonstrate this, net uptake of RbCl is followed, and it is found that Rb^+ uptake is Cl^- dependent and furosemide blocked. When RER vesicles are placed in KCl, furosemide blocks the ATP-dependent uptake of Ca^{2+}, showing that the KCl cotransport pathway is relevant to action of the Ca^{2+} pump (Fig. 10). Neither the K^+ conductance nor the KCl pathway is affected directly by IP_3.

After the vesicles have accumulated Ca^{2+} in the presence of KCl, the addition of IP_3 results in a transient release of Ca^{2+}. The transient nature results from the breakdown of IP_3 by the RER fraction (Fig. 11). The effect of IP_3 is to open a conductance, because in the presence of

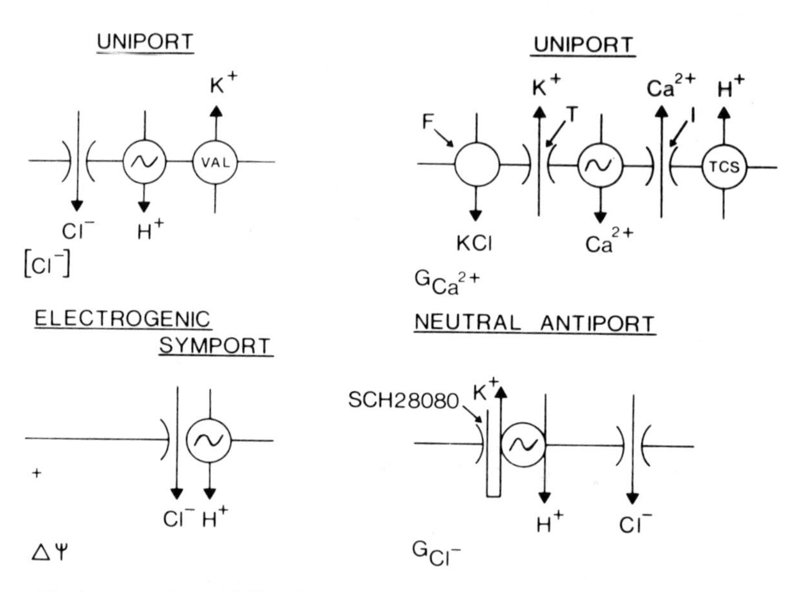

Figure 12. A composite modeling the lysosomal H^+ pump (upper left), the renal H^+ pump (lower left), the resting and stimulated gastric H^+ pump (lower right), and the pathways in liver RER (upper right).

IP_3, Ca^{2+}-loaded vesicles diluted into Ca^{2+} tracer take up Ca^{2+}, and the uptake is blocked by K^+ providing an alternate conductive pathway (data not shown). The efflux of Ca^{2+} that is IP_3 induced, if conductive, must require countertransport of a conductive species. Thus, an ion such as K^+ must be present for the IP_3-mediated release of Ca^{2+} to occur.

This Ca^{2+} pathway could be either pump associated or independent. Loading the RER with K^+ and diluting it into K^+-free medium results in the uptake of Ca^{2+}. In the absence of IP_3, however, this Ca^{2+} uptake is completely blocked by vanadate, and valinomycin does not overcome the vanadate inhibition, showing that it is both the Ca^{2+} and the K^+ pathways that are blocked by the pump inhibitor vanadate. In the presence of IP_3, the Ca^{2+} uptake remains vanadate sensitive and TEA sensitive, but when valinomycin is added, partial Ca^{2+} uptake is driven by the inward negative potential. Thus, the pump is reversed by K^+ gradients because of the K^+ diffusion potential, but the IP_3 pathway is not part of the pump. As above, this pathway is conductive, since K^+-gradient-driven uptake is TEA blocked and valinomycin reversed. This allows the model of Fig. 12 to be postulated, in which there is a uniport Ca^{2+} ATPase, a K^+ conductance that is TEA inhibited, a KCl pathway that is furosemide sensitive, and a Ca^{2+} channel that is opened by IP_3. A summary of the various ion pathways that are pump associated is given in Fig. 12.

6. *Pharmacology of the H^+–K^+ ATPase*

6.1. *Potassium Site Antagonists*

It is not clear what direct application there might be in controlling the H^+ pump of lysosomes or kidney. Regulation of the intracellular Ca^{2+} pump could be of major significance in the regulation of stimulus–secretion coupling in various organs. In the case of the gastric H^+–K^+ ATPase, inhibition of this enzyme would be a most effective means of dealing with acid-related diseases of the upper gastrointestinal tract. One can think of several means of doing this. Blockade of the Cl^- site or the K^+ site by selective agents would be one reasonable means of achieving inhibition of acid secretion.

To illustrate this, we first select compound SCH28080, a pyridyl-

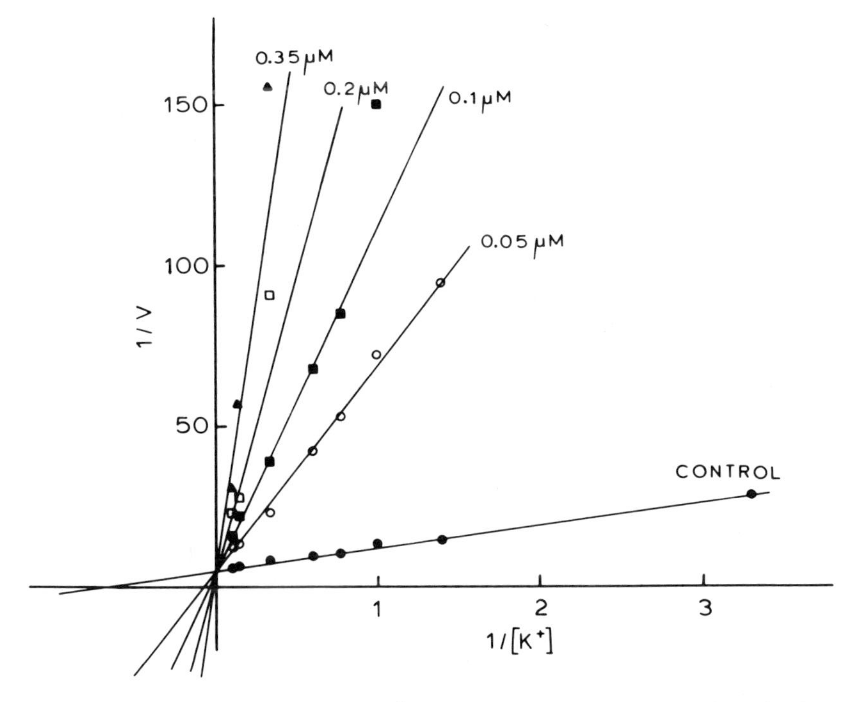

Figure 13. Lineweaver–Burke plots of H^+–K^+ ATPase activity as a function of K^+ concentration in the absence and presence of SCH 28080 at different concentrations.

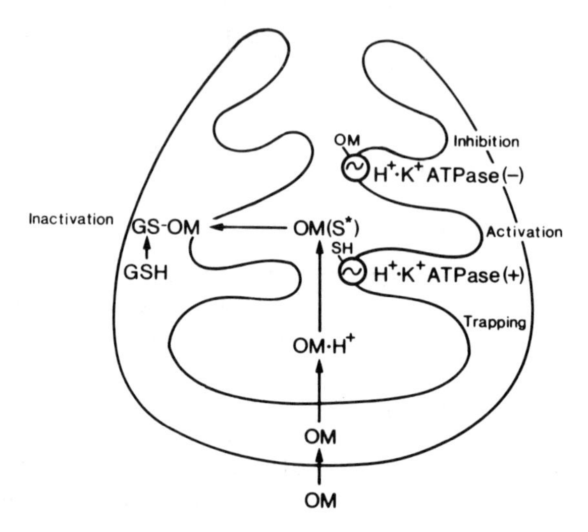

Figure 14. A model of the action of the acid secretion inhibitor omeprazole. This shows acid gradient-dependent accumulation of the weak base, acid activation, and reaction with an SH group of the H^+–K^+ ATPase. Any cellular access of acidic compound reacts with glutathione, rendering it harmless.

imidazole derivative (Kaminski *et al.*, 1985) shown to be an effective inhibitor of acid secretion in animals and man. In contrast to H_2 antagonists, it appears to block all forms of stimulus. It has been shown previously that hydrophobic amines competitively block H^+–K^+ ATPase by competition with K^+ (Im *et al.*, 1984). Inspection of the formula of this compound shows that when protonated it would also act as a hydrophobic amine. It should also be noted that because of its pK_a it would be expected to accumulate in the acid space of the parietal cell. Its effect on ATPase activity is competitive with K^+, as shown in Fig. 13. Thus, this compound would be expected to block the $K^+:K^+$ exchange catalyzed by the H^+–K^+ ATPase, and this in fact occurs. Interaction with the catalytic cycle shows that the compound blocks phosphorylation and competitively prevents the action of K^+. Thus, this compound mainly interacts with the E_2 form of the enzyme, preventing further cycling of the ATPase. It remains to be seen whether this reversible type of inhibition is superior to the irreversible inhibition achieved by omeprazole, an SH reagent (Fellenius *et al.*, 1981).

6.2. *Sulfhydryl Reagents*

Figure 14 is a model of the inhibitory mechanism of omeprazole. The compound is unreactive at neutral pH, but since it is a weak base it is

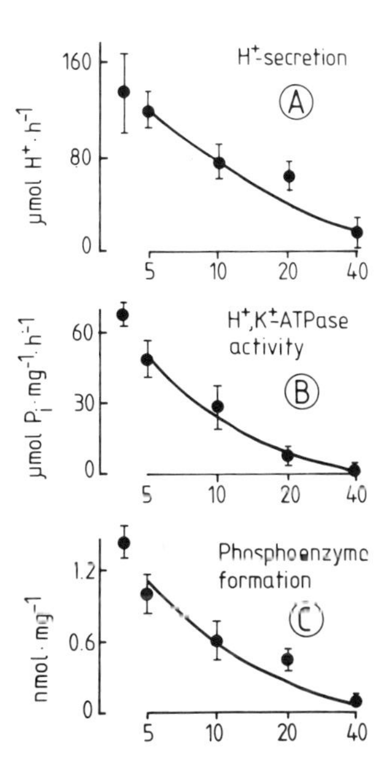

Figure 15. Dose–response relationship for the inhibition of acid secretion, H^+–K^+ ATPase activity, and phosphoenzyme formation in rats treated with omeprazole. Omeprazole was given as single oral doses. Measurement of the antisecretory effects and preparation of membranes were done 3 hr after administration. Control groups given vehicle (0.25% Methocel) were run in parallel at each dose group. A: Effect of omeprazole on supramaximal carbachol (110 nmole·kg^{-1}·hr^{-1})- plus pentagastrin (20 nmole·kg^{-1}·hr^{-1}, s.c.)-stimulated acid secretion in the chronic fistula rat. B,C: Effect on K^+-stimulated ATPase activity (B) and K^+-sensitive phosphoenzyme (C) in the density-gradient-purified microsomal fraction of the rat mucosa.

concentrated in the acid space of the parietal cell. The compound is activated in acid and becomes a highly reactive SH reagent that reacts with an SH group on the ATPase, irreversibly inhibiting the enzyme. This compound has already proved useful in the type of severe clinical hyperacidity that occurs in the Zollinger–Ellison syndrome.

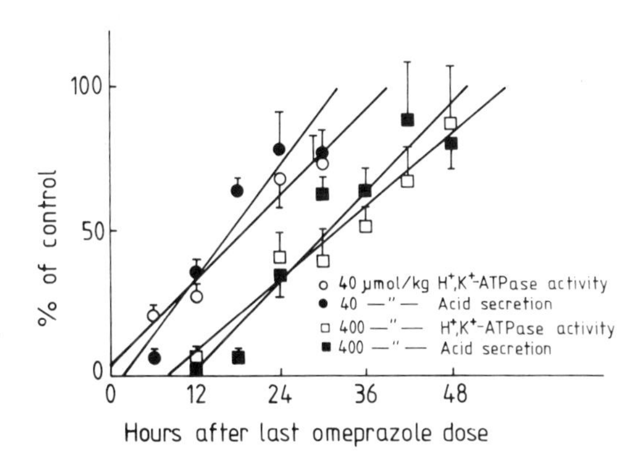

Figure 16. Recovery of acid secretion and H^+–K^+ ATPase activity following omeprazole treatment. Omeprazole was given once daily for 3 consecutive days in doses of 40 and 400 μmole/kg, respectively. The effects on supramaximally stimulated acid secretion and K^+-stimulated ATPase activity in the microsomal density gradient fraction were measured at the times indicated. The results are expressed as a percentage of the secretory rates and K^+-stimulated ATPase activities obtained in control rats.

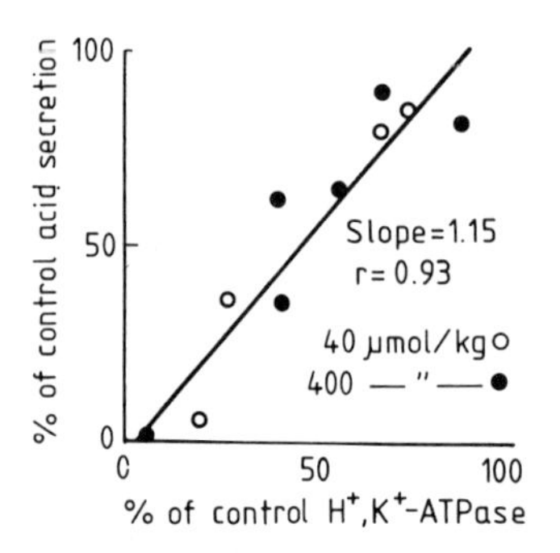

Figure 17. Correlation between inhibition of acid secretion and H^+–K^+ ATPase activity under the influence of omeprazole. Data were taken from Fig. 16. The correlation line was drawn according to the method of least squares.

When rats were given increasing doses of omeprazole, a progressive decrease in the rate of gastric acid secretion was found (ED_{50} = 10 μmole/kg) (Fig. 15A). In parallel to inhibition of acid secretion, decrease of both the K^+-stimulated ATPase and the K^+-sensitive phosphoenzyme levels of the H^+–K^+ ATPase occurred, with ED_{50} values of 12 and 10 μmole/kg, respectively (Fig. 15B,C). However, no effect on the basal Mg^{2+}-stimulated ATPase activity was seen after omeprazole treatment (data not shown).

The administration of supramaximal doses of omeprazole (40 and 400 μmole/kg, respectively) initially induced complete inhibition of both acid secretion and H^+–K^+ ATPase activity (Fig. 16). The recovery of both the rate of acid secretion and H^+–K^+ ATPase activity was found to follow the same time course (Fig. 16). When the degrees of inhibition for the rate of acid secretion and H^+–K^+ ATPase activity were correlated for different times following administration of omeprazole, a straight-line relationship was obtained with a slope close to unity (Fig. 17). Thus, a direct correlation between the rate of gastric acid secretion and the activity of the gastric H^+–K^+ ATPase was obtained irrespective of whether gastric acid secretion was progressively inhibited under submaximal omeprazole doses or the recovery of acid secretion was followed after inhibition had been completely established.

ACKNOWLEDGMENTS. This work is supported by National Institutes of Health grant AM17328 and the Veterans Administration SMI award. We gratefully acknowledge the contributions of Dr. Shmuel Muallem.

References

Berridge, M. J., 1984, Inositol triphosphate and diacylglycerol as second messengers, *Biochem. J.* **220:**345–360.

Cidon, S., and Nelson, N., 1983, A novel ATPase in the chromaffin granule membrane, *J. Biol. Chem.* **258:**2892–2898.

Cidon, S., Ben-David, H., and Nelson, N., 1983, ATP driven proton fluxes across membranes of secretory organelles, *J. Biol. Chem.* **258:**11684–11688.

Cuppole, H. J., and Sachs, G. 1984, Regulation of gastric acid secretion via modulation of a chloride conductance, *J. Biol. Chem.* **259:**14952–14957.

Fellenius, E., Berglindh, T., Sachs, G., Olbe, L., Elander, B., Sjöstrand, S. E., and Wallmark, B., 1981, Substituted benzimidazoles inhibit gastric acid secretion by blocking $(H^+ + K^+)$ ATPase, *Nature* **290:**159–161.

Gluck, S., and Al-Awqati, Q., 1984, An electrogenic proton translocating ATPase from bovine kidney medulla, *J. Clin. Invest.* **73:**1704–1710.

Heilmann, C., Spamer, C., and Gerok, W., 1984, The calcium pump in rat liver endoplasmic reticulum. Demonstration of the phosphorylated intermediate, *J. Biol. Chem.* **259:**11139–11144.

Heinz, E., 1981, *Electrical Potentials in Membrane Transport,* Springer, Berlin.

Hokin, M. R., and Hokin, L. E., 1952, Enzyme secretion and the incorporation of P^{32} into phospholipids of pancreas slices, *J. Biol. Chem.* **203:**967–977.

Im, W. B., Blakeman, D. P., Mendlein, J., and Sachs, G., 1984, Inhibition of $(H^+ + K^+)$ ATPase and H^+ accumulation in hog gastric membranes by trifluoperazine, verapamil and 8 (N,N-diethylamino)octyl-3,4,5-trimethoxybenzoate, *Biochim. Biophys. Acta* **770:**65–72.

Kaminski, J. J., Bristol, J. A., Puchalski, C., Lovey, R. G., Elliott, A. J., Guzik, H., Solomon, D. M., Conn. D. J., Domalski, M. S., Wong, S.-L., Gold, E. H., Long, J. F., Chiu, P. J. S., Steinberg, H., and McPhail, A. T., 1985, Antiulcer agents. I. Gastric antisecretory and cytoprotective properties of substituted imidazole (1-2-α) pyridines, *J. Med. Chem.* (in press).

Mitchell, P., 1966, Chemiosmotic coupling in oxidative and photosynthetic phosphorylation, *Biol. Rev.* **41:**445–502.

Rabon, E., Gunther, R. D., Soumarmon, A., Lewin, M. J. M., and Sachs, G., 1985, *J. Biol. Chem.* **260:**6641–6653.

Racker, E., 1976, *A New Look at Mechanisms in Bioenergetics,* Academic Press, New York.

Reeves, J. P., and Reames, T., 1981, ATP stimulates amino acid accumulation by lysosomes incubated with amino acid methyl esters. Evidence for a lysosomal proton pump, *J. Biol. Chem.* **256:**6047–6053.

Sachs, G., Chang, H. H., Rabon, E., Schackmann, R., Lewin, M., and Saccomani, G., 1976, A non-electrogenic H^+ pump in plasma membranes of hog stomach, *J. Biol. Chem.* **251:**7690–7698.

Schackmann, R., Schwartz, A., Saccomani, G., and Sachs, G., 1977, Cation transport by gastric H^+:K^+-ATPase, *J. Membr. Biol.* **32:**361–381.

Schatzman, H. J., 1982, The plasma membrane Ca^{2+} pump of erythrocytes and other animal cells, in: *Membrane Transport of Calcium* (E. Carafoli, ed.), Academic Press, New York, pp. 41–108.

Wolosin, J. M., and Forte, J. G., 1983, Kinetic properties of the KCl transport at the secreting apical membrane of the oxyntic cell, *J. Membr. Biol.* **71:**195–207.

III

CALCIUM CHANNELS AND THE CALCIUM PUMP

6

Shifts between Modes of Calcium Channel Gating as a Basis for Pharmacological Modulation of Calcium Influx in Cardiac, Neuronal, and Smooth-Muscle-Derived Cells

A. P. FOX, P. HESS, J. B. LANSMAN, B. NILIUS, M. C. NOWYCKY, and R. W. TSIEN

1. Introduction

When Ca^{2+} channels open in response to an appropriate change in membrane potential, they allow Ca^{2+} ions to move down their electrochemical gradient into the cytoplasm. This inflow of Ca^{2+} not only transfers depolarizing charge into excitable cells but also carries a specific message to be decoded by Ca^{2+} receptor proteins. The signal leads to the initiation of contraction in heart and smooth muscle cells, transmitter release from nerve cell synaptic terminals, hormone secretion by gland cells, and other important cellular responses (Hagiwara and Byerly, 1981, 1983; Reuter, 1983; Tsien, 1983). In linking membrane potential changes to the delivery of a messenger substance, Ca^{2+} channels perform a function that is vital and possibly unique (Tsien *et al.*, 1983; Hille, 1984).

The development of new methods for recording Ca^{2+} channel currents from single cells and single channels (Hamill *et al.*, 1981; Lee *et al.*, 1980) has led to a rapid growth in our basic understanding of Ca^{2+}

A. P. FOX, P. HESS, J. B. LANSMAN, B. NILIUS, M. C. NOWYCKY, and R. W. TSIEN • Department of Physiology, Yale University School of Medicine, New Haven, Connecticut 06510.

channels, with contributions from many groups (for reviews, see Reuter, 1983; Hagiwara and Byerly, 1983; Tsien, 1983). In our own laboratory, patch-clamp recordings have allowed us to study the mechanism of ion permeation and selectivity (Hess and Tsien, 1984; Hess *et al.*, 1985b; Lansman *et al.*, 1985) and to identify and characterize multiple types of Ca^{2+} channels in sensory neurons (Nowycky *et al.*, 1985b) and ventricular heart cells (Nilius *et al.*, 1985). These topics are not reviewed in this chapter, and the interested reader is referred to primary papers cited above.

The main focus in this chapter is on the mechanisms of Ca^{2+} channel modulation. A large variety of hormones, neurotransmitters, and drugs are known to alter Ca^{2+} entry into cells by increasing or decreasing Ca^{2+} channel activity (for review, see Reuter, 1983; Tsien, 1983; Siegelbaum and Tsien, 1983). Here we are mainly concerned with two important classes of Ca^{2+} channel modulator: (1) dihydropyridine (DHP) Ca^{2+} agonists and antagonists and (2) β-adrenergic agonists. Calcium agonists are exemplified by Bay K 8644 and CGPP 392. These compounds increase cardiac contractility and divalent cation influx even though they are very similar in structure to nifedipine and other DHP Ca^{2+} antagonists. Important differences in molecular conformation may help to explain why such similar compounds produce seemingly opposite effects (see Triggle *et al.*, Chapter 7, this volume).

We and others have used cell-attached patch recordings to study the effects of Ca^{2+} agonists and antagonists on the activity of single Ca^{2+} channels, both in heart cells (Ochi *et al.*, 1984; Kokubun and Reuter, 1984; Hess *et al.*, 1984) and in neurons (Nowycky *et al.*, 1985a). This chapter reviews this type of analysis and provides new information about basic similarities between large-conductance Ca^{2+} channels in smooth-muscle-derived cells and in neuronal and cardiac cells. It includes a more detailed description of Ca^{2+} channels in ventricular heart cells and examines the mechanisms of enhancement of Ca^{2+} channel activity by DHP Ca^{2+} agonists and β-adrenergic agonists and of inhibition of DHP Ca antagonists. The overall conclusion is that Ca^{2+} channels in a variety of cell types share the ability to express different modes of gating behavior; shifts between these gating modes can be an important mechanism in modulatory effects of neurochemicals and drugs.

2. *Experimental Methods*

Unitary calcium channel activity was recorded from cell-attached patches by the method of Hamill *et al.* (1981), using patch pipettes filled with 110 mM $BaCl_2$ and 10 mM HEPES (pH 7.3 or 7.5). In some cases, the resting potential of the cell was zeroed by exposing it to isotonic K^+-aspartate bathing solution. Drugs were administered by delivering a bolus of concentrated solution to the solution outside the cell. Smooth-muscle-cell-derived A10 cells were cultured by standard techniques (see Kimes and Brandt, 1976; Kongsamut *et al.*, 1985) (from stocks kindly provided by Drs. R. J. Miller and S. Kongsamut of the Department of Physiological and Pharmacological Sciences, University of Chicago). Heart cells were obtained from guinea pig ventricles by enzymatic dissociation (see Lee and Tsien, 1983, for references). Sensory neurons were isolated from dorsal root ganglia (DRG) of 8- to 12-day old chick embryos and maintained for 3–10 days in cell culture before recording (see Nowycky *et al.*, 1985a, for references). Methods for filtering, storing, and analyzing single-channel recordings are as previously described (Hess *et al.*, 1984).

3. *Large-Conductance Ca^{2+} Channels in a Smooth Muscle Cell Line*

Calcium channels in smooth muscle are of obvious importance in activation of contraction and as a target of action of therapeutic drugs, but they have received relatively little attention in studies using modern electrophysiological methods. This situation is changing with the application of whole-cell and single-channel recording techniques to isolated smooth muscle cells (e.g., Isenberg and Klockner, 1985; Caffrey *et al.*, 1985; Mitra and Morad, 1985).

We have used the patch-clamp method to study Ca^{2+} channel activity in A10 cells, a continuous cell line derived from rat aorta (Kimes and Brandt, 1976). Figure 1 shows a cell-attached patch recording made from a large-conductance Ca^{2+} channel in an A10 cell. The Ca^{2+} channel activity is evoked by applying depolarizing pulses from a holding potential of −70 mV to a test potential of 0 mV. With 110 mM Ba as the charge carrier in the pipette solution, the channel type under investigation activates with relatively strong depolarizations (−10 mV and upwards) and displays a relatively large slope conductance (~25 pS). In both of these respects, it bears a strong resemblance to the large-conductance Ca^{2+} channel in heart cells (Reuter *et al.*, 1982; Cavalie *et al.*, 1983; Nilius *et al.*, 1985) and in DRG neurons (Nowycky *et al.*, 1985a,b; see also Brown *et al.*, 1982).

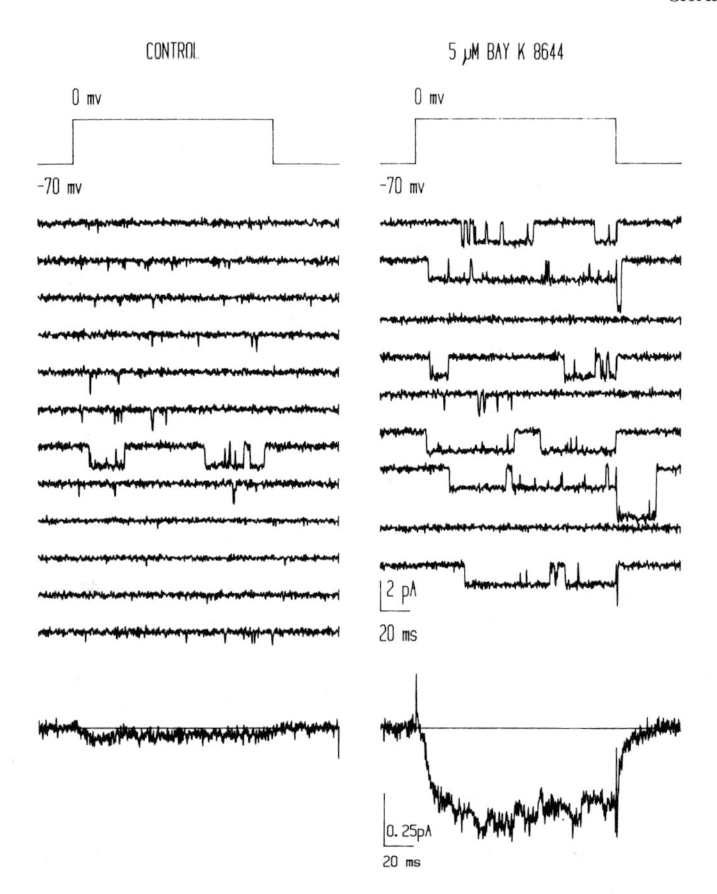

Figure 1. Calcium channel activity in an A10 cell (cell line derived from rat aortic smooth muscle). Cell-attached patch recordings with 110 mM $BaCl_2$ in the patch pipette in the absence of drug (A) and in the presence of 5 μM Bay K 8644 (B). Uppermost traces show the voltage-clamp pulse applied to the patch (the cell membrane potential was zeroed with a bathing solution containing 150 mM K-aspartate). The current records were taken consecutively with depolarizing pulses repeated once every 3 sec. (C,D) Averaged current records from the corresponding runs. A10 cells were kindly provided by Drs. Richard J. Miller and Sathapana Kongsamut, Departments of Physiological and Pharmacological Sciences, University of Chicago.

4. Modal Activity and Bay K 8644 Response in A10 Cells and Sensory Neurons

The records in Fig. 1 serve as a general illustration of different modes of Ca^{2+} channel gating (see also Hess *et al.*, 1984; Nowycky *et al.*, 1985a).

When depolarizing voltage clamp pulses (top traces) are applied once every 3 sec, a variety of patterns of channel activity are seen in the absence of drug (A). Some sweeps contain brief (~1 msec) channel openings ("mode 1"), whereas others contain no detectable openings whatsoever ("null mode" or "mode 0"). The group of consecutive records also includes one sweep that stands out in sharp contrast to the rest because it contains several long-lasting openings; we call this type of gating "mode 2" (Hess *et al.*, 1984).

The effect of Bay K 8644 is shown in Fig. 1B. The Ca^{2+} agonist greatly increases the proportion of sweeps that display mode 2 activity while decreasing the percentage showing mode 1 or mode 0 activity. Within each category, however, there is no obvious difference in the pattern of gating between records taken before or after the application of Ca^{2+} agonist. The overall result is a five- to sixfold increase in the averaged current in the presence of Ca^{2+} agonist (D) relative to control (C). The stimulatory effect of Ca^{2+} agonist on averaged single-channel current is in qualitative agreement with its promotion of $^{45}Ca^{2+}$ uptake in A10 cells as reported by Miller and colleagues (Kongsamut *et al.*, 1985).

Figure 2 illustrates a very similar set of recordings from a chick dorsal root ganglion cell maintained in primary culture (Nowycky *et al.*, 1985a). Here, as in Fig. 1, control runs in the absence of drug (A) consist of a mixture of the three modes of activity; here also, exposing the cell to Bay K 8644 (B) shifts the channel activity away from modes 0 and 1 and towards mode 2. The overall effect is a large increase in divalent cation influx (compare C and D). The increase in activity is shown as a function of time in panel E. Each vertical bar represents the degree of channel openness ($N_T \cdot p$) for a single sweep, where N_T is the total number of channels in the patch and p is the degree of openness per channel. Since $N_T = 3$ in this experiment, the full-scale value of 3 corresponds to 100% openness. As is evident, the response to Bay K 8644 begins within a few sweeps after its delivery to the bathing solution and is well maintained throughout the period of exposure. In DRG neurons, as in other cell types, Bay K 8644 has clear and prompt effects on channels under patch pipettes following its application to the rest of the cell. Since the seal between the membrane and the circumference of the pipette is very effective in preventing movement of small molecules (Hamill *et al.*, 1981), one can only presume that the high lipid solubility of DHPs allows them to enter the bilayer rapidly and to diffuse within the plane of the membrane in order to reach the channel protein within the patch (see also Kokubun and Reuter, 1984).

These neuronal responses are noteworthy because there is considerable disagreement in the literature about whether neuronal Ca^{2+} chan-

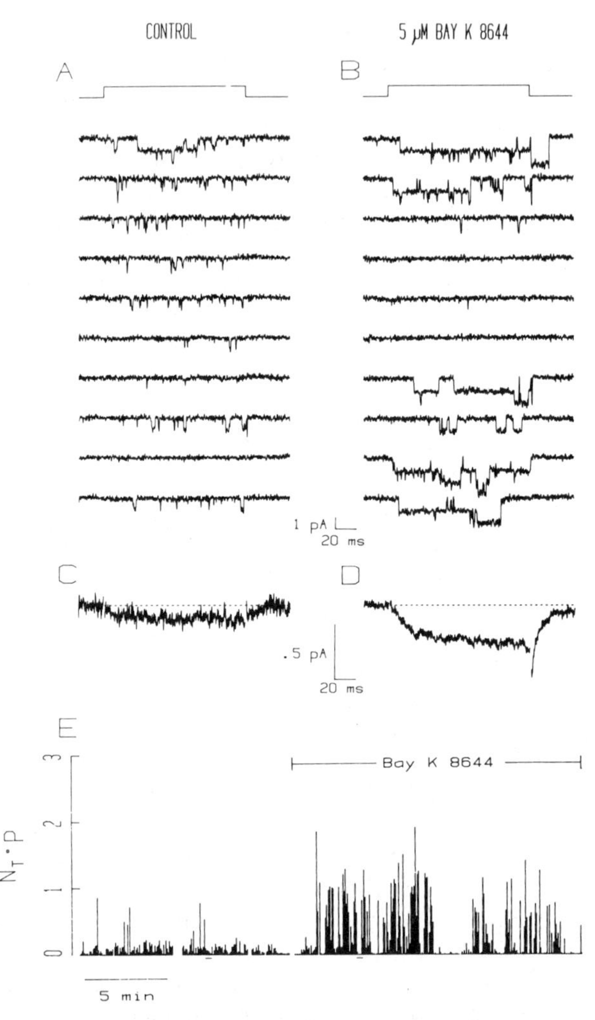

Figure 2. Effect of Bay K 8644 on channel activity and modal behavior in a chick dorsal root ganglion cell in dissociated culture. Cell-attached patch recording with 100 mM Ba^{2+} in the patch pitpette. Uppermost traces show voltage-clamp pulses from resting potential minus 60 mV to resting potential. (A,B) Sets of consecutive sweeps taken in the absence of drug (A) and in after 5 μM Bay K 8644 was applied to the bathing solution outside the cell (B). (C,D) Averaged currents from the corresponding runs. The scale bars in D refer to 20 msec horizontally and 2 pA vertically for A and B and 0.2 pA vertically for C and D. (E) Plot of channel openness (N_T) as a function of experimental time. (From Nowycky *et al.*, 1985a.)

nels are pharmacologically responsive to dihydropyridines (for review, see Miller and Freedman, 1984). It is important to mention here the coexistence of three types of Ca^{2+} channel in chick dorsal root ganglion cells (Nowycky *et al.*, 1985b). In addition to the L-type Ca^{2+} channel illustrated in Fig. 2, we find two additional kinds of Ca^{2+} channel with smaller unitary conductances that generate rapidly decaying currents. One of these is activated with weak depolarizations (type T), whereas the other requires strong depolarizations (type N). The T-type and N-type channels do not respond to Bay K 8644 at concentrations as high as 5 μM (Nowycky *et al.*, 1985b).

5. *Graded Shifts in the Balance between Modes*

Figure 3 shows our approach to the quantitative analysis of modal activity. The results come from a one-channel patch on a ventricular heart cell, which is the type of preparation in which the long openings promoted by Bay K 8644 were first seen (Ochi *et al.*, 1984; Kokubun and Reuter, 1984; Hess *et al.*, 1984; Brown *et al.*, 1984). Individual records from this experiment (Hess *et al.*, 1984, Fig. 2) exhibit the same kind of gating patterns as the neuronal and smooth-muscle-derived cells (Figs. 1 and 2), and the average current records show a correspondingly large enhancement of average current with exposure to the Ca^{2+} agonist (compare Fig. 3A,B).

Figure 3 describes a simple way of sorting out the different forms of gating behavior. This relies on measurements of the degree of openness (p) in individual sweeps, obtained by integrating over a period when the current signal is steady (see Fig. 3, legend). Panels C and D show p on a sweep-by-sweep basis, and panels E and F show the same data in the form of a histogram. In the absence of drug (E), the distribution of p values displays two distinguishable peaks, corresponding to the patterns of gating that we have labeled mode 0 and mode 1, In this particular experiment, mode 2 activity is represented by only a single sweep, but its p value ($\sim$0.85) is very far away from the sweeps in the peak labeled "mode 1," so there can be little doubt that it represents a qualitatively different kind of behavior. In the presence of Bay K 8644, the histogram of p values (F) is clearly trimodal; peaks corresponding to the patterns of gating that we have termed modes 0 and 1 remain in the same position along the p axis, and a clearly defined peak in the distribution of p values appears for the gating pattern we call mode 2.

Having shown that mode 2 sweeps are grouped as a peak along the spectrum of possible p values, we should point out that they are also

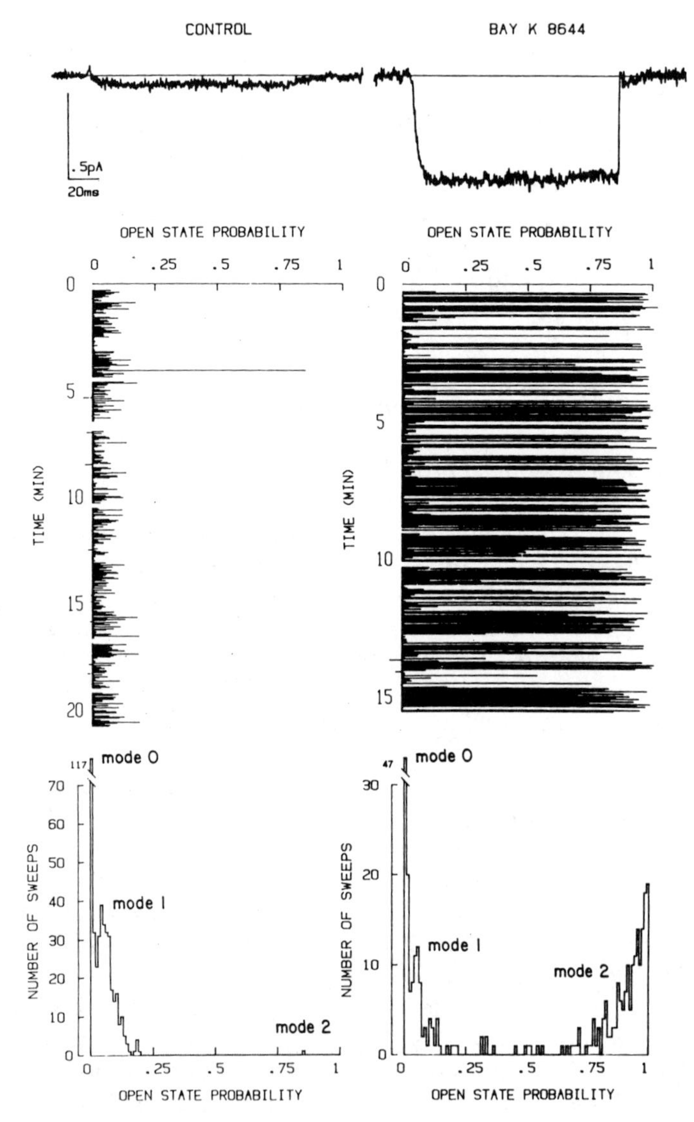

Figure 3. Analysis of the activity of a single Ca^{2+} channel of a guinea pig ventricular heart cell in the absence (left column) and presence (right column) of 5 μM Bay K 8644. Cell-attached patch depolarized from RP − 40 to RP + 70 mV, where RP was approximately −60 mV. (A,B) Averaged current records from all sweeps in each run. (C,D) Probability of channel openness (p) plotted sweep by sweep as a function of experimental time. For each sweep, steady-state values of p were determined as $p = \langle I \rangle / i$, where i is the unitary current and $\langle I \rangle$ is the steady-state value of current, averaged over an integration period from 22 msec after the depolarizing step until the end of the pulse. (E,F) Histograms of p values for the runs shown in C,D. Each bin is 0.008 wide. Sweeps with no detectable openings ("nulls") have near-zero values of p and show up in the first few bins. (From Hess *et al.*, 1984.)

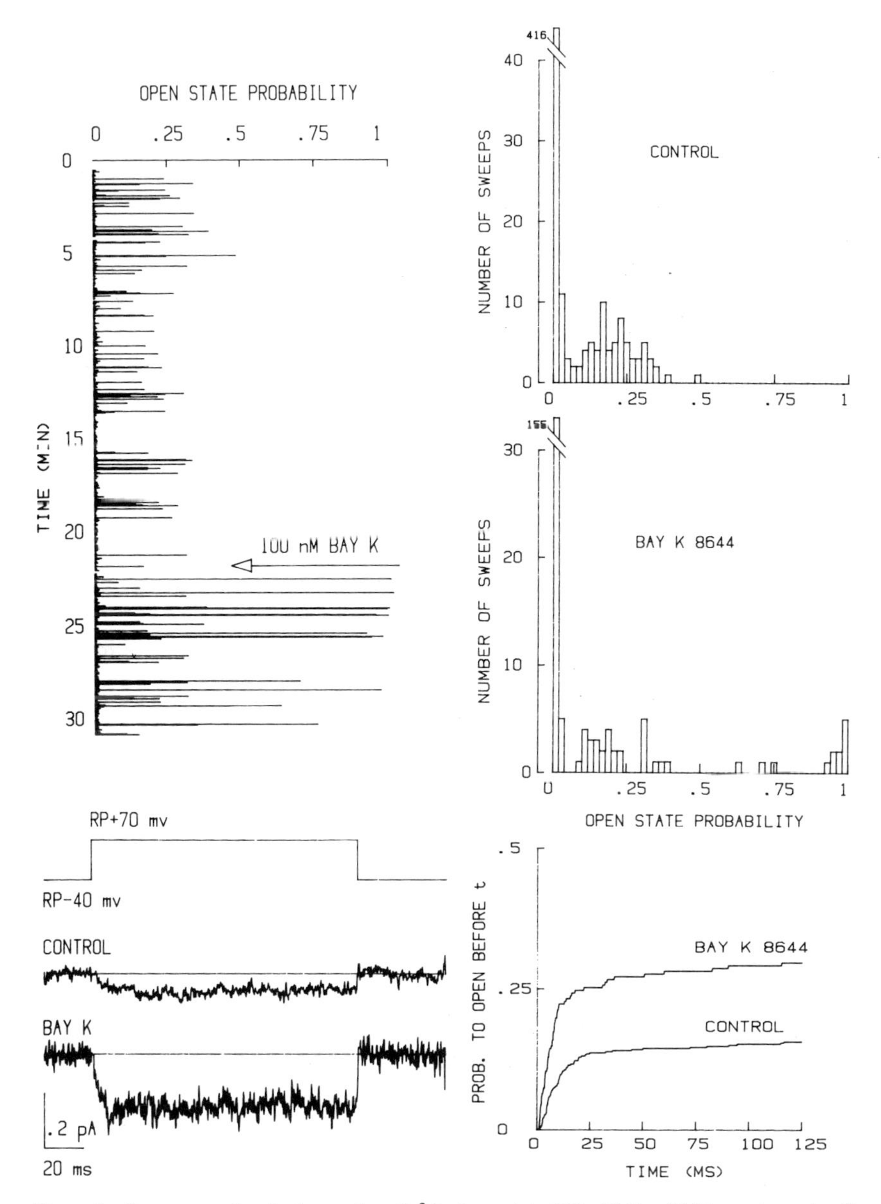

Figure 4. Response of a single cardiac Ca^{2+} channel to 100 nM Bay 8466, a submaximally effective dose for the Ca^{2+} agonist effect. (A) p values for each sweep plotted against time. (B,C) Histograms of p values in the absence and presence of Bay K 8644. (D) Averaged current records without and with the drug, shown below voltage clamp pulse. RP = −60 mV. (E) Superimposed cumulative distributions of the latency to first opening. The plateau levels correspond to the chances of the channel being available (i.e., not in mode 0).

grouped to follow each other. This can be shown quantitatively by constructing conditional probability tables. The probability of a given sweep showing mode 2 activity is much higher if the previous sweep contained mode 2 activity (see Hess *et al.*, 1984; Nowycky *et al.*, 1985a).

In the experiment illustrated in Fig. 3, the Ca^{2+} agonist concentration (5 μM) was sufficient to produce a maximal enhancement of Ca^{2+} channel activity. We view the results as showing graded changes in the amplitude of the modal peaks rather than changes in their positions along the probability axis. This interpretation is consistent with experiments with submaximally effective doses of Ca^{2+} agonist. For example, 100 nM Bay K 8644 (Fig. 4) produces an overall increase in Ca^{2+} channel activity of about half that seen with maximally effective doses. Here, as in Fig. 3, the probability histograms show a decrease in the number of blank sweeps and an increase in the number of mode 2 sweeps. The mode 2 peak lies between $p = 0.7$ and $p = 1.0$, much as it does with the maximally effective dose of Bay K 8644 in Fig. 3. The same point can be made by considering experiments conducted in the absence of Ca^{2+} agonist, where the incidence of mode 2 activity is sometimes as high as 11% (Hess *et al.*, 1984, Fig. 5B). In this case, mode 2 sweeps give rise to a peak in the p distribution that falls between $p = 0.7$ and $p = 1.0$, just as in Fig. 3 or Fig. 4.

The overall conclusion from these comparisons is that spontaneous or agonist-induced variations in overall activity arise from shifts in the relative weighting of the different forms of gating behavior.

6. *Modes of Gating as a Framework for Describing Modulatory Effects*

Figure 5 shows our working hypothesis for how variations in Ca^{2+} channel gating behavior may provide a basis for pharmacological modulation of Ca^{2+} influx. The scheme explicitly includes DHP Ca^{2+} agonists and Ca^{2+} antagonists, although we believe that it may also have usefulness in explaining effects of β-adrenergic agents and other drugs (see below). We propose that neither Ca^{2+} agonists nor antagonists block the Ca^{2+} channel pore. Instead, they bind with different affinities when the channel is in different modes and thereby stabilize one or more modes at the expense of the others. Within a given mode, the kinetics of gating transitions need not be changed when the DHP molecule binds to its receptor. Pure agonist effects arise when mode 2 is specifically favored, and pure antagonist effects result when mode 0 is preferentially stabilized. This framework bears some similarity to the modulated receptor hypoth-

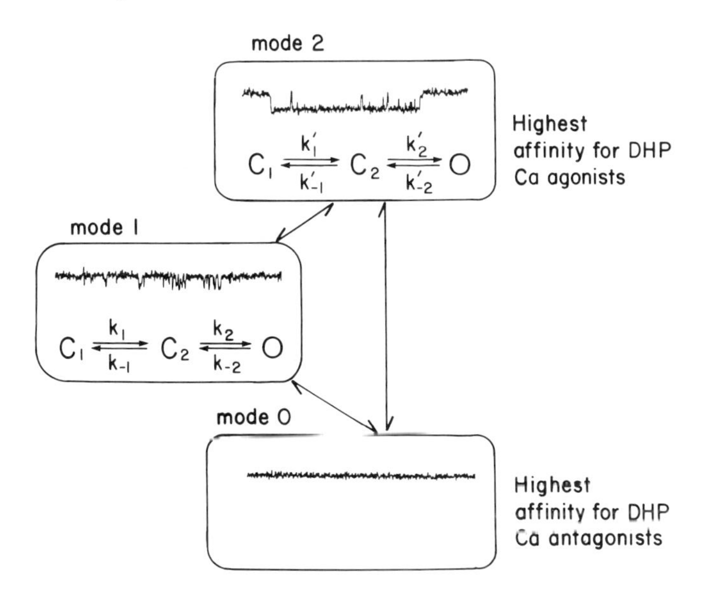

Figure 5. Modes of Ca^{2+} channel gating as a framework for describing modulatory effects of DHP Ca^{2+} agonists and Ca^{2+} antagonists. In the absence of drug, transitions between modes are assumed to be much slower than gating reactions within modes; details of intermode connections are left unspecified and require further study. At the test potentials under study, mode 1 and mode 2 are distinguished by qualitatively different patterns of opening and closing. This is put in terms of a kinetic model for Ca^{2+} channel gating with two closed (C_1, C_2) and one open (O) state. Rate constants for transitions between the states are assigned different values depending on the mode. No gating reactions are depicted for mode 0 because it represents condition(s) in which the channel is unavailable for opening at the test depolarization. Since voltage-dependent inactivation is one factor that biases the channel toward mode 0, the rate constants for transitions to and from mode 0 are presumably voltage dependent. The voltage dependence of transitions between mode 1 and mode 2 is not know. (From Hess *et al.*, 1984.)

esis for local anesthetic block of sodium channels (Hille, 1977; Hondeghem and Katzung, 1977). There are also important differences; in the modulated receptor hypothesis, (1) drug occupancy is equated with channel block, and there is no possibility of current enhancement, and (2) drug occupancy stabilizes individual states of the sodium channel, namely, resting, open, and/or inactivated, rather than modes of gating.

A mode of gating is different from an individual gating state. As Fig. 5 indicates, one can think of a mode of gating as a confederation of individual closed or open states interconnected in a kinetic scheme that gives rise to the characteristic pattern of gating appropriate for that mode. The concept of modes is useful because transitions between modes are slow, generally occurring over the course of several seconds, and transitions within modes are rapid, usually requiring only milliseconds. Because of the disparity in time scale, a channel may undergo hundreds or thousands of gating transitions between the closed state and the open state within mode 1 before switching to the pattern of gating characteristic of mode 2. Modes 1 and 2 will probably remain useful for some time as simplifications of a much more complicated multistate diagram with many transitions that cannot yet be described in detail.

7. *Nimodipine as an Example of Ca^{2+} Antagonists that Stabilize Mode 0*

The inhibitory actions of DHP Ca^{2+} antagonists are illustrated in Fig. 6, which shows the response of the large-conductance cardiac Ca^{2+} channel to nimodipine. A large concentration of drug (15 μM) is used because the Ca^{2+} channel is far less sensitive to Ca^{2+} antagonists at a high concentration of external Ba^{2+} than at a physiological concentration of external Ca^{2+} (see Lee and Tsien, 1983). The most dramatic effect of the drug on the single-channel records is that the proportion of sweeps with detectable openings is greatly decreased. In this particular experiment, roughly 65% of the sweeps contained activity in the absence of drug, and ~30% of the sweeps showed opening after nimodipine. The drug also promotes the disappearance of channel activity part way through the pulse. This results in an uneven distribution of openings strongly biased toward the beginning of the pulse (B) and a clear-cut decay of the averaged current record (D) not seen at this particular test potential in the absence of drug (C).

The increased proportion of blank sweeps and the accelerated disappearance of activity were seen with nifedipine and nitrendipine as well as nimodipine (Hess *et al.*, 1984). Both effects can be described as a promotion of a condition in which the channel is unavailable or inactivated—that is, mode 0. In this respect, our single-channel recordings fit very nicely with recent experiments in single ventricular heart cells (Bean *et al.*, 1984) or in cardiac Purkinje fibers (Sanguinetti and Kass, 1984). The common conclusion is that inactivation and the inhibitory effects of DHP Ca^{2+} antagonists reinforce each other. The lines of evidence are as follows: (1) DHP Ca^{2+} antagonists speed the decay of Ba^{2+} current in whole-cell (Lee and Tsien, 1983) as well as single-channel recordings (Hess *et al.*, 1984; Fig. 4); (2) the drugs slow the recovery from inactivation following a depolarizing pulse (Lee and Tsien, 1983; Sanguinetti and Kass, 1984); (3) steady depolarizations that inactivate Ca^{2+} channels

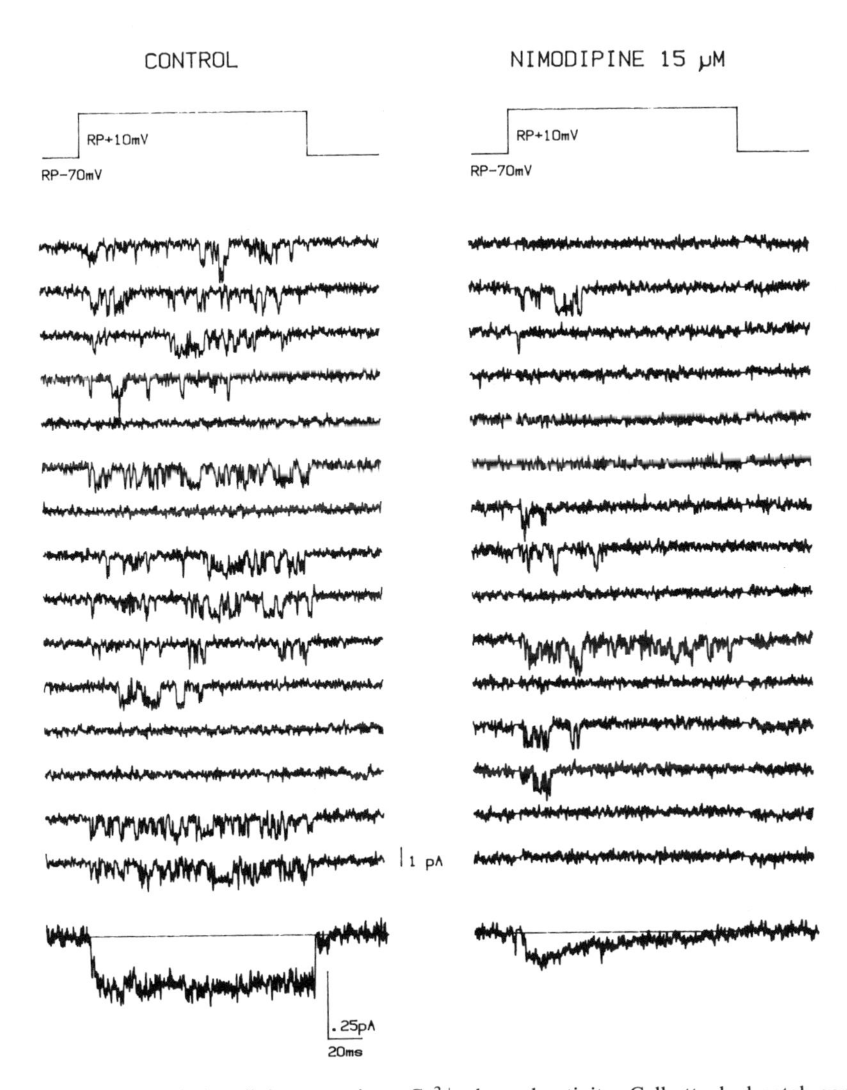

Figure 6. Effect of nimodipine on unitary Ca^{2+} channel activity. Cell-attached patch containing two Ca^{2+} channels as judged from the appearance of two nonzero conductance levels. (A,B) Representative groups of consecutive current records in the absence of drug (A) and in the presence of 15 μM nimodipine (B). Voltage-clamp protocol shown above current records. (C,D) Averaged current records from corresponding runs: 368 sweeps in the absence of drug (C) and 337 sweeps with nimodipine present (D). (From Hess *et al.*, 1985a.)

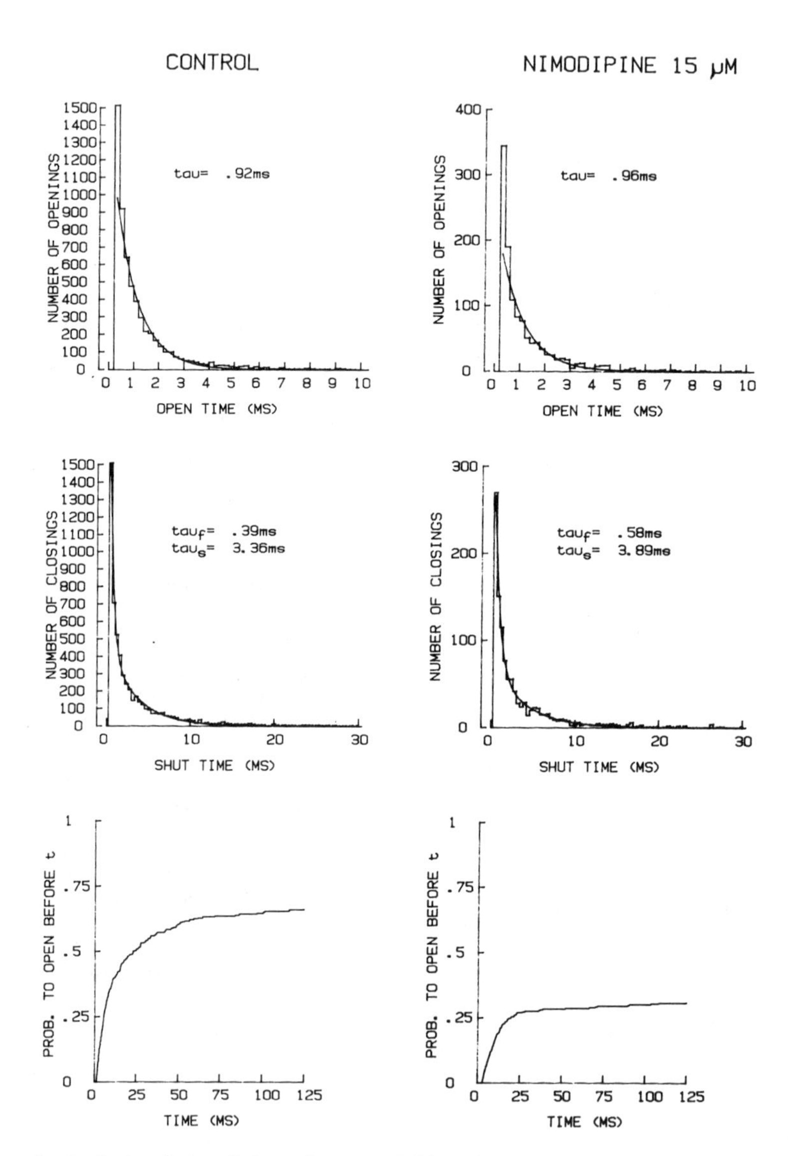

Figure 7. Analysis of nimodipine effects on Ca^{2+} channel kinetics. Same patch as Fig. 6. (A,B) Distributions of open times in the absence (A) and presence (B) of 15 μM nimodipine, fitted by single exponentials with parameters as indicated (C,D) Closed time distributions fitted by the sum of two exponentials. (E,F) Cumulative distributions of latency to first opening (corrected for the presence of two channels). (From Hess *et al.*, 1985a.)

increase the inhibitory potency of DHP Ca^{2+} antagonists (Sanguinetti and Kass, 1984; Bean 1985a).

In the case of nimodipine, the inhibitory effect seems limited to changes in the bias between availability and inactivation. Detailed kinetic analysis with the drug shows no significant change in the distributions of durations of individual opening and closing events (Fig. 7). The situation seems more complicated in the case of nitrendipine, which shows a mixture of nimodipine- and Bay K 8644-like effects (Hess *et al.*, 1984). It remains to be seen whether this reflects opposite effects of the individual enantiomers within the racemic mixture of nitrendipine (see Triggle *et al.*, 1985, for references). An alternative explanation attributes partial antagonist and partial agonist effects to an individual molecular species. We believe that this remains open as a possibility; within the framework depicted in Fig. 5, it seems conceivable that a single molecule could jointly favor mode 2 and mode 0 at the expense of mode 1.

Our single-channel studies of Ca^{2+} antagonist action have been carried out almost exclusively in guinea pig ventricular heart cells; more work is needed to characterize Ca^{2+} antagonist effects in neuronal and smooth-muscle-derived cells.

8. *β-Adrenergic Stimulation Promotes Mode 1 Rather Than Mode 2*

Increases in the rhythm and strength of the heartbeat during sympathetic stimulation result in large part from enhancement of Ca^{2+} channel activity. Norepinephrine, epinephrine, and other β-adrenergic agonists modulate Ca^{2+} channel activity by elevating intracellular cAMP and activating cAMP-dependent protein kinase (see Watanabe *et al.*, 1981; Reuter, 1983; Siegelbaum and Tsien, 1983; Sperelakis, 1985, for reviews). The mechanism of the modulatory action has been studied extensively with patch-clamp techniques (Reuter *et al.*, 1982; Cachelin *et al.*, 1983; Bean *et al.*, 1984; Brum *et al.*, 1984). There is general agreement that β stimulation of cAMP increases the opening probability p through the prolongation of openings and the abbreviation of closings; β stimulation also appears to increase the availability of channels for opening as judged by an increase in the proportion of nonblank sweeps in single-channel recording (Cachelin *et al.*, 1983; Tsien *et al.*, 1983; Brum *et al.*, 1984) or an increase in the functional number of channels in fluctuation analysis of whole-cell recordings (Bean *et al.*, 1984). Here we present new evidence on the nature of the β-adrenergic response with the specific aim of comparing it with the effects of DHP Ca^{2+} agonists and antagonists.

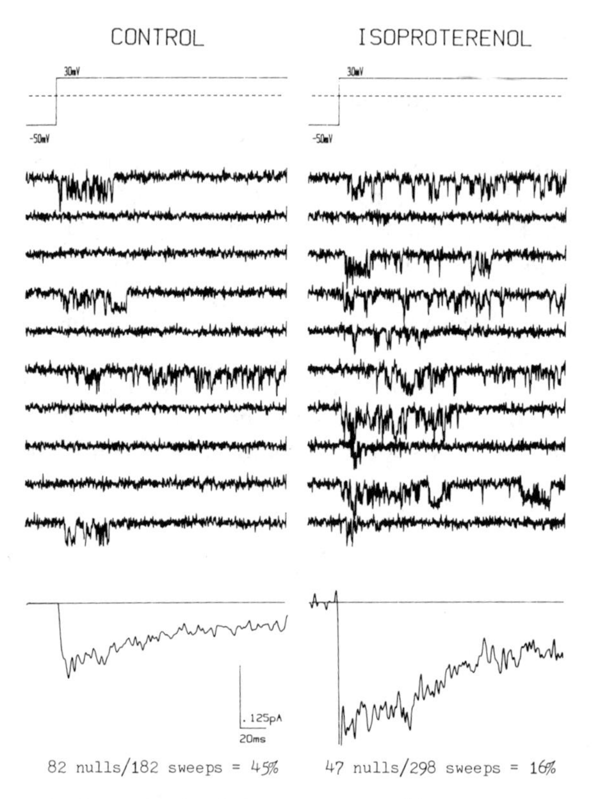

Figure 8. Effect of β-adrenergic stimulation on unitary Ca^{2+} channel activity. Cell-attached patch containing two Ca^{2+} channels as judged from the appearance of two nonzero conductance levels. (A,B) Representative groups of consecutive current records in the absence of drug (A) and in the presence of 14 μM isoproterenol (B). Voltage-clamp protocol shown above current records. (C,D) Averaged current records from corresponding runs. Inactivation is more prominent in C than in other experiments illustrated in this paper because the test polarization (+30 mV) is more positive. Cell B08E.

Figure 8 illustrates the effect of isoproterenol on a cell-attached patch on a guinea pig ventricular cell. As in other experiments, channel activity was evoked by applying a standard voltage-clamp depolarization once every 3 sec (upper traces in A,B). Representative groups of consecutive current records show unitary Ca^{2+} channel activity during runs taken in the absence of drug (A) and after introduction of isoproterenol into the bathing solution (B). The overall effect of the drug stimulation is a twofold increase of the peak inward current seen in the averaged current records from these runs (C,D). The mechanism of the β-adrenergic enhancement

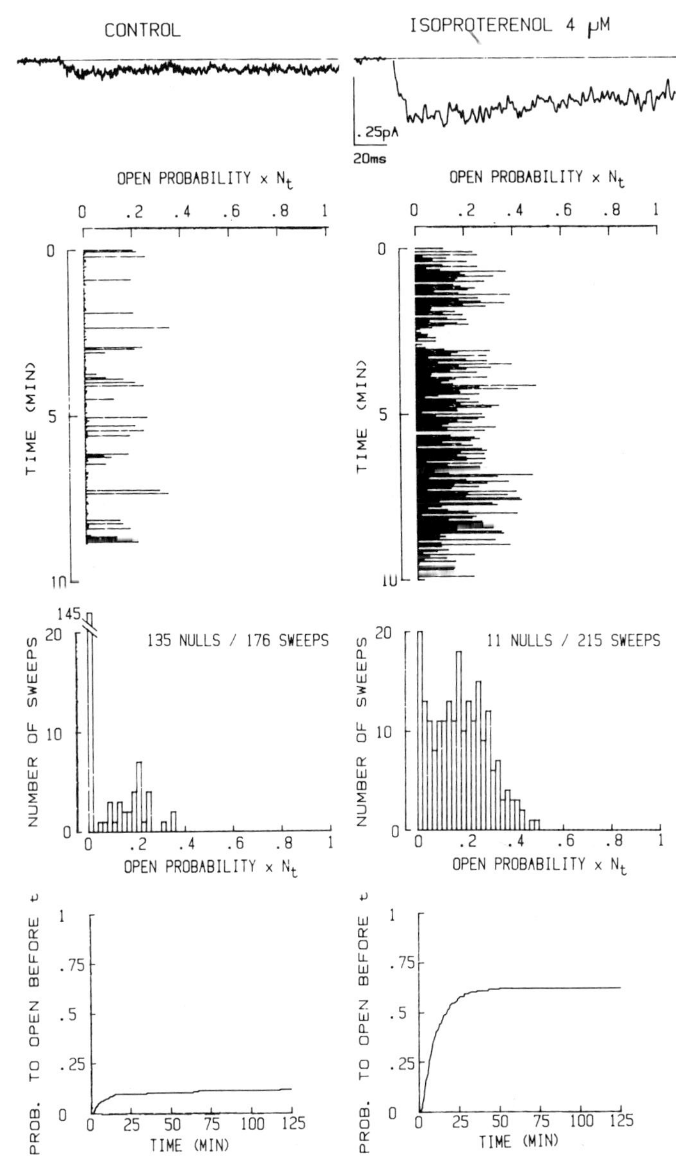

Figure 9. Response of unitary cardiac Ca^{2+} channel activity to β-adrenergic stimulation. Cell-attached patch containing two channels. (A,B) Averaged current records without drug (A) and after cell was exposed to 4 μM isoproterenol (B). Voltage-clamp pulse (not shown) from −60 mV to +20 mV. (C,D) p values for individual sweeps plotted against time in the absence and presence of isoproterenol. (E,F) Histograms of p values. (G,H) Cumulative distributions of the latency to first opening. The distribution saturates at levels corresponding to the chances of an individual channel being available (i.e., not in mode 0). Correction has been made for the presence of two channels in the patch. As a result, the saturating probability of opening is $1 - (135/176)^{1/2} = 0.124$ in G and $1 - (11/215)^{1/2} = 0.774$ in H. Cell B11E.

is different from that in the case of DHP Ca^{2+} agonists: unlike Bay K 8644, isoproterenol does not increase mode 2 activity. The most obvious change is an increase in mode 1 activity and a decrease in the percentage of null sweeps from 45% to 16%.

To what extent is the increase in average current accounted for by the increase in p_f, the probability of an individual channel being available (i.e., not in mode 0)? Taking into account the existence of two (presumably independent) channels in the patch (see legend to Fig. 8), we found that p_f rose from 0.33 in control to 0.60 after β stimulation. The 1.8-fold increase in availability is clearly the predominant factor in the overall enhancement of the peak inward current in this particular example.

In our previous study of the effect of isoproterenol on frog ventricular cells, we found a sixfold enhancement of Ca^{2+} channel current (Bean *et al.*, 1984), which is much larger than the responses to 8-bromo-cAMP or β-adrenergic stimulation found in mammalian cells, which averaged about 1.5- to twofold (e.g., Cachelin *et al.*, 1983; Brum *et al.*, 1984). This prompted us to look for particularly large responses in our unitary recordings from guinea pig ventricular cells to see if the mechanism of enhancement might be compatible with the results obtained with frog myocytes. As Fig. 9 illustrates, we have seen particularly dramatic increases in Ca^{2+} channel activity in patches in which the activity is relatively sparse in the absence of drug.

Figure 9 shows results from a two-channel patch in which the peak amplitude of the averaged inward current record increased by about fivefold (A,B). To show temporal variations in channel availability, we plotted the open-state probability on a sweep-by-sweep basis (C,D) as well as in histograms (E,F). The results document a dramatic difference between the behavior in the absence of drug and in the presence of β-adrenergic agonist. Isoproterenol largely eliminates nulls, raising the proportion of nonblank sweeps from 41/176 = 24% to 204/215 = 95%. After exposure to the drug, the nonblank sweeps tend to have a slightly higher value of open probability (perhaps because of the presence of two channels in the patch). However, changes in the channel availability are clearly the dominant factor in the overall increase in activity.

9. *Changes in Ca^{2+} Channel Inactivation with β-Adrenergic Stimulation*

When DHP Ca^{2+} antagonists decrease the peak Ca^{2+} channel current, they also accelerate the time course of decay. This leads to the question of whether β-adrenergic enhancement of Ca^{2+} current is ac-

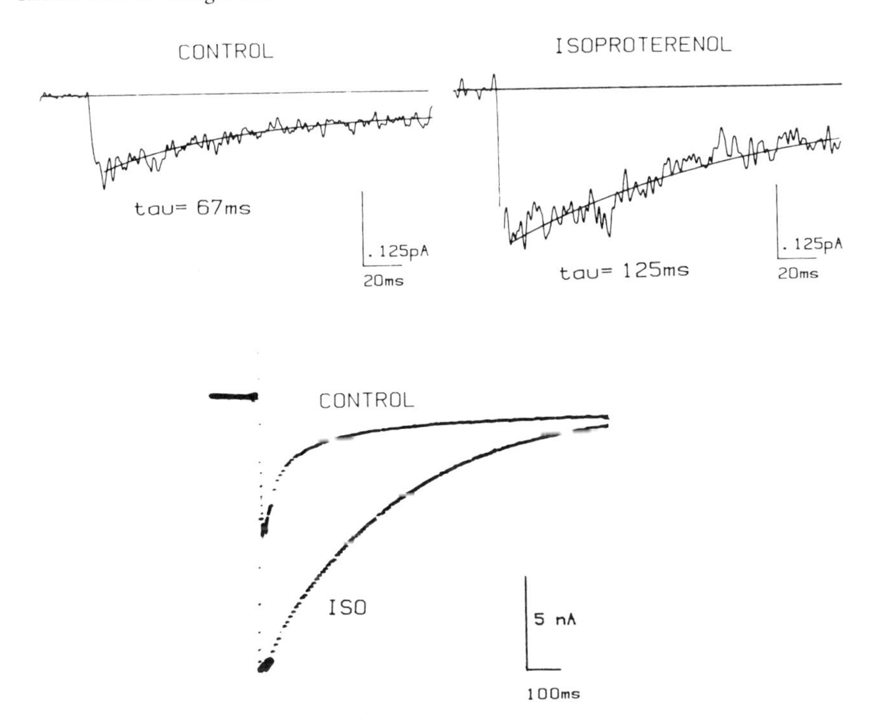

Figure 10. Changes in the time course of cardiac Ca^{2+} channel inactivation with β-adrenergic stimulation. (A,B) Averaged current records from Fig. 8 with single exponentials fitted to their decaying phase. (C) Records from a suction pipette recording from a guinea pig ventricular cell in the absence of drug and in the presence of 10 nM is isoproterenol for 6 min. Cell 106E. (Unpublished experiment of K. S. Lee and B. P. Bean.)

companied by an opposite effect on the time course of the Ca^{2+} channel current. Figure 10 show results at the level of single channels and single cells. Panels A and B show the averaged current records from Fig. 8, fitted with smooth curves decaying with exponential time constants of 67 msec in control and 125 msec in isoproterenol. Panel B shows whole-cell recordings with the suction pipette method (Lee *et al.*, 1980), with 10 mM external Ca^{2+} as the charge carrier. In this case, isoproterenol increases the peak current amplitude about twofold and increases the half-decay time from 50 msec to more than 200 msec. These changes in the inactivation time course are consistent with results reported previously in guinea pig ventricular myocytes (Brum *et al.*, 1984) and frog ventricular cells (Bean *et al.*, 1984). One obvious interpretation is that the slowing of inactivation and the decrease in the proportion of null sweeps are different expressions of a common mechanism that biases activity against mode 0.

10. How Similar Are Large-Conductance Ca^{2+} Channels in Various Cell Types?

One of the main purposes of this chapter is to provide a comparison among certain forms of Ca^{2+} channel activity in cardiac, neuronal, and smooth muscle cells. We have used very similar recording conditions and voltage-clamp protocols in guinea pig ventricular heart cells, chick dorsal root ganglion cells, and A10 cells derived from rat aortic smooth muscle. Each of these preparations displays a type of Ca^{2+} channel with 110 mM Ba^{2+} as the charge carrier that shares the following properties:

1. A large slope conductance ranging from 20 to 25 pS.
2. Steeply voltage-dependent activation between −20 and +30 mV.
3. Relatively slow inactivation with Ba^{2+} as the charge carrier.
4. A predominant form of gating (mode 1) characterized at moderate depolarizations by brief (~1-msec) openings, often grouped in brief clusters.
5. An occasional spontaneous appearance of a second form of gating (mode 2) characterized by relatively long-lasting openings (~10–20 msec) even in the absence of drug.
6. An increased occurrence of mode 2 sweeps with exposure to the Ca^{2+} agonist Bay K 8644.

We tentatively assign this kind of Ca^{2+} channel activity the label "type L" ("L" for its large conductance and its ability to generate a long-lasting average current with isotonic Ba^{2+} as the charge carrier).

11. Distinguishing Between L-Type Ca^{2+} Channels and Other Types of Ca^{2+} Channels

L-type Ca^{2+} channels are to be distinguished from other kinds of Ca^{2+} channels. T-type Ca^{2+} channels have already been found in various neuronal preparations (see Carbone and Lux, 1984; Nowycky *et al.*, 1985b, for references) and in heart cells of the atrium (Bean, 1985b) and ventricle (Nilius *et al.*, 1985). N-type Ca^{2+} channels have only been found so far in DRG neurons (Nowycky *et al.*, 1985b). In contrast to L-type Ca^{2+} channels, T- and N-type channels display smaller slope conductances, more rapidly decaying activity, and no responsiveness to Bay K 8644 at doses up to 10 μM.

The large-conductance Ca^{2+} channels we have described in cardiac, neuronal, and smooth-muscle-derived cells have much more in common with each other than with T-type or N-type channel activity. The list of generic properties of L-type Ca^{2+} channels may grow as studies of ion permeation and block of large-conductance cardiac Ca^{2+} channels (Hess *et al.*, 1986b; Lansman *et al.*, 1986) are extended to Ca^{2+} channels in other preparations. The evidence obtained so far with whole-cell recordings from smooth muscle cells (Isenberg and Klockner, 1985) indicates strong similarities to L-type Ca channels in heart cells (e.g., Hess and Tsien, 1984).

12. Possible Differences among L-Type Ca^{2+} Channels

It is also important to point out that within the general classification of L-type activity, significant differences may exist between cardiac and neuronal Ca^{2+} channels with respect to properties of modulation and inactivation. With regard to modulation, L-type Ca^{2+} channels in cardiac cells show a dramatic increase in activity with β-adrenergic stimulation and elevation of intracellular cAMP, as already discussed in Section 8 and in other papers. On the other hand, Ca^{2+} currents in chick DRG neurons are not enhanced by application of forskolin (Dunlap, 1985) or by internal dialysis with cAMP-containing solutions (Forscher and Oxford, 1985). With regard to inactivation, the decay of L-type cardiac Ca^{2+} channel activity is strongly accelerated by increased levels of intracellular Ca^{2+}. This is seen in cardiac cells under whole-cell voltage clamp as a much more rapid rate of inactivation, with Ca^{2+} rather than Ba^{2+} as the the external charge carrier (see Eckert and Chad, 1984; Lee *et al.*, 1985, for references). In chick DRG neurons, on the other hand, replacement of Ca^{2+} with Ba^{2+} leaves the rate of Ca^{2+} channel inactivation unaffected (Nowycky *et al.*, 1984; but see also Dunlap and Fischbach, 1981).

13. Calcium-Agonist-Promoted Mode 2 Gating: A General Property of L-Type Channels?

The idea of modes of Ca^{2+} channel gating was first developed as a framework for analyzing the effects of Bay K 8644 in heart cells (Hess *et al.*, 1984). Its usefulness extends to Ca^{2+} agonist effects in other cell types, including chick DRG neurons (Nowycky *et al.*, 1985a) and smooth-muscle-derived A10 cells (this chapter). Further observations of the spontaneous occurrence of long-lasting openings and of promotion of this type of activity by Bay K 8644 have been made in neuroblastoma cells (Fox *et al.*, 1984), adrenal chromaffin cells (T. Hoshi, J. Rothlein, and S. J. Smith, unpublished data), and single smooth muscle cells isolated from the stomach of *Amphiuma* (Josephson *et al.*, 1985). With similar results from six different preparations, it seems reasonable to expect that Ca^{2+}-agonist-modulated mode 2 activity will prove to be a general property of large-conductance Ca^{2+} channels.

We have seen similar modal patterns of gating for T-type and N-type Ca^{2+} channels in chick DRG neurons, but we do not yet know of a pharmacological agent for altering the balance between modes.

14. Questions about the Basis of Spontaneously Occurring Mode 2 Activity

The voltage dependence of the probability of channel openness is roughly 20 mV more negative in mode 2 than in mode 1 (Hess *et al.*, 1984, Fig. 1; Nowycky *et al.*, 1985a); at a given potential, the average amount of charge transfer per sweep can be at least an order of magnitude greater in mode 2 than in mode 1. It thus seems that even a small proportion of mode 2 activity can make a big difference to the overall Ca^{2+} entry per unit time.

The cause(s) of spontaneously occurring mode 2 activity are not known. The main possibilities are (1) an intrinsic substance with DHP-like activity (see Schramm *et al.*, 1983; Thayer *et al.*, 1984), and (2) an intrinsic property of the Ca^{2+} channel protein (see Coronado and Affolter, 1985). We see tremendous variability in the frequency of occurrence of spontaneous mode 2 activity, ranging from 0 to 11% of sweeps in guinea pig ventricular cells, for example (Hess *et al.*, 1984), and over an even wider range in the case of chick DRG neurons (Nowycky *et al.*, 1985a). We suggest that spontaneous variations in the proportion of channel activity in mode 2 might contribute to chronically elevated intracellular Ca^{2+} in certain disease states. Possible examples include Duchenne's muscular dystrophy, cardiomyopathy, and certain forms of hypertension. It is intriguing that calcium leakage has been implicated as a cause of the high resting tension in vascular smooth muscle from the spontaneously hypertensive rat (Noon *et al.*, 1978).

15. Altered Balance between Modes of Channel Gating during β Stimulation of Cardiac Ca^{2+} Channels

We have used the same kind of analysis to describe effects of β-adrenergic agonists and DHP Ca^{2+} agonists. Single-channel recording experiments leave no doubt that these agents enhance Ca^{2+} channel activity by very different mechanisms (Kokubun and Reuter, 1984; Hess *et al.*, 1984). Isoproterenol and Bay K 8644 both decrease the proportion of null sweeps; however, Bay K 8644 specifically increases the proportion of mode 2 activity (Fig. 3), whereas isoproterenol does not (Figs. 8,9).*

The difference in mechanisms of β agonists and DHP Ca agonists fits nicely with the finding that unlike isoproterenol, Bay K 8644 and CGP 28 392 fail to elevate intracellular cAMP in cultured rat ventricular cells (Kokubun and Reuter, 1984). Common effects of β stimulation and DHP Ca^{2+} agonists downstream of cAMP production are not ruled out but seem unlikely in view of the different effects on single Ca^{2+} channel activity. The contrasting actions of isoproterenol and Bay K 8644 at the single-channel level are also consistent with studies of myocardial contraction, which show that maximally effective doses of the Ca^{2+} agonist do not completely occlude the positive inotropic effect of the β agonist (Thomas *et al.*, 1985).

16. Comparison between Up-Modulation by β Agonists and Down-Modulation by Ca^{2+} Antagonists such as Nimodipine

In our experience, the predominant effect of β-adrenergic stimulation can be summarized as a shift away from mode 0 and in favor of mode 1. To a first approximation, the response to isoproterenol is opposite to the response to nimodipine, with isoproterenol increasing rather than decreasing the proportion of sweeps in which the channel is available and slowing rather than speeding the time course of decay (compare Figs. 6 and 10).

The effects of isoproterenol and nimodipine may not be diametrically opposed. Besides altering the balance between modes, β stimulation or cAMP administration is known to produce additional changes in the rapid

* In illustrating the β-adrenergic effect, Reuter (1983) showed a mode 2 sweep in the presence of isoproterenol and none in its absence. In light of our own results and those of Cachelin *et al.* (1983) and Kokubun and Reuter (1984), we suggest that the mode 2 sweep was an instance of spontaneously occurring mode 2 activity rather than a representative example of the specific action of β stimulation.

kinetics within mode 1, favoring longer openings and shorter closings (see Cachelin *et al.*, 1983; Brum *et al.*, 1984). Although such effects are not especially prominent in the experiments illustrated in this chapter, we have seen clear expression of fast kinetic effects in other guinea pig ventricular cells and in whole-cell recordings from frog ventricular cells in which kinetic changes appear as a slight increase in the half-time of the rising phase of Ca^{2+} channel current and as a significant increase in opening probability (Bean *et al.*, 1984).

In the case of nimodipine, no changes in fast kinetic parameters were detected (Fig. 7). The situation may be different for the phenylalkylamine Ca^{2+} antagonist D600 (methoxyverapamil). Pelzer and Trautwein (1985) have found that D600 not only increases the proportion of null sweeps but also shortens individual openings within mode 1. It remains to be seen whether the apparently diametric opposition between β stimulation and D600 is based on a common molecular mechanism.

REFERENCES

Bean, B. P., 1985a, Nitrendipine block of cardiac calcium channels: High-affinity binding to the inactivated state, *Proc. Natl. Acad. Sci. U.S.A.* **81:**6388–6392.

Bean, B. P., 1985b, Two kinds of calcium channels in atrial cells from dog and frog hearts, *Biophys. J.* **47:**496a.

Bean, B. P., Nowycky, M. C., and Tsien, R. W., 1984, β-Adrenergic modulation of calcium channels in frog ventricular heart cells, *Nature* **307:**371–375.

Brown, A. M., Camerer, H., Kunze, D. L., and Lux, H. D., 1982, Similarity of unitary Ca^{2+} currents in three different species, *Nature* **299:**156–158.

Brown, A. M., Kunze, D. L., and Yatani, A., 1984, The agonist effect of dihydropyridines on Ca channels, *Nature* **311:**570–572.

Brum, G., Osterrieder, W., and Trautwein, W., 1984, β-Adrenergic increase in the calcium conductance of cardiac myocyte studies with the patch clamp, *Pfluegers Arch.* **401:**111–118.

Cachelin, A. B., dePeyer, J. E., Kokubun, S., and Reuter, H., 1983, Calcium channel modulation by 8-bromo-cyclic AMP in cultured heart cells, *Nature* **304:**402–404.

Caffrey, J. M., Josephson, I. R., and Brown, A. M., 1985, Calcium channels in smooth muscle: Dihydropyridine action, *Biophys. J.* **47:**265a.

Carbone, E., and Lux, H. D., 1984b, A low voltage-activated, fully inactivating Ca channel in vertebrate sensory neurons, *Nature* **310:**501–502.

Cavalie, A., Ochi, R., Pelzer, D., and Trautwein, W., 1983, Elementary currents through Ca^{2+} channels in guinea pig myocytes, *Pfluegers Arch.* **398:**284–297.

Cavalie, A., Pelzer, D., and Trautwein, W., 1985, Modulation of the gating properties of single calcium channels by D600 in guinea-pig ventricular myocytes, *J. Physiol.* (*Lond.*) **258:**59P.

Coronado, R., and Affolter, H., 1985, Kinetics of dihydropyridine-sensitive single calcium channels from purified muscle transverse tubules, *Biophys. J.* **47:**434a.

Dunlap, K., 1985, Forskolin prolongs action potential duration and blocks potassium current in embryonic chick sensory neurons, *Pfluegers Arch.* **403:**170–174.

Eckert, R., and Chad, J. D., 1984, Inactivation of Ca channels. *Prog. Biophys. Mol. Biol.* **44**:215–267.

Forscher, P., and Oxford, G. S., 1985, Modulation of calcium channels by norepinephrine in internally dialyzed avian sensory neurons, *J. Gen. Physiol.* **85**:743–763.

Fox, A. P., Hess, P., Lansman, J. B., Nowycky, M. C., and Tsien, R. W., 1984, Slow variations in the gating properties of single calcium channels in guinea pig heart cells, chick neurons and neuroblastoma cells, *J. Physiol. (Lond.)* **353**:75P.

Franckowiak, G., Bechem, M., Schramm, M., and Thomas, G., 1985, The optical isomers of the 1,4-dihydropyridine Bay K 8644 show opposite effects on Ca channels, *Eur. J. Pharmacol.* **114**:223–226.

Hagiwara, S., and Byerly, L., 1981, Calcium channel, *Annu. Rev. Neurosci.* **4**:69–125.

Hagiwara, S., and Byerly, L., 1983, The calcium channel, *Trends Neurosci.* **6**:189–193.

Hamill, O., Marty, A., Neher, E., Sakmann, B., and Sigworth, F. J., 1981, Improved patch-clamp techniques for high-resolution current recording from cells and cell-free membrane patches, *Pfluegers Arch.* **391**:85–100.

Hess, P., and Tsien, R. W., 1984, Mechanism of ion permeation through calcium channels, *Nature* **309**:453–456.

Hess, P., Lansman, J. B., and Tsien, R. W., 1984, Different modes of Ca channel gating behavior favored by dihydropyridine Ca agonists and antagonists, *Nature* **311**:538–544.

Hess, P., Lansman, J. B., and Tsien, R. W., 1985a, Mechanism of calcium channel modulation by dihydropyridine agonists and antagonists, in: *Control and Manipulation of Calcium Movement* (J. R. Parratt, ed.), Raven Press, New York.

Hess, P., Lansman, J. B., and Tsien, R. W., 1986b, Calcium channel selectivity for divalent and monovalent cations. Voltage- and concentration-dependence of single channel current in guinea pig ventricular heart cells, *J. Gen. Physiol.* (in press).

Hille, B., 1977, Local anesthetics: Hydrophilic and hydrophobic pathways for the drug-receptor reaction, *J. Gen. Physiol.* **69**:497–515.

Hille, B., 1984, *Ionic Channels of Excitable Membranes*, Sinauer Associates, Sunderland, MA.

Hondeghem, L. M., and Katzung, B. G., 1977, Time- and voltage-dependent interactions of antiarrhythmic drugs with cardiac sodium channels, *Biochim. Biophys. Acta* **472**:373–398.

Isenberg, G., and Klockner, U., 1985, Calcium currents of smooth muscle cells isolated from the urinary bladder of the guinea-pig: Inactivation, conductance and selectivity is controlled by micromolar amounts of $[Ca]_o$, *J. Physiol. (Lond.)* **358**:60P.

Josephson, I. R., Caffrey, J. M., and Brown, A. M., 1985, Single Ca channels in isolated smooth muscle cells, *Science* (in press).

Kimes, B. W., and Brandt, B. L., 1976, Characterization of two putative smooth muscle cell lines from rat thoracic aorta, *Exp. Cell Res.* **89**:349–366.

Kokubun, S., and Reuter, H., 1984, Dihydropyridine derivatives prolong the open state of Ca channels in cultured cardiac cells, *Proc. Natl. Acad. Sci. U.S.A.* **81**:4824–4827.

Kongsamut, S., Freedman, S. B., and Miller, R. J., 1985, Dihydropyridine sensitive calcium channels in a smooth muscle cell line, *Biophys. Biochem. Res. Commun.* (in press).

Lansman, J. B., Hess, P., and Tsien, R. W., 1986, Blockade of current through single Ca channels from guinea-pig ventricular heart cells by Cd, Mg and Ca, *J. Gen. Physiol.* (in press).

Lee, K. S., and Tsien, R. W., 1983, Mechanism of calcium channel blockade by verapamil, D600, diltiazem and nitrendipine in single dialyzed heart cells, *Nature* **302**:790–794.

Lee, K. S., Akaike, N., and Brown, A. M., 1980, The suction pipette method for internal perfusion and voltage clamp of small excitable cells, *J. Neurosci. Methods* **2**:51–78.

Lee, K. S., Marban, E., and Tsien, R. W., 1985, Inactivation of calcium channels in mammalian heart cells. Joint dependence on membrane potential and intracellular calcium, *J. Physiol.* (*Lond.*) **364:**395–411.

Miller, R. J., and Freedman, S. B., 1984, Are dihydropyridine binding sites voltage sensitive calcium channels? *Life Sciences* **34:**1205–1221.

Mitra, R. L., and Morad, M., 1985, Gastric smooth muscle cells have only Ca^{2+} and Ca^{2+}-activated K^+ channels, *Biophys. J.* **47:**66a.

Nilius, B., Hess, P., Lansman, J. B., and Tsien, R. W., 1985, A novel type of cardiac calcium channel in ventricular cells, *Nature* **316:**443–446.

Noon, J. P., Rice, P. J., and Baldessarini, R. J., 1978, Calcium leakage as a cause of the high resting tension in vascular smooth muscle from the spontaneously hypertensive rat, *Proc. Natl. Acad. Sci. U.S.A.* **75:**1605–1607.

Nowycky, M. C., Fox, A. P., and Tsien, R. W., 1984, Two components of calcium channel current in chick dorsal root ganglion cells, *Biophys. J.* **45:**36a.

Nowycky, M. C., Fox, A. P., and Tsien, R. W., 1985a, Long-opening mode of gating of neuronal calcium channels and its promotion by the dihydropyridine calcium agonist Bay K 8644, *Proc. Natl. Acad. Sci. U.S.A.* **82:**2178–2182.

Nowycky, M. C., Fox, A. P., and Tsien, R. W., 1985b, Three types of neuronal calcium channel with different calcium agonist sensitivity, *Nature* **316:**440–443.

Ochi, R., Hino, N., and Niimi, Y., 1984, Prolongation of calcium channel open time by the dihydropyridine derivative Bay K 8644 in cardiac myocytes, *Proc. Jpn. Acad.* [B] **60:**153–156.

Reuter, H., 1983, Calcium channel modulation by neurotransmitters, enzymes and drugs, *Nature* **301:**569–574.

Reuter, H., Stevens, C. F., Tsien, R. W., and Yellen, G., 1982, Properties of single calcium channels in cultured cardiac cells, *Nature* **297:**501–504.

Sanguinetti, M. C., and Kass, R. S., 1984, Voltage-dependent block of calcium channel current in the calf cardiac Purkinje fiber by dihydropyridine calcium channel antagonists, *Circ. Res.* **55:**336–348.

Schramm, M., Thomas, G., Towart, R., and Franckowiak, G., 1983, Novel dihydropyridines with positive inotropic action through activation of Ca^{2+} channels, *Nature* **303:**535–537.

Siegelbaum, S. A., and Tsien, R. W., and 1983, Modulation of gated ion channels as a mode of transmitter action, *Trends Neurosci.* **6:**307–313.

Sperelakis, N., 1985, Phosphorylation hypothesis of the myocardial slow channels and control of Ca influx, in: *Cardiac Electrophysiology and Arrhythmias* (D. P. Zipes and J. Jalife, eds.) Grune and Stratton, Orlando, Florida, pp. 123–135.

Thayer, S. A., Stein, L., and Fairhurst, A. S., 1984, Endogenous modulator of calcium channels, in: *IUPHAR 9th International Congress of Pharmacology* Macmillan, London, p. 891P.

Thomas, G., Chung, M., and Cohen, C. J., 1985, A dihydropyridine (Bay K 8644) that enhances calcium currents in guinea pig and calf myocardial cells: A new type of positive inotropic agent, *Circ. Res.* **56:**87–96.

Tsien, R. W., 1983, Calcium channels in excitable cell membranes, *Annu. Rev. Physiol.* **45:**341–358.

Tsien, R. W., Bean, B. P., Hess, P., and Nowycky, M., 1983, Calcium channels: Mechanisms of β-adrenergic modulation and ion permeation, *Cold Spring Harbor Symp. Quant. Biol.* **48:**201–212.

Watanabe, A. M., Lindemann, J. P., Jones, L. R., Besch, H. R., Jr., and Bailey, J. C., 1981, Biochemical mechanisms mediating neural control of the heart, in: *Disturbances in Neurogenic Control of the Circulation*, American Physiological Society, Bethesda.

7

Chemical Pharmacology of Ca^{2+} Channel Ligands

D. J. TRIGGLE, A. SKATTEBOL, D. RAMPE, A. JOSLYN, and P. GENGO

1. Introduction

A variety of terms, including Ca^{2+} antagonist, Ca^{2+} channel blocker, and slow channel blocker, have been applied to a structurally heterogeneous group of agents including the clinically available verapamil, nifedipine, and diltiazem (Fig. 1; for general reviews see Fleckenstein, 1977, 1983). With the recent introduction of 1,4-dihydropyridine analogues of nifedipine that function as calcium channel activators (Fig. 2, Bay K 8644 and CGP 28 392; Schramm *et al.*, 1983), a more appropriate generic title might be calcium channel ligands.

The calcium channel ligands, both antagonists and activators, have provided a powerful stimulus for the elucidation of the biochemical and biophysical bases of calcium channel structure and function. This explosion of interest has followed the therapeutic exploitation of these compounds, primarily in the control of cardiovascular disease. (A therapeutic profile of verapamil, nifedipine, and diltiazem is presented in Table I.) It is noteworthy that the listed agents do not share a common profile or selectivity of action (Schwartz and Triggle, 1984; Opie, 1984); thus, consistent with their chemical heterogeneity and the absence of any all-inclusive structure–activity relationship, it is likely that these agents modulate channel function at different sites and by different mechanisms (Triggle and Swamy, 1983; Janis and Triggle, 1983; Janis and Scriabine,

D. J. TRIGGLE, A. SKATTEBOL, D. RAMPE, A. JOSLYN, and P. GENGO • Department of Biochemical Pharmacology, School of Pharmacy, State University of New York at Buffalo, Buffalo, New York 14260.

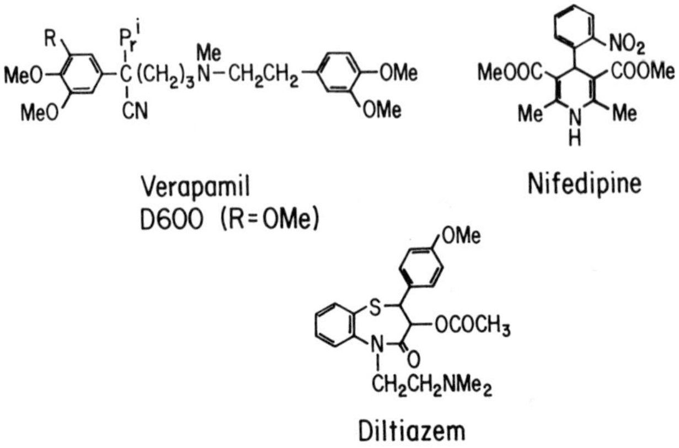

R, P_r^i, Me, MeO, MeO, C $(CH_2)_3$ N – CH_2CH_2, CN, OMe, OMe

Verapamil
D600 (R=OMe)

MeOOC, NO_2, COOMe, Me, N, H, Me

Nifedipine

OMe, S, $OCOCH_3$, N, O, $CH_2CH_2NMe_2$

Diltiazem

Figure 1. Structural formula of Ca^{2+} channel antagonists.

1983). In turn, the elucidation of sites and mechanisms of action will permit development of new members of this class with different and improved therapeutic profiles.

Many lines of evidence (reviewed in Triggle and Swamy, 1983; Janis and Triggle, 1983) derived from chemical (structure–activity relationships including stereoselectivity), pharmacological (tissue contractility and Ca^{2+} movements), and physiological (membrane potential and ion currents) measurements are consistent with the thesis that the Ca^{2+} channel

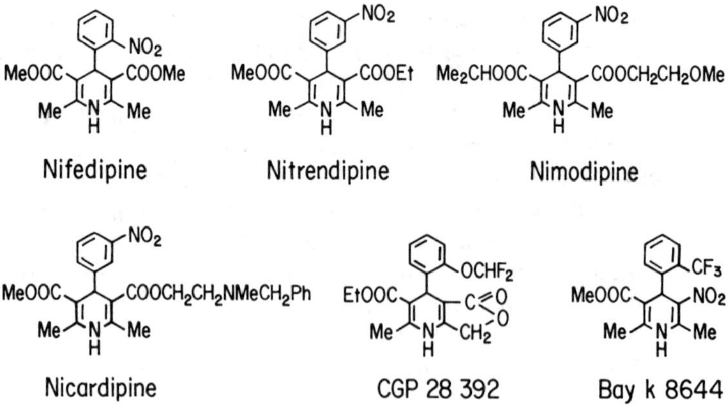

MeOOC, NO_2, COOMe, Me, N, H, Me — Nifedipine

MeOOC, NO_2, COOEt, Me, N, H, Me — Nitrendipine

Me_2CHOOC, NO_2, $COOCH_2CH_2OMe$, Me, N, H, Me — Nimodipine

MeOOC, NO_2, $COOCH_2CH_2NMeCH_2Ph$, Me, N, H, Me — Nicardipine

EtOOC, $OCHF_2$, C=O, O, CH_2, Me, N, H — CGP 28 392

MeOOC, CF_3, NO_2, Me, N, H, Me — Bay k 8644

Figure 2. Structural formula of 1,4-dihydropyridine Ca^{2+} channel ligands: CGP 28 392 and Bay K 8644 are activator species.

Table I. Calcium Channel Antagonists: Some Therapeutic Applications

	Verapamil	Nifedipine	Diltiazem
Cardiac arrhythmias			
PSVT	√	—	√
Atrial flutter	√	—	—
Myocardial ischemia			
Exertional angina	√	√	√
Variant angina	√	√	√
Arterial hypertension	√	√	—
Hypertrophic cardiomypathy	√	—	—

antagonists exert their primary actions at the potential-dependent Ca^{2+} channel. This specificity of action is seen very clearly with nifedipine and other 1,4-dihydropyridines, since they block channel function at concentrations two to five orders of magnitude less than those at which they affect other cellular processes. This specificity and high potency, together with the relatively easy availability of a large number of analogues of nifedipine, including activator species, has made the 1,4-dihydropyridine class of molecules a particularly useful probe of calcium channel structure and function. Specifically, the following questions may be raised:

1. What are the structural demands for ligand interaction with the 1,4-dihydropyridine site(s)? How may antagonist and activator structures be distinguished?
2. What are the relationships, topographical and functional, between the 1,4-dihydropyridine site and the permeation and gating machinery of the calcium channel?
3. Are 1,4-dihydropyridine receptors and Ca^{2+} channels regulated molecular species?

2. *Specific Sites of Interaction for 1,4-Dihydropyridines*

Structure–activity data derived from the pharmacological activities of 1,4-dihydropyridines in both *in vivo* and *in vitro* preparations led early to the postulation of a specific site or sites of action for these agents (Loev *et al.*, 1974; Rodenkirchen *et al.*, 1978; Rosenberger and Triggle, 1978). A schematic representation of the structural requirements for the expression of antagonist action is presented in Fig. 3. These requirements may be summarized as follows:

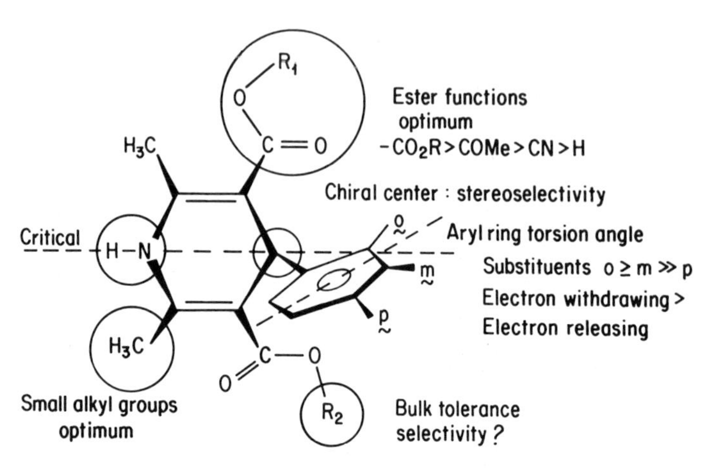

Figure 3. Summary of structural requirements for expression of antagonist activity in 1,4-dihydropyridines.

1. The presence of a 1,4-dihydropyridine ring with a free −NH group. Conversion to oxidized or reduced states abolishes activity.
2. Ester groups in the 3 and 5 positions are optimum. Replacement by other electron-withdrawing groups (−COMe, −CN) reduces antagonist activity. However, substantial tolerance to the size of the ester group alkyl substituent exists, and with increasing size there is evidence that partial selectivity of action may occur in different vascular beds.
3. A substituted phenyl group in the 4 position of the 1,4-dihydropyridine ring. Electron-withdrawing substituents in the ortho (*o*) or meta (*m*) positions of the phenyl ring confer optimum activity, but any substituent in the para (*p*) position greatly reduces activity, suggesting steric hindrance to binding.

It is also clear that the stereochemistry of the molecule is an important determinant of activity. Evidence from structure determinations of the solid state (Triggle *et al.*, 1980; Fossheim *et al.*, 1982; Langs and Triggle, 1985) and from the activities of rigid analogues of nifedipine (Seidel *et al.*, 1984) indicates that optimum activity is associated with an approximately orthogonal arrangement of the 1,4-dihydropyridine and phenyl rings (Fig. 3).

These conclusions, drawn from pharmacological studies, have been confirmed by the demonstration of specific, saturable, and high-affinity

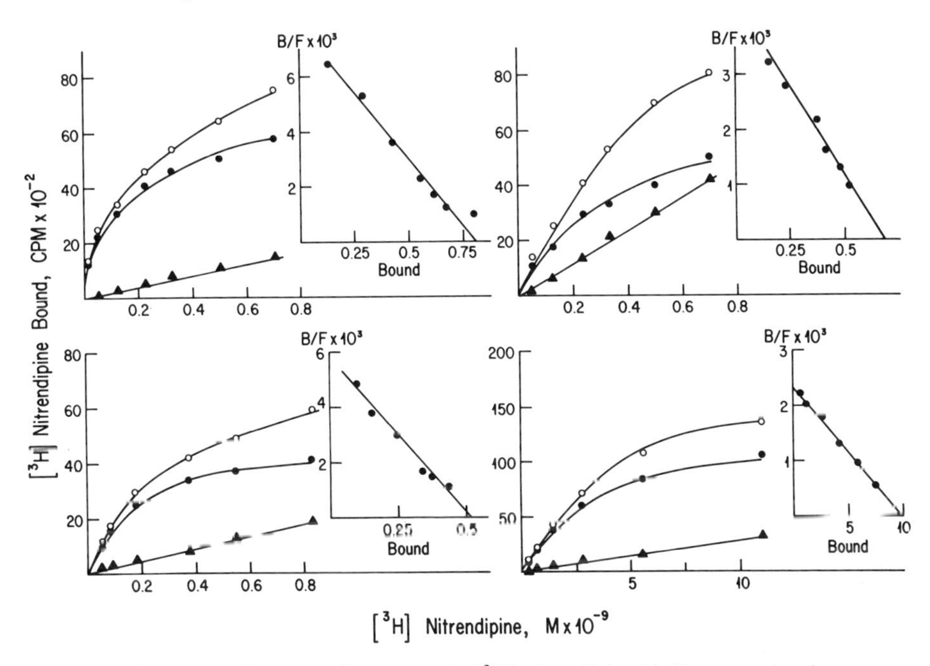

Figure 4. Representative saturation curves for [^{3}H]-nitrendipine binding to crude microsome preparations from smooth muscle (top left), cardiac muscle (top right), neuronal tissue (bottom left), and skeletal muscle (bottom right). Each saturation curve depicts total (○), nonspecific (▲), and specific binding (●). Each inset is a Scatchard transformation of the depicted binding data.

binding sites for several 1,4-dihydropyridines in smooth muscle, cardiac muscle, and brain (Fig. 4; for a general review see Triggle and Janis, 1984). In these tissues dissociation constants (K_D) for nitrendipine are approximately 1–3 $\times$ 10^{-10} M (25°), and capacities range from 50 to 1000 fmoles/mg protein, corresponding to approximate binding densities of 1–10 sites/μm^2 of plasmalemma (Bolger *et al.*, 1983; Janis *et al.*, 1984). Sites of low affinity but of very high capacity, associated with the t-tubules, are found in skeletal muscle.

The sensitivity of 1,4-dihydropyridine binding in smooth and cardiac muscle to inhibition by a series of 1,4-dihydropyridine antagonists is clearly consistent with interaction at a site at which Ca^{2+} antagonism is exerted (Fig. 5). The data of Fig. 5A show that there is a one-to-one correlation between binding and pharmacological activities for a series of 1,4-dihydropyridines in smooth muscle. In cardiac muscle an identical rank order of activities is exhibited, although binding affinities are some ten- to 100-fold higher than the corresponding pharmacological activities

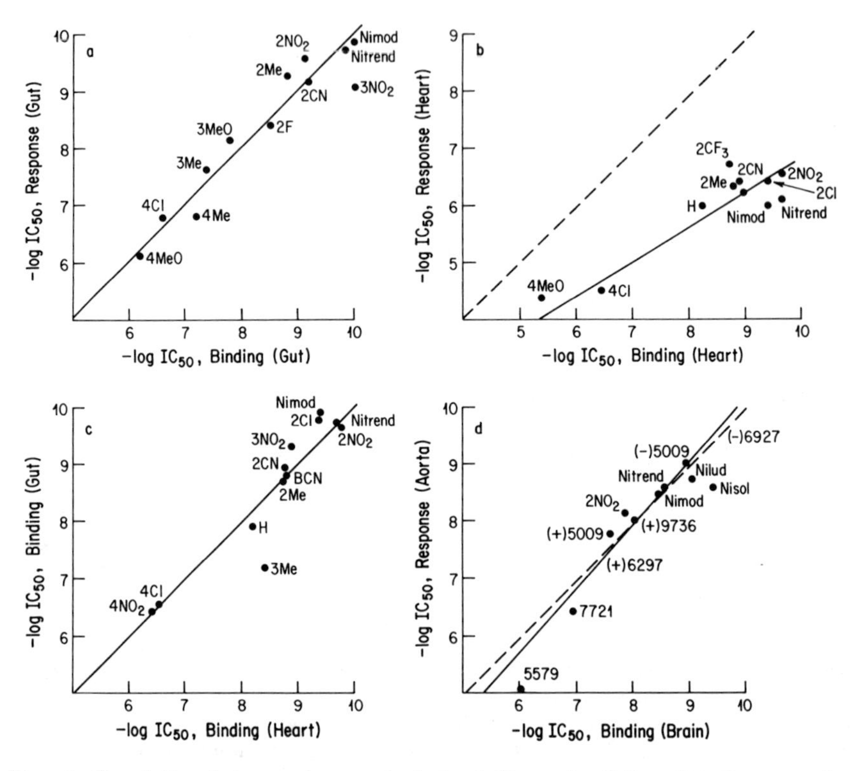

Figure 5. Correlations between pharmacological activities and radioligand-binding properties (ability to inhibit [^{3}H]-nitrendipine binding) in series of 1,4-dihydropyridines. Where substituents are shown, they refer to compounds in the series, 2,6-dimethyl-3,5-dicarbomethoxy-4-(substituted phenyl)1,4-dihydropyridines. A: K^+ depolarization-induced tension responses in guinea pig ileal smooth muscle (data from Bolger *et al.*, 1983). B: Tension responses in electrically driven cat papillary muscles (data from Rodenkirchen *et al.*, 1978) and binding in rabbit ventricle membranes (data from Janis *et al.*, 1983). C: Binding in guinea pig ileal smooth muscle and rabbit ventricle membranes. D: Binding in rat braian membranes and K^+ depolarization-induced tension responses in rabbit aorta (data from Bellemann *et al.*, 1983).

(Fig. 5B); binding in smooth and cardiac muscle membranes is very similar (Fig. 5C). 1,4-Dihydropyridine binding to mammalian neuronal tissue is fundamentally similar to that observed in smooth and cardiac muscle, and a very good correlation can thus be derived between neuronal binding and smooth muscle pharmacology (Fig. 5D). However, with the exception of a number of clonal cell lines, it is difficult to demonstrate high-affinity pharmacology of 1,4-dihydropyridines (or other Ca^{2+} channel antagonists) in neuronal tissue. Thus, an apparent discrepancy of some mag-

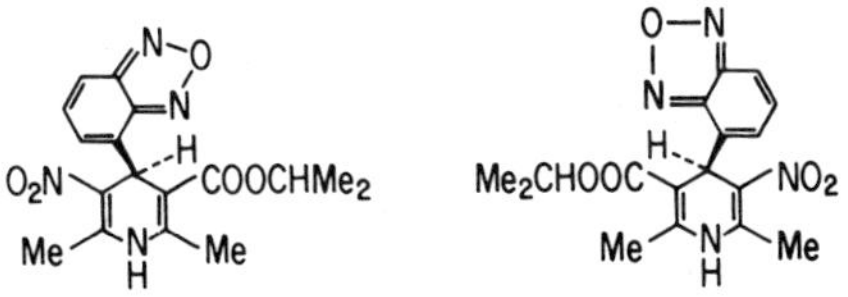

Figure 6. Structural formulas of enantiomers of isopropyl-4-(2,1,3-benzoxadiazol-4-yl)-1,4-dihydro-2,6-dimethyl-5-nitro-3-pyridine carboxylate (Hof *et al.*, 1985; Sandoz 202-791).

nitude exists whereby high-affinity binding in smooth, cardiac, and neuronal membranes is accompanied respectively by high-, medium-, and low-affinity pharmacology. The resolution of this anomaly, and a similar anomaly in skeletal muscle, will greatly facilitate our understanding of the mechanism(s) of action of this class of compounds (Triggle, 1983; Triggle and Janis, 1984; Miller and Freedman, 1984).

It is now clear that the basic 1,4-dihydropyridine structure embraces both antagonist and activator properties, with the latter being expressed in CGP 28 392 and Bay K 8644 (Fig. 2). The structural requirements for the relative expression of antagonist and activator properties have not been elucidated. However, pharmacological and radioligand-binding studies indicate that antagonists and activators have one binding site in common (Schramm *et al.*, 1983; Janis *et al.*, 1984; Rampe *et al.*, 1984; Su *et al.*, 1984). The solid-state structure of CGP 28 392 and Bay K 8644 reveal the anticipated general similarity to the antagonist structures previously described (Langs and Triggle, 1985). This is consistent with interaction at a common binding site. Subtle differences do, however, exist in geometry. Both activators show, relative to antagonists, enhanced 1,4-dihydropyridine ring planarity and increased acidity of the –NH protein. In addition, the orientation of the 3 and 5 substituents may be of importance: in known antagonist structures the ester groups at C_3 and C_5 enjoy either a *cis,cis* or *cis,trans* disposition relative to the C=C bonds of the 1,4-dihydropyridine ring. The activator structures thus far determined are the only ones that do not have at least one ester group in a *cis* orientation (Langs and Triggle, 1985).

Much further work will be necessary to determine the structural clues that indicate activator or antagonist properties. Undoubtedly, the chiral nature of the interactions will be important, as shown by the opposing properties of the *R*- and *S*- enantiomers of isopropyl-4-(2,1,3-benzoxadiazol-4-yl)-1,4-dihydro-2,6-dimethyl-5-nitro-3-pyridine carboxylate, where the *S*- and *R*- enantiomers are activator and antagonist, respectively (Hof *et al.*, 1985; Fig. 6).

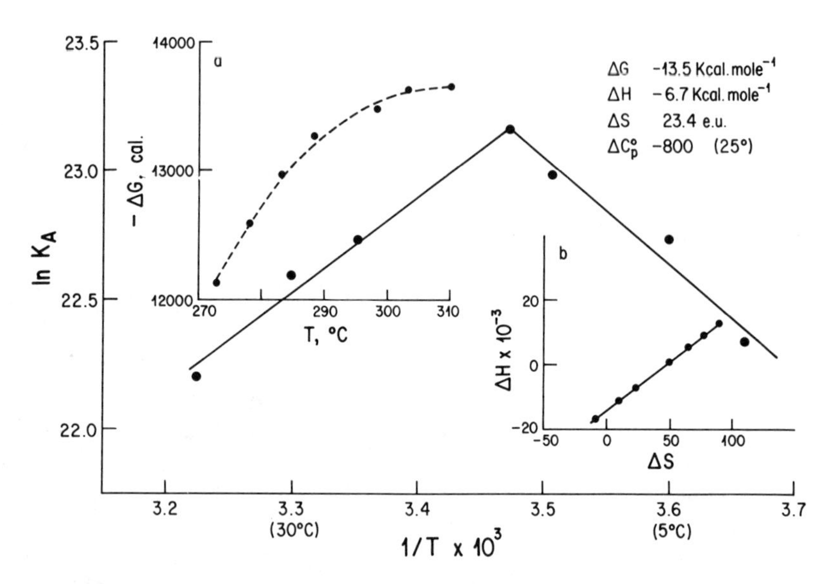

Figure 7. Temperature dependence of [^{3}H]-nitrendipine binding to a microsomal fraction from guinea pig ventricle muscle. Major representation is an Arrhenius plot of ln K_A against T^{-1}. Inset a: Variations of free energy of binding with temperature; the points represent experimental data, and the line is drawn to a polynomial fit. Inset b: Relationship between enthalpy and entropy of binding in the temperature range 0° to 37°.

Differences in the detail of antagonist and agonist interactions are also suggested from the thermodynamics of activator and antagonist binding. The dependence of nitrendipine binding on temperature is modest, but the Arrhenius plot is nonlinear, and the variation of free energy of binding with temperature is fit by a second-order expression with compensating enthalpy and entropy changes (Fig. 7). Enthalpy and entropy changes become more positive with decreasing temperature, which is consistent with a hydrophobic association process. In contrast, binding of Bay K 8644 is markedly temperature dependent and over the range of 15°–37° is an enthalpically driven process ($\Delta H = -22$ kcal mole^{-1} and $\Delta S = -34$ e.u.; R. A. Janis and D. J. Triggle, unpublished observations). Similar differences have been observed for agonist–antagonist pairs in other systems including β-adrenoceptors (Weiland *et al.*, 1980). Such differences should be interpreted with caution, since they reflect binding to crude membrane fractions, but they may correspond to activator and antagonist stabilization of different states of the Ca^{2+} channel.

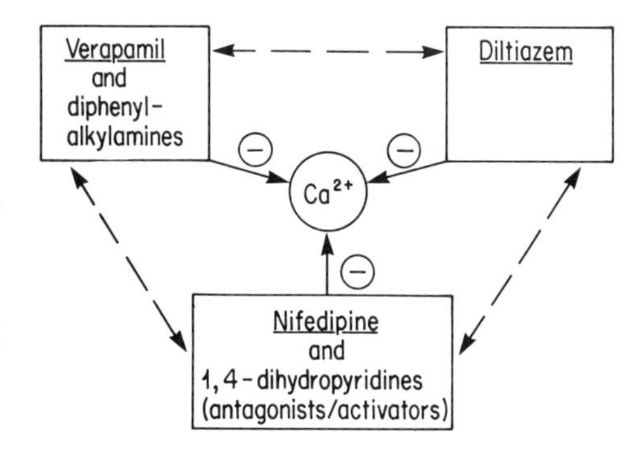

Figure 8. Schematic representation of interrelationships between proposed binding sites for drugs active at Ca^{2+} channels. The sites are regarded as a set of linked allosteric receptors (for further details see Glossmann *et al.*, 1982; Janis and Triggle, 1984).

3. Topographic and Functional Relationship of Ligand-Binding Sites to Ca^{2+} Channels

There is little doubt that the ligand binding studies, conducted mostly with 1,4-dihydropyridines but also with verapamil, D600, and diltiazem, probe binding sites associated with Ca^{2+} channels despite some quantitative discrepancies between pharmacological and binding behavior. However, the more difficult problem of determination of the topographic disposition of such binding sites in terms of Ca^{2+} channel function remains to be solved. Two-dimensional relationships between various binding sites such as those indicated in Fig. 8 have been suggested by many studies.

The existence of activator and antagonist species makes it tempting to locate the 1,4-dihydropyridine binding site close to the permeation machinery of the channel. This may be premature, since activator analogues in other ligand classes may await discovery. Nonetheless, the divalent-cation dependence of 1,4-dihydropyridine binding (Glossmann *et al.*, 1982; Gould *et al.*, 1982; Luchowski *et al.*, 1984) is suggestive of an association between the ligand-binding sites and the metal coordination sites of the Ca^{2+} channel. The high sensitivity of this process to divalent cations, including Ca^{2+}, suggests that this may represent an intracellular process. Furthermore, the good correlation existing in a series of calmodulin antagonists between their activities as inhibitors of 1,4-dihydropyridine binding and of calmodulin function (Fig. 9) suggests that a calmodulinlike protein may be involved in the regulation of 1,4-dihydropyridine binding. It is thus of particular interest to note that models of Ca^{2+} channel permeation and inactivation have been described that invoke as an integral feature high-affinity divalent cation binding sites (Hess and Tsien, 1984; Almers *et al.*, 1984; Standen and Stanfield, 1982). In the description by Hess and Tsien (1984), two high-affinity Ca^{2+} bind-

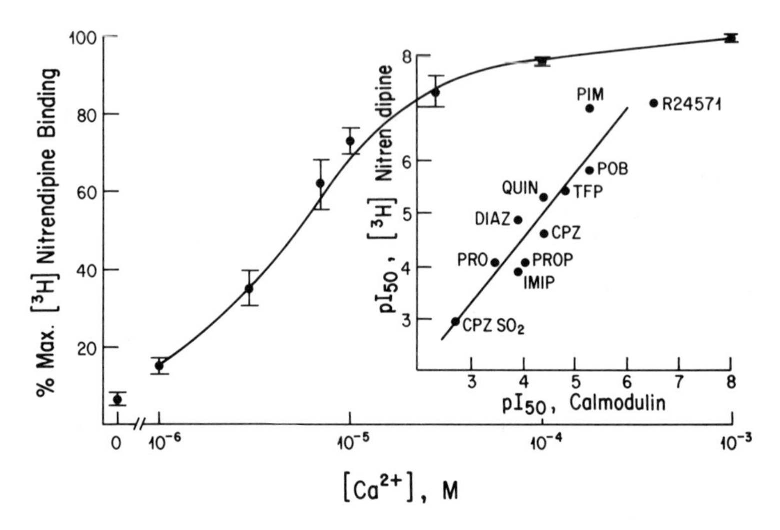

Figure 9 The dependence of specific [3H]-nitrendipine binding in Ca^{2+}-depleted guinea pig ileal longitudinal smooth muscle microsomal preparations on added Ca^{2+}. Inset: The ability of a series of calmodulin antagonists, including R24571 (calmidazolium), pimozide (PIM), phenoxybenzamine (POB), trifluoperazine (TFP), quinacrine (QUIN), diazepam (DIAZ), chlorpromazine (CPZ), propranolol (PROP), procaine (PRO), imipramine (IMIP), and chlorpromazine sulfoxide (CPZ SO_2) to inhibit specific nitrendipine binding and to inhibit calmodulin-dependent phosphodiesterase activity. (Data from Luchowski *et al.*, 1984.)

ing sites are present in the channel, and Ca^{2+} flux increases with double occupancy of these binding sites. 1,4-Dihydropyridine ligands could function as antagonist or activator species by mediating appropriate changes in the binding affinities and saturation curves of these Ca^{2+} binding sites.

Any description of the relationship of channel ligand-binding sites to the function and structural machinery of the Ca^{2+} channel must recognize the necessarily dynamic character of channel configuration and conformation. Channels exist minimally in open, closed, or inactivated states, and the affinity of a ligand for a channel component may vary according to state (Fig. 10; Hille, 1977; Hondeghem and Katzung, 1977). The marked frequency dependence of verapamil and D600 action is consistent with their binding to Ca^{2+} channels only after they open. It is likely that they stabilize an inactivated state via the pathway (Fig. 10) $R \rightarrow O \rightarrow O^* \rightarrow I^*$ (*inter alia*, Pelzer *et al.*, 1982; Lee and Tsien, 1983; McDonald *et al.*, 1984). 1,4-Dihydropyridine antagonists show marked voltage-dependent activity in the absence of repetitive pulsing, and they probably also stabilize an inactivated state via a different pathway, $R \rightarrow I \rightarrow I^*$ (Lee and Tsien, 1983; Sanguinetti and Kass, 1984a; Bean, 1984). In contrast, 1,4-

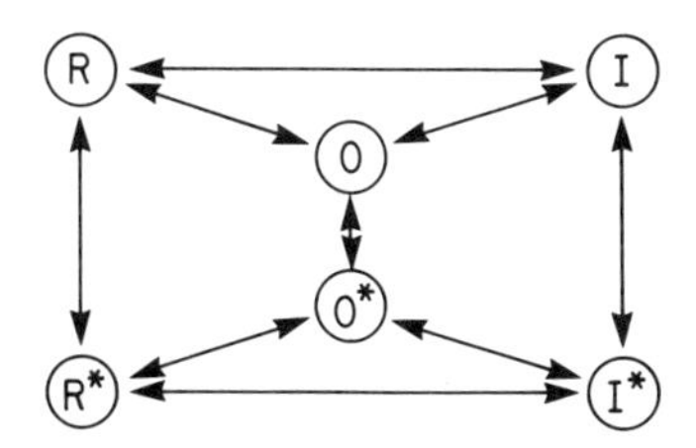

Figure 10. Schematic representation of equilibrium among resting (R), open (O), and inactivated (I) states of the Ca^{2+} channel. * Indicates channels with bound drug (see text for further details).

dihydropyridine activators require a polarized membrane potential and function by the stabilization of an open channel state, R → O → O* (Fig. 10; Hess *et al.*, 1984; Sanguinetti and Kass, 1984b).

These state-dependent interactions are likely to be of considerable importance in determining the tissue selectivity patterns of different categories of antagonist and in correlating pharmacological and radioligand-binding data. The several hundredfold discrepancy between binding and pharmacology seen in cardiac tissue for a series of 1,4 dihydropyridines (Fig. 5) is in accord with differences in affinity between resting and inactivated states, with high-affinity binding to membrane fragments presumably measuring dominantly the inactivated state (Sanguinetti and Kass, 1984b; Bean, 1984). An additional clue to the importance of the voltage-dependent interactions of the Ca^{2+} channel antagonists lies in apparently competitive interactions seen in tissue but not in radioligand-binding experiments (for reviews see Fleckenstein, 1977; Janis and Triggle, 1983, 1984). This effect of Ca^{2+} and other polyvalent cations arises at least in part from their ability to bind to or screen surface negative change in excitable cells (Frankenhaueser and Hodgkin, 1957; Hille, 1984). Changes in Ca^{2+} concentration will alter the effective transmembrane potential and channel-gating processes. Increases in Ca^{2+} should shift the voltage–channel availability relationship in a depolarizing direction: for a given stimulus more channels will be available, and a competition between Ca^{2+} ions and Ca^{2+} channel antagonists will be realized.

Electrophysiological evidence for modulated ligand binding to Ca^{2+} channels has been derived largely from cardiac tissue. It is possible that different behavior occurs in other tissues, including neurons and skeletal muscle, that could accommodate the very obvious discrepancies in these tissues between radioligand binding and pharmacological activity.

The conformational changes of the channel proteins that necessarily accompany channel state interconversions are also important to structure–activity relationships in the several groups of Ca^{2+} channel ligands. These structure–activity relationships must be interpreted in terms of the channel state(s) being measured. It is conceivable that both qualitative and quantative differences in structural requirements for optimum activity exist between the several channel states and that explanations of such differences will be of value to the design of state- and tissue-specific channel ligands.

4. *Regulation of Ligand-Binding Sites and Function at the Calcium Channel*

Elucidation of the mechanisms by which channel ligand-binding sites are regulated will be of particular value in determining the relationships between these binding sites and channel function. Only limited information is currently available.

The ontogeny of 1,4-dihydropyridine binding sites in central and peripheral tissues (Erman *et al.*, 1983; Kazazoglou *et al.*, 1983; Renaud *et al.*, 1984) reveals generally low levels of sites in fetal stages, which increase during development. These increases in binding sites do differ in amounts and time courses between tissues but show an approximate accord with the development of other excitable functions.

Some studies of Ca^{2+} channel ligand binding (Renaud *et al.*, 1984) suggest similar observations with other receptor systems (*inter alia*, Renaud *et al.*, 1981; Halvorsen and Hathanson, 1984): at the earliest stages the 1,4-dihydropyridine binding sites may not be coupled or may be cou-

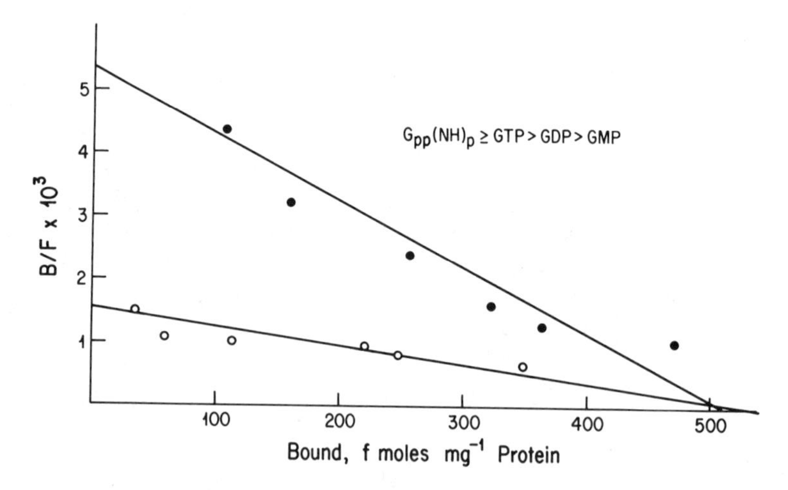

Figure 11. The effect of 5′-guanylimidodiphosphate [Gpp(NH)p, 10^{-4} M] on [^{3}H]-nitrendipine binding to a microsomal preparation from guinea pig ventricle cardiac muscle. Control binding (●); binding in presence of Gpp(NH)p (○). Control, B_{max} 511 ± 38 fmole/mg protein, K_D 99.4 ± 15 × 10^{-12} M; treated, B_{max} 571 ± 128 fmole/mg protein, K_D 372 ± 116 × 10^{-12} M. Medium contained Mg^{2+} (1.3 mM) and Ca^{2+} (1.2 mM).

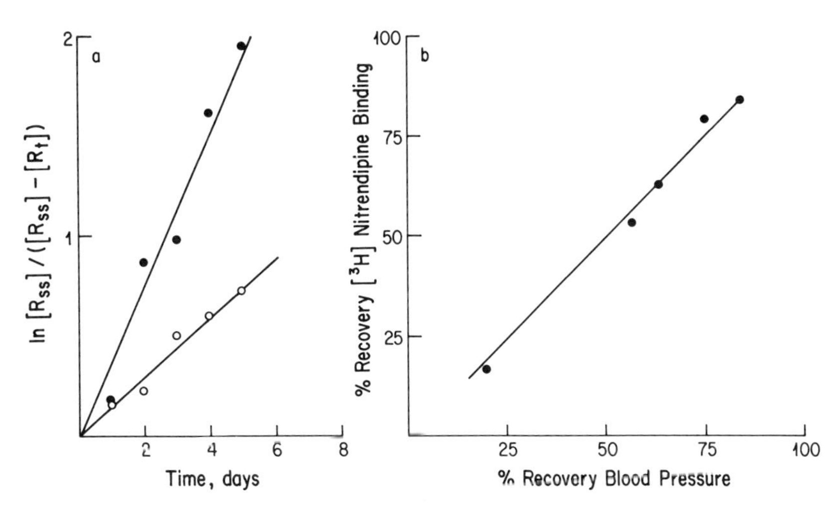

Figure 12. Recovery of [^{3}H]-nitrendipine binding sites in rat heart and Ca^{2+} channel function following *in vivo* treatment of rats with 2,6-dimethyl-3,5-dicarbomethoxy-4-(2-isothiocyanophenyl)-1,4-dihydropyridine (25 mg/kg, i.v.). A: Recovery of [^{3}H]-nitrendipine binding sites plotted according to the equation $\ln [R_{ss}]/([R_{ss}] - [R_t]) = kt$ (Mauger *et al.*, 1982), where $[R_{ss}]$ = receptor concentration at steady state, $[R_t]$ = receptor concentration at time t, and k = rate constant for degradation, 0.02 hr^{-1}. This corresponds at steady state to an incorporation of approximately 5.0 fmole/mg protein per hr. ○, Control; ○, after treatment with cycloheximide. B: Recovery of blood pressure (tail cuff determination) as a function of recovery of maximum density of nitrendipine binding sites.

pled inefficiently to functional channels. The nature and even existence of coupling systems linking recognition sites and functional machinery in Ca^{2+} channels await discovery and definition. However, the presence of coupling systems would support a role for endogenous ligands that are active at the channel and serve to link physiologically the recognition and activation sites (Schramm *et al.*, 1983; Triggle and Janis, 1985).

A candidate coupling system for Ca^{2+} channels is the guanine nucleotide binding proteins (G proteins), which mediate excitatory and inhibitory effects through hormonally sensitive adenylate cyclase (Gilman, 1984; Hughes, 1983) but which may also be involved in coupling other systems including Ca^{2+} movements. Tentative support for this postulate is provided by reports that guanine nucleotides potentiate Ca^{2+}-dependent exocytosis in mast cells (Gomperts, 1983) and that pertussis toxin inhibits chemotactic factor-induced Ca^{2+} mobilization in neutrophils (Molski *et al.*, 1984) as well as by the inhibitory effects of guanine nucleotides on [^{3}H]-verapamil and [^{3}H]-nitrendipine binding in skeletal mus-

Table II. Turnover of Membrane Receptor and Ion Channel Complexes

System	Tissue	$t_{1/2}$ (hr)	Reference
Acetylcholine receptor	Chick skeletal muscle	17	Gardner and Fambrough (1979)
Benzodiazepine receptor	Chick brain cultures	4 32	Borden *et al.* (1984)
Dopamine receptor	Rat brain	45 120	Leff *et al.* (1984)
Epidermal growth factor	Fibroblasts A 431 cells	9 16	Krupp *et al.* (1982)
Adrenergic receptor			
α	BC_3H_1 cells	23	Mauger *et al.* (1982)
α_1	Spleen	38	Hamilton *et al.* (1984)
α_2	Spleen	84	

cle t-tubules (Galizzi *et al.*, 1984) and cardiac muscle microsomes, respectively (Fig. 11).

By analogy to other membrane proteins, including receptors and ion channels, it may be assumed that Ca^{2+} channels and the associated ligand-binding sites have a finite lifetime and that they are replenished by newly incorporated material. An estimate of the turnover of the Ca^{2+} channel in cardiac tissue has been obtained by measuring the rate of appearance of [^{3}H]-nitrendipine binding sites following the *in vivo* administration of 2,6-dimethyl-3,5-dicarbomethoxy-4-(2-isothiocyanophenyl)1,4-dihydropyridine, a persistent analogue of nifedipine (Fig. 12). The rate of recovery of binding sites ($t_{1/2}$ 41 hr) parallels the rate of restoration of blood pressure (Fig. 12), consistent with the ligand-binding sites representing recovery of a functional state. This observed rate accords with those reported for other receptor systems (Table II).

It may be assumed that the concentration and localization of nitrendipine binding sites are controlled by trophic factors. The nature of any such factors remains unknown. However, there is an increase in [^{3}H]-nitrendipine binding sites in skeletal muscle following denervation (Fig. 13). More extensive studies by Schmid *et al.* (1984) in chick skeletal muscle have also documented an increase in [^{3}H]-nitrendipine binding sites in denervated chick skeletal muscle. Similarly, in chick hearts, 6-hydroxydopamine treatment produces an increase in nitrendipine binding site density (Renaud *et al.*, 1984).

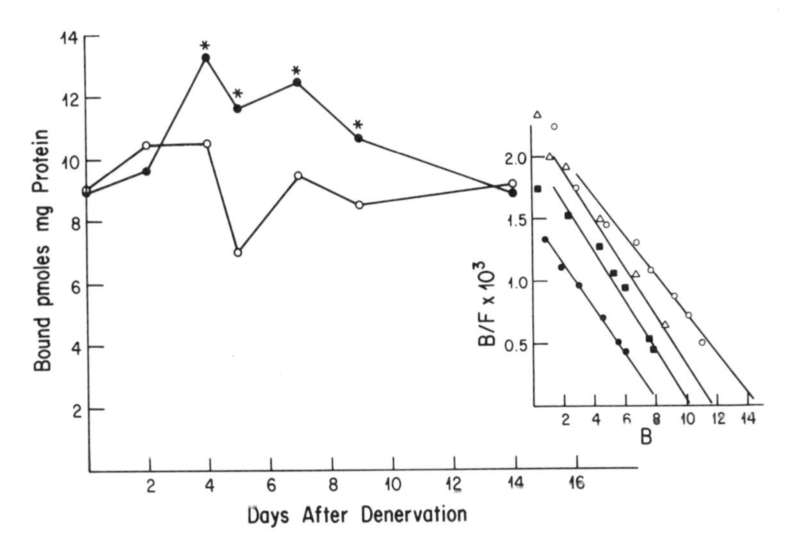

Figure 13. Effect of denervation (nerve section) on [^{3}H]-nitrendipine binding density in rat gastrocnemius skeletal muscle: ●, denervated; ○, control (contralateral). Inset: Scatchard plots of [^{3}H]-nitrendipine binding to denervated muscle at days 2 (●), 4 (○), 9 (△), and 15 (■). * Indicates significant difference from control.

5. Conclusions and Prospective

The Ca^{2+} channel ligands are a new and exciting group of structures that have generated very considerable interest as both therapeutic agents and molecular probes. They have proved to be of considerable utility in the management of a number of serious cardiovascular disorders and have also considerably extended our knowledge of Ca^{2+} channel function. It seems very probable that progress in both areas will pursue common paths, at least in the immediate future. New structures will indicate new therapeutic areas, and the same structures will probe more subtly the architecture and function of the Ca^{2+} channel. For patient, clinician, and scientist alike, these are likely to be rewarding and exciting times.

ACKNOWLEDGMENTS. This work was supported in part by a grant from the National Institute of Health (HL 16003). Additional support from the Miles Institute for Preclinical Pharmacology is gratefully acknowledged.

References

Almers, W., McCleskey, E. W., and Palade, P. T., 1984, A non-selective cation conductance in frog muscle membrane blocked by micromolar external calcium ions, *J. Physiol. (Lond.)* **353:**565–583.

Bean, B. P., 1984, Nitrendipine block of cardiac calcium channels: High affinity binding to the inactivated state, *Proc. Natl. Acad. Sci. U.S.A.* **81:**6388–6392.

Bellemann, P., Schade, A., and Towart, R., 1983, Dihydropyridine receptor in rat brain labelled with [^{3}H]nimodipine, *Proc. Natl. Acad. Sci. U.S.A.* **80:**2356–2360.

Bolger, G. T., Gengo, P. T., Klockowski, R., Luchowski, E., Siegel, H., Janis, R. A., Triggle, A. M., and Triggle, D. J., 1983, Characterization of binding of the Ca^{2+} channel antagonist, [^{3}H]nitrendipine, to guinea pig ileal smooth muscle, *J. Pharmacol. Exp. Ther.* **225:**291–309.

Borden, L. A., Czajkowski, C., Chan, C. Y., and Farb, D. H., 1984, Benzodiazepine receptor synthesis and degradation by neurons in culture, *Science* **226:**857–860.

Edelhoch, H., and Osborne, J. C., Jr., 1976, The thermodynamic basis of the stability of proteins, nucleic acids and membranes, *Adv. Protein Chem.* **30:**183–250.

Erman, R. D., Yamamura, H. I., and Roeske, W. R., 1983, The ontogeny of specific binding sites for the calcium channel antagonist, nitrendipine, in mouse heart and brain, *Brain Res.* **278:**327–331.

Fleckenstein, A., 1977, Specific pharmacology of calcium in myocardium, cardiac pacemakers and vascular smooth muscle, *Annu. Rev. Pharmacol. Toxicol.* **17:**149–166.

Fleckenstein, A., 1983, *Calcium Antagonism in Heart and Smooth Muscle. Experimental Facts and Therapeutic Prospects*, Wiley Interscience, New York.

Fossheim, R., Svarteng, K., Mostad, A., Rømming, C., Shefter, E., and Triggle, D. J., 1982, Crystal structures and pharmacologic activities of calcium channel antagonists: 2,6-Dimethyl-3,5-dicarbomethoxy-4-(unsubstituted, 3-methyl, 4-methyl, 3-nitro, 4-nitro, and 2,4-dinitrophenyl)-1,4-dihydropyridine, *J. Med. Chem.* **25:**126–131.

Frankenhaueser, B., and Hodgkin, A. L., 1957, The action of calcium on the electrical properties of squid axons, *J. Physiol. (Lond.)* **137:**218–244.

Galizzi, J. P., Fosset, M., and Lazdunski, M., 1984, Properties of receptors for the Ca^{2+}-channel blocker verapamil in transverse-tubule membranes of skeletal muscle, *Eur. J. Biochem.* **144:**211–215.

Gardner, J. M., and Fambrough, D. M., 1979, Acetylcholine receptor degradation measured by density labelling: Effects of cholinergic ligands and evidence against recycling, *Cell* **16:**661–674.

Gilman, A. G., 1984, G Proteins and dual control of adenylate cyclase, *Cell* **36:**577–579.

Glossman, H., Ferry, D. R., Lübbecke, F., Mewes, R., and Hofmann, F., 1982, Calcium channels: Direct identification with radioligand binding studies, *Trends Pharm. Sci.* **3:**431–437.

Gomperts, B. D., 1983, Involvement of guanine nucleotide-binding protein in the gating of Ca^{2+} by receptors, *Nature* **306:**64–66.

Gould, R. J., Murphy, K. M. M., and Snyder, S. H., 1982, [^{3}H]Nitrendipine-labeled calcium channels discriminate inorganic calcium agonists and antagonists, *Proc. Natl. Acad. Sci. U.S.A.* **79:**3656–3660.

Halvorsen, S. W., and Nathanson, N. M., 1984, Ontogenesis of physiological responsiveness and guanine nucleotide sensitivity of cardiac muscarinic receptors during chick embryonic development, *Biochemistry* **23:**5813–5821.

Hamilton, C. A., Dalrymple, H. W., Reid, J. L., and Summer, D. J., 1984, The recovery of α-adrenoceptor function and binding sites after phenoxybenzamine. An index of receptor turnover, *Naunyn Schmiedebergs Arch. Pharmacol.* **325:**34–41.

Hess, P., and Tsien, R. W., 1984, Mechanism of ion permeation through calcium channels, *Nature* **309:**453–456.

Hess, P., Lansman, J. B., and Tsien, R. W., 1984, Different modes of Ca channel gating behavior favored by dihydropyridine Ca agonists and antagonists, *Nature* **311:**538–544.

Hille, B., 1977, Local anesthetics: Hydrophilic and hydrophobic pathways for the drug–receptor reaction, *J. Gen. Physiol.* **69:**497–515.

Hille, B., 1984, Modifiers of Gating, in: *Ionic Channels of Excitable Membranes* Sinauer Associates, Sunderland, A, pp. 303–328.

Hof, R. P., Rüegg, U. T., Hof, A., and Vogel, A., 1985, Stereoselectivity at the calcium channel: Opposite action of the enantiomers of a 1,4-dihydropyridine, *J. Cardiovasc. Pharmacol.* **7:**689–693.

Hondeghem, L. M., and Katzung, B. G., 1977, Time- and voltage-dependent interactions of antiarrhythmic drugs with cardiac sodium channels, *Biochim. Biophys. Acta* **472:**373 398.

Hughes, S. M., 1983, Are guanine nucleotide binding proteins a distinct class of regulatory proteins? *FEBS Lett.* **164:**1–8.

Janis, R. A., and Scriabine, A., 1983, Sites of action of Ca^{2+} channel inhibitors, *Biochem. Pharmacol.* **32:**3499–3507.

Janis, R. A., and Triggle, D. J., 1983, New developments in Ca^{2+} channel antagonists, *J. Med. Chem.* **26:**775–785.

Janis, R. A., Rampe, D., Sarmiento, J. G., and Triggle, D. J., 1984a, Specific binding of a calcium channel activator, [^{3}H]Bay K 8644 to membranes from cardiac muscle and brain, *Biochem. Biophys. Res. Commun.* **121:**317–323.

Janis, R. A., Sarmiento, J. G., Maurer, S. C., Bolger, G. T., and Triggle, D. J., 1984b, Characteristics of the binding of [^{3}H]nitrendipine to rabbit ventricular membranes: Modification by other Ca^{2+} channel antagonists and by the Ca^{2+} channel agonist, Bay K 8644, *J. Pharmacol. Exp. Ther.* **231:**8–15.

Kazazoglou, T., Schmid, A., Renaud, J. R., and Lazdunski, M., 1983, Ontogenic appearance of Ca^{2+} channels characterized as binding sites for nitrendipine during development of nervous, skeletal and cardiac muscle systems in the rat, *FEBS Lett.* **164:**75–79.

Krupp, M. N., Connolly, D. T., and Lane, M. D., 1982, Synthesis, turnover and down-regulation of epidermal growth factor receptors in human A431 epidermoid carcinoma cells and skin fibroblasts, *J. Biol. Chem.* **257:**11489–11496.

Langs, D. A., and Triggle, D. J., 1985, Conformational features of calcium channel agonist and antagonist analogs of nifedipine, *Mol. Pharmacol.* **27:**544–548.

Lee, K. S., and Tsien, R. W., 1983, Mechanism of calcium channel blockade by verapamil, D600, diltiazem and nitrendipine in single dialyzed heart cells, *Nature* **302:**790–794.

Leff, S. E., Gariano, R., and Creese, I., 1984, Dopamine receptor turnover rates in rat striatum are age-dependent, *Proc. Natl. Acad. Sci. U.S.A.* **81:**3910–3914.

Loev, B., Goodman, M. M., Snader, K. M., Tedeschi, R., and Macko, E., 1974, "Hantzsch-type" dihydropyridine hypotensive agents, *J. Med. Chem.* **17:**956–965.

Luchowski, E., Yousif, F., Triggle, D. J., Maurer, S. C., Sarmiento, J. G., and Janis, R. A., 1984, Effects of metal cations and calmodulin antagonists on [^{3}H]nitrendipine binding in smooth and cardiac muscle, *J. Pharmacol. Exp. Ther.* **230:**607–613.

Mauger, J. P., Sladeczek, F., and Bockaert, J., 1982, Characteristics and metabolism of α_1-adrenergic receptors in a nonfusing muscle cell line, *J. Biol. Chem.* **257:**875–879.

McDonald, T. F., Pelzer, D., and Trautwein, W., 1984, Cat ventricular muscle treated with D600: Characteristics of calcium channel block and unblock, *J. Physiol.* (*Lond.*) **352:**217–241.

Miller, R. J., and Freedman, S. B., 1984, Are dihydropyridine binding sites voltage-sensitive calcium channels? *Life Sci.* **34:**1205–1221.

Molski, T. F. P., Naccache, P. H., Marsh, M. L., Kermode, T., Becker, E. L., and Sha'afi, R. I., 1984, Pertussis toxin inhibits the rise in the intracellular concentration of free calcium that is induced by chemotactic factors in rabbit neutrophils: Possible role of the "G proteins" in calcium mobilization, *Biochem. Biophys. Res. Commun.* **124:**664–650.

Opie, L. H., 1984, Calcium ions, drug action and the heart—with special reference to calcium antagonist drugs, *Pharmacol. Ther.* **25:**271–295.

Pelzer, D., Trautwein, W., and McDonald, T. F., 1982, Calcium channel block and recovery from block in mammalian ventricular muscle treated with organic channel inhibitors, *Pflüegers Arch.* **394:**97–105.

Rampe, D., Janis, R. A., and Triggle, D. J., 1984, Interaction of Bay K 8644, a 1,4-dihydropyridine Ca^{2+} channel activator: Dissociation of binding and functional effects in brain synaptosomes, *J. Neurochem.* **43:**1688–1692.

Renaud, J. F., Romey, G., Lomget, A., and Lazdunski, M., 1981, Differentiation of the fast Na^{+} channel in embryonic heart cells: Interaction of the channel with neurotoxins, *Proc. Natl. Acad. Sci. U.S.A.* **78:**5348–5352.

Renaud, J. F., Kazazoglou, T., Schmid, A., Romey, G., and Lazdunski, M., 1984, Differentiation of receptor sites for [^{3}H]nitrendipine in chick hearts and physiological relation to the slow Ca^{2+} channel and to excitation–contraction coupling, *Eur. J. Biochem.* **139:**673–681.

Rodenkirchen, R., Bayer, R., Steiner, R., Bossert, F., Meyer, H., and Moller, E., 1979, Structure–activity studies on nifedipine in isolated cardiac muscles, *Naunyn Schmiedebergs Arch. Pharmacol.* **310:**69–78.

Rosenberger, L. B., and Triggle, D. J., 1978, Calcium, calcium translocation and specific calcium antagonists, in: *Calcium and Drug Action* (G. B. Weiss, ed.), Plenum Press, New York, pp. 3–31.

Sanguinetti, M. C., and Kass, R. S., 1984a, Voltage-dependent block of calcium channel current in the calf cardiac Purkinje fiber by dihydropyridine calcium channel antagonists, *Circ. Res.* **55:**336–348.

Sanguinetti, M. C., and Kass, R., 1984b, Regulation of cardiac calcium current and contractile activity by the dihydropyridine Bay K 8644 is voltage-dependent, *J. Mol. Cell. Cardiol.* **16:**667–670.

Schmid, A., Renaud, J. F., Fosset, M., Meaux, J. P., and Lazdunski, M., 1984, The nitrendipine-sensitive Ca^{2+} channel in chick muscle cells and its appearance during myogenesis *in vitro* and *in vivo*, *J. Biol. Chem.* **259:**11366–11372.

Schramm, M., Thomas G., Towart, F., and Franckowiak, G., 1983, Novel dihydropyridines with positive inotropic action through activation of Ca^{2+} channels, *Nature* **303:**535–537.

Schwartz, A., and Triggle, D. J., 1984, Cellular action of calcium channel blocking drugs, *Annu. Rev. Med.* **35:**325–339.

Seidel, W., Meyer, H., Born, L., Kazda, S., and Dompert, W., 1984, Rigid calcium antagonists of the nifedipine type: Geometrical requirements for the dihydropyridine receptor. *Abstr. Am. Chem. Soc.* **187.**

Standen, N. B., and Stanfield, P. R., 1982, A binding-site model for calcium channel inactivation that depends on calcium entry, *Proc. R. Soc. Lond.* [*Biol.*] **217:**101–110.

Su, C. M., Swamy, V. C., and Triggle, D. J., 1984, Calcium channel activation in vascular smooth muscle by Bay K 8644, *Can. J. Physiol. Pharmacol.* **62:**1401–1410.

Triggle, A. M., Shefter, E., and Triggle, D. J., 1980, Crystal structures of calcium channel antagonists: 2,6-Dimethyl-3,5-dicarbomethoxy-4[2-nitro,-3-cyano-, 4-(dimethylamino)- and 2,3,4,5,6,-pentafluorophenyl]-1,4-dihydropyridine, *J. Med. Chem.* **23:**1442–1445.

Triggle, D. J., 1984, Ca^{2+} channels revisited: Problems and promises, *Trends Pharmacol. Sci.* **5:**4–5.

Triggle, D. J., and Janis, R. A., 1984, Calcium channel antagonists: New perspectives from the radioligand binding assay, in: *Modern Methods in Pharmacology*, Vol. II (N. Back and S. Spector, eds.), Alan R. Liss, New York, pp. 1–28.

Triggle, D. J., and Janis, R. A., 1985, The 1,4-dihydropyridine receptor: A regulatory component of the Ca^{2+} channel, *J. Cardiovasc. Pharmacol.* **6:**S949–955.

Triggle, D. J., and Swamy, V. C., 1983, Calcium antagonists: Some chemical–pharmacological aspects, *Circ. Res.* **52**(Suppl. I)**:**17–28.

Weiland, G. A., Minneman, K. P., and Molinoff, P. B., 1980, Thermodynamics of agonist and antagonist interactions with mammalian β-adrenergic receptors, *Mol. Pharmacol.* **18:**341–347.

8

Information Flow in the Calcium Messenger System

HOWARD RASMUSSEN, ITARU KOJIMA, and PAULA BARRETT

1. Introduction

Work in the past 20 years has led to an ever-expanding appreciation of the intracellular messenger function of calcium ion [Ca^{2+}].* What began as an interest in the role of Ca^{2+} in stimulus–response coupling in skeletal muscle has expanded to the point where Ca^{2+} is recognized as one of the few universal intracellular messengers involved in coupling a wide variety of stimuli to an equally diverse range of responses: Ca^{2+} participates in the control of neuro-, exocrine, and endocrine secretion, the contraction of all forms of muscle, fluid and electrolyte transport, energy and fuel metabolism, and growth and development. The control of these diverse tissue responses requires a considerably more elaborate system than simply a rise and fall in intracellular free Ca^{2+} concentration—the original model of the way Ca^{2+} was thought to serve its messenger function. Hence, it is more appropriate to introduce the concept of the calcium messenger system. Encompassed within this term are the cellular components involved in generating, terminating, transmitting, and receiving the Ca^{2+} message.

* Selected references include Berridge, 1975, 1982, 1984; Borle, 1981; Campbell, 1983; Cheung, 1980; Exton, 1981; Joseph and Williamson, 1983; Kaibuchi *et al.*, 1983; Means, 1980; Nishizuka, 1983, 1984; Putney, 1979, 1983; Rasmussen, 1970, 1981, 1983; Rasmussen and Barrett, 1984; Rasmussen *et al.*, 1984; Rubin, 1982; Schulz, 1980; Wang and Waisman, 1979; Williamson *et al.*, 1981.

HOWARD RASMUSSEN, ITARU KOJIMA, and PAULA BARRETT • Departments of Cell Biology and Internal Medicine, Yale University School of Medicine, New Haven, Connecticut 06510.

Our discussion focuses on our present knowledge of how information flows in this system from cell surface to intracellular calcium receptor proteins and response elements. In particular, we consider the flow of information in cells that display sustained responses to the sustained increases in extracellular messenger concentration.

Two concepts are stressed. First, when the calcium messenger system couples stimulus to sustained cellular response, the flow of information from cell surface to cell interior commonly occurs via two distinct branches having distinct temporal roles: a calmodulin branch involved in initiating the response and a C-kinase branch in sustaining the response. Secondly, the magnitude of the sustained response is determined by two functionally linked events at the plasma membrane: the amount of the activated, calcium-sensitive form of the calcium-activated, phospholipid-dependent protein kinase or C-kinase associated with the membrane and the rate of Ca^{2+} cycling across the membrane.

2. *Calcium as Messenger*

Although our focus is on the calcium messenger system in sustained cellular responses, the original work on messenger Ca^{2+} was in systems displaying either brief and repetitive or cycling responses such as neurosecretion and skeletal and cardiac muscle contraction (Fig. 1). In the case of neurosecretion, the early model of Ca^{2+} messenger function was one in which Ca^{2+} influx across the presynaptic membrane led to a rise of Ca^{2+} in a restricted subcellular domain that was sufficient to trigger neurotransmitter release (Katz, 1966). Cessation of response was associated with an efflux of Ca^{2+} back out of the cell: Ca^{2+} cycled across the plasma membrane.

The model of Ca^{2+} messenger function in skeletal muscle was quite different. Muscle contracts and relaxes in response to a rise and fall in cytosolic free Ca^{2+}, $[Ca^{2+}]_c$, but the source of and sink for this cytosolic Ca^{2+} is a bound but rapidly exchangeable calcium pool in the endoplasmic (sarcoplasmic) reticulum (Ebashi *et al.*, 1978; Weber and Murray, 1973). Changes in Ca^{2+} fluxes across the plasma membrane make little or no contribution to the process. There is a simple intracellular Ca^{2+} cycle between calcium bound to the endoplasmic reticulum and cytosolic free Ca^{2+} (Chapman, 1979). Strength of contraction depends on the magnitude of change in $[Ca^{2+}]_c$.

The situation in cardiac muscle is quite similar. The bulk of Ca^{2+} for contraction comes from the pool in the endoplasmic reticulum. However, there is one important difference. In skeletal muscle, Ca^{2+} release from

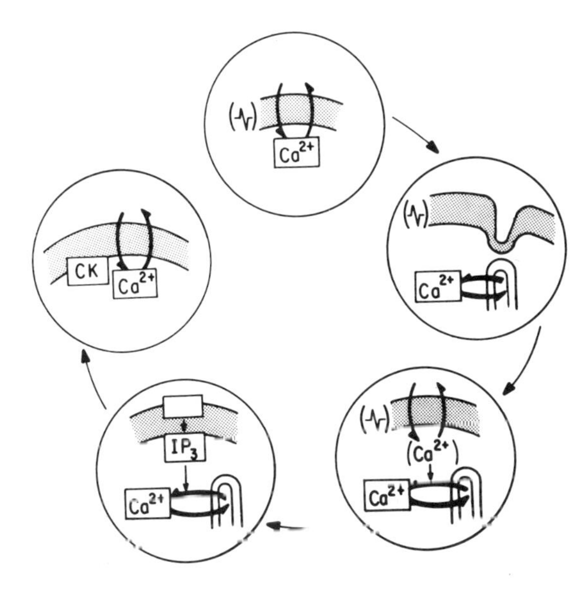

Figure 1. Models of calcium cycling and cellular response. Beginning at the top and moving clockwise: neurosecretion—cycling of Ca^{2+} across the synaptic membrane controls neurotransmitter secretion; skeletal muscle contraction—cycling of Ca^{2+} across the membrane of the sarcoplasmic reticulum controls response and rate of cycling is controlled by a direct link between plasma membrane and sarcoplasmic reticulum; cardiac muscle contraction—cycling of Ca^{2+} across the sarcoplasmic reticulum is the major regulator of response, but Ca^{2+} cycling across the plasma membrane controls the Ca^{2+} cycling across the sarcoplasmic reticulum; hepatic glycogenolysis—cycling across the endoplasmic reticulum controls the rate of glucose production, and cycling across the endoplasmic reticulum is regulated by the production of IP_3 at the plasma membrane; sustained response in the adrenal glomerulosa cell—Ca^{2+} cycling across the plasma membrane controls the activity of a specific membrane-bound enzyme, the C-kinase.

this pool is controlled by a direct link between plasma membrane and endoplasmic reticulum at a site called the T-system. In cardiac muscle, Ca^{2+} release from endoplasmic reticulum is triggered by an influx of Ca^{2+} across the plasma membrane. During each systole this trigger Ca^{2+} enters the cell, and during each diastole it is pumped back out. There are two Ca^{2+} cycles: one in and out of the endoplasmic reticulum, the other into and out of the cell. Thus, the model in the cardiac cell incorporates the major features of the other two models. The model developed most recently to account for sustained cellular response returns to an emphasis on the role of Ca^{2+} cycling across the plasma membrane and its link to membrane-associated protein kinase as a means of controlling cell response (Rasmussen and Barrett, 1984). The data in support of this model form the focus of the present discussion.

2.1. *Calcium as Messenger during Sustained Cellular Response*

When it was realized that Ca^{2+} played a role in the regulation of more sustained cellular responses such as pancreatic exocrine secretion, insulin secretion, and smooth muscle contraction, it was tacitly assumed that in these cases response was also a function of $[Ca^{2+}]_c$ and, therefore, that the $[Ca^{2+}]_c$ rose when the cell was activated and remained elevated as long as response continued. However, the introduction of methods for measuring $[Ca^{2+}]_c$ in small cells has led to a different conclusion (Albert and Tashjian, 1984; Borle and Snowdowne, 1982; Charesi *et al.*, 1983; Gennaro *et al.*, 1984; Morgan and Morgan, 1983, 1984; Pozzan *et al.*, 1982; Snowdowne and Borle, 1984; Tsien *et al.*, 1981). When such cells are activated by appropriate agonists, the $[Ca^{2+}]_c$ rises promptly, as had been predicted, but it then falls back to, or nearly to, its original basal value even though response continues. There is no simple relationship between the change in $[Ca^{2+}]_c$ and the magnitude of the cellular response during the sustained phase of this response.

These results raise two questions. First, why does the cell not simply employ a sustained elevation of $[Ca^{2+}]_c$ as a means of maintaining a response; and second, if $[Ca^{2+}]_c$ does not remain elevated, how does a cell display a sustained response to an apparently transient message?

2.2. *Autoregulation of Cytosolic Calcium Concentration*

Progress in our understanding of cellular calcium homeostasis provides insights into the first of these issues. A simple experiment done in human red cells (cells without internal organelles or calcium pools) illustrates the major role played by the plasma membrane in the maintenance of cellular calcium homeostasis (Scharff *et al.*, 1983). Addition of a Ca^{2+} ionophore, A23187, in a low concentration sufficient to produce a sustained increase in the rate of Ca^{2+} influx into these cells leads to a prompt but transient increase in total cellular Ca^{2+} and $[Ca^{2+}]_c$. However, within 1–3 min, both fall, and a new steady state is reached in which both $[Ca^{2+}]_c$ and total calcium levels are only slightly greater than before ionophore addition even though the Ca^{2+} influx rate remains high.

This remarkable response is mediated by the Ca^{2+} pump in the red cell membrane. The activity of the pump is regulated by calmodulin (CaM). Hence, when $[Ca^{2+}]_c$ rises, Ca-CaM associates with the pump and brings about two changes in its behavior: an increase in V_{max} and a

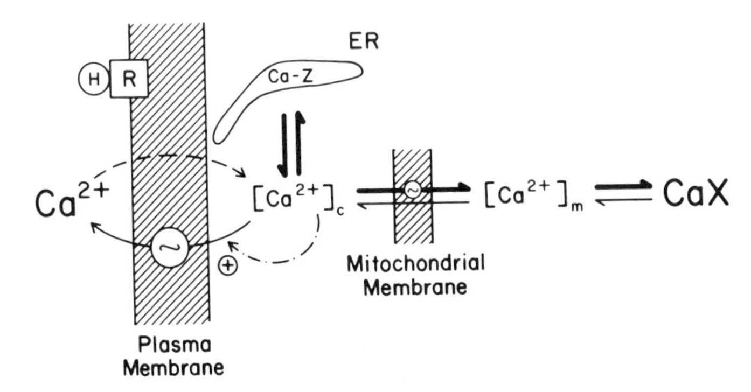

Figure 2. A schematic representation of cellular calcium metabolism that emphasizes the key role of the plasma membrane in the autoregulation of the Ca^{2+} concentration in the cell cytosol, $[Ca^{2+}]_c$. Four intracellular calcium pools are depicted: $[Ca^{2+}]_c$; free Ca^{2+} in the mitochondrial matrix space, $[Ca^{2+}]_m$; bound calcium in the matrix space, CaX; bound calcium in the endoplasmic reticulum, Ca-Z. The pool in the endoplasmic reticulum provides a trigger pool of Ca^{2+} that can, when released, bring about a transient increase in $[Ca^{2+}]_c$; the $[Ca^{2+}]_m$ and $[Ca^{2+}]_c$ are in rapid exchange and are stabilized by CaX, which is large (50–100 μmole/kg cell water) compared to $[Ca^{2+}]_m$ (0.2–0.5 μM) and $[Ca^{2+}]_c$ (0.1–0.3 μM). When $[Ca^{2+}]_c$ rises, part of the rise is damped by this mitochondrial stabilizing function. In addition, a rise in $[Ca^{2+}]_c$ leads to the calmodulin-dependent activation of the plasma membrane Ca^{2+} pump (~). This system operates as an autoregulatory device to minimize and damp any change in $[Ca^{2+}]_c$.

decrease in its K_m for Ca^{2+}, its substrate. This means that the cell has an autoregulatory system in its plasma membrane that immediately compensates for any increase in $[Ca^{2+}]_c$ by an increase in Ca^{2+} efflux (Fig. 2). This system operates so that any sustained increase in Ca^{2+} influx is balanced by a sustained increase in efflux with very little change in $[Ca^{2+}]_c$. Since the only calcium pool capable of supporting a sustained increase in $[Ca^{2+}]_c$ over long periods of time is the extracellular pool, the presence of this autoregulatory system dictates that under physiological circumstances it is very unlikely that $[Ca^{2+}]_c$ ever remains elevated for significant periods of time. If this statement is correct, it leads logically back to the second question: how does a cell display a sustained response to a transient message? To answer this question, it is necessary to introduce the concept of sensitivity modulation (Rasmussen, 1983; Rasmussen and Waisman, 1982).

2.3. Sensitivity Modulation

In contrast to the above models of Ca^{2+} messenger function are the more recent ones, in which Ca^{2+} serves as messenger even under conditions in which the $[Ca^{2+}]_c$ does not change appreciably. In order to understand these models, it is necessary to understand the concept of sensitivity modulation (Rasmussen and Waisman, 1982). By definition, this is a situation in which information flow changes as a consequence of a change in the sensitivity of the response system (calcium receptor protein and the response element it associates with) to activation by Ca^{2+}. Sensitivity modulation can be either positive or negative. If it is positive, there is an increased sensitivity to Ca^{2+}; if it is negative, there is a decreased sensitivity. A common way to bring about sensitivity modulation in components of the calcium messenger system is by the cAMP-dependent phosphorylation of response elements in this system. The classic example is the enzyme phosphorylase kinase, the key enzyme in the glycogenolytic cascade (Brostrom *et al.*, 1971; Cohen, 1979). This enzyme is a calmodulin-dependent protein kinase that catalyzes the phosphorylation of phosphorylase *b* and glycogen synthetase. Its activity is controlled by the $[Ca^{2+}]_c$. Hence, when $[Ca^{2+}]_c$ rises from 0.2 to 1.0 μM, the enzyme goes from virtual inactivity to full activity—phosphorylase becomes phosphorylated. As a consequence, phosphorylase becomes active, and glycogen breakdown is accelerated—a case of amplitude modulation.

Phosphorylase kinase has an additional characteristic. Since it is a substrate for the cAMP-dependent protein kinase, it can exist in either a phosphorylated or a nonphosphorylated form. Each form is a Ca^{2+}-calmodulin-dependent protein kinase capable of catalyzing the phosphorylation of phosphorylase, but there is an essential difference between the two forms. The activity is controlled by changes of $[Ca^{2+}]_c$ in the range of approximately 0.05 to 0.25 μM for the phosphorylated form and 0.2 to 1.0 μM for the nonphosphorylated form; that is, the phosphorylation of phosphorylase kinase brings about its positive sensitivity modulation.

From a practical point of view, this modality of information flow is of considerable importance because it means that a Ca^{2+}-sensitive process can increase in rate without a change occurring in $[Ca^{2+}]_c$. It also means that different Ca^{2+}-calmodulin-regulated processes can be differentially activated or inhibited, depending on their states of sensitivity to activation by Ca^{2+}.

3. *Two-Branch Model of Cell Activation*

A knowledge of the existence of sensitivity modulation and the fact that during a sustained response the $[Ca^{2+}]_c$ does not remain elevated led us to consider the possibility that some type of sensitivity modulation mediated the sustained phase of such cellular response. In particular, the discovery of the calcium-activated, phospholipid-dependent protein kinase (C-kinase) by Nishizuka and co-workers led us to consider the possibility that this enzyme played a key role in the sustained phase of cellular response (Kaibuchi *et al.*, 1983; Nishizuka, 1983; Kikkawa *et al.*, 1983; Takai *et al.*, 1979, 1981, 1984). This possibility was considered for two reasons. First, it was shown that C-kinase undergoes positive sensitivity modulation with respect to Ca^{2+} when it associates with phosphatidylserine and a specific type of diacylglycerol. In its calcium-sensitive form, the enzyme is controlled by changes in $[Ca^{2+}]$ in the range of 0.05 to 0.5 μM. Second, it has been shown that during platelet activation by thrombin, both a calmodulin-dependent protein kinase and the C-kinase are activated and that both are necessary in order to observe a maximal effect, that is, events in the two pathways act synergistically. However, platelet activation is a brief event, so that these studies did not address the issue of possible specific temporal roles for the two pathways.

Since the activated, Ca^{2+}-sensitive form of the C-kinase is thought to be bound to the endoplasmic face of the plasma membrane and is sensitive to changes in $[Ca^{2+}]$ in the range of 0.05 to 0.05 μM, it seemed possible that when this enzyme is in its activated state *in situ*, it could continue to operate in cells in which the $[Ca^{2+}]_c$ was close to its basal value of 0.1–0.3 μM.

Our first approach to determining whether the calmodulin and C-kinase branches of the calcium messenger system might have distinct temporal roles was to carry out experiments in which receptor-mediated events were bypassed and the two postulated branches were activated separately or in unison by specific drugs. The agents employed to activate the calmodulin branch were calcium ionophores, in particular, A23187; the agents employed to activate the C-kinase branch were either phorbol esters such as TPA (Castagna *et al.*, 1982) or synthetic diacylglycerols such as 1-oleoyl-2-acetylglycerol (OAG), each of which is known to bring about the activation of the C-kinase in intact cells.

Studies of this type were carried out in four different cells or tissues: adrenal glomerulosa cells (Kojima *et al.*, 1983; Kojima *et al.*, 1984, 1985 a,b,c,d; Barrett *et al.*, 1985); cultured pituitary cells (Delbeke *et al.*, 1984); vascular smooth muscle (Rasmussen *et al.*, 1984a; Forder *et al.*, 1985); and pancreatic islets (Zawalich *et al.*, 1983, 1984). The data obtained have provided comparable results. Treatment of each of these cell types or tissues with small doses of A23187 (or other agents that increase $[Ca^{2+}]_c$) leads to a transient, Ca^{2+}-dependent response; treatment with TPA leads to a very slowly developing, sustained, but submaximal response; but treatment with a combination of TPA and A23187 leads to responses that

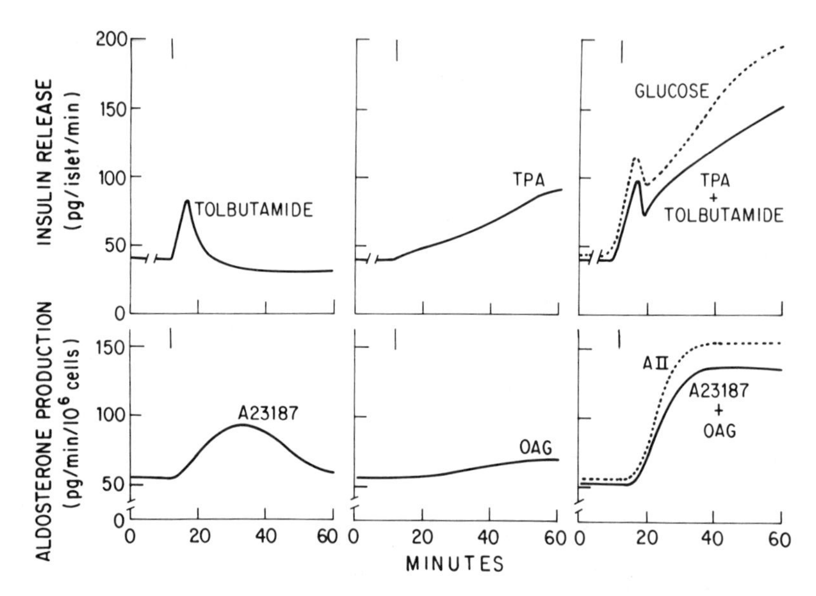

Figure 3. The time course of change in insulin secretion and aldosterone secretion in response to agonists. Insulin secretory responses to 7.5 mM glucose, 200 nM TPA, tolbutamide, or combine TPA and tolbutamide are shown in the upper part of the figure, and aldosterone secretory responses to 1 mM angiotensin II, 60 μM OAG, 0.5 μM A23187, or combined OAG and A23187 are shown in the lower portion of the figure.

are qualitatively and quantitatively similar to those induced by the natural agonists.

Two examples of this type of data are shown in Fig. 3. The first compares glucose-induced insulin release from β cells in isolated rat islets with the patterns of insulin secretion induced by tolbutamide, TPA, or combined tolbutamide and TPA. Addition of glucose to the perifusate leads to a biphasic pattern of insulin secretion: an early brief peak occurs within the first 8–10 min, followed by a slowly developing and sustained increase in secretory rate. These response are Ca^{2+} dependent. Addition of tolbutamide, which causes a transient increase in $[Ca^{2+}]_c$ and in Ca^{2+} influx, leads to only a phase 1 or brief insulin secretory response. Addition of TPA leads to no immediate increase but to a slowly developing, sustained, but submaximal increase in insulin secretory rate. Addition of combined TPA and tolbutamide leads to a biphasic pattern of insulin secretion similar to that seen after glucose addition (Zawalich *et al.*, 1984).

The second system to be employed was the control of aldosterone secretion from adrenal glomerulosa cells by the peptide hormone angiotensin II. When angiotensin II is added to a system in which isolated glomerulosa cells are being continuously perifused, there is a prompt (within 3–5 min) monotonic increase to a sustained plateau in the rate of aldosterone secretion. Addition of low doses of A23187 (0.5 μM) leads to a prompt, transitory secretory response, so that by 30 min the rate of secretion is only slightly above the basal rate; and addition of TPA leads to a very slowly developing, sustained, but submaximal (15–25%) secretory response. The combined addition of A23187 and TPA leads to a secretory response that is similar in time course and magnitude to that seen after the addition of angiotensin II. Comparable data are obtained if OAG rather than TPA is employed to activate C-kinase.

These data and a variety of data from other cellular systems have been interpreted in terms of a model in which the flow of information from cell surface to cell interior occurs via two branches that serve distinct temporal functions: a calmodulin branch, which is transiently activated by a transient rise in $[Ca^{2+}]_c$ and is responsible for initiating the response, and a C-kinase branch, which is activated in a sustained manner by diacylglycerol and is responsible for sustaining the response. The integrated cellular response depends on the temporal integration of the events occurring in these two pathways. These pathways can be activated separately by bypassing the receptor-mediated events and employing pharmacologic agents to cause an increase just in $[Ca^{2+}]_c$ (A23187) or just in the activated form of C-kinase (TPA or OAG).

4. *Validation of the Two-Branch System*

To validate this model, we have performed an extended series of experiments employing the adrenal glomerulosa cells. These included a detailed analysis of early receptor-mediated events (Berridge, 1982, 1984; Aub and Putney, 1984; Downs and Michell, 1982; Irvine *et al.*, 1982; Joseph *et al.*, 1984; Kirk *et al.*, 1981; Michell, 1975), an analysis of the effects of hormone on cellular calcium metabolism (Berridge and Irvine, 1984; Borle, 1973, 1981; Mauger *et al.*, 1984; Schulz, 1980; Shears and Kirk, 1984; Streb and Schulz, 1983; Streb *et al.*, 1983; Studer and Borle, 1983; Suematsu *et al.*, 1983), and an analysis of the pattern of cellular protein phosphorylation seen after the addition of hormone and drug (Garrison, 1983).

4.1. *Phosphatidylinositide Turnover*

The results of the experiments concerned with early receptor-mediated events show that the addition of angiotensin II to these cells causes

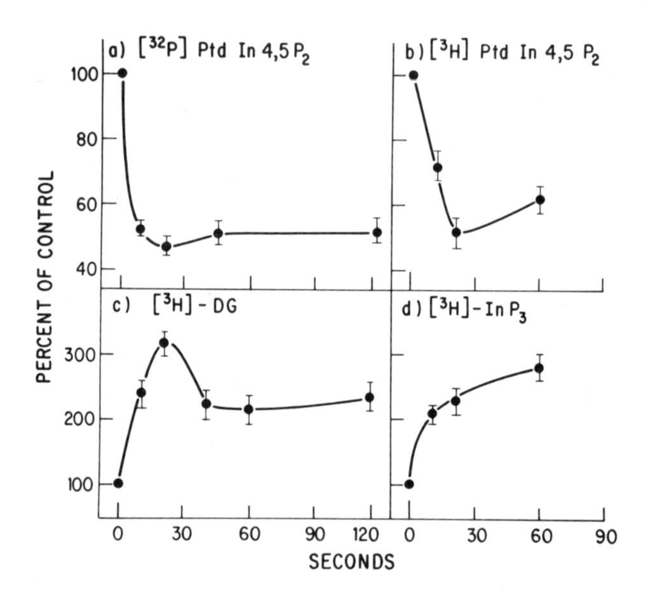

Figure 4. Early biochemical events following angiotensin II–receptor interaction in the adrenal glomerulosa cell. Addition of hormone leads to the rapid hydrolysis of phosphatidylinositol-4,5-bisphosphate ($PtdIn4,5P_2$) whether this compound is prelabeled with [^{32}P]-HPO_4^{2-} (a) or [^{3}H]-inositol (b). The breakdown of this compound leads to the production of diacylglycerol (c) and inositol trisphosphate (d).

the immediate (10–20 sec) reduction in the contents of phosphatidylinositol-4-phosphate (PIP) and phosphatidylinositol-4,5-bisphosphate (PIP_2) and the production of inositol trisphosphate (IP_3) and diacylglycerol rich in arachidonic acid (DG) (Fig. 4). Thus, this hormone stimulates the hydrolysis of PIP_2, resulting in the production of both postulated messengers, IP_3 and DG.

4.2. Cellular Calcium Metabolism

Angiotensin II has multiple effects on cellular calcium metabolism. It causes both an immediate increase in the mobilization of Ca^{2+} from a dantrolene-sensitive intracellular (endoplasmic reticulum) pool and an immediate and sustained increase in the rate of plasma membrane Ca^{2+} influx (Kojima *et al.*, 1984, 1985a,b,c). These events lead to a transient rise in $[Ca^{2+}]_c$ as measured by the photoprotein aequorin. The rise in $[Ca^{2+}]_c$ leads to the activation (via CaM) of the plasma membrane Ca^{2+} pump and thus to a new efflux of Ca^{2+} out of the cell. As a consequence, total cell calcium falls initially, then rises slightly, and remains stable thereafter. During the period of the sustained response, the $[Ca^{2+}]_c$ is near basal, total cell calcium is stable, and the rate of calcium influx across the plasma membrane increases 2.0–2.5 times. Hence, the rate of plasma membrane efflux must also be increased 2.0 to 2.5-fold.

In saponin-permeabilized cells, IP_3 induces the efflux of Ca^{2+} from a nonmitochondrial, dantrolene-sensitive, ATP-dependent calcium pool that is thought to be located in the endoplasmic reticulum (Kojima *et al.*, 1984). When the hormone acts, the receptor-mediated rise in IP_3 is thought to lead to the mobilization of Ca^{2+} from this pool, and it is this Ca^{2+} that is pumped out of the cell and accounts for a decrease in total cell calcium.

4.3. Patterns of Protein Phosphorylation

The effects of angiotensin, ionophore, and ionophore plus TPA on the patterns of protein phosphorylation were determined at 1 min and 30 min after agonist addition (Barrett *et al.*, 1985). One minute after A23187 is added to prelabeled cells, there is an increase in ^{32}P labeling of at least ten cellular proteins. Thirty minutes after A23187 is added, none of these proteins is labeled, even though the A23187 is present throughout and causes a sustained threefold increase in the rate of Ca^{2+} influx into these cells. The transient nature of the A23187 effect on protein phosphorylation coincides with the transient nature of A23187 action on aldosterone secretion. Our interpretation of these data is that A23187 leads to a transient increase in $[Ca^{2+}]_c$ and a transient activation of CaM-dependent enzymes including CaM-dependent protein kinases, which in turn catalyze the transient phosphorylation of ten cellular proteins that mediate a transient increase in aldosterone production.

One minute after the addition of A23187 plus TPA, there is an increase in the labeling of 11 proteins, the same ten seen after adding A23187 alone plus one new protein, which is presumably phosphorylated as a result of the action of C-kinase. An increase in the labeling of the same 11 proteins is seen 1 min after the addition of angiotensin II. Thus, at this level of analysis, the initial effects of A23187 plus TPA are identical to those of the natural agonist, angiotensin II. Thirty minutes after the addition of either angiotensin II or TPA plus A23187, there is an increase in the labeling of at least seven proteins, four of the original 11 plus three new ones. In other words, seven of the original 11 are no longer phosphorylated. One of the original four proteins that remains phosphorylated is the C-kinase substrate protein, which was phosphorylated at 1 min. The other three proteins that are phosphorylated at 1 min after treatment with either angiotensin II or A23187 alone and that are phosphorylated at 30 min after treatment with angiotensin II or combined A23187 and TPA but

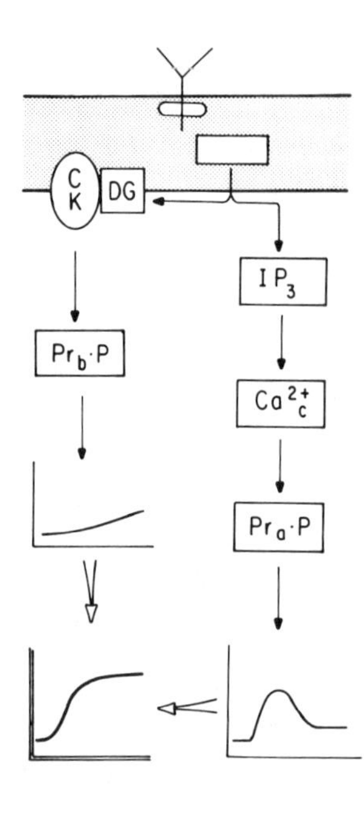

Figure 5. Model A: The flow of information in the calcium messenger system during angiotensin II action. Activation of the hormone receptor (Y) leads to the breakdown of PIP_2 ▭ with the production of DG and IP_3. The increase in IP_3 concentration brings about the mobilization of Ca^{2+} from the endoplasmic reticulum, resulting in a rise in $[Ca^{2+}]_c$. This rise in $[Ca^{2+}]_c$ leads to the activation of calmodulin-dependent protein kinase resulting in the phosphorylation of a distinct subset of cellular proteins, $Pr_a \cdot P$. However, the rise in $[Ca^{2+}]_c$ and the phosphorylation of these proteins is transient, so only a rapid but transient physiological response occurs. The rise in DG leads to the activation of C-kinase (CK), and this brings about a slower activation of a second subset of cellular proteins, $Pr_b \cdot P$, which, if they occur without prior phosphorylation of $Pr_a \cdot P$, lead to a slowly developing and submaximal response. The temporal integration of events in these two branches brings about the normal physiological response (lower left corner).

not after treatment with A23187 alone are assumed to be substrates for both the CaM-dependent protein kinase and the C-kinase. Thus, seven proteins are unique substrates for calmodulin-dependent protein kinase; four are unique substrates for C-kinase, and three are common substrates for both kinases.

These protein phosphorylation studies provide a major validation of our original model and allow us to expand the model as depicted in Fig. 5.

5. *Role of Ca^{2+} in the Operation of the C-Kinase Branch*

Even though the model shown in Fig. 5 provides a satisfactory explanation for much of our data, it does not address an important issue; what role, if any, do changes in $[Ca^{2+}]_c$ and/or changes in Ca^{2+} fluxes play in controlling the sustained response? It is known, for example, that many of these sustained cellular responses require extracellular Ca^{2+}, but it is not known whether this requirement reflects a specific role for Ca^{2+} or a more global role in maintaining membrane and cell integrity (Schulz, 1980; Borle, 1981). Likewise, it is possible that when C-kinase undergoes positive sensitivity modulation during its association with the plasma membrane, it becomes sufficiently sensitive to Ca^{2+} that it becomes, in fact, fully activated at the basal $[Ca^{2+}]_c$.

An examination of these issues in adrenal cells has led to the conclusion that Ca^{2+} plays two important roles in controlling information flow through the C-kinase branch of the calcium messenger system: it helps to determine the amount of C-kinase that is shifted to the activated form, and it determines the turnover rate of the activated enzyme.

5.1. Calcium and Activation of C-Kinase

Three types of experiments indicate that even though $[Ca^{2+}]_c$ is close to its basal value during the sustained phase of the response, the transient increase in $[Ca^{2+}]_c$ that occurs within the first few minutes after hormone addition helps to determine the magnitude of the sustained response.

When glomerulosa cells are pretreated with dantrolene, this blocks the mobilization of Ca^{2+} from an intracellular pool (endoplasmic reticulum) that normally occurs after the addition of angiotensin II. As a consequence, the rise in $[Ca^{2+}]_c$ is considerably lower, and the aldosterone secretory response develops more slowly and is diminished. The plateau value is approximately 50% of the normal increase. If dantrolene is added 20 min or more after adding angiotensin II, at a time when the sustained secretory response is established, the drug has no effect on the production rate of aldosterone. When glomerulosa cells are exposed simultaneously to A23187 and OAG, the subsequent secretory response is qualitatively and quantitatively similar to that seen after the addition of angiotensin II. However, if A23187 is added first, and the transient increases in $[Ca^{2+}]_c$ and aldosterone secretion are allowed to occur and wane before OAG is added, then OAG induces an increase of approximately 50% of maximal in the rate of aldosterone secretion.

Experiments with the calcium channel agonist Bay K 8644 provide additional evidence. When Bay K 8644 alone is added to glomerulosa cells, there is a nearly twofold increase in rate of Ca^{2+} influx but little change in rate of aldosterone secretion. If Ca^{2+} influx is increased by Bay K 8644 and OAG is added simultaneously, the aldosterone secretion rate increases to approximately 50% of maximal. If at the time of addition of Bay K 8644 and OAG, the cells are also exposed to 8 mM K^+ for 5 min (a condition known to increase $[Ca^{2+}]_c$ transiently) and then returned to a normal-K^+ buffer, the sustained rate of aldosterone secretion is maximal and remains so.

Our interpretation of these data is that the transient rise in $[Ca^{2+}]_c$ that occurs during the initial phase of cellular response acts with the increase in the DG content of the plasma membrane to bring about the

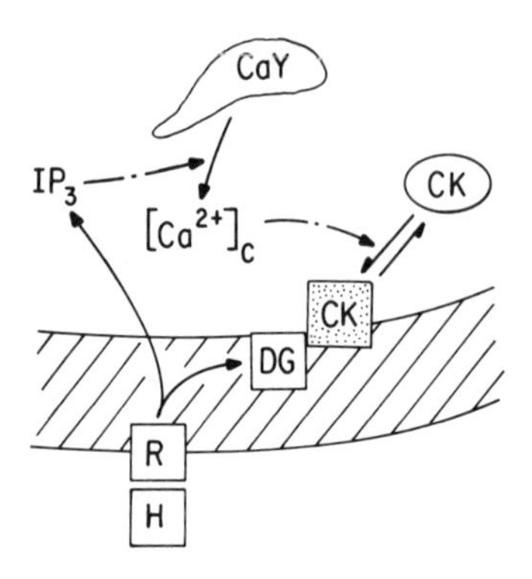

Figure 6. A schematic representation of the events occuring during the initial phase of cellular response (phase 1) that determine the amount of C-kinase that is switched from its nonactivated, calcium-insensitive form CK to its activated, calcium sensitive form CK . Both the increase in the DG content of the plasma membrane and the IP_3-dependent increase in $[Ca^{2+}]_c$ play complementary roles in the activation process.

conversion of the nonactivated form of C-kinase to the activated form (Fig. 6).

5.2. *Calcium and C-Kinase Turnover Rate*

The final aspects of this system to be considered are the changes in plasma membrane Ca^{2+} influx and their role during the sustained phase of the response. As noted above, when angiotensin II acts there is an immediate 2.0 to 2.5-fold increase in the rate of calcium influx across the plasma membrane (Kojima *et al.*, 1985b). This increase in Ca^{2+} influx rate is sustained as long as angiotensin II is present. In spite of this sustained increase, the $[Ca^{2+}]_c$ is near its original basal value during the sustained phase of the response, and there is no net gain of total cell calcium during this phase. One can calculate that if increased influx were not compensated for by an increase in efflux rate, the total cell calcium would more than double over a period of 30 min. This clearly does not occur. Hence, one is led to the conclusion that during the sustained phase of cellular response, the sustained increase in Ca^{2+} influx across the plasma membrane is balanced by a sustained rate of Ca^{2+} efflux under conditions in which $[Ca^{2+}]_c$ is near its basal value.

Two kinds of experiments were done to examine the informational role of this plasma membrane Ca^{2+} cycling during the sustained phase of the response. These experiments were performed with media containing a normal Ca^{2+} concentration (1.25 mM). These cells display no increase in Ca^{2+} influx rate in response to angiotensin II on either the addition of 1 μM nitrendipine, a calcium channel blocker, or the reduction of the external $[K^+]$ to 2.0 mM. (The basal Ca^{2+} influx rate is near normal in the presence of nitrendipine and is only reduced about 15% in the low-K^+ medium.) In each of these circumstances, the addition of angiotensin

Figure 7. A schematic representation of the factors determining the rate of turnover of C-kinase during the sustained phase of cellular response (phase 2). Both the amount of activated, calcium-sensitive C-kinase associated with the endoplasmic face of the plasma membrane and the rate of calcium cycling across the membrane determine the C-kinase turnover rate. It is not clear how calcium cycling is linked to C-kinase turnover rate, but a logical postulate is that Ca^{2+} cycling determines the Ca^{2+} concentration in a specific submembrane domain $[Ca^{2+}]_{sm}$, and this Ca^{2+} pool regulates C-kinase activity.

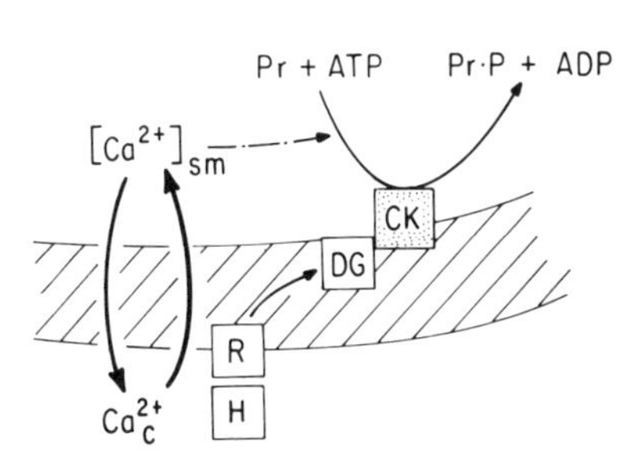

II leads to a mobilization of intracellular Ca^{2+} and a nearly normal initial phase of aldosterone secretion. In both cases, however, the secretion rate declines progressively after reaching 75–80% of the normal peak value, so that after 30–45 min it has reached 10–15% of the normal sustained rate seen after angiotensin II. Studies of protein phosphorylation under these conditions show that 1 min after angiotensin II is added to nitrendipine-treated cells, all 11 proteins that are normally phosphorylated become phosphorylated. In the presence of nitrendipine, however, none of the seven proteins normally phosphorylated at 30 min after the addition of angiotensin II become phosphorylated.

These data indicate that the increase in plasma membrane Ca^{2+} cycling plays a critical informational role during the sustained phase of cel-

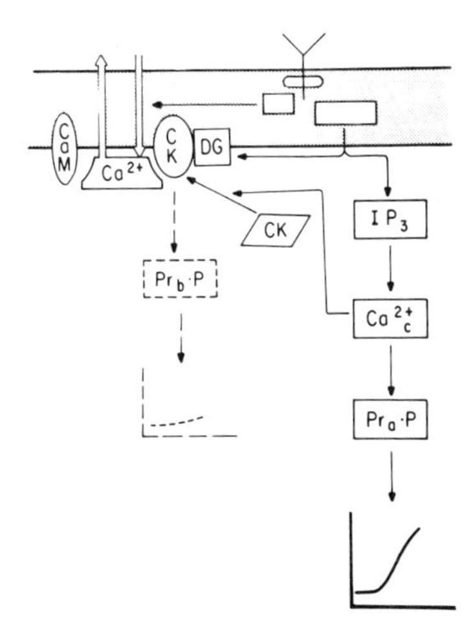

Figure 8. A model of the flow of information through the calcium messenger system during the initial phase of cellular response. The branch of major importance is the calmodulin branch. It is activated by the IP_3-induced rise in $[Ca^{2+}]_c$ and leads to the phosphorylation of a specific subset of cellular proteins, $Pr_a \cdot P$, responsible for the initiation of cellular response. During this initial phase of response there is also a shift of C-kinase from its nonactivated to its activated, Ca^{2+}-sensitive form as a result of the combined effects of a rise in $[Ca^{2+}]_c$ and an increase in the DG content of the plasma membrane. This phase of the response is transient because the increase in $[Ca^{2+}]_c$ is transient. This model is an elaboration of that shown in Fig. 5.

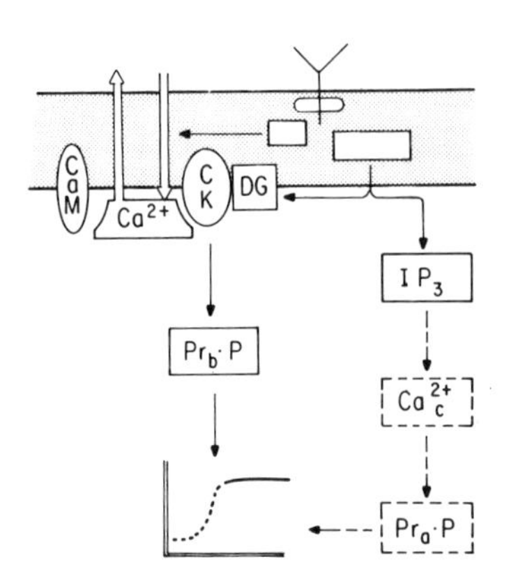

Figure 9. The flow of information through the calcium messenger system during the sustained phase of cellular response. The branch of major importance is the C-kinase branch. The rate at which this enzyme, C-kinase, phosphorylates a specific subset of proteins, $Pr_b \cdot P$, is determined both by the amount of calcium-sensitive protein kinase associated with the plasma membrane and by the $[Ca^{2+}]$ in a specific submembrane domain, which is determined by the rate of Ca^{2+} cycling across the plasma membrane. During this phase Ca^{2+} influx and efflux are balanced because of the Ca^{2+}-calmodulin (CaM)-dependent activation of the plasma membrane Ca^{2+} efflux pathway, the Ca^{2+} pump. See Figs. 5 and 8 for comparison.

lular response. We postulate that this controls the turnover rate of the activated form of C-kinase (Fig. 7). In this view, the flow of information in the C-kinase branch of the calcium messenger system during the sustained phase of the response is determined by two interrelated factors in the plasma membrane: the amount of the activated form of C-kinase associated with the membrane and the rate of Ca^{2+} cycling across the membrane. The C-kinase "reads" this cycling rate and responds to it by an increase or decrease in its turnover rate. It is not clear how the enzyme senses this cycling. It is most likely that the increase in Ca^{2+} cycling leads to an increase in the concentration of Ca^{2+} in a restricted domain in, at, or near the membrane, and the activated form of the C-kinase senses this domain. Activated C-kinase and plasma membrane Ca^{2+} cycling form a functional unit that determines the rate and/or extent of phosphorylation of a specific subset of cellular proteins that regulate the rate of cellular response.

5.3. Temporal Integration of Response

These data lead to a model of cellular response that can be considered to have two phases. During the initial phase (Fig. 8), the flow of information through the calmodulin branch of the system plays the predominant role and is responsible for the rapid initiation of response. Events in the C-kinase pathway contribute little, but a key event occurs in terms of subsequent response: there is a Ca^{2+}- and DG-dependent shift in C-kinase from its Ca^{2+}-insensitive to its Ca^{2+}-sensitive form. During the sustained phase of response, the flow of information through the C-kinase branch is of predominant importance (Fig. 9). The response is a function of the extent of phosphorylation of a specific subset of cellular proteins. The extent of their phosphorylation is controlled by the $[Ca^{2+}]_{sm}$ and the amount of Ca^{2+}-sensitive C-kinase in a specific membrane domain.

6. *General Applicability of the Model*

It is important to know whether this model applies only to the adrenal glomerulosa cell or to situations in other cells as well. Work in at least six other systems suggests that the model applies in a general way to a number of cellular responses. Data obtained from studies of peptide hormone (insulin and prolactin) secretion (Albert and Tashjian, 1984; Delbeke *et al.*, 1984; Martin, 1983; Martin, 1983; Martin and Kowalchyk, 1984a,b; Zawalich *et al.*, 1983, 1984), pancreatic exocrine secretion (dePont and Fleurer-Jakobs, 1984), neutrophil activation (Robinson *et al.*, 1984; Dale and Penfield, 1984), histamine release from mast cells (Katakami *et al.*, 1984), lymphocyte activation (Trunek *et al.*, 1984), and smooth muscle contraction (Forder *et al.*, 1985; Rasmussen *et al.*, 1984a) support the notion that the general organization of the calcium messenger system is similar in all of these systems. In addition, recent studies of the mode of action of growth factors such as platelet-derived growth factor (PDGF) and epidermal growth factor (EGF) indicate that these agents act via the calcium messenger system (Balls *et al.*, 1984; Cooper *et al.*, 1982; Heiden and Westermark, 1984; Huang *et al.*, 1984; Moolenaar *et al.*, 1984; Sawyer and Cohen, 1981; Sawyer and Kantz, 1984). It is already clear that in doing so they activate both branches of the system, but the issues of the temporal roles of these branches and the importance of plasma membrane Ca^{2+} cycling in these cases have not yet been addressed.

6.1. *Peptide Hormone Secretion*

The major distinction between insulin secretion and the secretion of other peptide hormones is that extracellular glucose, the primary regulator of insulin secretion, does not activate a glucoreceptor. Instead, it undergoes metabolic conversion leading to a change in concentration of a specific (but unidentified) metabolite and/or to a change in the redox potential. One of these changes then brings about the activation of the calcium messenger system (Hedeskov, 1980).

In the case of TRH-regulated prolactin secretion, it seems clear that the separate branches of the calcium messenger system control, respectively, the initial and the sustained phases of peptide hormone secretion, just as they control the secretion of steroid hormone in angiotensin-II-regulated aldosterone secretion. In addition, there is evidence that the sustained phase of secretion is associated with an increase in rate of plasma membrane calcium influx and is dependent on extracellular Ca^{2+} (Albert and Tashjian, 1984). A point of controversy is whether the $[Ca^{2+}]_c$ is near its basal value during the sustained phase of the response, as we postulate in the adrenal cell, or whether it remains significantly above basal during this phase. Evidence in support of this latter view has been obtained in cells treated with very high concentrations of TRH in a static incubation system in which intracellular free Ca^{2+} is measured with the fluorescent probe Quin-2, a calcium chelator and buffering agent. Any of these facts could lead to an apparent or real elevation of $[Ca^{2+}]_c$ that is not relevant to the normal physiological situation.

It is our belief that when $[Ca^{2+}]_c$ is measured with a nonbuffering calcium indicator in response to physiologically relevant hormone concentrations in a flow-through system, the $[Ca^{2+}]_c$ during the sustained phase of the response will be found to be close to the basal value. However, this finding alone would not resolve the issue. As we developed this model, we have emphasized that the sources of Ca^{2+} for the two phases of the response are different and have postulated that there is a direct link between plasma membrane Ca^{2+} influx (and/or a submembrane domain of Ca^{2+}) and the turnover rate of C-kinase. It is equally possible that Ca^{2+} influx and Ca^{2+} cycling bring about subtle changes in $[Ca^{2+}]_c$ that are of sufficient magnitude to control C-kinase turnover rate but are not sufficient to activate CaM-dependent protein kinase activity. From an operational point of view, these alternative models are equivalent in the sense that in each the rate of plasma membrane Ca^{2+} influx determines the $[Ca^{2+}]$ in a cellular pool that regulates the turnover rate of the C-kinase. There are several reasons why we currently favor the model of a distinct submembrane Ca^{2+} pool or domain: (1) the C-kinase branch can be activated without activating the CaM branch; (2) Ca^{2+} cycling across the plasma membrane does not lead to a net uptake of Ca^{2+} by the cell; (3) the CaM-activated plasma membrane Ca^{2+} pump behaves as though the $[Ca^{2+}]$ is higher than the measured $[Ca^{2+}]_c$; and (4) in cells in which $[Ca^{2+}]_c$ is clamped, C-kinase activation requires higher $[Ca^{2+}]$ than predicted from measurements of $[Ca^{2+}]_c$ in nonclamped cells responding to angiotensin II.

6.2. *Smooth Muscle Contraction*

Using the model developed from out studies in adrenal glomerulosa cells, we have sought to determine the role of the C-kinase branch of the calcium messenger system in the regulation of vascular smooth muscle contraction, a second target tissue for angiotensin II (Bolton, 1979; Smith *et al.*, 1984).

Previous work by a number of investigators had led to a model in which the flow of information through the calmodulin branch is responsible for initiating a contractile response by bringing about the calcium-calmodulin-dependent activation of enzymatic myosin light chain phosphorylation (Aksoy *et al.*, 1982; Adelstein *et al.*, 1980, Gerthoffer and Murphy, 1983; Kerrick *et al.*, 1980; Silver and Stull, 1982). If this model is to account for a sustained contraction, it requires a sustained increase in $[Ca^{2+}]_c$ and/or a sustained increase in the content of phosphorylated myosin light chain (MLC·P) following agonist addition. However, neither of these conditions is met. Within the first 3 min following the addition of angiotensin II to vascular smooth muscle, the $[Ca^{2+}]_c$ rises transiently and then falls to (or very near to) its basal value even though contraction increases and is then maintained (Morgan and Morgan, 1982, 1984). The content of MLC·P rises rapidly with a longer half-time but then gradually declines over a period of 15–45 min (Silver and Stull, 1982).

These data lead to the postulate that a second Ca^{2+}-sensitive system other than myosin light chain kinase plays a role in regulating the sustained phase of contraction. Our studies suggest that this second system is the C-kinase branch of the calcium messenger system. At present this conclusion is based on the pharmacological evidence that TPA will induce a very slowly developing, sustained, calcium-dependent contraction of smooth muscle that is rapidly and completely reversible by forskolin, an activator of adenylate cyclase that rapidly reverses agonist-induced contraction. Most importantly, the time after TPA addition when contraction begins and the rate at which it increases are both enhanced by agents that alter plasma membrane Ca^{2+} influx and/or cycling (Fig. 10). It is possible to initiate a contraction within 2–5 min of the addition of TPA in vascular or tracheal smooth muscles pretreated with agents that alter plasma membrane calcium cycling without themselves inducing a contraction. In addition, it has been shown that agonists that activate smooth muscle contraction also stimulate the turnover of phosphatidylinositols (Smith *et al.*, 1984).

Further work is necessary to validate our conclusion, but our working hypothesis is that the calmodulin branch of the calcium messenger system, acting via myosin light chain kinase, is responsible for initiating a rapid but relatively brief contractile response and that the C-kinase branch is responsible for sustaining the contractile response at a time when $[Ca^{2+}]_c$ and [MLC·P] are close to their basal values. This model implies that the Ca^{2+}-sensitive site at which the sustained contraction is regulated is the

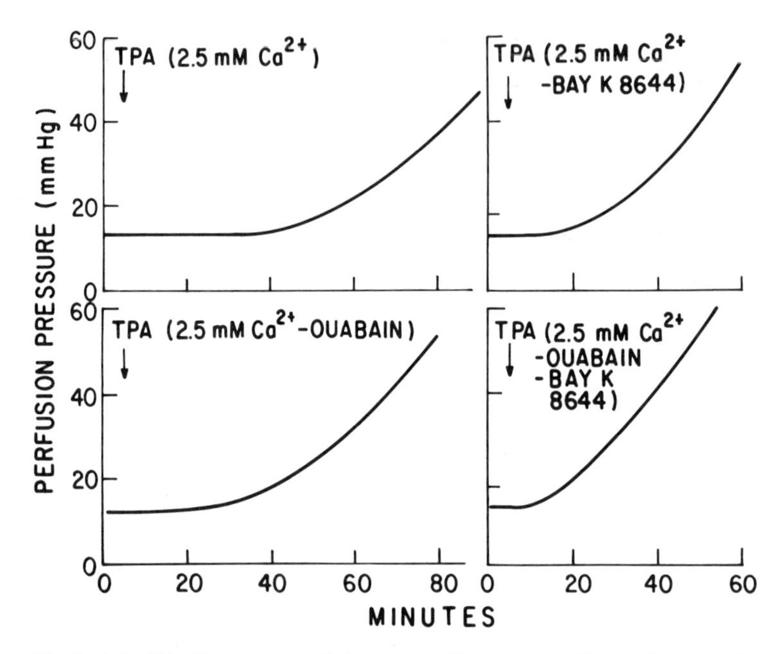

Figure 10. A plot of the time course of the contractile response of vascular smooth muscle to 200 nM TPA alone (upper left); when combined with Bay K 8644 (upper right); with ouabain (lower left); and with both Bay K 8644 and ouabain (lower right). Note that each of these agents singly reduces the latent period and increases the rate of increase of the contractile response to TPA, and that when combined they act to further reduce the latent period and increase the rate of increase in tension. The fact that none of the agents causes a contractile response in the absence of TPA is noteworthy.

C-kinase associated with the plasma membrane and that one or more of the phosphoprotein products of this enzyme regulate actin–myosin interaction, or actin–actin interactions.

7. Alternative Modalities of Gain Control

One can view the function of the C-kinase branch of the calcium messenger system as a means of providing a type of gain control in this system. Even though this function of the C-kinase branch is relatively common, it is not the only means of achieving gain control in cells. There is an alternative pathway in the adrenal glomerulosa cell, the same cell in which angiotensin II acts via the two branches of the calcium messenger system. These cells also respond to small changes in extracellular $[K^+]$

(Kojima *et al.*, 1985d). A rise in [K^+] from 3.5 to 7.0 mM leads to a two- to threefold increase in the rate of aldosterone production. The effect of K^+ is Ca^{2+} dependent. However, an increase in [K^+] does not activate PIP_2 breakdown, cause the production of IP_3 or DG, or induce the release of Ca^{2+} from an intracellular pool. The major (if not the sole) effect of K^+ on cellular Ca^{2+} metabolism is to bring about a fourfold increase in the rate of plasma membrane Ca^{2+} influx.

One might argue that this change alone is sufficient to bring about a sustained increase in the rate of aldosterone production by causing a sustained increase in $[Ca^{2+}]_c$. However, the addition of A23187, a calcium ionophore, in concentrations sufficient to induce an equal or greater increase in the rate of Ca^{2+} influx leads to a transient aldosterone secretory response rather than a sustained one. This difference between the effect of K^+ and A23187 in spite of a common and similar effect on Ca^{2+} influx is explained by the fact that K^+ causes a small rise in the $[cAMP]_c$, whereas A23187 causes a small fall. If A23187 is administered with a small dose of forskolin (sufficient to raise $[cAMP]_c$ to the value seen in K^+-treated cells), the aldosterone secretory responses are similar. In this case, the activation of the cAMP messenger system provides a second type of gain control in the calcium messenger system.

Our knowledge of the molecular mechanism(s) by which this gain control is achieved is not complete. One attractive possibility is that cAMP activates a protein kinase that catalyzes the phosphorylation of enzymes activated by Ca^{2+}-calmodulin and, by doing so, brings about their positive sensitivity modulation, as it does with the enzymes phosphorylase kinase (see above).

8. Set Point Control by cAMP

The preceding discussion of K^+ action touches on only one aspect of the modulatory effects of cAMP on the calcium messenger system (Rasmussen, 1983). In the case of K^+ action in adrenal glomerulosa cells, a working model proposes that the K^+-induced increase in cAMP brings about the positive sensitivity modulation of one or more calmodulin-regulated enzymes so that these enzymes respond to lower concentrations of $[Ca^{2+}]_c$. In this case, the effects of cAMP are confined to the calmodulin branch of the calcium messenger system.

It is evident from work in pituitary cells, pancreatic islets, and smooth muscle that cAMP also modulates events in the C-kinase branch of the calcium messenger system. The effects of activating adenylate cyclase with forskolin in islets perifused with combined TPA and tolbutamide

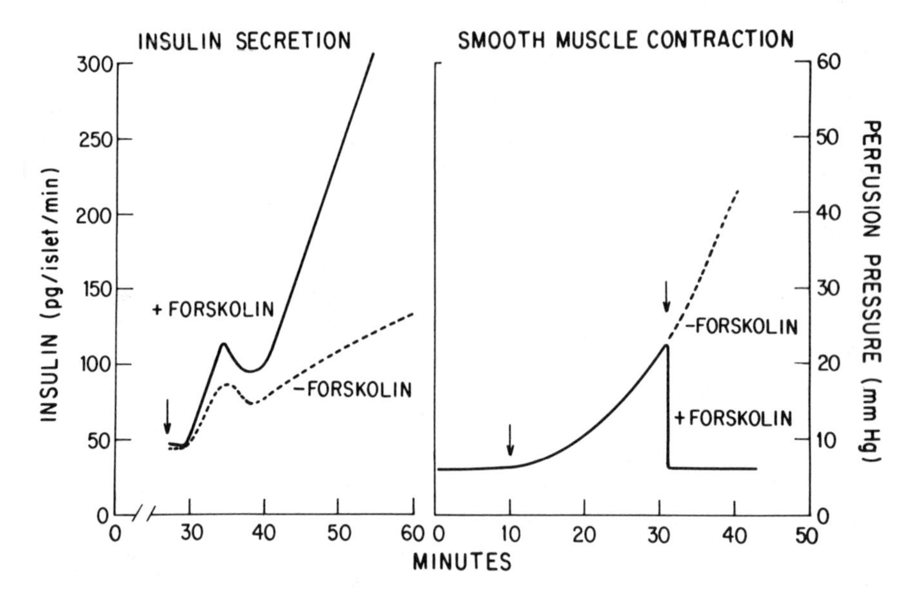

Figure 11. Positive (left) and negative (right) set-point control of cellular response brought about by the activation of adenylate cyclase in pancreatic rat islets (left) and vascular smooth muscle (right). Insulin secretion (left) was induced by the combined addition of TPA and tolbutamide in the presence (——) or absence (– – – –) of forskolin. Smooth muscle contraction was induced by adding TPA to muscle preincubated with combine ouabain–Bay K 8644. The addition of forskolin leads to a rapid and complete relaxation of the muscle.

contrast sharply with those achieved in vascular smooth muscle perifused with TPA and a combination of ouabain and Bay K 8644 (Fig. 11).

The addition of forskolin to the islets leads to an enhancement of the rate of insulin secretion during both phases of the response (Zawalich *et al.*, 1983; 1984). The effect of forskolin (mediated by an increase in [cAMP]) cannot be simply to raise the $[Ca^{2+}]_c$ since when forskolin and TPA are combined there is no first phase of insulin secretion but the second phase of secretion is two times that seen with TPA alone. It is our working hypothesis the cAMP has two effects: it leads to an enhanced rise in $[Ca^{2+}]_c$ during phase 1 so that more of the C-kinase is converted into its activated, calcium-sensitive form, and it brings about an increase in plasma membrane Ca^{2+} cycling rate during phase 2 (and hence the $[Ca^{2+}]$ in our hypothetical submembrane domain) so that both the amount of activated C-kinase and its turnover rate are increased during phase 2 (Fig. 12). It remains possible that cAMP may also alter the function of the phosphoprotein products of C-kinase.

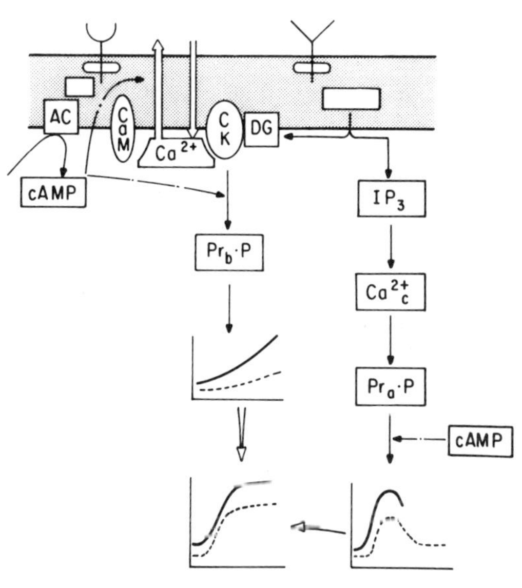

Figure 12. Positive set-point control by cAMP of the flow of information in the calcium messenger system. Cellular response to a standard concentration of agonist working via the calcium messenger system in the absence (----) or presence (——) of another agent that activates adenylate cyclase. The rise in $[cAMP]_c$ acts at several sites to increase the magnitude of the eventual response. It increases Ca^{2+} cycling and $[Ca^{2+}]$ in the submembrane domain and increases the initial increase in $[Ca^{2+}]_c$ (not shown) so that both the amount of the calcium-sensitive C-kinase and its turnover rate are increased, leading to an increase of information flow via the C-kinase branch. Likewise, the greater increase in $[Ca^{2+}]_c$ plus positive sensitive modulation of Ca^{2+}-sensitive enzymes in the calmodulin branch increase information flow via the calmodulin branch. As a consequence, both phases of cellular response are enhanced.

The effect of activating adenylate cyclase with forskolin in vascular smooth muscle contracted by combined treatment with TPA and ouabain plus Bay K 8644 is exactly the opposite of the effect seen in islets (Rasmussen *et al.*, 1984; Forder *et al.*, 1985): the rise in $[cAMP]_c$ leads to an immediate and complete inhibition of the contractile response. It is our working hypothesis that a rise in $[cAMP]_c$ lowers the $[Ca^{2+}]$ in the submembrane domain by increasing both the Ca^{2+} uptake into the sarcoplasmic reticulum and the rate of Ca^{2+} efflux across the plasma membrane.

These two contrasting responses provide a framework within which to consider the relationship between the cAMP messenger system and the C-kinase branch of the calcium messenger system. The simplest way to relate these two pathways of information flow is to postulate that in many cellular responses the cAMP system determines the set point around which the calcium messenger system operates. A rise in $[cAMP]_c$ can lead to either a negative or a positive change in set point, depending on the particular tissue (Fig. 12).

9. *Membrane Ca^{2+} Cycling as a General Regulatory Device*

Given the evidence that plasma membrane Ca^{2+} cycling may control the response of the activated cell, the question arises as to whether Ca^{2+} cycling across other subcellular membranes, specifically those of the endoplasmic reticulum and mitochondria, plays regulatory roles. Several facts suggest that such might be the case.

Recent work on the endoplasmic reticulum has shown that the rate of protein synthesis is a function of $[Ca^{2+}]$ and that the S6 ribosomal protein is a substrate for a calmodulin-dependent protein kinase (Brostrom *et al.*, 1983). It is also known that a common event in the activated cell is an increase in the rate of protein synthesis. It seems likely on the basis of present evidence that IP_3 concentration in those cells in which hormone–receptor interaction leads to an increase in IP_3 production remains high as long as hormone is present and that as a consequence the rate of Ca^{2+} efflux from the endoplasmic reticulum remains high. Therefore, it is quite possible that the $[Ca^{2+}]_c$ in a submembrane domain of this organelle is increased in the activated cell and plays a role in controlling the rate of protein synthesis. Other metabolic events that occur in this organelle, such as steroid hydroxylations and glucose 6-phosphate hydrolysis, may also be controlled in this fashion.

There is also evidence to suggest that an increase in Ca^{2+} cycling across the inner mitochondrial membrane could be of regulatory importance (Denton and McCormack, 1980). It is now quite clear that the total calcium content of mitochondria in intact normal cells is considerably lower than had been previously supposed; that it is possible to alter the rate of calcium cycling across the inner membrane (Nichitta and Williamson, 1984); and that several Ca^{2+}-sensitive intramitochondrial dehydrogenases are switched on when the cell is activated by an appropriate hormone (McCormack and Denton, 1984). Yet if these enzymes are to remain activated during the sustained phase of the cellular response, then either the total intramitochondrial free $[Ca^{2+}]$ or the $[Ca^{2+}]$ in a restricted submembrane domain must remain high. The latter might occur as a consequence of an increased rate of Ca^{2+} cycling across the inner mitochondrial membrane.

These concepts require experimental validation, but they do raise the interesting possibility that altered rates of Ca^{2+} cycling across each of three subcellular membranes (the endoplasmic reticulum, the plasma membrane, and the inner mitochondrial membrane) act to coordinate metabolic events in three separate intracellular domains in the activated cell under conditions in which total cellular Ca^{2+} and free $[Ca^{2+}]$ in the cytosol are not significantly different from those found in the nonactivated cell. The flow information in the calcium messenger system appears to occur through pathways considerably more subtle than had heretofore been recognized.

ACKNOWLEDGMENTS. This work is supported by grants from the National Institutes of Health (AM19813), the Muscular Dystrophy Association, and the National Dairy Council. Part of this work was performed during the period when Dr. Rasmussen held a John Simon Guggenheim Fellowship. The authors are indebted to Ms. Anne Levine and Ms. Nancy Canetti for expert secretarial and editorial assistance.

References

Adelstein, R. S., Conti, M. A., and Pato, M. D., 1980, Regulation of myosin light chain kinase by reversible phosphorylation and calcium–calmodulin, *Ann. N.Y. Acad. Sci.* **356**:142–150.

Aksoy, M. O., Murphy, R. A., and Kamm, K. E., 1982, Role of Ca^{2+} and myosin light chain phosphorylation in regulation of smooth muscle, *Am. J. Physiol.* **242**:C109–C116.

Albert, P. R., and Tashjian, A. H., Jr., 1984, Relationship of thyrotropin-releasing hormone-induced spike and plateau phases in cytosolic free Ca^{2+} concentrations to hormone secretion, *J. Biol. Chem.* **259**:15350–15363.

Aub, D. L., and Putney, J. W., Jr., 1984, Metabolism of inositol phosphates in parotid cells: Implication for the pathway of the phosphoinositide effect and for the possible messenger role of inositol triphosphate, *Life Sci.* **34**:1347–1355.

Balls, S. D., Morisis, A., and Gunther, H. S., 1984, Phorbol 12-myristate 13-acetate, ionomycin or ouabain, and raised extracellular magnesium induce proliferation of chicken heart mesenchymal cells, *Proc. Natl. Acad. Sci. U.S.A.* **81**:6418–6421.

Barrett, P. Q., Kojima, I., Kojima, K., and Rasmussen, H., 1985, Patterns of protein phosphorylation induced by angiotensin II in adrenal glomerulosa cells, *Biochem. J.* (in press).

Berridge, M. J., 1975, The interaction of cyclic nucleotides and calcium in the control of cellular activity, *Adv. Cyclic Nucleotide Res.* **6**:1–98.

Berridge, M. J., 1982, A novel signalling system based on the integration of phospholipid and calcium metabolism, in: *Calcium and Cell Function*, Vol. III (W. H. Cheung, ed.), Academic Press, New York, pp. 1–37.

Berridge, M. J., 1984, Inositol trisphosphate and diacylglycerol as second messengers, *Biochem. J.* **220**:345–360.

Berridge, M. J., and Irvine, R. F., 1984, Inositol trisphosphate, a novel second messenger in cellular signal transduction, *Nature* **213**:315–321.

Bolton, T. B., 1979, Mechanisms of action of transmitters and other substances on smooth muscle, *Physiol. Rev.* **59**:606–718.

Borle, A. B., 1973, Calcium metabolism at the cellular level, *Fed. Proc.* **32**:1944–1950.

Borle, A. B., 1981, Control, modulation and regulation of cell calcium, *Rev. Physiol. Biochem. Pharmacol.* **90**:13–153.

Borle, A. B., and Snowdowne, K. W., 1982, Measurement of intracellular free calcium in monkey kidney cells with aequorin, *Science* **217**:252–254.

Brostrom, C. O., Hunkeler, F. L., and Krebs, E. G., 1971, The regulation of skeletal muscle phosphorylase kinase by Ca^{2+}, *J. Biol. Chem.* **246:**1961–1967.

Brostrom, C., Bocckino, S. B., and Brostrom, M. A., 1983, Identification of a Ca^{2+} requirement for protein synthesis in eukaryotic cells, *J. Biol. Chem.* **258:**14390–14399.

Campbell, A. K., 1983, *Intracellular Calcium*, John Wiley & Sons, New York.

Castagna, M., Takai, Y., Kaibuchi, K., Sano, K., Kikkawa, U., and Nishizuka, Y., 1982, Direct activation of calcium-activated, phospholipid dependent protein kinase by tumor promoting phorbol esters, *J. Biol. Chem.* **257:**7847–7851.

Chapman, R. A., 1979, Excitation–contraction coupling in cardiac muscle, *Prog. Biophys. Mol. Biol.* **35:**1–52.

Charesi, R., Blackmore, P. F., Berthon, B., and Exton, J. H., 1983, Changes in free cytosolic Ca^{2+} in hepatocytes following alpha$_1$-adrenergic stimulation, *J. Biol. Chem.* **258:**8769–8773.

Cheung, W. Y., 1980, Calmodulin plays a pivotal role in cellular regulation, *Science* **207:**19–27.

Cohen, P., 1979, The hormonal control of glycogen metabolism in mammalian muscle by multivalent phosphorylation, *Biochem. Soc. Trans.* **7:**459–480.

Cooper, J. A., Bowen-Pope, D. F., Raines, E., Ross, R., and Hunter, T., 1982, Similar effects of platelet-derived growth factor and epidermal growth factor on the phosphorylation of tyrosine in cellular proteins, *Cell* **31:**263–273.

Dale, M. M., and Penfield, A., 1984, Synergism between phorbol ester and A23187 in superoxide production by neutrophils, *FEBS Lett.* **175:**170–172.

Denton, R. M., and McCormack, J. G., 1980, On the role of the calcium transport cycle in heart and other mammalian mitochondria, *FEBS Lett.* **119:**1–8.

Delbeke, D., Kojima, I., Dannies, P. A., and Rasmussen, H., 1984, Synergistic stimulation of prolactin release by phorbol ester, A23187 and forskolin, *Biochem. Biophys. Res. Commun.* **123:**735–741.

dePont, J. J. H. H. M., and Fleurer-Jakobs, A. M. M., 1984, Synergistic effect of A23187 and a phorbol ester on amylase secretion from rabbit pancreatic acini, *FEBS Lett.* **170:**64–68.

Downes, P., and Michell, R. H., 1982, Phosphatidylinositol 4-phosphate and phosphatidylinositol 4,5-bis-phosphate: Lipids in search of a function, *Cell Calcium* **3:**467–502.

Ebashi, S., Mikawa, T., Hirata, M., and Nonomura, Y., 1978, The regulatory role of calcium in muscle, *Ann. N.Y. Acad. Sci.* **307:**451–461.

Exton, J. H., 1981, Molecular mechanisms involved in α-adrenergic responses, *Mol. Cell. Endocrinol.* **23:**233–264.

Forder, J., Scriabine, A., and Rasmussen, H., 1985, TPA-induced contractions of vascular smooth muscle, *J. Pharmacol. Exp. Ther.* **235:**267–273.

Garrison, J. C., 1983, Role of Ca^{2+}-dependent protein kinases in the response of hepatocytes to α-agonists, angiotensin II and suppressants, in: *Isolation, Characterization and Use of Hepatocytes* (R. A. Harris and N. W. Cornell, eds.), Elsevier, Amsterdam, pp. 551–559.

Gennaro, R., Pozzan, T., and Romeo, D., 1984, Monitoring of cytosolic free Ca^{2+} and CSa-stimulated neutrophils, loss of receptor-modulated Ca^{2+} stores and Ca^{2+} uptake in granule-free cytoplasts, *Proc. Natl. Acad. Sci. U.S.A.* **81:**1416–1420.

Gershengorn, M. C., and Thaw, C., 1983, Calcium influx is not required for TRH to elevate free cytoplasmic calcium in GH_3 cell, *Endocrinology* **113:**1522–1524.

Gerthoffer, W. T., and Murphy, R. A., 1983, Ca^{2+}, myosin phosphorylation and relaxation of arterial smooth muscle, *Am. J. Physiol.* **245:**C271–C277.

Hedeskov, C. J., 1980, Mechanism of glucose-induced insulin secretion, *Physiol. Rev.* **60:**442–509.

Heiden, C.-H., and Westermark, B., 1984, Growth factors: Mechanism of action and relation to oncogenes, *Cell* **37:**9–20.

Huang, J. S., Huang, S. S., and Deuel, T. F., 1984, Transforming protein of simian sarcoma virus stimulates autocrine growth of SSV-transformed cells through PDGF cell-surface receptors, *Cell* **39:**79–87.

Irvine, R. F., Dawson, R. M. C., and Freinkel, N., 1982, Stimulated phosphatidylinositol turnover. A brief appraisal, in: *Contemporary Metabolism*, Vol. 2 (N. R. Freinkel, ed.), Plenum Press, New York, pp. 301–342.

Joseph, S. K., and Williamson, J. R., 1983, The origin, quantitation and kinetics of intracellular calcium mobilization by vasopressin and phenylephrine in hepatocytes, *J. Biol. Chem.* **258:**10425–10432.

Joseph, S. K., Thomas, A. P., Williams, R. J., Irvine, R. F., and Williamson, J. R., 1984, Myoinositol 1,4,5-triphosphate: A second messenger for the hormonal mobilization of intracellular Ca^{2+} in liver, *J. Biol. Chem.* **259:**3077–3081.

Kaibuchi, K., Takai, Y., Sawamura, M., Hoshijima, M., Fujikura, T., and Nishizuka, Y., 1983, Synergistic functions of protein phosphorylation and calcium mobilization in platelet activation, *J. Biol. Chem.* **258:**6701–6704.

Katakami, Y., Kaibuchi, K., Sawamura, M., Takai, Y., and Nishizuka, Y., 1984, Synergistic action of protein kinase C and calcium for histamine release from rat peritoneal mast cells, *Biochem. Biophys. Res. Commun.* **121:**573–578.

Katz, B., 1966, *Nerve, Muscle, Synapse*, McGraw-Hill, New York.

Kerrick, W. G. L., Hoar, P. E., and Cassidy, P. S., 1980, Calcium-activated tension: The role of myosin light chain phosphorylation, *Fed. Proc.* **39:**1558–1563.

Kikkawa, U., Minakuchi, R., Takai, Y., and Nishizuka, Y., 1983, Calcium-activated phospholipid-dependent protein kinase (protein kinase C) from rat brain, *Methods Enzymol.* **99:**288–299.

Kirk, C. J., Michell, R. H., and Hems, D. A., 1981, Phosphatidylinositol metabolism in rat hepatocytes stimulated by vasopressin, *Biochem. J.* **194:**155–165.

Kojima, I., Lippes, H., Kojima, K., and Rasmussen, H., 1983, Aldosterone secretion: Effect of phorbol ester and A23187 and TPA, *Biochem. Biophys. Res. Commun.* **116:**555–562.

Kojima, I., Kojima, K., Kreutter, D., and Rasmussen, H., 1984, The temporal integration of the aldosterone secretory response to angiotensin II occurs via two intracellular pathways, *J. Biol. Chem.* **259:**14448–14457.

Kojima, I., Kojima, K., and Rasmussen, H., 1985a, Characteristics of angiotensin III-, K^+- and ACTH-induced calcium influx in adrenal glomerulosa cells, *J. Biol. Chem.* **260:**9171–9176.

Kojima, I., Kojima, K., and Rasmussen, H., 1985b, Role of calcium fluxes in the sustained phase of angiotensin II-mediated aldosterone secretion from adrenal cells, *J. Biol. Chem.* **260:**9177–9184.

Kojima, I., Kojima, K., and Rasmussen, H., 1985c, Role of calcium and cAMP in the action of adrenocorticotropin on aldosterone secretion, *J. Biol. Chem.* **260:**4248–4256.

Kojima, I., Kojima, K., and Rasmussen, H., 1985d, Intracellular calcium and adenosine 3′,5′-cyclic monophosphate as mediators of potassium-induced aldosterone secretion, *Biochem. J.* **228:**69–76.

Martin, T. J. F., 1983, Thyrotropin-releasing hormone rapidly activates the phosphodiesterase hydrolysis of polyphosphoinositides in GH_3 pituitary cells, *J. Biol. Chem.* **258:**14816–14822.

Martin, T. J. F., and Kowalchyk, T. A., 1984a, Evidence for the role of calcium and diacylglycerol as dual second messengers in thyrotropin-releasing hormone action: Involvement of diacylglycerol, *Endocrinology* **115:**1517–1526.

Martin, T. F. J., and Kowalchyk, T. A., 1984b, Evidence for the role of calcium and diacylglycerol as dual second messengers in thyrotropin-releasing hormone action: Involvement of Ca^{2+}, *Endocrinology* **115:**1527–1536.

Mauger, J. P., Poggioli, J., Guesdon, F., and Claret, M., 1984, Noradrenaline, vasopressin and angiotensin increase Ca^{2+} influx by opening a common pool of Ca^{2+} channels in isolated rat liver cells, *Biochem. J.* **221:**121–127.

McCormack, J. G., and Denton, R. M., 1979, The effects of calcium ions and adenine nucleotides on the activity of pig heart 2-oxoglutarate dehydrogenase complex, *Biochem. J.* **180:**533–544.

Means, A. R., and Dedman, J. R., 1980, Calmodulin—an intracellular calcium receptor, *Nature* **285:**73–77.

Michell, R. H., 1975, Inositol phospholipids and cell surface receptor function, *Biochim. Biophys. Acta* **415:**81–147.

Moolenaar, W. H., Tertoolen, L. G. J., and de Laat, S. W., 1984, Growth factors immediately raise cytoplasmic free Ca^{2+} in human fibroblasts, *J. Biol. Chem.* **259:**8066–8069.

Morgan, J. P., and Morgan, K. G., 1983, Vascular smooth muscle: The first recorded Ca^{2+} transients, *Pflügers Arch.* **395:**75–77.

Morgan, J. P., and Morgan, K. G., 1984, Stimulus-specific patterns of intracellular calcium levels in smooth muscle of ferret portal vein, *J. Physiol.* (*Lond.*) **351:**156–167.

Nichitta, C. V., and Williamson, J. R., 1984, Spermine, a regulator of mitochondrial calcium cycling, *J. Biol. Chem.* **259:**12978–12983.

Nishizuka, Y., 1983, A receptor-linked cascade of phospholipid turnover in hormone action, in: *Endocrinology* (K. Shizume, H. Imura, and H. Shimizu, eds.), Excerpta Medica, Amsterdam, pp. 15–24.

Nishizuka, Y., 1984. The role of protein kinase C in cell surface signal transduction and tumour promotion, *Nature* **308:**693–698.

Pozzan, T. P., Arslan, P., Tsien, R. Y., and Rink, T. J., 1982, Antiimmunoglobulin, cytoplasmic free calcium, and capping in lymphocytes, *J. Cell Biol.* **94:**335–340.

Putney, J. W., 1979, Stimulus–permeability coupling: Role of calcium in the receptor regulation of membrane permeability, *Pharmacol. Rev.* **30:**209–245.

Putney, J. W., 1982, Inositol lipids and cell stimulation in mammalian salivary gland, *Cell Calcium* **3:**369–383.

Rasmussen, H., 1970, Cell communication, calcium ion and cyclic adenosine monophosphate, *Science* **170:**404–412.

Rasmussen, H., 1981, *Calcium and cAMP as Synarchic Mesengers*, John Wiley & Sons, New York.

Rasmussen, H., 1983, Pathways of amplitude and sensitivity modulation in the calcium messenger system, in: *Calcium and Cell Function*, Vol. 4 (W. Y. Cheung, ed.), Academic Press, New York, pp. 1–61.

Rasmussen, H., and Barrett, P. W., 1984, Calcium messenger system: An integrated view, *Physiol. Rev.* **64:**938–984.

Rasmussen, H., and Waisman, D. M., 1982, Modulation of cell function in the calcium messenger system, *Rev. Physiol. Biochem. Pharmacol.* **95:**111–148.

Rasmussen, H., Forder, J., Kojima, I., and Scriabine, A., 1984a, TPA-induced contraction of isolated rabbit vascular smooth muscle, *Biochem. Biophys. Res. Commun.* **122:**776–784.

Rasmussen, H., Kojima, I., Kojima, K., Zawalich, W., and Apfeldorf, W., 1984b, Calcium as intracellular messenger: Sensitivity modulation, C-kinase pathway and sustained cellular response, *Adv. Cyclic Nucleotide Protein Phosphor. Res.* **18:**159–193.

Robinson, J. M., Badwey, J. A., Karnovsky, M. L., and Karnovsky, M. J., 1984, Superoxide release by neutrophils: Synergistic effects of a phorbol ester and a calcium ionophore, *Biochem. Biophys. Res. Commun.* **122:**734–739.

Rubin, R. P., 1982, *Calcium and Cellular Secretion*, Plenum Press, New York.

Sawyer, S. T., and Cohen, S., 1981, Enhancement of calcium uptake and phosphatidylinositol turnover by epidermal growth factor in A-431 cells, *Biochemistry* **20:**6280–6286.

Sawyer, S. T., and Krantz, S. B., 1984, Erythropoietin stimulates $^{45}Ca^{2+}$ uptake in Friend virus-infected erythroid cells, *J. Biol. Chem.* **259:**2769–2774.

Scharff, O., Foder, B., and Skibsted, U., 1983, Hysteretic activation of the Ca^{2+} pump revealed by calcium transients in human red cells, *Biochim. Biophys. Acta* **730:**295–305.

Schulz, I., 1980, Messenger role of calcium in function of pancreatic acinar cells, *Am. J. Physiol.* **239:**G335–G347.

Shears, S. B., and Kirk, C. J., 1984, Determination of mitochondrial calcium content in hepatocytes by a rapid cellular fractionation technique: Vasopressin stimulates mitochondrial Ca^{2+} uptake, *Biochem. J.* **219:**383–389.

Silver, P. J., and Stull, J. T., 1982, Regulation of myosin light chain and phosphorylase phosphorylation in tracheal smooth muscle, *J. Biol. Chem.* **257:**6145–6150.

Smith, J. B., Smith, L., Brown, E. R., Barnes, D., Sabir, M. A., Davis, J. S., and Farese, R. V., 1984, Angiotensin II rapidly increases phosphatidic-phosphoinositide synthesis and phosphoinositide hydrolysis and mobilizes intracellular calcium in cultured arterial muscle cells, *Proc. Natl. Acad. Sci. U.S.A.* **81:**7812–7816.

Snowdowne, K. W., and Borle, A. B., 1984, Changes in cytosolic ionized calcium induced by activators of secretion in GH_3 cells, *Am. J. Physiol.* **246:**E198–E201.

Streb, H., and Schulz, I., 1983, Regulation of cytosolic free Ca^{2+} concentration in acinar cells of rat pancreas, *Am. J. Physiol.* **245:**G347–G357.

Streb, H., Irvine, R. F., Berridge, M. J., and Schulz, I., 1983, Release of Ca^{2+} from a nonmitochondrial store in pancreatic acinar cells by inositol-1,4,5-trisphosphate, *Nature* **306:**67–69.

Studer, R. K., and Borle, A. B., 1983, Sex differences in cellular calcium metabolism of rat hepatocytes and in α-adrenergic activation of glycogen phosphorylase, *Biochim. Biophys. Acta* **762:**302–314.

Suematsu, E., Hirata, M., Hashimoto, T., and Kuriyama, H., 1983, Inositol 1,3,4-trisphosphate releases Ca^{2+} from intracellular storage sites in skinned single cells of porcine coronary arteries, *Biochem. Biophys. Res. Commun.* **120:**481–485.

Takai, Y., Kishimoto, A., Iwasa, Y., Kawahara, Y., Mori, T., and Nishizuka, Y., 1979, Calcium-dependent activation of a multifunctional protein kinase by membrane phospholipids, *J. Biol. Chem.* **254:**3692–3695.

Takai, Y., Kishimoto, A., Kawahara, Y., Minakuchi, R., Sano, K., Kikkawa, U., Mori, T., Yu, B., Kaibuchi, K., and Nishizuka, Y., 1981, Calcium and phosphatidylinositol turnover as signalling for transmembrane control of protein phosphorylation, *Adv. Cyclic Nucleotide Res.* **14:**301–313.

Takai, Y., Kikkawa, U., Kaibuchi, K., and Nishizuka, Y., 1984, Membrane phospholipid metabolism and signal transduction for protein phosphorylation, *Adv. Cyclic Nucleotide Protein Phosphor. Res.* **18:**119–158.

Trunek, A., Albert, F., Goldstein, P., and Schnitt-Verhulst, A.-M., 1985, Early steps of lymphocyte activation bypassed by synergy between calcium ionophores and phorbol ester, *Nature* **313:**318–320.

Tsien, R. Y., Pozzan, T., and Rink, T. J., 1981, Calcium homeostasis in intact lymphocytes: Cytoplasmic free calcium monitored with a new intracellularly trapped fluorescent indicator, *J. Cell Biol.* **94:**325–334.

Wang, J. H., and Waisman, D. M., 1979, Calmodulin and its role in the second messenger system, *Curr. Top. Cell Regul.* **15**:47–107.

Weber, A., and Murray, J. M., 1973, Molecular control mechanisms in muscle contraction, *Physiol. Rev.* **53**:612–673.

Williamson, J. R., Cooper, R. H., and Hoek, J. R., 1981, Role of calcium in the hormonal regulation of liver metabolism, *Biochim. Biophys. Acta* **639**:243–295.

Zawalich, W., Brown, C., and Rasmussen, H., 1983, Insulin secretion: Combined effect of phorbol ester and A23187, *Biochem. Biophys. Res. Commun.* **117**:448–455.

Zawalich, W., Zawalich, K., and Rasmussen, H., 1984, Insulin secretion: Combined tolbutamide, forskolin and TPA mimic action of glucose, *Cell Calcium* **5**:551–558.

9

Neutrophil Activation, Polyphosphoinositide Hydrolysis, and the Guanine Nucleotide Regulatory Proteins

PAUL H. NACCACHE and RAMADAN I. SHA'AFI

1. Introduction

Mechanochemical transduction is an intimate component of several functions that neutrophilic polymorphonuclear leukocytes (neutrophils) perform as part of their role in the body's first line of defense against foreign pathogens. Thus, the neutrophils represent a convenient model system for the study of the activation mechanisms in nonmuscle cells.

The general outline of the excitation–response coupling sequence in the neutrophils has been adequately described. Neutrophils recognize specific extracellular signals (termed chemotactic factors because of their ability to evoke a locomotory response from these cells) by means of membrane-located receptors. Several reactions are then initiated that culminate in a rise in the level of intracellular calcium. The latter, in conjunction with other cytoplasmic events, initiates a coordinated, localized, and transient contractile response from the neutrophils' actin-based microfilaments (Southwick and Stossel, 1983), which results in the displacement of the appropriate part of the cell.

Significant gaps remain in our knowledge of the excitation–response

PAUL H. NACCACHE and RAMADAN I. SHA'AFI • Departments of Pathology and Physiology, University of Connecticut Health Center, Farmington, Connecticut 06032.
Present address of P.H.N: Department of Immunology and Rheumatism, Centre Hospitalier de l'Université Laval, Ste. Foy, Quebec G1V 4G2, Canada.

coupling sequence in the neutrophils, as in other nonmuscle cells. A partial and nonexhaustive list of the latter would include (1) a complete cataloging of the various neutrophil stimuli and of the soluble factors that regulate their activity, (2) a definite understanding of the sequence of events that leads from receptor occupation to the observed increases in the cytoplasmic levels of ionized calcium, (3) an elucidation of the targets of the raised levels of calcium and of the consequences of their activation by the divalent cation, (4) a further understanding of the biochemical pathways activated by chemotactic factors that appear to be either independent of a rise in calcium or that occur in parallel with the latter, and (5) an integration of the *in vitro* data concerning the behavior of the constituents of the cytoskeletons of these cells into the *in situ* conditions. The present review is concerned with a discussion of the current concepts related to the mechanism of signal transduction in the neutrophils, i.e., with points 2, 3, and 4 of the above list.

2. *Biochemical Events Associated with Cellular Activation*

In the past few years several lines of investigation directed at the elucidation of the mechanism of activation of nonmuscle cells have coalesced, and the stimulated hydrolysis of the polyphosphoinositides has emerged as their common element (reviewed in Berridge and Irvine, 1984; Berridge, 1984; Nishizuka, 1984). Two critical compounds are generated as a result of the activation of the polyphosphoinositide phosphodiesterase: the unsaturated fatty acid rich 1,2-diacylglycerol (DG) and the water-soluble inositides, of which inositol 1,4,5-trisphosphate (IP_3) appears to be the compound most active physiologically, although it may not be the only one (Irvine *et al.*, 1984).

As presently understood, the physiological role of IP_3 is to serve as an intracellular mediator of calcium mobilization from internal stores. The calcium-mobilizing properties of IP_3 have accordingly been demonstrated in a variety of permeabilized cells including liver, pancreas, and other tissues (see Berridge and Irvine, 1984). Very recently, IP_3 has also been shown to release intracellular calcium from the *Limulus* ventral photoreceptor cells following its injection into intact cells (Brown and Rubin, 1984). Wherever characterized, IP_3 releases calcium from a nonmitochondrial, smooth-endoplasmic-reticulum-related storage pool (Streb *et al.*, 1984). In several instances (hepatocytes, *Limulus* photoreceptor cells, platelets, and pancreatic acinar cells), the latter structures can be observed to lie in close proximity to the plasma membrane, thereby facilitating communication between the source of IP_3 and its target. The phosphatases responsible for the degradation of IP_3 have also been found to be localized in the plasma membrane (Storey *et al.*, 1984), further emphasizing the local nature of the mediatory effects of IP_3. The physiological significance of IP_3 has been underscored recently by the demonstration that intracellular injection of IP_3 in intact *Xenopus* oocytes mimics the muscarinic responses of these cells, inducing a polarizing chloride current (Oron *et al.*, 1985). An additional common feature of several cells in which the actions of IP_3 have been studied is that their major physiological role is secretion (the *Limulus* photoreceptor cells, however, are a clear exception).

In addition to generating IP_3, the activation of the polyphosphoinositide phosphodiesterase also results in the liberation of 1,2-diacylglycerol (DG), a compound capable of activating and/or potentiating several processes that play apparently central roles in nonmuscle cell functions and are of possibly critical relevance to cell proliferation and differentiation (Moolenar *et al.*, 1984; Rozengurt *et al.*, 1984) 1,2-Diacylglycerol has been found to be a potent activator of protein kinase C (C-kinase), lowering this enzyme's requirement for calcium to the level found in resting cells (10^{-7} M) (Kaibuchi *et al.*, 1981; DiVirgilio *et al.*, 1984). The evidence linking the activation of C-kinase to the initiation of cellular responsiveness has recently been reviewed by Nishizuka (1984).

It should be pointed out, however, that with the recent exception of the C-kinase stimulation of the phosphorylation of guanylate cyclase (Zwiller *et al.*, 1985), the nature and function of the *in situ* C-kinase substrates are still unknown. The realization that phorbol esters directly activate C-kinase by substituting for DG (Castagna *et al.*, 1982; Niedel *et al.*, 1983) has provided a basis for the interpretation of the pleiotropic effects of these compounds as being mediated by C-kinase. The currently prevailing assumption that most, if not all, of the effects of the phorbol ester can be attributed to the activation of C-kinase remains to be critically examined. Additional ties between C-kinase and polyphosphoinositide metabolism were recently uncovered, and it was found that phorbol esters (Halenda *et al.*, 1984; de Chaffoy de Courcelles *et al.*, 1984), mitogens (Taylor *et al.*, 1984), and the products of the *src* (Sugimoto *et al.*, 1984) and *ros* (Macara *et al.*, 1984) viral oncogenes all stimulated the phosphorylation of phosphatidylinositol.

The stimulated hydrolysis of the polyphosphoinositides may also be related to the commonly observed liberation of arachidonic acid. Two mechanisms for this link have been proposed and substantiated to a certain extent. A DG-specific phospholipase A_2 has been detected in platelets (Rittenhouse-Simmons, 1979; Bell *et al.*, 1979); thus, the hydrolysis of PIP_2 would conceivably provide a substrate for this enzyme. Secondly,

Dawson *et al.* (1983) have demonstrated that unsaturated-fatty-acid-rich DG caused physical changes in the organization of liposomes that render them more susceptible to attack by phospholipases. According to this scheme, the endogenous liberation of DG would thus stimulate the activity of other lipid-remodeling enzymes. The attraction of the second model is that it requires receptor occupation to initiate a single event, namely, the activation of phospholipase C. The other observed changes would result from the liberation of the active metabolites DG and IP_3. Finally, direct fusogenic properties have also been attributed to DG (Allan and Michell, 1977), possibly as a result of this compound's ability to lower the surface pressure of the phospholipid monolayers and thus to promote their breakdown (Demel *et al.*, 1975).

3. Relevance of Polyphosphoinositide Hydrolysis to the Activation of the Neutrophils

In view of the apparent generality of the above indices of cell activation, several recent studies have focused on the examination of their occurrence in response to the addition of chemotactic factors and other stimuli to neutrophils. The results obtained thus far have established that stimulated polyphosphoinositide hydrolysis and C-kinase activation are indeed detectable in the neutrophils but have at the same time raised questions about their physiological significance.

3.1. Effect of Chemotactic Factors on Polyphosphoinositide Metabolism in the Neutrophils

The addition of fMet-Leu-Phe but not of calcium ionophores to rabbit and human neutrophils and to HL60 human promyelocytic leukemia cells results in rapid changes in the levels of the various phosphoinositides, including:

1. A rapid decrease (less than 0.5 min) in the level of phosphatidylinositol-4,5-bisphosphate (PIP_2) and especially in the arachidonate-rich pool (Yano *et al.*, 1983a), which is followed by a return to basal values within the next minute (Volpi *et al.*, 1983; Yano *et al.*, 1983a; Dougherty *et al.*, 1984; Bradford and Rubin, 1985).
2. Somewhat delayed although more sustained decreases in the levels of phosphatidylinositol-4-monophosphate (PIP) (Volpi *et al.*, 1983; Yano *et al.*, 1983a).
3. Increases in the levels of DG and phosphatidic acid (Volpi *et al.*, 1983, 1984; Yano *et al.*, 1983b).

4. Increases in the levels of water-soluble inositol phosphate compounds including IP_3, inositol-4,5-bisphosphate (IP_2) and inositol-4-monophosphate (IP) in cytochalasin-B-treated cells (Dougherty *et al.*, 1984; Bradford and Rubin, 1985).

The above results indicate that the occupation of the formyl peptide receptors leads to the activation of a PIP_2-specific phospholipase C. However, they do not adequately allow us to determine whether a phospholipase D (see Cockroft, 1982) plays a role in the above observations. The lack of stimulatory activity of calcium ionophores (Bradford and Rubin, 1985) supports the argument that the hydrolysis of PIP_2 is causally linked to the mobilization of calcium and no vice versa.

The physiological relevance of the generation of IP_3 appears to be similar to that described in other systems, since its addition to permeabilized neutrophils has been shown to release calcium from a nonmitochondrial internal pool (Prentki *et al.*, 1984; Burgess *et al.*, 1984) that presumably is the same as that mobilized by fMet-Leu-Phe, although this point has not been established at present.

Despite the apparent agreement between the results presented above and those obtained in other cells, there are several observations that indicate that mechanisms other than IP_3 must be invoked to account for all aspects of calcium mobilization in neutrophils. First, significantly less fMet-Leu-Phe is required to mobilize calcium and to elicit a chemotactic response than to initiate the hydrolysis of PIP_2 (Fig. 1). The ED_{50} of the chemotactic responses of rabbit neutrophils to fMet-Leu-Phe is 7×10^{-11} M (Showell *et al.*, 1976), that for calcium mobilization (as derived from quin-2 data) is $1–2 \times 10^{-10}$ M (White *et al.*, 1983; Volpi *et al.*, 1984), and those for phosphatidic acid and IP_3 production are 1×10^{-9} M (Volpi *et al.*, 1984) and 0.7×10^{-9} M (Bradford and Rubin, 1985), respectively. In addition, Petroski *et al.* (1979) used radioisotopic methods to detect internal calcium redistribution at concentrations of fMet-Leu-Phe equal to or lower than 1×10^{-11} M. On the other hand, the concentrations of fMet-Leu-Phe that induce the hydrolysis of PIP_2 are similar to those that evoke oxidative and secretory responses from the neutrophils, which suggests a more direct interrelationship between PIP_2 breakdown and these latter two functional responses than with chemotaxis.

Secondly, the stimulation of the turnover of PIP_2 may not be a characteristic shared by all chemotactic factors, as would be expected from the hypothesis that IP_3 is the mediator of calcium mobilization. Leukotriene B_4 is a potent neutrophil chemoattractant that has been shown to raise the intracellular level of calcium in the neutrophils using part of the same internal store of calcium that is affected by fMet-Leu-Phe (Sha'afi

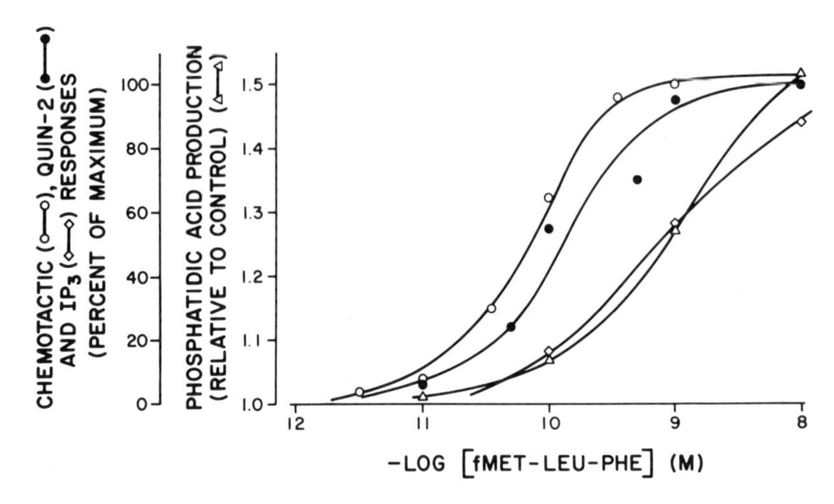

Figure 1. Comparison of the dependence of the chemotactic, quin-2, phosphatidic acid, and IP_3 responses on the concentration of fMet-Leu-Phe in rabbit neutrophils. The data were extracted from the following references: chemotaxis, Becker *et al.*, 1978; quin-2, Volpi *et al.*, 1984, and White *et al.*, 1983; phosphatidic acid, Volpi *et al.*, 1984; IP_3, Bradford and Rubin, 1984.

et al., 1981; Naccache *et al.*, 1984). However, when its effects on polyphosphoinositide turnover were examined, it was found that leukotriene B_4 did not stimulate the breakdown of PIP_2 (Volpi *et al.*, 1984) or only stimulated it to a small extent (Bradford and Rubin, 1985). Similarly, leukotriene B_4 did not induce the formation of phosphatidic acid or lysophosphatidic acid (Volpi *et al.*, 1984), two additional indices of polyphosphoinositide turnover.

The basis for these results is for the moment unknown. However, they strongly imply that a mode of calcium release independent of IP_3 is also available to the neutrophils. Although fMet-Leu-Phe utilizes the two mechanisms, at least at relatively high concentrations, leukotriene B_4 only accesses one of them (the mechanism independent of IP_3). A similar conclusion was recently reached concerning the epidermal growth factor (EGF)-stimulated 3T3 fibroblasts; it was found that EGF causes increases in calcium and C-kinase activation in the absence of detectable PIP_2 breakdown (quoted in Hesketh *et al.*, 1985). This interpretation is consistent with the previous observations that whereas leukotriene B_4 raises the level of intracellular calcium as much as fMet-Leu-Phe in the presence of external calcium, the fatty acid is only partly as efficient a stimulus for the release of internal calcium as is the formyl peptide (Naccache *et al.*, 1984; Lew *et al.*, 1984). These results imply that calcium mobilization by leukotriene B_4 is largely dependent on increases in the permeability of the plasma membrane to calcium and thus on a net influx of calcium. The relatively low activity of leukotriene B_4 as an activator of the oxidative burst (Gay *et al.*, 1984) and as a secretagogue (Showell *et al.*, 1982) (two neutrophil functions that require both increases in cytoplasmic free calcium and activation of C-kinase for their optimization) can be similarly explained in terms of these findings, emphasizing that the primary function of leukotriene B_4 is that of a chemoattractant.

It remains to be seen whether a similar dual control of calcium release is available to other cells and what relationship, if any, exists between these two aspects of calcium mobilization. It is intriguing to postulate that the mechanism of calcium release utilized by leukotriene B_4 may be related to the pool of "trigger" calcium, the existence of which had previously been postulated in pancreatic acinar cells, among others (Schulz, 1980).

3.2. Protein Kinase C and the Neutrophils

On the basis of the concepts and results developed in several cells and in the platelets in particular (Nishizuka, 1984), several studies have investigated the possible involvement of C-kinase in the mechanism of stimulation of the neutrophils by chemotactic factors.

C-kinase accounts for the majority of the kinase activity of the neutrophils (Huang *et al.*, 1983; Helfman *et al.*, 1983). Its intracellular distribution is to some extent dependent on the state of activation of the cells. As previously observed in the parietal yolk sac (Kraft and Anderson, 1983), phorbol myristate acetate (PMA) causes a large decrease in cytosolic C-kinase activity and a concomitant increase in particulate activity (McPhail *et al.*, 1984). On the other hand, fMet-Leu-Phe was found not to affect the soluble activity, although it did increase the activity of the particulate C-kinase in the presence of cytochalasin B, suggesting that the formyl peptide did not induce a shift in enzyme distribution towards the plasma membrane. It is also noteworthy that fMet-Leu-Phe by itself did not affect the activity of the extracted C-kinase.

Evidence for the involvement of C-kinase in neutrophil activation has been derived mostly from studies of the effects of phorbol esters, particularly PMA. This is necessary because the nature and function of the neutrophils' C-kinase substrates are not sufficiently defined at present to allow a direct study of their characteristics. These investigations are also based on the previously described agonist properties of PMA towards the neutrophils.

Phorbol myristate acetate is a potent aggregatory stimulus (Camussi *et al.*, 1981) and an activator of the respiratory burst of the neutrophils

(Repine *et al.*, 1974; De Chatelet *et al.*, 1976) as a result of the activation of the NADPH oxidase (Suzuki and Lehrer, 1980). The phorbol ester also possesses significant secretory activity (Estensen *et al.*, 1974), although it is somewhat selective towards the specific granules. The demonstration that PMA stimulates neutrophils without raising the cytosolic level of calcium (Sha'afi *et al.*, 1983) clearly distinguishes the mode of action of the phorbol ester from that of surface-receptor-dependent chemotactic factors. This result is in line with the known properties of PMA as an activator of C-kinase (Kaibuchi *et al.*, 1981). Phorbol myristate acetate has also been shown to activate the Na^+/H^+ antiporter in the neutrophils (Volpi *et al.*, 1985) in accord with the results obtained in lymphocytes by Rosoff *et al.* (1984) and to stimulate the ATPase-driven Ca^{2+} pump of a plasma membrane fraction derived from the neutrophils (Lagast *et al.*, 1984). In intact cells, PMA only marginally lowers the resting levels of cytoplasmic free calcium (Sha'afi *et al.*, 1983; Schell-Frederick, 1984; DiVirgilio *et al.*, 1984; Naccache *et al.*, 1985), although it does stimulate the rate of return to basal of the fMet-Leu-Phe- (Schell-Frederick, 1984) or A23187-induced rises in the intracellular levels of calcium (Lagast *et al.*, 1984).

The addition of PMA to neutrophils in suspension or in an adherent monolayer has been shown to stimulate the phosphorylation of several proteins with an apparent molecular mass (as deduced from SDS polyacrylamide gel electrophoresis) ranging from 13 to 130 kD (Schneider *et al.*, 1981; Andrews and Babior, 1983, 1984; White *et al.*, 1984). On the other hand, the phosphorylation level of a 20-kD polypeptide was found to be diminished in adherent neutrophils following stimulation by PMA (Andrews and Babior, 1984). The physiological significance of this finding is unknown. The addition of the synthetic diglyceride 1-oleyl-2-acetylglycerol (OAG), an activator of C-kinase (Kaibushi *et al.*, 1981), also induces the phosphorylation of several proteins and decreases the labelling of a 19-kD band (Fujita *et al.*, 1984). Several of the proteins whose phosphorylation is stimulated by PMA in intact cells also show increased labeling in broken-cell preparations on the addition of phosphatidylserine and calcium (i.e., the cofactors of C-kinase) (Huang *et al.*, 1983; Helfman *et al.*, 1983), indicating that the *in situ* phosphorylation may similarly be directly mediated by C-kinase.

No direct data are available at present concerning the possible phosphorylation of cytoskeletal proteins by PMA in the neutrophils (the possibility that the 130-kD protein mentioned above is vinculin has not yet been adequately addressed) or of the indirect activation of tyrosine kinase by the phorbol ester. Some of the phosphoproteins appear to respond similarly to fMet-Leu-Phe and to PMA (in particular, a peptide with an apparent molecular mass of 45–50 kD, whereas others are selectively activated by one but not the other of these two stimuli [11 kD by PMA and 69 kD by fMet-Leu-Phe (Andrews and Babior, 1984)]. The site of phosphorylation of the 45- to 50-kD protein was found to reside on the same proteolytic fragment whether the protein was phosphorylated in response to fMet-Leu-Phe or PMA in intact cells or to phosphatidylserine plus calcium in a broken-cell preparation (White *et al.*, 1984).

The PMA-induced phosphorylation of several proteins is inhibitable by trifluperazine in whole cells (White *et al.*, 1984; Andrews and Babior, 1984) as well as in broken-cell preparations (Huang *et al.*, 1983; Helfman *et al.*, 1983). On the other hand, Andrews and Babior (1984) have reported that the PMA-stimulated phosphorylation observed in adherent cells is unaffected by cyclic nucleotides, theophylline, aspirin, colchicine, or corticosteroids. The sensitivity of the phosphorylation to trifluperazine has also served in part as a basis for a positive correlation between the phosphorylation of the 45- to 50-kD substrate and the secretory responses of the cells (White *et al.*, 1984). In contrast to the results reported in cultured fibroblasts (Gilmore and Martin, 1983), the phosphorylation of the 45- to 50-kD protein in the neutrophils does not appear to be mediated by a tyrosine kinase, since it is sensitive to alkali treatment (C.-K. Huang, personal communication).

In accord with the concept originally put forward by Nishizuka (1984), it has also been demonstrated that neutrophils could be synergistically activated on the simultaneous addition of PMA (or of DG) and of a calcium ionophore such as A23187. These effects have been observed in studies of neutrophil degranulation (Kajikawa *et al.*, 1983; White *et al.*, 1984; O'Flaherty *et al.*, 1984; Naccache *et al.*, 1985) and of superoxide production (Dale and Penfield, 1984; Robinson *et al.*, 1984). In addition, White *et al.* (1984) have shown that the functional synergism described above occurred in parallel with a similar effect on the time course of the phosphorylation of the previously mentioned 45- to 50-kD substrate.

The synergistic effects of the simultaneous addition of PMA and A23187 are commonly interpreted as a reflection of the bifurcating model of cell activation in which calcium and PMA stimulate parallel series of posttransductional events. It is, however, plausible that part of the synergism is caused by the enhanced generation of an endogenously produced mediator. Preliminary data supportive of this interpretation are presented in Table I, which shows that subthreshold concentrations of PMA and A23187 induce the liberation of significantly larger than additive amounts of arachidonic acid and also, to a smaller extent of phosphatidic acid. The free arachidonic acid would be expected to be metabolized rapidly via 5-lipoxygenase and to generate leukotriene B_4 among others, a potent che-

Table I. Synergistic Effects of PMA and A23187 on the Liberation of Arachidonic Acid and the Production of Phosphatidic Acid[a]

Conditions	Arachidonate release	Phosphatidate production
PMA (20 ng/ml)	709 ± 200	320 ± 18
A23187 (5×10^{-7} M)	7,318 ± 618	4,415 ± 320
PMA + A23187	12,694 ± 950	6,640 ± 890

[a] The values are the means ± S.E.M. of the increases over the control in the radioactivity associated with free arachidonic acid and with phosphatidic acid in four determinations made 2 min after the additions of PMA and/or A23187. The average counts in the controls were 5515 and 800 cpm for arachidonic acid and phosphatidic acid, respectively.

moattractant that may be partially responsible for the increased biological activity of the cells. Since local mediators derived from arachidonic acid have been described in other cells (e.g., thromboxane A_2 in the platelets), the applicability of the present results to other systems becomes an important element in the interpretation of the commonly observed synergistic effects of calcium ionophores and phorbol esters.

Although preliminary in several important aspects, the results summarized in the preceding paragraphs are nevertheless in overall agreement with the bifurcating activation pathways observed in other cells, in which the hydrolysis of PIP_2 simultaneously activates C-kinase (through the production of DG) and mobilizes calcium (though IP_3) and indicate that C-kinase may play significant roles in the initiation of the oxidative burst and the degranulation of the neutrophils. The relationship of the activation of C-kinase to the chemotactic responses of the neutrophils remains unexamined, however. The examination of this relationship is of more than trivial or confirmatory interest, as indicated by the observations that the hydrolysis of PIP_2 can only be demonstrated at concentrations of fMet-Leu-Phe in excess of those required to elicit a chemotactic response from these cells (see Fig. 1), that leukotriene B_4 does not induce a significant turnover of the polyphosphoinositides, and that fMet-Leu-Phe at concentrations as high as 10^{-7} M does not shift the distribution of C-kinase to the plasma membrane, a process apparently mediated by the accumulation of DG in the membranes (Nishizuka, 1984).

The complexity of the consequences of C-kinase activation has been highlighted by the recent demonstration of the inhibitory properties of PMA towards neutrophil stimulation. If the cells are preexposed to the phorbol ester for a short period of time (1–5 min) (Sha'afi *et al.*, 1983; Lagast *et al.*, 1984; Schell-Frederick, 1984; Naccache *et al.*, 1985), the phorbol ester has been found to be a potent antagonist of neutrophil activation by a variety of chemotactic factors: fMet-Leu-Phe, leukotriene B_4, and AGEPC (P. H. Naccache, unpublished observations). The phorbol ester 4-phorbol didecanoate, which does not stimulate C-kinase (Castagna *et al.*, 1982), is inactive in this respect, indicating that the above effect of PMA is specific and is probably mediated by the ability of the latter to interact with C-kinase.

Phorbol myristate acetate exerts its inhibitory activity at a postreceptor site, as evidenced by its antagonism to several stimuli that interact with distinct sets of receptors and by the observation that the binding of fMet-Leu-Phe is unaffected by the phorbol ester. The secretory activity of A23187, on the other hand, is stimulated by PMA as discussed above. Preincubation with PMA also inhibits the ability of the above chemoattractants to raise the level of intracellular calcium. These two sets of results clearly indicate that the site of action of PMA lies between the binding of the stimuli and the rise in calcium; i.e., PMA uncouples the chemotactic receptors from the mobilization of calcium. The characteristics of the PMA-induced reduction of the reactivity of the neutrophils are also suited to providing a molecular mechanism for the previously described heterologous desensitization of these cells in which preincubation with relatively high concentrations of one chemotactic factor inhibits the ability of the cells to respond to another chemotactic factor (Henson *et al.*, 1978; Showell *et al.*, 1979).

The inhibitory properties of PMA are not limited to the neutrophils or to calcium-mobilizing systems. The phorbol ester has recently been shown to antagonize, among other things, the increases of cytosolic calcium induced by antigen in histamine-secreting rat basophil leukemia cells (Sagi-Eisenberg *et al.*, 1985) and by thrombin in human platelets (Zavoico *et al.*, 1985). It can also block the contraction of guinea pig ileum and rat uterus induced by neurotransmitters (Baraban *et al.*, 1985) and uncouple the β receptors from the adenylate cyclase in epidermis cells (Garte and Belman, 1980). Several recent observations indicate that PMA affects one or more steps preceding the activation of the PIP_2-specific phospholipase C. Phorbol myristate acetate has thus been shown to stimulate the phosphorylation of the insulin receptor in cultured hepatoma cells and to block the growth-factor-stimulated tyrosine kinase activity of the insulin receptor without altering its binding characteristics (Takayama *et al.*, 1984).

These results raise the possibility that the inhibitory site of action of the phorbol ester is similarly located in the neutrophils, i.e., that PMA induces the phosphorylation of the chemotactic receptor or, alternatively, that it so affects one or more of the coupling factors, for example, the guanine-nucleotide-binding regulatory proteins (see Section 4). Phorbol myristate acetate has also been found to inhibit the stimulated hydrolysis of the polyphosphoinositides in the platelets (Zavoico *et al.*, 1984) and rat hippocampal sices (Labarca *et al.*, 1984). The inhibition of the generation of IP_3 could thus account, at least in part, for the decreased mobilization of calcium. The determination of the effects of PMA on the stimulation of C-kinase by chemotactic factors is complicated by the potent stimulatory activity of the phorbol ester towards that enzyme. The ability of PMA to modulate either negatively (Jolles *et al.*, 1981; Aloyo *et al.*, 1983) or positively (Halenda *et al.*, 1984; DeChaffoy de Courcelles *et al.*, 1984) the phosphorylation level of inositol phosphatides further illustrates the sensitivity of this pathway to the phorbol ester.

In spite of the inhibitory properties of PMA demonstrated *in vitro*, the physiological relevance of these findings remains to be evaluated in the context of the short half-life of diglycerides generated on cell stimulation. Thus, whether C-kinase is sufficiently activated under physiological conditions to generate the negative feedback signals detected with the use of PMA remains to be determined.

4. *The Role of the Guanine-Nucleotide-Binding Regulatory Proteins in Signal Transduction and Calcium Mobilization in Neutrophils*

One of the main "black boxes" remaining in the understanding of the mechanism of calcium mobilization in neutrophils, as well as in other cells, concerns the nature of the membrane components involved in signal transduction across the plasma membrane. Based on the extensive data related to the role of the guanine-nucleotide-binding regulatory proteins (the "G proteins") in the hormonal modulation of the activity of adenylate cyclase (for a recent review, see Gilman, 1984), recent studies have focused on the possibility that the latter elements may also be involved in signal transmission in calcium-mobilizing systems. This hypothesis is supported, albeit indirectly, by the demonstrated ability of individual G proteins to interact *in vitro* with receptors other than those to which they are physiologically coupled. The stimulation by light-activated rhodopsin of the GTPase of the α subunit of G_i when reconstituted with the β subunit of G_i or of transducin (Kanaho *et al.*, 1984) is one example of the cross reactivity of the nucleotide-binding regulatory proteins. The neutrophil has been at the vanguard of these studies, whose major conclusions, although still at a tentative stages, are nevertheless quite enticing.

Investigations of the possible role of the G proteins are greatly facilitated, at least in their initial stages, by the availability of bacterial toxins known to interact specifically with these proteins. Cholera toxin

ADP-ribosylates the α component of G_s and irreversibly activates it. Pertussis toxin (islet-activating protein), on the other hand, ADP-ribosylates α_i and prevents its dissociation from the β subunit, thereby blocking the effects of hormones that inhibit the activity of adenylate cyclase.

With these toxins, it has recently been shown that the α subunits of both G_s and G_i are present in the neutrophils' membrane (Lad *et al.*, 1984; Okajima and Ui, 1984). As in other systems, G_s is apparently primarily involved in the stimulation of adenylate cyclase (Lad *et al.*, 1984) and is thus not of direct relevance to the mechanism of chemotactic-factor-generated signal transmission. On the other hand, the studies performed in the past year have placed G_i (possibly its α subunit) at one of the earliest and most essential positions in the excitation–response coupling sequence in neutrophils, mast cells, and possibly other cells.

Pertussis toxin inhibits in a dose- and time-dependent fashion the mobilization of calcium in neutrophils stimulated by fMet-Leu-Phe, leukotriene B_4 and C5a (Molski *et al.*, 1984; Okajima and Ui, 1984; Goldman *et al.*, 1984; Shefcyk *et al.*, 1985). As expected from these results, pertussis toxin inhibits several chemotactic-factor-stimulated functional responses of the neutrophils including chemotaxis, degranulation, and superoxide production (Okajima and Ui, 1984; Bokoch and Gilman, 1984; Goldman *et al.*, 1984; Shecfyk *et al.*, 1985; Becker *et al.*, 1985) and interferes with the stimulated polymerization of actin (Shefcyk *et al.*, 1985). The effects of pertussis toxin appear to be noncompetitive in nature, as they are not reversed by concentrations of stimuli as much as 100 times larger than those required for optimal functional responses and are not mediated at the binding-site level (Okajima and Ui, 1984; Becker *et al.*, 1985). In addition, the inhibitory effects of pertussis toxin are not mediated by increases in cAMP (Bokoch and Gilman, 1984; Becker *et al.*, 1985). Cholera toxin, on the other hand, does not inhibit the chemotactic-factor-induced increases in cytoplasmic calcium and only partially and competitively antagonizes the secretory responses to these stimuli (Shefcyk *et al.*, 1985). The selective inhibitory effects displayed by pertussis toxin imply that it is the α subunit of G_i and not that of G_s that mediates the activation of the neutrophils.

Additional information concerning the mode (and site) of action of pertussis toxin is provided by the results of several sets of experiments in which its effects on the biochemical events discussed in the preceding sections have been investigated. Pertussis toxin has been found to inhibit the breakdown of PIP_2 and of PIP and the generation of phosphatidic acid (Shefcyk *et al.*, 1985; Volpi *et al.*, 1985), thus strongly implying that its site of action precedes the stimulated activation of phospholipase C. Pertussis toxin has accordingly been found to leave unaffected the responses of the neutrophils to A23187 and to PMA (Bokoch and Gilman, 1984; Becker *et al.*, 1985). The cytoskeletal response (increased actin association with Triton X-100-insoluble structures) to A23187 is, on the other hand, inhibited by pertussis toxin (Shefcyk *et al.*, 1985). This last set of results suggests that whereas the secretory activity of A23187 can be ascribed directly to its ionophoretic properties, its effects on actin polymerization are mediated by the stimulated generation of a receptor-dependent local mediator such as AGEPC.

Pertussis-toxin-induced inhibition of calcium mobilization can, like that of PMA, be attributed at least in part to a decreased generation of IP_3 (see, however, the above discussion of the lack of dependence of the leukotriene-B_4-induced calcium mobilization on PIP_2 hydrolysis). Two sets of observations indicate that the inhibition of protein kinase C stimulation that would be expected to result from decreased phospholipase C activity is indeed demonstrable: (1) the fMet-Leu-Phe-stimulated phosphorylation of the 45- to 50-kD protein that is apparently a substrate of protein kinase C is severely curtailed by pertussis toxin (Volpi *et al.*, 1985), and (2) the fMet-Leu-Phe-induced activation of the Na^+/H^+ antiport (as judged either from the stimulation of $^{22}Na^+$ influx or from the rise in pH_i), another index of protein kinase C stimulation (Rosoff *et al.*, 1984), is similarly inhibited by pertussis toxin (Volpi *et al.*, 1985).

In addition to the above effects of pertussis toxin on chemotactic-factor-induced phospholipid remodeling, the toxin also inhibits the stimulated liberation of arachidonic acid (Okajima and Ui, 1984; Bokoch and Gilman, 1984). The recent observations of Dawson *et al.* (1983) concerning the potentiating effects of unsaturated-fatty-acid-rich diglycerides on the activity of phospholipases (and phospholipase A_2 in particular) may tie together the effects of pertussis toxin on both phospholipase C and A_2.

However incomplete, the present data demonstrate that the pertussis toxin target(s) play an essential role in signal transduction in the neutrophils. The sensitivity of the mast cells to the toxin (Nakamura and Ui, 1984) indicates that this membrane component may hold a similar role in other calcium-mobilizing systems. The demonstrated ADP-ribosylation of a 41-kD protein by pertussis toxin in both cell types (Nakamura and Ui, 1984; Okajima and Ui, 1984) strongly suggests that the pertussis toxin target is indeed the α subunit of G_i. The conclusion that the toxin interferes with a guanine-nucleotide-binding protein is supported by the recently reported inhibition of the fMet-Leu-Phe-stimulated GTPase (presumably that of the α subunit of G_i) of guinea pig neutrophil membranes (Okajima *et al.*, 1985) and by the potentiating effects of nonhydrolyzable GTP analogues on the calcium-dependent degranulation of mast cells (Gomperts, 1983) and platelets (Haslam and Davidson, 1984).

The results presented in the above section raise the possibility that one of the major physiological roles of the α subunit of G_i is to mediate signal transmission in calcium-mobilizing systems, whereas its β subunit, as previously proposed (Gilman, 1984), is primarily involved in the hormonally induced inhibition of the adenylate cyclase. The guanine-nucleotide-binding regulatory proteins would thus serve as "universal coupling factors" for the generation of the two major second messengers in nonmuscle cells.

5. Conclusions: A Model for Neutrophil Activation

The following model of neutrophil activation is suggested by the data presented in the previous sections (see Fig. 2 for a diagrammatic representation):

1. Stimulus (chemotactic factor) recognition is mediated by distinct sets of membrane-bound receptors, the characteristics and regulation of which have been summarized in several recent reviews (e.g., Snyderman and Pike, 1984; Sklar *et al.*, 1984).
2. Transmembrane signal transduction is mediated in part by the guanine-nucleotide-binding regulatory proteins G_s and G_i. G_s is primarily involved in the modulation of the levels of cAMP, a compound with inhibitory properties towards neutrophil activation, via regulation of adenylate cyclase (exerted possibly at the level of calcium reuptake by the endoplasmic reticulum) (see Feinstein *et al.*, 1983). G_i is intimately involved in the initiation of all reactions presently known to be critical to neutrophil activation. This conclusion is based on the inhibitory effects of pertussis toxin on calcium mobilization, phospholipase C and C-kinase activation, and on the neutrophil's chemotactic, secretory, and oxidative responses to chemotactic factors. It should be pointed out that the nature of the pertussis toxin target remains to be conclusively identified.
3. The postulated chemotactic factor receptor and G_i complex initiates two parallel series of events leading, respectively, to a motile or to secretory/oxidative responses. At low concentrations of stimuli (either on occupation of the high-affinity receptors or on the occupation of a small percentage of the total receptor population), calcium is released, possibly from the plasma membrane,

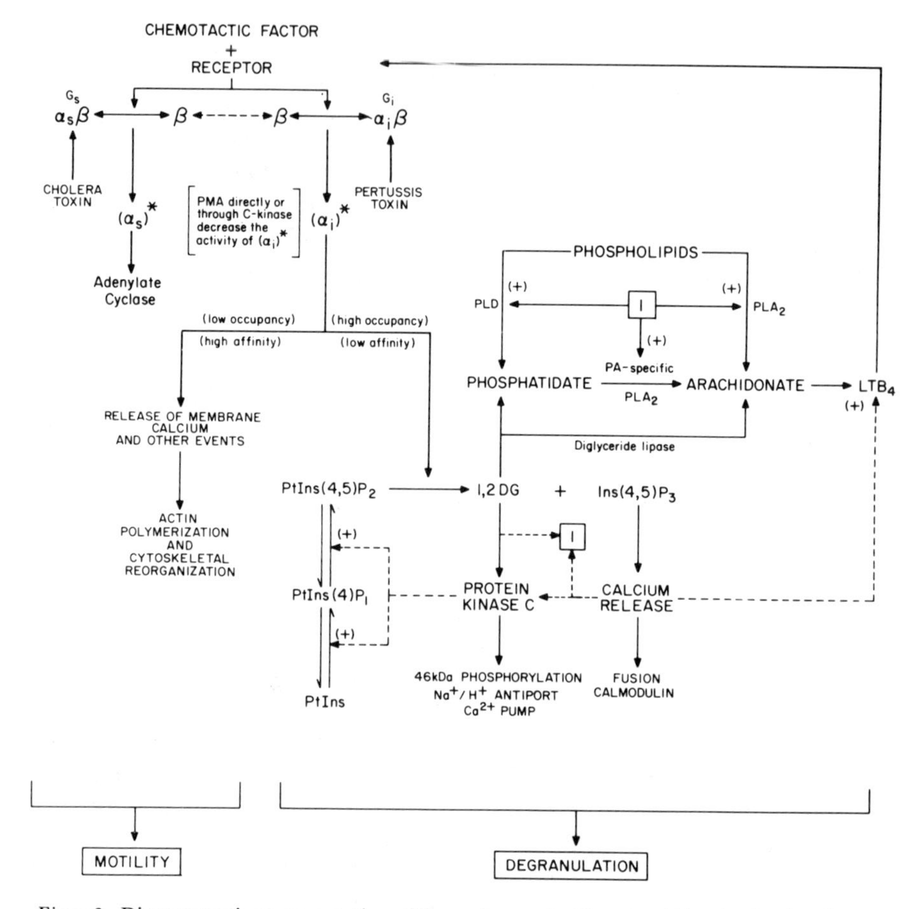

Figure 2. Diagrammatic representation of the early events of neutrophil activation by chemotactic factors. PLA_2 and PLD stand for phospholipases A_2 and D, respectively. PtIns, PtIns(4)P, and PtIns(4,5)P represent phosphatidylinositol and its mono- and bisphosphate forms, respectively.

and actin filaments form and reorganize. Petroski *et al.* (1979) have presented evidence for internal calcium redistribution at concentrations of fMet-Leu-Phe as low as 10^{-12} M, and the data of Torres and Coates (1984) indicate that at least one index of internal calcium release—decreased cell-associated chlortetracycline fluorescence—is preserved essentially intact in the neutrophil's cytoplasts. The latter are anucleated, granule-free cell fragments that would be expected to be comparatively deficient, at the very least, in their endoplasmic reticulum content. The lack of effect of leukotriene B_4 on the hydrolysis of PIP_2 and the dissociation between the fMet-Leu-Phe dose dependencies of the neutrophils' chemotactic and PIP_2 responses further support this parallel model of neutrophil activation. It remains to be established how the redistribution of calcium and the cytoskeletal reorganization are coordinated to support the series of contractile cycles necessary for cell motility, and also which other reactions are involved in this pathway of neutrophil activation. At high concentrations of stimuli (i.e., high occupancy rate/interaction with the low-affinity receptors), the PIP_2 cycle is initiated. The two products of this pathway, DG and IP_3, have been detected following the stimulation of the neutrophils by fMet-Leu-Phe and carry out the same mediator functions ascribed to them in other cells. 1,2-Diacylglycerol serves as an activator of C-kinase, and IP_3 releases calcium from an endoplasmic-reticulum associated pool. The chemotactic factor stimulation of C-kinase is evidenced by the similarity between the pattern of protein phosphorylation (at least in the case of the 46-kD protein) induced by fMet-Leu-Phe and PMA in intact cells and by phosphatidylserine and calcium in broken cells. The activation of C-kinase also underlies the observed stimulation of the Na^+/H^+ antiporter. In the presence of cytochalasin B, fMet-Leu-Phe has been shown to induce the formation of IP_3, and the former releases calcium from permeabilized cells. Working synergistically, PMA (or DG) and IP_3 (or calcium) initiate secretory and oxidative responses from the neutrophils. The mechanisms employed by the neutrophils to selectively control and/or initiate these two functions remain to be investigated.

4. Arachidonic acid is rapidly liberated on the stimulation of the neutrophils by chemotactic factors (its possible sources—phospholipids, diglycerides, or phosphatidic acid—are yet to be defined) and metabolized via the 5-lipoxygenase to generate, among others, leukotriene B_4. Apparently the latter predominantly activates the PIP_2-independent pathway, and its major biological activity towards the neutrophils is accordingly that of a chemoattractant.
5. Several feedback pathways that modulate the main lines of the excitation–response coupling sequence sketched above have been uncovered. C-kinase directly or indirectly stimulates the phosphorylation of PI and its conversion to PIP and PIP_2, thus potentially increasing the availability of the substrate for the PIP_2-specific phospholipase C. By contrast, the sustained activation of C-kinase, as achieved by incubation with PMA, results in a reduced ability of chemotactic factors to stimulate the PIP_2-specific phos-

pholipase C. On the other hand, it has previously been demonstrated that PMA potentiates the chemotactic responses of the neutrophils (Hill, 1978), and preliminary evidence indicates that the stimulation of the incorporation of actin into the neutrophils' cytoskeletons induced by fMet-Leu-Phe is not as sensitive to the phorbol ester as the responses described above (J. Shefcyk, unpublished observations). These two pieces of data, although preliminary, indicate that incubation with PMA interferes predominantly with the receptor–PIP_2 coupling process. The combined presentation of PMA and A23187 to the neutrophils also results in a synergistic effect on the liberation of arachidonic acid (M. Volpi, unpublished observations; see also Table I), thereby raising the possibility that the previously demonstrated functional synergism between these two stimuli may result at least in part from the generation of leukotriene B_4 or other lipid mediators.

The above model of neutrophil activation, however incomplete, incorporates most of the presently available data and, more importantly, suggests several relatively straightforward tests of its validity, the examination of which should allow further refinement.

ACKNOWLEDGMENTS. This work was supported in part by NIH grants AM 31000 and AI 13734.

References

Allan, D., and Michell, R. H., 1977, Calcium-ion-dependent diacylglycerol accumulation in erythrocytes is associated with microvesiculation but not with efflux of potassium ions, *Biochem. J.* **166:**495–499.

Aloyo, V. J., Zwiers, H., and Gispen, W. H., 1983, Phosphorylation of B-50 protein by calcium activated, phospholipid dependent protein kinase and B-50 protein kinase, *J. Neurochem.* **41:**649–653.

Andrews, P. C., and Babior, B. M., 1983, Endogenous protein phosphorylation by resting and activated human neutrophils, *Blood* **61:**333–340.

Andrews, P. C., and Babior, B. M., 1984, Phosphorylation of cytosolic proteins by resting and activated neutrophils, *Blood* **64:**883–890.

Baraban, J. M., Gould, R. J., Peroutka, S. J., and Snyder, S. H., 1985, Phorbol ester effects on neurotransmission: Interaction with neurotransmitters and calcium in smooth muscle, *Proc. Natl. Acad. Sci. U.S.A.* **82:**604–607.

Becker, E. L., Showell, H. J., Naccache, P. H., and Sha'afi R. I., 1978, Enzymes in granulocyte movement; preliminary evidence for the involvement of Na^+/K^+, in: *Leukocyte Chemotaxis* (J. I. Gallin and P. G. Quie, eds.), Raven Press, New York, pp. 113–121.

Becker, E. L., Kermode, J. C., Naccache, P. H., Yassin, R., Marsh, M. L., Munoz, J. J., and Sha'afi, R. I., 1985, The inhibition of neutrophil granule enzyme secretion and chemotaxis by pertussis toxin, *J. Cell Biol.* **100:**1641–1646.

Bell, R. L., Kennedy, D. A., Stanford, N., and Majerus, P. W., 1979, Diglyceride lipase: A pathway for arachidonate release from human platelets, *Proc. Natl. Acad. Sci. U.S.A.* **76:**3238–3241.

Berridge, M. J., 1984, Inositol trisphosphate and diacylglycerol as second messengers, *Biochem. J.* **220:**345–360.

Berridge, M. J., and Irvine, R. F., 1984, Inositol trisphosphate, a novel second messenger in cellular signal transduction, *Nature* **312:**315–321.

Bokoch, G. M., and Gilman, A. G., 1984, Inhibition of receptor-mediated release of arachidonic acid by pertussis toxin, *Cell* **39:**301–308.

Bradford, P. G., and Rubin, R. P., 1985, Characterization of formyl-methionyl-leucyl-phenylalanine stimulation of inositol trisphosphate accumulation in rabbit neutrophils, *Mol. Pharmacol.* **27:**74–78.

Brown, J. E., and Rubin, L. J., 1984, A direct demonstration that inositol trisphosphate induces an increase in intracellular calcium in *Limulus* photoreceptors, *Biochem. Biophys. Res. Commun.* **125:**1137–1142.

Burgess, G. M., McKinney, J. S., Irvin, R. F., Berridge, M. J., Hoyle, P. C., and Putney, J. W., 1984, Inositol 1,4,5-trisphosphate may be a signal for fMet-Leu-Phe induced intracellular Ca mobilization in human leukocytes (HL 60 cells), *FEBS Lett.* **176:**193–196.

Camussi, G., Tetta, C., Bussolino, F., Caligaris Cappio, F., Coda, R., Masera, C. and Segoloni, G., 1981, Mediators of immune-complex induced aggregation of polymorphonuclear neutrophils. II. Platelet-activating factor as the effector substance of immune induced aggregation. *Int. Arch. Allergy Appl. Immunol.* **64:**25–41.

Castagna, M., Takai, Y., Kaibuchi, K., Sano, K., Kikkawa, V., and Nishizuka, Y., 1982, Direct activation of calcium activated phospholipid dependent protein kinase by tumor promoting phorbol esters, *J. Biol. Chem.* **257:**7847–7851.

Cockroft, S., 1982, Phosphatidylinositol metabolism in mast cells and neutrophils, *Cell Calcium* **3:**337–349.

Dale, M. M., and Penfield, A., 1984, Synergism between phorbol ester and A23187 in superoxide production by neutrophils, *FEBS Lett.* **175:**170–172.

Dawson, R. M. C., Hemington, N. L., and Irvine, R. F., 1983, Diacylglycerol potentiates phospholipase attack upon phospholipid bilayers: Possible connection with cell stimulation, *Biochem. Biophys. Res. Commun.* **117:**196–201.

De Chaffoy de Courcelles, D., Roevens, P., and van Belle, H., 1984, 12-O-Tetradecanoylphorbol 13-acetate stimulates inositol lipid phosphorylation in intact human platelets, *FEBS Lett.* **173:**389–393.

De Chatelet, L. R., Shirley, P. S., and Johnston, R. B., Jr., 1976, Effect of phorbol myristate acetate on the oxidative metabolism of human polymorphonuclear leukocytes, *Blood* **47:**545–554.

Demel, R. A., Geurts Van Kessel, W. S. M., Zwale, R. F. A., Roelofsen, B., and Van Deenen, L. L. M., 1975, Relation between various phospholipase actions on human red cell membranes and the interfacial phospholipid pressure in monolayers, *Biochim. Biophys. Acta* **406:**97–107.

DiVirgilio, F., Lew, P. D., and Pozzan, T., 1984, Protein kinase C activation of physiological processes in human neutrophils at vanishingly small cytosolic Ca^{2+} levels, *Nature* **310:**691–693.

Dougherty, R. W., Godfrey, P. P., Hoyle, P. C., Putney, J. W., Jr., and Freer, R. J., 1984, Secretagogue induced phosphoinositide metabolism in human leukocytes, *Biochem. J.* **222:**307–314.

Estensen, R. D., White, J. G., and Holmes, B., 1974, Specific degranulation of human polymorphonuclear leukocytes, *Nature* **248:**347–348.

Feinstein, M. B., Egan, J. J., Sha'afi, R. I., and White, J., 1983, The cytoplasmic concentration of free calcium in platelets is controlled by stimulators of cyclic AMP production (PGD_2, PGE_1, forskolin), *Biochem. Biophys. Res. Commun.* **113:**598–604.

Fujita, I., Irita, K., Takeshige, K., and Minakami, S., 1984, Diaglycerol, 1-oleyl-2-acetylglycerol stimulates superoxide generation from human neutrophils, *Biochem. Biophys. Res. Commun.* **120:**318–324.

Garte, S. J., and Belman, S., 1980, Tumor promoter uncouples beta-adrenergic receptor from adenylate cyclase in mouse epidermis, *Nature* **284:**171–173.

Gay, J. G., Beckman, J. K., Brash, A. R., Oates, J. A., and Lukens, J. N., 1984, Enhancement of chemotactic factor-stimulated neutrophil oxidative metabolism by leukotriene B_4, *Blood* **64:**780–785.

Gilman, A. G., 1984, G proteins and dual control of adenylate cyclase, *Cell* **36:**577–579.

Gilmore, T., and Martin, G. S., 1983, Phorbol ester and diacylglycerol induce protein phosphorylation at tyrosine, *Nature* **306:**487–490.

Goldman, D. W., Gifford, C. A., Bourne, H. R., and Goetzl, E. J., 1984, Pertussis toxin inhibits the activation of human neutrophils by chemotactic factors, *J. Cell Biol.* **99:**278.

Gomperts, B. D., 1983, Involvement of guanine nucleotide binding protein in the gating of Ca^{2+} by receptors, *Nature* **306:**64–66.

Halenda, S. P. and Feinstein, M. B. 1984, Phorbol myristate acetate stimulates formation of phosphatidylinositol 4-phosphate and phosphatidylinositol 4,5-bisphosphate in human platelets. *Biochem. Biophys. Res. Commun.* **124:**507–513.

Haslam, R. J., and Davidson, M. M. L., 1984, Guanine nucleotides decrease the free [Ca^{2+}] required for secretion of serotonin from permeabilized blood platelets. Evidence of a role for a GTP binding protein in platelet activation, *FEBS Lett.* **174:**90–95.

Helfman, D. M., Appelbaum, B. D., Vogler, W. R., and Kuo, J. F., 1983, Phospholipid-sensitive Ca^{2+}-dependent protein kinase and its substrates in human neutrophils, *Biochem. Biophys. Res. Commun.* **111:**847–853.

Henson, P. M., Zanolari, B., Schwartzman, N. A., and Hong, S. R., 1978, Intracellular control of human neutrophil secretion. 1. C5a induced stimulus specific desensitization and the effects of cytochalasin, *Br. J. Immunol.* **121:**851–855.

Hesketh, T. R., Moore, J. P., Morris, J. D. H., Taylor, M. V., Rogers, J., Smith, G. A., and Metcalf, J. C., 1985, A common sequence of calcium and pH signals in the mitogenic stimulation of eukaryotic cells, *Nature* **313:**481–484.

Hill, H. R., 1978, Cyclic nucleotides as modulators of leukocyte chemotaxis, in: *Leukocyte Chemotaxis* (J. I. Gallin and P. G. Quie, eds.), Raven Press, New York, pp. 179–193.

Huang, C.-K., Hill, J. M., Jr., Bormann, B. J., Mackin, W. M., and Becker, E. L., 1983, Endogenous substrates for cyclic-AMP dependent and calcium dependent protein phosphorylation in rabbit peritoneal neutrophils, *Biochim. Biophys. Acta* **760:**126–135.

Irvine, R. F., Letcher, A. J., Lander, D. J., and Downes, C. P., 1984, Inositol trisphosphates in carbachol stimulated rat parotid glands, *Biochem. J.* **223:**237–243.

Jolles, J., Zwiers, H., Dekker, A., Wirtz, K. W. A., and Gispen, W. H., 1981, Corticotropin (1–24)-tetracosapeptide affects protein phosphorylation and polyphosphoinositide metabolism in rat brain, *Biochem. J.* **194:**283–291.

Kaibuchi, K., Takai, Y., and Nishizuka, Y., 1981, Cooperative roles of various membrane phospholipids in the activation of calcium activated phospholipid dependent protein kinase, *J. Biol. Chem.* **256:**7146–7149.

Kajikawa, N., Kaibuchi, K., Matsubara, T., Kikkawa, U., Takai, Y., and Nishizuka, Y., 1983, A possible role of protein kinase C in signal induced lysosomal enzyme release, *Biochem. Biophys. Res. Commun.* **116:**743–750.

Kanaho, Y., Tsai, S.-C., Adamik, R., Hewlett, E. L., Moss, J., and Vaughan, M., 1984,

Rhodopsin enhanced GTPase activity of the inhibitory GTP binding protein of adenylate cyclase, *J. Biol. Chem.* **259:**7378–7381.

Kraft, A. S., and Anderson, W. B., 1983, Phorbol esters increased the amount of Ca^{2+}, phospholipid dependent protein kinase associated with plasma membrane, *Nature* **304:**621–623.

Labarca, R., Janowsky, A., Patel, J., and Paul, S. M., 1984, Phorbol esters inhibit agonist induced ^{3}H-inositol-1-phosphate accumulation in rat hippocampal slices, *Biochem. Biophys. Res. Commun.* **123:**703–709.

Lad, P. M., Glovsky, M. M., Smiley, P. A., Klempner, M., Reisinger, D. M., and Richards, J. H., 1984, The beta-adrenergic receptor in the human neutrophil plasma membrane: Receptor–cyclase uncoupling is associated with amplified GTP activation, *J. Immunol.* **132:**1466–1471.

Lagast, H., Pozzan, T., Waldvogel, F. A., and Lew, P. D., 1984, Phorbol myristate acetate stimulates ATP dependent calcium transport by the plasma membrane of neutrophils, *J. Clin. Invest.* **73:**878–883.

Lew, P. D., Dayer, J.-M., Wollheim, C. B., and Pozzan, T., 1984, Effect of leukotriene B_4, prostaglandin E_2 and arachidonic acid on cytosolic free calcium in human neutrophils, *FEBS Lett.* **166:**44–48.

Macara, I. G., Marinetti, G. V., and Balduzzi, P. C., 1984, Transforming protein of avian sarcoma virus UR2 is associated with phosphatidylinositol kinase activity: Possible role in tumorigenesis, *Proc. Natl. Acad. Sci. U.S.A.* **81:**2728–2732.

McPhail, L. C., Wolfson, M., Clayton, C., and Snyderman, R., 1984, Protein kinase C and neutrophil (PMN) activation: Differential effects of chemoattractants and phorbol myristate acetate (PMA), *Fed. Proc.* **43:**1661.

Molski, T. F. P., Naccache, P. H., Marsh, M. L., Kermode, J., Becker, E. L., and Sha'afi, R. I., 1984, Pertussis toxin inhibits the rise in intracellular concentration of calcium that is induced by chemotactic factors in rabbit neutrophils: Possible role of the "G proteins" in calcium mobilization, *Biochem. Biophys. Res. Commun.* **124:**644–650.

Moolenar, W. H., Tertoolen, L. G. J., and deLaat, S. W., 1984, Phorbol ester and diacylglycerol mimic growth factors in raising cytoplasmic pH, *Nature* **312:**371–374.

Naccache, P. H., Molski, T. F. P., Borgeat, P., and Sha'afi, R. I., 1984, Mechanism of action of leukotriene B_4: Intracellular calcium redistribution in rabbit neutrophils, *J. Cell. Physiol.* **118:**13–18.

Naccache, P. H., Molski, T. F. P., Borgeat, P., and Sha'afi, R. I., 1985, Phorbol esters inhibit the fMet-Leu-Phe and leukotriene B_4 stimulated calcium mobilization and enzyme secretion in rabbit neutrophils, *J. Biol. Chem.* **260:**2125–2131.

Nakamura, T., and Ui, M., 1984, Islet activating protein, pertussis toxin, inhibits Ca^{2+} induced and guanine nucleotide dependent releases of histamine and arachidonic acid from rat mast cells, *FEBS Lett.* **173:**414–418.

Niedel, J. E., Kuhn, L. J., and Vandenbark, G. R., 1983, Phorbol diester copurifies with the protein kinase C, *Proc. Natl. Acad. Sci. U.S.A.* **80:**36–40.

Nishizuka, Y., 1984, The role of protein kinase C in cell surface signal transduction and tumour promotion, *Nature* **308:**693–698.

O'Flaherty, J. T., Schmitt, J. D., McCall, C. E., and Wykle, R. L., 1984, Diacylglycerols enhance human neutrophil degranulation responses: Relevancy to a multiple mediator hypothesis of cell function, *Biochem. Biophys. Res. Commun.* **123:**64–70.

Okajima, F., and Ui, M., 1984, ADP-ribosylation of the specific membrane protein by islet activating protein, pertussis toxin, associated with inhibition of a chemotactic peptide induced arachidonate release in neutrophils. A possible role of the toxin substrate in Ca^{2+} mobilizing biosignaling, *J. Biol. Chem.* **259:**13863–13871.

Okajima, F., Katada, T., and Ui, M., 1985, Coupling of the guanine nucleotide regulatory protein to chemotactic peptide receptors in neutrophil membranes and its uncoupling by islet-activating protein, pertussis toxin. A possible role of the toxin substrate in Ca^{2+} mobilizing receptors mediated signal transduction, *J. Biol. Chem.* **260:**6761–6768.

Oron, Y., Dascal, N., Nadler, E., and Lupu, M., 1985, Inositol 1,4,5-trisphosphate mimics muscarinic response in *Xenopus* oocytes, *Nature* **313:**141–143.

Petroski, R. J., Naccache, P. H., Becker, E. L., and Sha'afi, R. I., 1979, Effect of chemotactic factors on calcium levels of rabbit neutrophils, *Am. J. Physiol.* **237:**C43–C49.

Prentki, M., Wollheim, C. B., and Lew, P. D., 1984, Ca^{2+} homeostasis in permeabilized human neutrophils. Characterization of Ca^{2+} sequestering pools and the action of inositol 1,4,5-trisphosphate, *J. Biol. Chem.* **259:**13777–13782.

Repine, J. E., White, J. G., Clawson, C. C., and Holmes, B. M., 1974, Effects of phorbol myristate acetate on the metabolism and ultrastructure of neutrophils in chronic granulomatous disease, *J. Clin. Invest.* **54:**83–90.

Rittenhouse-Simmons, S., 1979, Production of diglyceride from phosphatidylinositol in activated human platelets, *J. Clin. Invest.* **63:**580–587

Robinson, J. M., Badwey, J. A., Karnovsky, M. L., and Karnovsky, M. J., 1984, Superoxide release by neutrophils: Synergistic effects of a phorbol ester and a calcium ionophore, *Biochem. Biophys. Res. Commun.* **122:**734–739.

Rosoff, P. M., Stein, L. F., and Cantley, L. C., 1984, Phorbol esters induce differentiation of a pre B lymphocyte cell line by enhancing Na^+/H^+ exchange, *J. Biol. Chem.* **259:**7056–7060.

Rozengurt, E., Rodriguez-Pena, A., Coombs, M., and Sinnett-Smith, J., 1984, Diacylglycerol stimulates DNA synthesis and cell division in mouse 3T3 cells: Role of Ca^{2+}-sensitive phospholipid dependent protein kinase, *Proc. Natl. Acad. Sci. U.S.A.* **81:**5748–5752.

Sagi-Eisenberg, R., Lieman, H., and Pecht, I., 1985, Protein kinase C regulation of the receptor-coupled calcium signal in histamine secreting rat basophilic leukemia cells, *Nature* **313:**59–60.

Schneider, C., Zanetti, M., and Romeo, D., 1981, Surface reactive stimuli selectively increase protein phosphorylation in human neutrophils, *FEBS Lett.* **127:**4–8.

Sha'afi, R. I., Naccache, P. H., Molski, T. F. P., Borgeat, P., and Goetzl, E. J., 1981, Cellular regulatory role of leukotriene B_4: Its effects on cation homeostasis in rabbit neutrophils, *J. Cell. Physiol.* **108:**401–408.

Sha'afi, R. I., White, J. R., Molski, T. F. P., Shefcyk, J., Volpi, M., Naccache, P. H., and Feinstein, M. B., 1983, Phorbol 12-myristate 13-acetate activates rabbit neutrophils without an apparent rise in the level of intracellular free calcium, *Biochem. Biophys. Res. Commun.* **114:**638–645.

Shefcyk, J., Yassin, R., Volpi, M., Molski, T. F. P., Naccache, P. H., Munoz, J. J., Becker, E. L., Feinstein, M. D., and Sha'afi, R. I., 1985, Pertussis but not cholera toxin inhibits the stimulated increase in actin association with the cytoskeleton in rabbit neutrophils: Role of the "G proteins" in stimulus response coupling, *Biochem. Biophys. Res. Commun.* **126:**1174–1181.

Schell-Frederick, E. 1984, A comparison of the effects of soluble stimuli on free cytoplasmic and membrane bound calcium in human neutrophils. *Cell Calcium* **5:**237–251.

Showell, H. J., Freer, R. J., Zigmond, S. H., Schiffmann, E., Aswanikumar, S., Corcoran, B., and Becker, E. L., 1976, The structure activity relations of synthetic peptides as chemotactic factors and inducers of lysosomal enzyme secretion for neutrophils, *J. Exp. Med.* **143:**1154–1169.

Showell, H. J., Williams, D., Becker, E. L., Naccache, P. H., and Sha'afi, R. I., 1979,

Desensitization and deactivation of the secretory responsiveness of rabbit neutrophils induced by the chemotactic peptide, formyl-methionyl-leucyl-phenylalanine, *J. Reticuloendothel. Soc.* **25:**139–150.

Showell, H. J., Naccache, P. H., Borgeat, P., Picard, S., Vallerand, P., Becker, E. L., and Sha'afi, R. I., 1982, Characterization of the secretory activity of leukotriene B_4 towards rabbit neutrophils, *J. Immunol.* **128:**811–816.

Shulz, I., 1980, Messenger role of calcium in function of pancreatic acinar cells, *Am. J. Physiol.* **239:**G335–G347.

Sklar, L. A., Jesaitis, A. J., and Painter, R. G., 1984, The neutrophil N-formyl peptide receptor: Dynamics of ligand receptor interactions and their relationship to cellular responses, *Contemp. Top. Immunobiol.* **14:**29–82.

Snyderman, R., and Pike, M. C., 1984, Transductional mechanisms of chemoattractant receptor on leukocytes, *Contemp. Top. Immunobiol.* **14:**1–28.

Southwick, F. S., and Stossel, T. P., 1983, Contractile proteins in leukocyte function, *Semin. Hematol.* **20:**305–321.

Storey, D. J., Shears, S. B., Kirk, C. J., and Michell, R. H., 1984, Stepwise enzymatic dephosphorylation of inositol 1,4,5-trisphosphate to inositol in liver, *Nature* **312:**374–376.

Streb, H., Bayerdorffer, H., Haage, W., Irvine, R. F., and Shulz, I., 1984, Effect of inositol 1,4,5-trisphosphate on isolated subcellular fractions of rat pancreas, *J. Membr. Biol.* **81:**241–253.

Sugimoto, Y., Whitman, M., Cantley, L. C., and Erikson, R. L., 1984, Evidence that the Rous sarcoma virus transforming gene product phosphorylates phosphatidylinositol and diacylglycerol, *Proc. Natl. Acad. Sci. U.S.A.* **81:**2117–2121.

Suzuki, Y., and Lehrer, R. I., 1980, NAD(P)H oxidase activity in human neutrophils stimulated by phorbol myristate acetate, *J. Clin. Invest.* **66:**1409–1418.

Takayama, S., White, M. F., Lauris, V., and Kahn, R. C., 1984, Phorbol esters modulate insulin receptor phosphorylation and insulin action in cultured hepatoma cells, *Proc. Natl. Acad. Sci. U.S.A.* **81:**7797–7801.

Taylor, M. V., Metcalf, J. C., Hesketh, T. R., Smith, G. A., and Moore, J. P., 1984, Mitogen increase phosphorylation of phosphoinositides in thymocytes, *Nature* **312:**462–465.

Torres, M., and Coates, T. D., 1984, Neutrophil cytoplasts: Relationship of superoxide release and calcium pools, *Blood* **64:**891–895.

Volpi, M., Yassin, R., Naccache, P. H., and Sha'afi, R. I., 1983, Chemotactic factor causes rapid decreases in phosphatidylinositol 4,5-bisphosphate and phosphatidylinositol 4-monophosphate in rabbit neutrophils, *Biochem. Biophys. Res. Commun.* **112:**957–964.

Volpi, M., Yassin, R., Tao, W., Molski, T. F. P., Naccache, P. H., and Sha'afi, R. I., 1984, Leukotriene B_4 mobilizes calcium without the breakdown of polyphosphoinositides and the production of phosphatidic acid in rabbit neutrophils, *Proc. Natl. Acad. Sci. U.S.A.* **81:**5966–5969.

Volpi, M., Naccache, P. H., Molski, T. F. P., Shefcyk, J., Huang, C.-K., Marsh, M. L., Munoz, J., Becker, E. L., and Sha'afi, R. I., 1985, Pertussis toxin inhibits the formylmethionyl-leucyl-phenylalanine but not the phorbol ester stimulated changes in ion fluxes, protein phosphorylation and phospholipid metabolism in rabbit neutrophils: Role of the "G proteins" in excitation response coupling, *Proc. Natl. Acad. Sci. U.S.A.* **82:**2708–2712.

White, J. R., Naccache, P. H., Molski, T. F. P., Borgeat, P., and Sha'afi, R. I., 1983, Direct demonstration of increased intracellular concentration of free calcium in rabbit and human neutrophils following stimulation by chemotactic factor, *Biochem. Biophys. Res. Commun.* **113:**44–50.

White, J. R., Huang, C.-K., Hill, J. M., Jr., Naccache, P. H., Becker, E. L., and Sha'afi, R. I., 1984, Effect of phorbol 12-myristate 13-acetate and its analogue 4-α on protein phosphorylation, and phorbol 12,13-didecanoate lysosomal enzyme release in rabbit neutrophils, *J. Biol. Chem.* **259**:8605–8611.

Yano, K., Nakashima, S., and Nozawa, Y., 1983a, Coupling of polyphosphoinositide breakdown with calcium efflux in formyl-methionyl-leucyl-phenylalanine-stimulated rabbit neutrophils, *FEBS Lett.* **161**:296–300.

Yano, K., Hattori, H., Imai, A., and Nozawa, Y., 1983b, Modification of positional distribution of fatty acids in phosphatidylinositol of rabbit neutrophils stimulated with formylmethionyl-leucyl-phenylalanine, *Biochim. Biophys. Acta* **752**:137–144.

Zavoico, G. B., Halenda, S. P., Sha'afi, R. I., and Feinstein, M. B., 1985, PMA inhibits thrombin stimulated Ca^{2+} mobilization and phosphatidylinositol 4,5-bisphosphate hydrolysis in human platelets, *Proc. Natl. Acad. Sci. U.S.A.* **82**:3859–3862.

Zwiller, J., Revel, M.-O., and Malviya, A. N., 1984, Protein kinase C catalyzes phosphorylation of guanylate cyclase *in vitro, J. Biol. Chem.* **260**:1350–1353.

IV

INOSITOL LIPIDS AND THE CALCIUM-MEDIATED CELLULAR RESPONSES

10

Agonist-Dependent Phosphoinositide Metabolism

A Bifurcating Signal Pathway

MICHAEL J. BERRIDGE

1. Introduction

The arrival of a chemical signal at the surface of a cell initiates a profound change in cellular activity providing it can gain access to one of the intracellular signal pathways. These signal pathways begin at the cell surface with a specific receptor, which detects the external signal and relays the information to a limited number of transducing mechanisms, which encode the message into various second messengers. A classic example of an intracellular second messenger is cAMP. Recently there has been rapid progress in the identification of two new second messengers, inositol trisphosphate and diacylglycerol, which are related to each other in that they are both derived from a common precursor, which is a unique inositol lipid located within the inner leaflet of the plasma membrane (Berridge, 1984; Berridge and Irvine, 1984; Nishizuka, 1984a,b).

The first indication that an inositol lipid might play a role in signal transduction emerged from studies on the pancreas in which acetylcholine was found to stimulate the incorporation of ^{32}P into phosphatidylinositol (PI) (Hokin and Hokin, 1953). Over the years it became apparent that this alteration in PI metabolism was a common feature of the action of many hormones, neurotransmitters, and growth factors (Michell, 1975; Berridge, 1981). The lipid-labeling phenomenon originally described by Hokin and Hokin (1953) turned out to be an indirect effect following a prior

MICHAEL J. BERRIDGE • AFRC Unit, Department of Zoology, University of Cambridge, Cambridge CB2 3EJ, England.

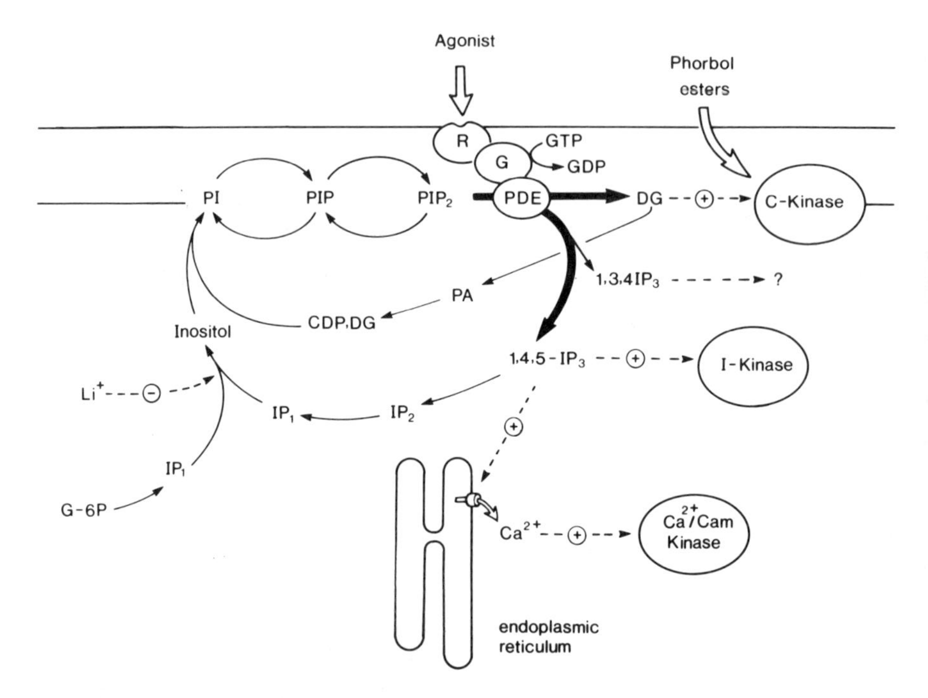

Figure 1. Summary of agonist-dependent inositol lipid hydrolysis as part of a bifurcating signal pathway. Occupation of a receptor (R) appears to activate a G protein (G) responsible for stimulating the phosphodiesterase (PDE) that cleaves PIP_2 to diacylglycerol (DG) and inositol-1,4,5-trisphosphate (1,4,5,-IP_3). While DG acts to stimulate C-kinase, 1,4,5-IP_3 can either mobilize calcium or stimulate an I-kinase.

agonist-dependent hydrolysis of the inositol lipids. For a long time, the assumption was that the substrate used by the receptor mechanism was PI, but recent evidence favors the view that the primary substrate used by the receptor is one of the quantitatively minor lipids phosphatidylinositol-4,5-bisphosphate (PIP_2) (Downes and Michell, 1982; Berridge, 1984; Berridge and Irvine, 1984).

The key event in any receptor mechanism is the transduction event, whereby an external signal is transduced into an internal signal that can be recognized by effector systems within the cell. For those receptors that use inositol lipids, signal transduction depends on the hydrolysis of PIP_2 to diacylglycerol (DG) and inositol trisphosphate (IP_3), both of which function as second messengers by stimulating a family of protein kinases (Fig. 1). The neutral DG stimulates C-kinase (Nishizuka, 1983, 1984a,b), whereas IP_3 acts indirectly by mobilizing intracellular calcium (Berridge, 1984; Berridge and Irvine, 1984), which then stimulates a Ca^{2+}-calmodulin-dependent protein kinase. There are now indications that IP_3 might also act directly by stimulating an I-kinase (Whitman *et al.*, 1984).

There are four major components to this signal pathway:

1. The biochemical pathways responsible for providing the receptor mechanism with PIP_2.
2. The coupling mechanism whereby surface receptors stimulate the PIP_2 phosphodiesterase.
3. The enzymatic mechanisms that regulate the levels of the two second messengers.
4. The mechanisms employed by the second messengers to bring about a change in cellular activity. This final step is complicated by the two second messengers acting synergistically with each other.

These four aspects are now considered, paying particular attention to recent developments and to some of the outstanding questions that remain to be answered.

2. *Formation of PIP_2*

Phosphatidylinositol (PI) is unique among membrane phospholipids in that it can be further phosphorylated to phosphatidylinositol-4-phosphate (PIP) and PIP_2 (Fig. 1). Since there are phosphomonoesterases that remove these additional phosphates, thus converting these polyphosphinositides back to the parent molecule (PI), these three inositol lipids are linked together through two futile cycles (Fig. 1). The turnover of these two linked cycles will determine the rate at which PIP_2 can be supplied to the receptor mechanism. Within seconds of stimulation, the demand for PIP_2 will fluctuate enormously, which means that the PIP_2 synthetic pathway must be sufficiently flexible to supply PIP_2 at widely varying rates.

There are numerous hints from labeling experiments that formation of PIP_2 may be enhanced during cell stimulation (Agranoff *et al.*, 1983; Billah and Lapetina, 1982a). There are now indications that this increase in PIP_2 formation might be regulated through C-kinase (DeChaffoy *et al.*, 1984; Taylor *et al.*, 1984). One of the products of the receptor mechanism thus forms a positive feedback loop, thereby enhancing the supply of PIP_2. It is not yet clear whether C-kinase acts to stimulate the kinases that convert PI to PIP_2 or whether it inhibits the phosphomonoesterases that dephosphorylate PIP_2 back to PI (Fig. 1). Nevertheless, the observation that the C-kinase pathway can enhance the supply of PIP_2, thereby increasing the formation of IP_3 in response to an agonist (Taylor *et al.*, 1984), indicates that the operation of these two linked futile cycles is extremely important with regard to the efficiency of the transduction mechanism. It suggests that the rate-limiting step for second messenger formation depends not only on the operation of the agonist-dependent transduction mechanism (discussed in Section 3) but also on the rate of supply of PIP_2.

The realization that the inositol lipid futile cycles may contribute to the efficiency of the transduction mechanism raises the intriguing possibility that the supply of PIP_2 might be regulated through other mechanisms in addition to the positive feedback loop operated by C-kinase. A particularly interesting possibility is that ACTH (Jolles *et al.*, 1980, 1981; Van Dongen *et al.*, 1984) and dopamine (Jork *et al.*, 1984) may function in the brain by regulating the supply of PIP_2. These neuromodulators act on a protein kinase that regulates the state of phosphorylation of a B50 protein that, in turn, regulates the activity of PIP kinase, which converts PIP to PIP_2. The end result of this phosphorylation cascade is that the neuromodulators enhance the supply of PIP_2 and thus could sensitize those receptors that use the inositol lipids to generate second messengers. A possible link with the mechanisms discussed earlier lies in the suggestion that the protein kinase that regulates the phosphorylation state of the putative regulator of PIP kinase might be C-kinase (Aloyo *et al.*, 1983).

So far we have considered the futile cycles as a closed system, but in reality PIP_2 will be hydrolyzed and hence withdrawn by the PIP_2 phosphodiesterase. Such loss of lipid is replaced by the synthesis of PI, which is obtained either by recycling the products of hydrolysis or by *de novo* synthesis. This rather simplistic interpretation of PI metabolism does not take into account experimental evidence that only a small proportion of the total cellular content of PI is immediately accessible to the receptor mechanism (Fain and Berridge, 1979; Monaco, 1982; Billah and Lapetina, 1982b). A small (10–20%) hormone-sensitive pool, which may be localized within the plasma membrane, is responsible for supplying the PIP_2 used by the receptor mechanism. The size of this pool will depend on how the rates of PIP_2 removal and replacement are balanced. The size of this pool may determine how effectively PIP_2 can be supplied to the receptor mechanism, whose excessive demands can rapidly reduce the cellular level of PI. For example, in blood platelets 30–35% of cellular PI is lost within the first minute of stimulation with thrombin (Broekman *et al.*, 1980; Rittenhouse-Simmons, 1979). Similarly, approximately 10% of liver PI is lost within minutes of stimulation with vasopressin (Kirk *et al.*, 1981). If one assumes that all this PI must be channeled through a small hormone-sensitive pool, it follows that during intense stimulation there may be a marked decrease in pool size, which might account for the biphasic appearance of DG and IP_3 in certain cells.

The only way the size of the pool can remain constant is for the lipid backbone to be rapidly recycled back to PI. An additional input of PI by *de novo* synthesis would function to enlarge this hormone-sensitive pool and could lead to enhanced efficiency of receptor function. It is of some interest, therefore, to find that some hormones such as insulin appear to stimulate *de novo* synthesis of PI (Farese, 1985).

The resynthesis of PI, whether by recycling or *de novo* synthesis, depends on a ready supply of free *myo*-inositol (Fig. 1), and there are conditions under which this supply might become limiting. The formation of *myo*-inositol through *de novo* synthesis from glucose-6-phosphate or from the inositol phosphates released by the receptor depends on a *myo*-inositol 1-phosphatase, which is inhibited by lithium (Hallcher and Sherman, 1980). Stimulation of cells in the presence of lithium results in an accumulation of IP_1 and a decline in the level of free inositol (Berridge *et al.*, 1982). When GH_3 cells are stimulated chronically in the presence of lithium, there is a 50% reduction in the total cellular level of PI (Drummond and Raeburn, 1984). Of particular interest was the finding that despite this drastic reduction in the level of PI, the PIP_2 level remained constant, suggesting that the level of this key substrate, perhaps that located within the hormone-sensitive pool, was kept constant by drawing on PI in the hormone-insensitive pool. It would be interesting to know what happens to PIP_2 levels when the cellular level of PI falls below 50%. At some stage the level of PIP_2 must decline, resulting in receptor desensitization as a result of substrate starvation. It has been proposed that such lithium-mediated substrate starvation might explain the therapeutic action of lithium in controlling manic–depressive illness (Berridge *et al.*, 1982).

We need to know more about the hormone-sensitive pool and the way in which turnover within the futile cycles is adapted to satisfy the varying demands of the receptor mechanism for the PIP_2 required to generate second messengers.

3. *Receptor Coupling to PIP_2-Specific Phosphodiesterase: Possible Role of a G Protein*

The fundamental event in signal transduction concerns the way in which external receptors act to stimulate the PIP_2 phosphodiesterase. The first indication that a G protein might play a role in coupling receptors to IP_3 and DG formation was the finding that GTP inhibited the binding of calcium-mobilizing agonists to their receptors (Goodhardt *et al.,* 1982; Koo *et al.,* 1983; Snyderman and Pike, 1984). Another clue was provided by Gomperts (1983), who showed that the sensitivity of mast cells to the stimulatory effect of calcium was heightened if they had taken up nonhydrolyzable GTP analogues such as GTPγS or GppNHp. When Haslam and Davidson (1984a) found that the latter were also able to increase the calcium sensitivity of permeabilized blood platelets, they proposed that these guanine nucleotides were acting to stimulate phosphoinositide breakdown, since they already knew that a similar sensitization was achieved when thrombin stimulated the formation of diacylglycerol (Haslam and Davidson, 1984b). Confirmation that a G protein might be involved in coupling receptors to the breakdown of inositol lipids has followed quickly. The nonhydrolyzable analogue GTPγS was found to stimulate the formation of diacylglycerol in permeabilized blood platelets (Haslam and Davidson, 1984c). Likewise, GTP (10 μM) stimulates the breakdown of PIP_2 in neutrophil membranes (Smith *et al.,* 1985), whereas GppNHp and GTPγS increase the formation of inositol phosphates in blowfly salivary gland homogenates (Litosch *et al.,* 1985).

Further evidence for a G-protein role in inositol lipid breakdown has been gathered by studying the effect of islet-activating protein (IAP), which is one of the toxins produced by *Bordetella pertussis*. The action of the chemotactic factor fMet-Leu-Phe on various processes in neutrophils, an effect that depends on the hydrolysis of inositol lipids (Volpi *et al.,* 1983; Yano *et al.,* 1983), is blocked by IAP (Bokach and Gilman, 1984; Molski *et al.,* 1984; Okajima and Ui, 1984; Volpi *et al.,* 1984). However, the ability of the calcium ionophore A 23187 to stimulate many of these processes was not blocked by IAP, suggesting that the toxin was interfering with some receptor-mediated event. This event is likely to be inositol lipid hydrolysis, since IAP was found to inhibit the fMet-Leu-Phe-stimulated decrease of PIP and PIP_2 in neutrophils (Volpi *et al.,* 1985). Using the Gompert technique (1983) of introducing guanine nucleotides into intact mast cells, Nakamura and Ui (1984) found that IAP inhibited subsequent calcium-induced histamine release. Since these inhibitory effects of IAP are associated with the ADP-ribosylation of a 41,000-dalton protein similar to the α subunit of the inhibitory G protein N_1, it has been proposed that a G protein may be involved in signal transduction at calcium-mobilizing receptors (Bokoch and Gilman, 1984; Molski *et al.,* 1984; Okajima and Ui, 1984; Volpi *et al.,* 1984).

It remains to be seen whether this coupling is mediated by the same inhibitory G protein N_1 already known to function in inhibiting adenylate cyclase. Since the activation of α_2 receptors, which use N_1 to inhibit adenylate cyclase, can be distinguished from the α_1-receptor-mediated hydrolysis of inositol lipids, it appears that there is a separate G protein responsible for the latter. Taken together with the stimulatory effects of guanine nucleotides discussed earlier, these inhibitory effects of IAP are consistent with the idea that a G protein transmits the stimulatory signal from receptors to the phosphodiesterase that cleaves PIP_2 to IP_3 and DG (Fig. 1).

4. Inositol Trisphosphate and DG Formation

The levels of IP_3 and DG are determined by the balance between their rates of formation by a single enzymatic step, the PIP_2 phosphodiesterase, and their rates of removal by separate enzymatic pathways (Fig. 1). An inositol trisphosphatase hydrolyzes IP_3 to IP_2, whereas DG is either phosphorylated to phosphatidic acid (PA) or deacylated to monoacylglycerol. Agonists alter the balance by stimulating the phosphodiesterase, perhaps through a G protein as discussed earlier, to increase the rate of formation of IP_3 and DG. The properties of the PIP_2 phosphodiesterase, particularly its sensitivity to calcium, have been a matter of considerable controversy (Michell *et al.*, 1981). However, a general consensus has emerged that it is a calcium-requiring enzyme that is not activated by changes in calcium concentration within the normal physiological range of 10^{-7} to 10^{-5} M (Irvine *et al.*, 1984a). Studies on intact platelets (Rittenhouse and Horne, 1984; Simon *et al.*, 1984) and basophils (Beaven *et al.* 1984) provide further evidence that an increase in intracellular calcium *per se* is not able to activate the breakdown of inositol lipids.

Another important property of the phosphodiesterase concerns its inositol lipid specificity. When studied *in vitro*, the enzyme appears to prefer the polyphosphoinositides over PI, but an unresolved question concerns whether or not it can discriminate between PIP and PIP_2. This is an interesting question, since the hydrolysis of either substrate will give DG, but only PIP_2 will give the calcium-mobilizing second messenger IP_3. In GH_3 cells, TRH results in the formation of both IP_3 and IP_2, but A23187 gives only IP_2, suggesting that high calcium concentrations stimulate the hydrolysis of PIP but not of PIP_2 (Kolesnick and Gershengorn, 1984).

A further complication concerning IP_3 formation is the existence of two separate IP_3 isomers (Irvine *et al.*, 1984b). In addition to the previously described inositol-1,4,5-trisphosphate, Irvine *et al.* (1984b) have described an inositol-1,3,4-trisphosphate that accumulates in large quantities in parotid gland. The source of this new isomer is unknown, but it presumably originates either from a corresponding phosphatidylinositol-

3,4-bisphosphate or by isomerization of 1,4,5-IP_3. There is also no information concerning the function of this new isomer, but the fact that its turnover is much slower than that of the 1,4,5 isomer raises some interesting possibilities concerning a role in longer-term adaptive changes.

The action of the 1,4,5 isomer is terminated by an inositol trisphosphatase (Downes *et al.*, 1982; Seyfred *et al.*, 1984), which specifically removes the phosphate from the 5 position to give inositol-1,4-bisphosphate (IP_2) and is relatively inactive in hydrolyzing the 1,3,4 isomer (Irvine *et al.*, 1984b). The activity of this enzyme requires magnesium and is inhibited by 2,3-phosphoglyceric acid (Downes *et al.*, 1982) and by spermine (Seyfred *et al.*, 1984). The latter observation is particularly interesting because spermine has been implicated in calcium mobilization (Koenig *et al.*, 1983). Instead of releasing calcium through a cation-exchange reaction as proposed by Koenig *et al.* (1983), spermine could act indirectly to inhibit the trisphosphatase, leading to an accumulation of IP_3.

Just how the 1,3,4 isomer is metabolized within the cell is unknown, but it does appear that the enzyme that degrades it is sensitive to lithium. This means that the IP_3 that accumulates in cells when stimulated in the presence of lithium is likely to be the 1,3,4 isomer rather than the 1,4,5 isomer. A preferential effect of lithium on the 1,3,4 isomer may explain why lithium could not enhance the effect of vasopressin on phosphorylase activation even though it resulted in a potentiation of IP_3 levels (Thomas *et al.*, 1984). The ability of high levels of lithium to specifically amplify the level of 1,3,4-IP_3 is particularly interesting with regard to the hypothesis that it might have long-term effects, because lithium is known to stimulate cell growth in a number of cell types (Gelfand *et al.*, 1979; Rybak and Stockdale, 1981; Tomooka *et al.*, 1983).

The other second messenger, DG, is rapidly phosphorylated to phosphatidic acid by a DG kinase, or it can function as a substrate for a DG lipase, which removes one of the fatty acid chains. Despite the importance of DG as an intracellular signal, not much is known about how these degradative pathways are controlled. In order for DG to act, it must form a quarternary complex in the plasma membrane with calcium, phosphatidylserine, and C-kinase. There is little information on the stability of this active complex, which raises an interesting question concerning the turnover of free versus bound DG and whether the latter is also susceptible to degradative enzymes.

5. *Second Messenger Mode of Action*

The final aspect of the bifurcating signal pathway concerns the way in which the two second messengers function to regulate various cellular processes. The neutral DG acts within the plane of the membrane to stimulate C-kinase, which phosphorylates a whole host of proteins (Nishizuka, 1984a,b). One function of the water-soluble IP_3 is to enter the cytosol, where it acts on the endoplasmic reticulum to mobilize calcium (Berridge, 1984; Berridge and Irvine, 1984), which then stimulates a Ca^{2+}-calmodulin protein kinase to phosphorylate a separate set of proteins. In addition, 1,4,5-IP_3 may act directly through an I-kinase to phosphorylate a membrane protein (Whitman *et al.*, 1984).

The task before us is to identify the protein substrates being phosphorylated and to determine how they contribute to the final response. Studies on blood platelets indicate that these signal pathways may interact with each other in a synergistic manner (Kaibuchi *et al.* 1983). The experimental basis for demonstrating such synergism depends on having pharmacological tools for stimulating each pathway independently of the other. An increase in intracellular calcium can be induced by using calcium ionophores such as A23187 or ionomycin, whereas C-kinase can be activated independently of calcium by adding 1-oleoyl-2-acetylglycerol (OAG) or a phorbol ester. In the first experiments on blood platelets, a maximal increase in platelet secretion was obtained by combining a calcium ionophore and an activator of C-kinase at concentrations at which neither was capable of stimulating secretion when added alone (Kaibuchi *et al.* 1983). By activating the two separate pathways, the ionophores and the activators of C-kinase were capable of mimicking the action of thrombin, thus supporting the notion that both arms of the bifurcating signal pathway contribute to platelet activation. A combination of ionophores and activators of C-kinase has now been found to mimic the action of many different agonists on a wide range of cellular processes (Table I). It appears, therefore, that the two second messengers IP_3 and DG function synergistically with each other to control many different cellular processes.

In order for synergism to occur, the two pathways must interact with each other at some stage during their course of action. Somehow the stimulation of one pathway must facilitate the action of the other. Some evidence for this has already come from studies on permeabilized adrenal cells (Knight and Baker, 1983) and blood platelets (Haslam and Davidson, 1984b), where the calcium activation curve for secretion is shifted to the left by treatments that activate C-kinase. Somehow the C-kinase pathway makes the exocytotic pathway more sensitive to the stimulatory effect of calcium. This phenomenon, which has been called "positive sensitivity modulation" (Rasmussen, 1981), might explain how OAG or phorbol esters can activate secretion in blood platelets without increasing the intracellular level of calcium (Rink *et al.*, 1983). Such a mechanism would

Table I. Summary of Cellular Processes That Can Be Stimulated by Combining the Actions of a Calcium Ionophore and an Activator of C-Kinase Such as 1-Oleoyl-2-acetylglycerol (OAG) or a Phorbol Ester

Tissue	Response	Reference
Blood platelet	5-HT secretion	Kaibuchi *et al.* (1983)
Mast cells	Histamine secretion	Nishizuka (1984a)
Neutrophils	Superoxide release	Robinson *et al.* (1984)
Lymphocytes	DNA synthesis	Mastro and Smith (1983)
Adrenal	Aldosterone secretion	Kojima *et al.* (1983)
Islet cells	Insulin secretion	Zawalich *et al.* (1983)
Liver	Glycogenolysis	Fain *et al.* (1984)
Pancreas	Amylase secretion	de Pont and Fleuren-Jakobs (1984)
Smooth muscle	Contraction	Rasmussen *et al.* (1984)
Parotid	Protein secretion	Putney *et al.* (1984)
Parasympathetic nerve	Acetylcholine secretion	Tanaka *et al.* (1984)
Pituitary tumor cell	Prolactin secretion	Delbeke *et al.* (1984)

be analogous to the control of phosphorylase kinase in which cAMP is able to sensitize the enzyme such that it can be activated by resting levels of calcium. Sensitivity modulation may represent one way whereby the two signal pathways interact, but other mechanisms may exist, and we need to find out more about the mode of action of each second messenger.

One way of trying to track down how second messengers act is to identify the protein substrates that are phosphorylated during a response. In the case of blood platelets, the C-kinase pathway phosphorylates a 40-kD protein, whereas a 20-kD protein is the substrate for the calcium pathway, but the way in which these proteins contribute to the final response has not been determined. Likewise, activation of liver by vasopressin results in the phosphorylation of ten cytosolic proteins, three by C-kinase and seven by the calcium pathway (Garrison *et al.*, 1984), but how all these phosphorylation events are integrated into a change in metabolic rate is unknown.

A particularly challenging problem is to unravel the apparent contribution of the bifurcating signal pathway in the control of cell growth. In the case of 3T3 cells, the action of certain growth factors results in the hydrolysis of PIP_2 to form DG (Habenicht *et al.*, 1981) and IP_3 (Berridge *et al.*, 1984). These two signal pathways may then play an important role in regulating the two major ionic events (increases in intracellular calcium and pH) that appear to be linked to the onset of DNA synthesis (Berridge, 1984; Berridge and Irvine, 1984). Further support for the participation of this bifurcating pathway in cell proliferation is provided by studies on lymphocytes, where DNA synthesis can be induced through the combined action of a calcium ionophore and a phorbol ester (Table I). During the action of growth factors, there are increases in the intracellular level of calcium (Moolenaar *et al.*, 1984; Hesketh *et al.*, 1985). Since these calcium signals are only slightly diminished by removing external calcium, much of the signal is derived by mobilizing calcium from intracellular stores, which is consistent with the observation that IP_3 can mobilize calcium from permeabilized Swiss 3T3 cells (Berridge *et al.*, 1984; Irvine *et al.*, 1984c). One limb of the signal pathway is thus concerned with elevating intracellular calcium, i.e., one of the important ionic signals for controlling growth.

The other major ionic signal for cell proliferation, an increase in intracellular pH, may be regulated by the other limb of the signal pathway (Berridge, 1984). The first clue that this might be the case was the observation that phorbol esters stimulated sodium entry into 3T3 cells (Burns and Rozengurt, 1983). Subsequent studies have provided further evidence that C-kinase may function to stimulate the Na^+/H^+ exchanger, which extrudes protons in exchange for sodium (Besterman and Cuatrecasas, 1984; Rosoff *et al.*, 1984; Whiteley *et al.*, 1984). The subsequent alkalinization of the cytosol seems to be an important stimulus for the onset of DNA synthesis. Interest in this pathway is likely to grow because of the observation that phorbol esters result in the rapid appearance of transcripts for the *fos* and *myc* oncogene (Greenberg and Ziff, 1984; Kruijer *et al.*, 1984). A tenuous link is beginning to emerge between events at the cell surface and the early transcription of specific genes. Further studies will have to concentrate on establishing the precise link between second messengers such as DG and calcium and the subsequent onset of DNA synthesis.

References

Agranoff, B. W., Murthy, P., and Seguin, E. B., 1983, Thrombin-induced phosphodiesteratic cleavage of phosphatidylinositol bisphosphate in human platelets, *J. Biol. Chem.* **258:**2076–2078.

Aloyo, V. J., Zwiers, H., and Gispen, W. H., 1983, Phosphorylation of B-50 protein by calcium-activated, phospholipid-dependent protein kinase and B-50 protein kinase, *J. Neurochem.* **41:**649–653.

Beaven, M. A., Moore, J. P., Smith, G. A., Hesketh, T. R., and Metcalfe, J. C., 1984, The calcium signal and phosphatidylinositol breakdown in 2H3 cells, *J. Biol. Chem.* **259:**7137–7142.

Berridge, M. J., 1981, Phosphatidylinositol hydrolysis: A multifunctional transducing mechanism, *Mol. Cell. Endocrinol.* **24:**115–140.

Berridge, M. J., 1984, Inositol trisphosphate and diacylglycerol as second messengers, *Biochem. J.* **220**:345–360.

Berridge, M. J., and Irvine, R. F., 1984, Inositol trisphosphate, a novel second messenger in cellular signal transduction, *Nature* **312**:315–321.

Berridge, M. J., Downes, M. J., and Hanley, M. R., 1982, Lithium amplifies agonist-dependent phosphatidylinositol responses in brain and salivary glands, *Biochem. J.* **206**:587–595.

Berridge, M. J., Heslop, J. P., Irvine, R. F., and Brown, K. D., 1984, Inositol trisphosphate formation and calcium mobilization in Swiss 3T3 cells in response to platelet-derived growth factor, *Biochem. J.* **222**:195–201.

Besterman, J. M., and Cuatrecasas, P., 1984, Phorbol esters rapidly stimulate amiloride-sensitive $Na^{+}H^{+}$ exchange in a human leukemic cell line, *J. Cell Biol.* **99**:340–343.

Billah, M. M., and Lapetina, E. G., 1982a, Rapid decrease of phosphatidylinositol 4,5-biphosphate in thrombin-stimulated platelets, *J. Biol. Chem.* **257**:12705–12708.

Billah, M. M., and Lapetina, E. G., 1982b, Evidence for multiple metabolic pools of phosphatidylinositol in stimulated platelets, *J. Biol. Chem.* **257**:11856–11859.

Bokoch, G. M., and Gilman, A. G., 1984, Inhibition of receptor-mediated release of arachidonic acid by pertussis toxin, *Cell* **39**:301–308.

Broekman, M. J., Ward, J. W., and Marcus, A. J., 1980, Phospholipid metabolism in stimulated human platelets: Changes in phosphatidylinositol, phosphatidic acid, and lysophospholipids, *J. Clin. Invest.* **66**:275–283.

Burns, C. P., and Rozengurt, E., 1983, Serum, platelet-derived growth factor, vasopressin and phorbol esters increase intracellular pH in Swiss 3T3 cells, *Biochem. Biophys. Res. Commun.* **116**:931–938.

DeChaffoy de Courcelles, D., Roevens, P., and Van Belle, H., 1984, 12-O-Tetradecanoylphorbol 13-acetate stimulates inositol lipid phosphorylation in intact human platelets, *FEBS Lett.* **173**:389–393.

Delbeke, D., Kojima, I. Dannies, P. S., and Rasmussen, H., 1984, Synergistic stimulation of prolactin release by phorbol ester, A23187 and forskolin, *Biochem. Biophys. Res. Commun.* **123**:735–741.

de Pont, J. J. H. H. M., and Fleuren-Jacobs, A. M. M., 1984, Synergistic effect of A23187 and a phorbol ester on amylase secretion from rabbit pancreatic acini, *FEBS Lett.* **170**:64–68.

Downes, C. P., and Michell, R. H., 1982, Phosphatidylinositol 4-phosphate and phosphatidylinositol 4,5-bisphosphate: Lipids in search of a function, *Cell Calcium* **3**:467–502.

Downes, C. P. Mussat, M. C., and Michell, R. H., 1982, The inositol trisphosphate phosphomonoesterase of the human erythrocyte membrane, *Biochem. J.* **203**:169–177.

Drummond, A. H., and Raeburn, C. A., 1984, The interaction of lithium with thyrotropin-releasing hormone-stimulated lipid metabolism in GH_3 pituitary tumour cells, *Biochem. J.* **223**:129–136.

Fain, J. N., and Berridge, M. J., 1979, Relationship between phosphatidylinositol synthesis and recovery of 5-hydroxytryptamine-responsive Ca^{2+} flux in blowfly salivary gland, *Biochem. J.* **180**:655–661.

Fain, J. N., Li, S.-Y., Litosch, I., and Wallace, M., 1984, Synergistic activation of rat hepatocyte glycogen phosphorylase by A23187 and phorbol ester, *Biochem. Biophys. Res. Commun.* **119**:88–94.

Garrison, J. C., Johnsen, D. E., and Campanile, C. P., 1984, Evidence for the role of phosphorylase kinase, protein kinase C, and other Ca^{2+} sensitive protein kinases in the response of hepatocytes to angiotensin II and vasopressin, *J. Biol. Chem.* **259**:3283–3292.

Gelfand, E. W., Dosh, H.-M., Hastings, D., and Shore, A., 1979, Lithium: A modulator of cyclic AMP-dependent events in lymphocytes, *Science* **203**:365–367.

Gomperts, B. D., 1983, Involvement of guanine nucleotide-binding protein in the gating of Ca^{2+} by receptors, *Nature* **306**:64–66.

Goodhardt, M., Ferry, N., Geynet, P., and Hanoune, J., 1982, Hepatic α_1-adrenergic receptors show agonist regulation by guanine nucleotides, *J. Biol. Chem.* **257**:11577–11583.

Greenberg, M. E., and Ziff, E. B., 1984, Stimulation of 3T3 cells induces transcription of the c-*fos* proto-oncogene, *Nature* **311**:433–438.

Habenicht, A. J. R., Glomset, J. A., King, W. C., Nist, C., Mitchell, C. D., and Ross, R., 1981, Early changes in phosphatidylinositol and arachidonic acid metabolism in quiescent Swiss 3T3 cells stimulated to divide by platelet-derived growth factor, *J. Biol. Chem.* **256**:12329–12335.

Hallcher, L. M., and Sherman, W. R., 1980, The effects of lithium ion and other agents on the activity of *myo*-inositol-1-phosphatase from bovine brain, *J. Biol. Chem.* **255**:10896–10901.

Haslam, R. J., and Davidson, M. M. L., 1984a, Guanine nucleotides decrease the free [Ca^{2+}] required for secretion of serotonin from permeabilized blood platelets, *FEBS Lett.* **174**:90–95.

Haslam, R. J., and Davidson, M. M. L., 1984b, Potentiation by thrombin of the secretion of serotonin from permeabilized platelets equilibrated with Ca^{2+} buffers, *Biochem. J.* **222**:351–361.

Haslam, R. J., and Davidson, M. M. L., 1984c, Receptor-induced diacylglycerol formation in permeabilized platelets: Possible role for a GTP-binding protein, *J. Receptor Res.* **4**:605–629.

Hesketh, T. R., Moore, J. P., Morris, J. D. H., Taylor, M. V., Rogers, J., Smith, G. A., and Metcalfe, J. C., 1985, A common sequence of calcium and pH signals in the mitogenic stimulation of eukaryotic cells, *Nature* **313**:481–484.

Hokin, M. R., and Hokin, L. E., 1953, Enzyme secretion and the incorporation of p^{32} into phospholipids of pancreas slices, *J. Biol. Chem.* **203**:967–977.

Irvine, R. F., Letcher, A. J., and Dawson, R. M. C., 1984a, Phosphatidylinositol-4,5-bisphosphate phosphodiesterase and phosphomonoesterase activities of rat brain, *Biochem. J.* **218**:177–185.

Irvine, R. F., Letcher, A. J., Lander, D. J., and Downes, C. P., 1984b, Inositol trisphosphate in carbachol-stimulated rat parotid glands, *Biochem. J.* **223**:237–243.

Irvine, R. F., Brown, K. D., and Berridge, M. J., 1984c, Specificity of inositol trisphosphate-induced calcium release from permeabilized Swiss-mouse 3T3 cells, *Biochem. J.* **222**:269–272.

Jolles, J., Zwiers, H., van Dongen, C. J., Schotman, P., Wirtz, K. W. A., and Gispen, W. H., 1980, Modulation of brain polyphosphoinositide metabolism by ACTH-sensitive protein phosphorylation. *Nature* **286**:623–625.

Jolles, J., Zwiers, H., Dekker, A., Wirtz, K. W. A., and Gispen, W. H., 1981, Corticotropin-(1–24)-tetracosapeptide affects protein phosphorylation and polyphosphoinositide metabolism in rat brain, *Biochem. J.* **194**:283–291.

Jork, R., De Graan, P. W. E., Van Dongen, C. J., Zwiers, H., Matthias, H., and Gispen, W. H., 1984, Dopamine-induced changes in protein phosphorylation and polyphosphoinositide metabolism in rat hippocampus, *Brain Res.* **291**:73–81.

Kaibuchi, K., Takai, Y., Sawamura, M., Hoshijima, M., Fujikura, T., and Nishizuka, Y., 1983, Synergistic functions of protein phosphorylation and calcium mobilization in platelet activation, *J. Biol. Chem.* **258**:6701–6704.

Kirk, C. J., Michell, R. H., and Hems, D. A., 1981, Phosphatidylinositol metabolism in rat hepatocytes stimulated by vasopressin, *Biochem. J.* **194:**155–165.

Knight, D. E., and Baker, P. F., 1983, The phorbol ester TPA increase the affinity of exocytosis for calcium in 'leaky' adrenal medullary cells, *FEBS Lett.* **160:**98–100.

Koenig, H., Goldstone, A., and Lu, C. Y., 1983, Polyamines regulate calcium fluxes in a rapid plasma membrane response, *Nature* **305:**530–534.

Kojima, I., Lippes, H., Kojima, K., and Rasmussen, H., 1983, Aldosterone secretion: Effect of phorbol ester and A23187, *Biochem. Biophys. Res. Commun.* **116:**555–562.

Kolesnick, R. N., and Gershengorn, M. C., 1984, Ca^{2+} ionophores affect phosphoinositide metabolism differently than thyrotropin-releasing hormone in GH_3 pituitary cells, *J. Biol. Chem.* **259:**9514–9519.

Koo, C., Lefkowitz, R. J., and Snyderman, R., 1983, Guanine nucleotides modulate the binding affinity of the oligopeptide chemoattractant receptor on human polymorphonuclear leukocytes, *J. Clin. Invest.* **72:**748–753.

Kruijer, W., Cooper, J. A., Hunter, T., and Verma, I. M., 1984, Platelet-derived growth factor induces rapid but transient expression of the c-*fos* gene and protein, *Nature* **312:**711–716.

Litosch, I., Wallis, C., and Fain, J. N., 1985, 5-Hydroxytryptamine stimulates inositol phosphate production in cell-free system from blowfly salivary glands: Evidence for a role of GTP in coupling receptor activation to phosphoinositide breakdown, *J. Biol. Chem.* **260:**5464–5471.

Mastro, A. M., and Smith, M. C., 1983, Calcium-dependent activation of lymphocytes by ionophore, A23187 and a phorbol ester tumor promotor, *J. Cell. Physiol.* **116:**51–56.

Michell, R. H., 1975, Inositol phospholipids and cell surface receptor function, *Biochim. Biophys. Acta* **415:**81–147.

Michell, R. H., Kirk, C. J., Jones, L. M., Downes, C. P., and Creba, J. A., 1981, The stimulation of inositol lipid metabolism that accompanies calcium mobilization in stimulated cells: Defined characteristics and unanswered questions. *Phil. Trans. R. Soc.* [*Biol.*] **296:**123–137.

Molski, T. F. P., Naccache, P. H., Marsh, M. L., Kermode, J., Becker, E. L., and Sha'afi, R. I., 1984, Pertussis toxin inhibits the rise in intracellular concentration of free calcium that is induced by chemotactic factors in rabbit neutrophils: possible role of the 'G proteins' in calcium mobilization, *Biochem. Biophys. Res. Commun.* **124:**644–650.

Monaco, M. E., 1982, The phospatidylinositol cycle in WRK-1 cells, *J. Biol. Chem.* **257:**2137–2139.

Moolenaar, W. H., Tertoolen, L. G. J., and de Laat, S. W., 1984, Growth factors immediately raise cytoplasmic free Ca^{2+} in human fibroblasts, *J. Biol. Chem.* **259:**8066–8069.

Nakamura, T., and Ui, M., 1984, Islet-activating protein, pertussis toxin, inhibits Ca^{2+}-induced and guanine nucleotide-dependent releases of histamine and arachidonic acid from rat mast cells, *FEBS Lett.* **173:**414–418.

Nishizuka, Y., 1983, Phospholipid degradation and signal translation for protein phosphorylation, *Trends Biochem. Sci.* **8:**13–16.

Nishizuka, Y., 1984a, The role of protein kinase C in cell surface signal transduction and tumor promotion, *Nature* **308:**693–697.

Nishizuka, Y., 1984b, Turnover of inositol phospholipids and signal transduction, *Science* **255:**1365–1370.

Okajima, F., and Ui, M., 1984, ADP-ribosylation of the specific membrane protein by islet-activating protein, pertussis toxin, associated with inhibition of a chemotactic peptide-induced arachidonate release in neutrophils, *J. Biol. Chem.* **259:**13863–13871.

Putney, J. W., McKinney, J. S., Aub, D. L., and Leslie, B. A., 1984, Phorbol ester-induced protein secretion in rat parotid gland, *Mol. Pharmacol.* **26:**261–266.

Rasmussen, H., 1981, *Calcium and cAMP as Synarchic Messengers*, John Wiley & Sons, New York.

Rasmussen, H., Forder, J., Kojima, I., and Scriabine, A., 1984, TPA-induced contraction of isolated rabbit vascular smooth muscle, *Biochem. Biophys. Res. Commun.* **122:**776–784.

Rink, T. J., Sanchez, A., and Hallam, T. J., 1983, Diacylglycerol and phorbol ester stimulate secretion without raising cytoplasmic free calcium in human platelets, *Nature* **305:**317–319.

Rittenhouse-Simmons, S., 1979, Production of diglyceride from phosphatidylinositol in activated human platelets, *J. Clin. Invest.* **63:**580–587.

Rittenhouse, S. E., and Horne, W. C., 1984, Ionomycin can elevate intraplatelet Ca^{2+} and activate phospholipase A without activating phospholipase C, *Biochem. Biophys. Res. Commun.* **123:**393–397.

Robinson, J. M., Badwey, J. A., Karnovsky, M. L., and Karnovsky, M. J., 1984, Superoxide release by neutrophils: Synergistic effects of a phorbol ester and a calcium ionophore, *Biochem. Biophys. Res. Commun.* **122:**734–739.

Rosoff, P. M., Stein, L. F., and Cantley, L. C., 1984, Phorbol esters induce differentiation in a pre-B-lymphocyte cell line by enhancing Na^+/H^+ exchange, *J. Biol. Chem.* **259:**7056–7060.

Rybak, S. M., and Stockdale, F. E., 1981, Growth effects of lithium chloride in BALB/c3T3 fibroblasts and Madin–Darby canine kidney epithelial cells, *Exp. Cell Res.* **136:**263–270.

Seyfred, M. A., Farrell, L. E., and Wells, W. W., 1984, Characterization of D-*myo* inositol 1,4,5-trisphosphate phosphatase in rat liver plasma membrane, *J. Biol. Chem.* **259:**13204–13208.

Simon, M.-F., Chap, H., and Douste-Blazy, L., 1984, Activation of phospholipase C in thrombin-stimulated platelets does not depend on cytoplasmic free calcium concentration, *FEBS Lett.* **170:**43–48.

Smith, C. D., Lane, B. C., Kusaka, I., Verghese, M. W., and Snyderman, R., 1985, Chemoattractant receptor induced hydrolysis of phosphatidylinositol 4,5-bisphosphate (PIP_2) in human polymorphonuclear leukocyte membranes. *J. Biol. Chem.* **260:**5875–5878.

Snyderman, R., and Pike, M. C., 1984, Chemoattractant receptors on phagocytic cells, *Annu. Rev. Immunol.* **2:**257–281.

Tanaka, C., Taniyama, K., and Kusunoki, M., 1984, A phorbol ester and A23187 act synergistically to release acetylcholine from the guinea pig ileum, *FEBS Lett.* **175:**165–169.

Taylor, M. V., Metcalfe, J. C., Hesketh, T. R., Smith, G. A., and Moore, J. P., 1984, Mitogens increase phosphorylation of phosphoinositides in thymocytes, *Nature* **312:**462–465.

Thomas, A. P., Alexander, J., and Williamson, J. R., 1984, Relationship between inositol polyphosphate production and the increase of cytosolic free Ca^{2+} induced by vasopressin in isolated hepatocytes, *J. Biol. Chem.* **259:**5574–5584.

Tomooka, Y., Imagawa, W., Nandi, S., and Bern, H. A., 1983, Growth effect of lithium on mouse mammary epithelial cells in serum-free collagen gel culture, *J. Cell. Physiol.* **117:**290–296.

Van Dongen, C. J., Zwiers, H., and Gispen, W. H., 1984, Purification and partial characterization of the phosphatidylinositol 4-phosphate kinase from rat brain, *Biochem. J.* **223:**197–203.

Volpi, M., Yassin, R., Naccache, P. H., and Sha'afi, R. I., 1983, Chemotactic factor causes rapid decreases in phosphatidylinositol 4,5-bisphosphate and phosphatidylinositol 4-monophosphate in rabbit neutrophils, *Biochem. Biophys. Res. Commun.* **112:**957–964.

Volpi, M., Naccache, P. H., Molski, T. F. P., Shefcyk, J., Huang, C.-K., Marsh, M. L., Munoz, J., Becker, E. L., and Sha'afi, R. I., 1985, Pertussis toxin inhibits the formyl-methionyl-leucyl-phenylalanine but not the phorbol ester stimulated changes in ion fluxes, protein phosphorylation and phospholipid metabolism in rabbit neutrophils: Role of the "G-proteins" in excitation response coupling, *Proc. Natl. Acad. Sci. U.S.A.* **82:**2708–2712.

Whitaker, M., and Irvine, R. F., 1984, Inositol 1,4,5-trisphosphate microinjection activates sea urchin eggs, *Nature* **312:**636–639.

Whiteley, B., Cassel, D., Zhuang, Y.-X., and Glaser, L., 1984, Tumour promotor phorbol 12-myristate 13-acetate inhibits mitogen-stimulated Na^+/H^+ exchange in human epidermoid carcinoma A431 cells, *J. Cell Biol.* **99:**1162–1166.

Whitman, M. R., Epstein, J., and Cantley, L., 1984, Inositol 1,4,5-trisphosphate stimulates phosphorylation of a 62,000-dalton protein in monkey fibroblasts and bovine brain cell lysates, *J. Biol. Chem.* **259:**13652–13655.

Yano, K., Nakashima, S., and Nazawa, Y., 1983, Coupling of polyphosphoinositide breakdown with calcium influx in formylmethionyl-leucyl-phenylalanine-stimulated rabbit neutrophils, *FEBS Lett.* **161:**296–300.

Zawalich, W., Brown, C., and Rasmussen, H., 1983, Insulin secretion: Combined effects of phorbol ester and A23187, *Biochem. Biophys. Res. Commun.* **117:**448–455.

11

Hormone-Induced Inositol Lipid Breakdown and Calcium-Mediated Cellular Responses in Liver

JOHN R. WILLIAMSON, SURESH K. JOSEPH, KATHLEEN E. COLL, ANDREW P. THOMAS, ARTHUR VERHOEVEN, and MARC PRENTKI

1. Introduction

Although it has been recognized for many years that changes of the intracellular free Ca^{2+} concentration by a variety of agonists form an important signaling device for regulation of cell function, the source of the Ca^{2+} and the molecular events regulating receptor-mediated changes of cellular calcium homeostasis have remained recalcitrant problems despite much effort directed towards their elucidation. However, advances made along a number of different lines have contributed towards the rapid increase of knowledge in this area. These include on the one hand the development of fluorescent Ca^{2+} indicators such as Quin 2 (Tsien, 1983) and more recently Fura 2 (Grynkiewicz *et al.*, 1985), which allow kinetic measurements of changes in the cytosolic free Ca^{2+} concentration of isolated cells, and on the other hand the elucidation of the signaling roles of two new intracellular second messengers, namely, inositol trisphosphate and diacylglycerol (for reviews see Nishizuka *et al.*, 1984; Nishizuka, 1984a; Berridge and Irvine, 1984; Williamson *et al.*, 1985; Williamson, 1986).

The connecting link between hormone or ligand interactions with cell surface receptors and the subsequent increase of cytosolic free Ca^{2+} was discovered as an extension of earlier work demonstrating hormone-me-

JOHN R. WILLIAMSON, SURESH K. JOSEPH, KATHLEEN E. COLL, ANDREW P. THOMAS, ARTHUR VERHOEVEN, and MARC PRENTKI • Department of Biochemistry and Biophysics, University of Pennsylvania School of Medicine, Philadelphia, Pennsylvania 19104.

diated increases of phosphatidylinositol turnover in a variety of tissues where intracellular signaling was thought to occur via calcium (for reviews see Michell, 1975, 1979). The relationship between these events, however, remained obscure for a number of years because of (1) the slow rate of phosphatidylinositol breakdown relative to Ca^{2+}-mediated functional responses, (2) poor correlations of hormone dose–response relationships among receptor binding, phosphatidylinositol turnover, and Ca^{2+}-mediated protein phosphorylations, and (3) inconsistencies of experimental findings relating to whether hormone-activated inositol lipid metabolism was Ca^{2+} dependent or Ca^{2+} independent (Exton, 1980; Williamson *et al.*, 1981; Abdel-Latif and Akhtar, 1982). Moreover, there was considerable uncertainty in the literature about whether the primary event resulting in an increase of cytosolic Ca^{2+} involved an activation of Ca^{2+} entry into the cell or a release of Ca^{2+} from an intracellular storage site (Michell *et al.*, 1981; Exton, 1981; Williamson *et al.*, 1981, 1985; Hawthorne, 1983).

Finally, the water-soluble products of phosphatidylinositol hydrolysis, *myo*-inositol-1,2-cyclic phosphate and *myo*-inositol-l-phosphate, were found to be inactive as putative Ca^{2+}-mobilizing second messengers (Irvine *et al.*, 1982). However, further studies showed that the hormone-mediated breakdown of two relatively minor inositol lipids, phosphatidylinositol-4-phosphate (PIP) and phosphatidylinositol-4,5-bisphosphate (PIP_2), preceded phosphatidylinositol breakdown (Michell *et al.*, 1981; Thomas *et al.*, 1983; Creba *et al.*, 1983) and that the products of PIP and PIP_2 breakdown accumulated rapidly in a Ca^{2+}-independent manner (Berridge, 1983; Thomas *et al.*, 1984). Further studies showed a direct Ca^{2+}-mobilizing effect of *myo*-inositol-1,4,5-trisphosphate (Ins-1,4,5-P_3) in permeabilized cells (Streb *et al.*, 1983; Joseph *et al.*, 1984a; Burgess *et al.*, 1984) and isolated microsomes (Prentki *et al.*, 1984; Joseph *et al.*, 1984b), thereby establishing a second messenger role for this compound.

Even more direct proof of its role as a physiological signaling messenger has been obtained by injection of Ins-1,4,5-P_3 into single cells. The elicitation of electrical events similar to those produced by light were obtained when Ins-1,4,5-P_3 was injected into *Limulus* photoreceptors (Brown *et al.*, 1984; Fein *et al.*, 1984); and injection of Ins-1,4,5-P_3 caused a Ca^{2+}-mediated activation of sea urchin eggs (Whitaker and Irvin, 1984) and *Xenopus* (frog) eggs (Oron *et al.*, 1985; Busa *et al.*, 1985). Various studies have also established that the primary source of the mobilized calcium responsible for the rise of cytosolic free Ca^{2+} is an intracellular nonmitochondrial calcium store, probably the endoplasmic reticulum (for reviews see Berridge and Irvine, 1984; Williamson *et al.*, 1985; Williamson, 1986).

Receptor-mediated breakdown of PIP_2 also produces a second intracellular signal, 1,2-diacylglycerol, which is formed simultaneously with Ins-1,4,5-P_3 and activates a specific Ca^{2+}- and phospholipid-dependent protein kinase, termed protein kinase C (Nishizuka, 1984a,b). This kinase is also activated by tumor-promoting phorbol esters such as phorbol myristate acetate (PMA), which, like diacylglycerol, increase the affinity of the enzyme for Ca^{2+} (Castagna *et al.*, 1982).

The mechanisms by which activation of protein kinase C affects cell function are presently poorly understood. Although many target proteins and enzymes that can be phosphorylated *in vitro* by this kinase have been described (for reviews see Nishizuka, 1984a; Williamson *et al.*, 1985), the number of recognized target proteins in intact cells is relatively small, and their functional roles have not yet been elucidated (e.g., Garrison *et al.*, 1984; Lapetina and Siegel, 1983; Fujita *et al.*, 1984, Drust and Marten, 1984. In some cells, notably platelets (Rink *et al.*, 1983; Kaibuchi *et al.*, 1983), pancreatic islets (Zawalich *et al.*, 1983), adrenal glomerulosa cells (Kojima *et al.*, 1983), and rat parotid gland (Putney *et al.*, 1984), the combined use of Ca^{2+} ionophores and PMA has shown that activation of protein kinase C and Ca^{2+} mobilization may act synergistically to elicit the full secretory response. However, recent studies on hepatocytes by Lynch *et al.* (1985) and Cooper *et al.* (1985) have shown that PMA had no synergistic effect with Ca^{2+} ionophores in increasing phosphorylase *a* levels and did not affect plasma membrane Ca^{2+} permeability, indicating that the Ins-1,4,5-P_3 signaling role for Ca^{2+} mobilization is independent of effects mediated by protein kinase C activation.

Phorbol esters also have inhibitory effects on receptor binding and hormone-mediated functional responses. Thus, PMA has been shown to inhibit α_1-adrenergic and to a lesser extent vasopressin-induced Ca^{2+} mobilization responses in liver (Cooper *et al.*, 1985; Lynch *et al.*, 1985), chemotactic factor-induced Ca^{2+} mobilization in neutrophils (Naccache *et al.*, 1985), muscarinic cholinergic effects in transformed smooth muscle (PC12) cells (Vincentini *et al.*, 1985), as well as agonist-induced formation of inositol phosphates or Ca^{2+} mobilization in platelets (MacIntyre *et al.*, 1985; Watson and Lapetina, 1985), hippocampal slices (Labarca *et al.*, 1984), and astrocytoma cells (Orellana *et al.*, 1985). In smooth muscle, either synergistic (rat vas deferens and dog basilar artery) or inhibitory (guinea pig ileum and rat uterus) effects on neurotransmitter actions on contractility were obtained with PMA (Baraban *et al.*, 1985). Protein kinase C activation may thus be involved in receptor internalization or desensitization as a general phenomenon for termination of agonist-mediated transduction processes (see also Macara, 1985).

Interest in the inositol lipid dual signal-generating pathway and recognition of its general importance for a variety of cell functions have been heightened over the last few years by a number of recent observations. Various growth factors that previously had been shown to stimulate inositol lipid metabolism (for review see Michell, 1982) cause an elevation of the cytosolic free Ca^{2+} (Moolenaar *et al.*, 1984a), which is thought to be a requirement for cell proliferation (for reviews see Berridge and Irvine, 1984; Macara, 1985). However, in most cells an increase of Ca^{2+} alone is an insufficient signal for cell growth, indicating that other factors must also be involved. Of particular interest is the demonstration that phorbol esters promote Na^{+}/H^{+} exchange across the plasma membrane with a consequent small alkalinization of the cell interior (Rosoff *et al.*, 1984; Besterman and Cuatrecasas, 1984; Moolenaar *et al.*, 1984b; Burns and Rozengurt, 1983; L'Allemain *et al.*, 1984; Kaibuchi *et al.*, 1985), an effect thought to be important for cell proliferation. Phorbol esters have also recently been shown to increase the levels of PIP and PIP_2 in platelets (de Chaffoy de Courcelles *et al.*, 1984; Halenda and Feinstein, 1984; Watson and Lapetina, 1985) and thymocytes (Taylor *et al.*, 1984), thereby increasing the amount of substrate available for receptor-stimulated hydrolysis. Certain oncogene products (*src* and *ros*) have PI and PIP kinase activities (Sugimoto *et al.*, 1984; Macara *et al.*, 1984), suggesting the possibility that viral transformation may promote a continuous production of second messengers independently of growth factor stimulation. It must be stressed, however, that the functional significance of a possible activation of PI(P) kinase activities and an increase of cellular PIP_2 levels on hormonal or growth factor-induced Ca^{2+}-signaling mechanisms remains to be established.

In this chapter we briefly review experimental evidence in support of the Ca^{2+}-signaling role of Ins-1,4,5-P_3 in liver, and some of its properties in causing Ca^{2+} release from permeabilized hepatocytes, RIN m5F insulinoma cells, and a nonmitochondrial microsomal fraction derived from Syrian hamster insulinoma cells. The hormone-induced increase of cytosolic free Ca^{2+}, as observed directly with Quin 2 and indirectly by changes of phosphorylase *a* in hepatocytes, is described in relation to the different phases of (1) intracellular Ca^{2+} mobilization and resequestration, (2) net efflux of Ca^{2+} from the cell, and (3) enhanced permeability of the plasma membrane to Ca^{2+}.

2. *Inositol-1,4,5-Trisphosphate as a Calcium-Mobilizing Second Messenger*

Figure 1 illustrates the basic reactions involved in receptor-mediated activation of PIP_2 hydrolysis in the plasma membrane by phospholipase

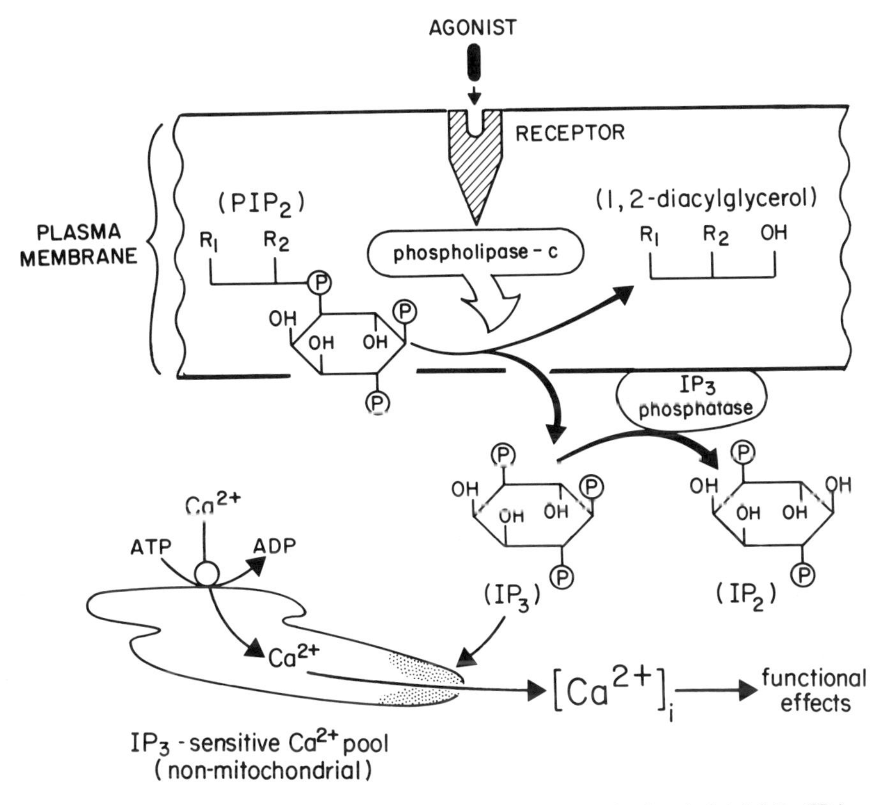

Figure 1. Schematic representation of second messenger role for inositol 1,4,5-P_3 (IP_3) as an intracellular Ca^{2+}-releasing agent.

C to yield the two intracellular second messengers Ins-1,4,5-P_3 and diacylglycerol. The molecular mechanism involved in the coupling between binding of Ca^{2+}-mobilizing agonists to the receptor and activation of phospholipase C is presently unknown, although there is growing evidence that a specific GTP-binding protein may be involved (Gomperts, 1983; Gilman, 1984; Haslam and Davidson, 1984).

Of the products of PIP_2 hydrolysis, the diacylglycerol moiety is highly lipophilic and presumably remains in the membrane bilayer, where it accomplishes its signaling role by activation of protein kinase C. The water-soluble product Ins-1,4,5-P_3 (IP_3) is depicted as activating efflux of Ca^{2+} from an intracellular, nonmitochondrial, vesicular calcium storage site into which cytosolic Ca^{2+} is pumped by a Ca^{2+}-ATPase. Therefore, the signaling role of Ins-1,4,5-P_3 causes a rapid increase of Ca^{2+} in the cytosol. Inositol trisphosphate is hydrolyzed by a phosphomonoesterase to *myo*-inositol-1,4-bisphosphate (Ins-1,4-P_2), which is located primarily on the plasma membrane, requires Mg^{2+} for activity, and is inhibited by Cd^{2+} but not by Li^+ (Seyfred *et al.*, 1984; Storey *et al.*, 1984; Joseph and Williams, 1985). Unlike Ins-1,4,5-P_3, Ins-1,4-P_2 is essentially inactive in causing Ca^{2+} release from permeabilized cells, and further hydrolysis of Ins-1,4-P_2 to *myo*-inositol-1-phosphate occurs in the cytosol (Joseph and Williams, 1985).

In liver, the major Ca^{2+}-mobilizing agonists are α_1-adrenergic agents (e.g., phenylephrine), vasopressin, and angiotensin II. Addition of glucagon, 8-bromo-cAMP, or forskolin to hepatocytes will also mobilize Ca^{2+} from the Ins-1,4,5-P_3-sensitive calcium pool, and the dose–response curve for this effect is similar to those for phosphorylase *a* increase and the stimulation of urea production from glutamine. (A. Verhoeven and J. R. Williamson, unpublished observations). Typical Ca^{2+}-mediated functional responses in liver are an increase of glycogenolysis by activation of phosphorylase kinase and the subsequent phosphorylation and activation of phosphorylase *b* to phosphorylase *a*; decreased glycogen synthesis by phosphorylation and inactivation of glycogen synthase; and increases of gluconeogenesis and ureogenesis (Pilkis *et al.*, 1978; Hems and Whitton, 1980; Williamson *et al.*, 1981; Kraus-Friedman, 1984).

Although this basic mechanism for the initiation of Ca^{2+} signaling appears to operate in many different cell types (for review see Berridge and Irvine, 1984), there are a number of aspects that require further elucidation. If PIP_2 is the sole substrate for phospholipase C, as depicted in Fig. 1, there will be an equal and simultaneous production of Ins-1,4,5-P_3 and diacylglycerol. However, studies with liver (Thomas *et al.*, 1983; Creba *et al.*, 1983) and other cells (Berridge, 1984) have shown an approximately equal rate of hormone-stimulated breakdown of both PIP and PIP_2, with a later but much larger breakdown of PI. Studies by Wilson *et al.* (1984) using purified phospholipase C and radiolabeled substrates reconstituted in phospholipid vesicles also showed a simultaneous hydrolysis of all the inositol lipids in a Ca^{2+}-dependent manner, with maximum hydrolysis rates in the order $PIP_2 > PIP > PI$. However, hydrolysis of polyphosphoinositides was favored over PI hydrolysis at low Ca^{2+} concentrations, suggesting that the Ins-1,4,5-P_3-mediated rise of cytosolic Ca^{2+} may be involved in regulating the substrate specificity of phospholipase C, thereby causing an increase in the ratio of diacylglycerol to Ins-1,4,5-P_3 formed after the initial phase of hormone stimulation.

Recent studies have indicated that in some systems, including the salivary gland (Irvine *et al.*, 1984; 1985) and guinea pig hepatocytes (Burgess *et al.*, 1985), two isomers of inositol trisphosphate are formed during

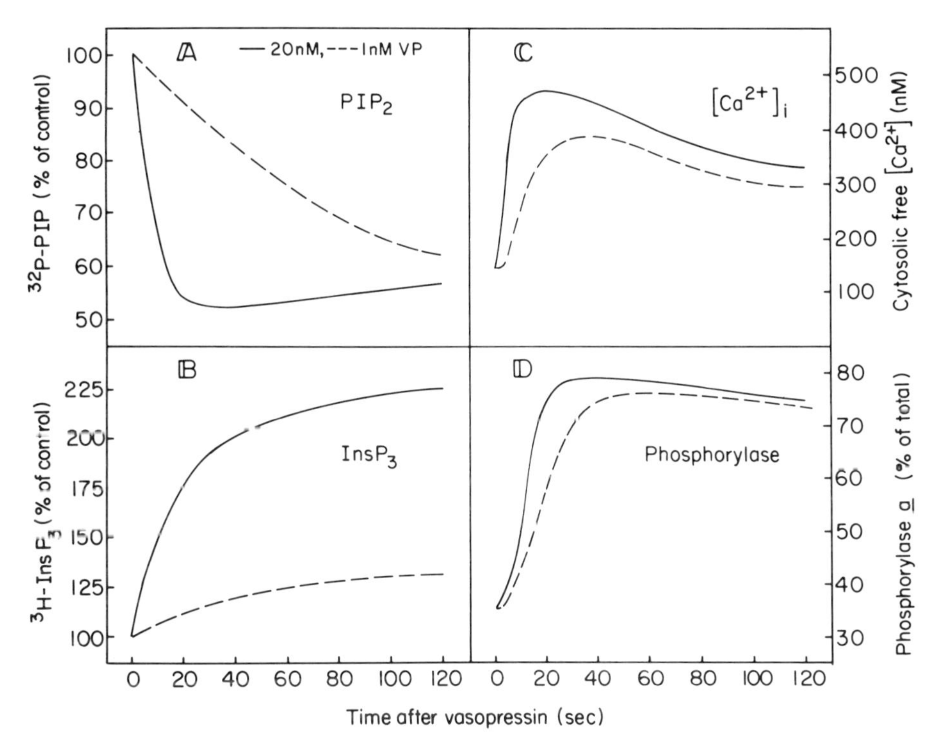

Figure 2. Kinetics of vasopressin-stimulated breakdown of [^{32}P]-labeled phosphatidylinositol-4,5-bisphosphate (PIP_2) (A). Formation of inositol trisphosphate (IP_3) from [^{3}H]-inositol-prelabeled cells (B). Increase of cytosolic $[Ca^{2+}]_i$ as measured by Quin 2 (C). Increase of phosphorylase *a* (D) in isolated hepatocytes.

hormonal stimulation, namely, D- or L-*myo*-inositol-1,3,4-trisphosphate as well as Ins-1,4,5-P_3. Ins-1,3,4-P_3 is produced after Ins-1,4,5-P_3 has reached a peak level (10–15 sec) and continues to accumulate with time, whereas Ins-1,4,5-P_3 falls to a new steady-state level threefold above control values. The slow kinetics of accumulation of Ins-1,3,4-P_3 suggests that it may be a relatively poor substrate for Ins-1,4,5-P_3 phosphatase. Its function is presently unknown, but it is important to note that most studies reporting the formation of radiolabeled IP_3 in hormone-stimulated cells include both isomers. The formation of Ins-1,3,4-IP_3 during vasopressin stimulation of hepatocytes may account for the fact that the total inositol trisphosphate pool reaches peak levels much later than the peak increase of cytosolic free Ca^{2+} (see Fig. 2) and is augmented by pretreatment of the cells with Li^+ (Thomas *et al.*, 1984). A possible explanation of the latter finding is that the enzyme responsible for degradation of Ins-1,3,4-P_3 may be Li^+ sensitive. Studies by Batty *et al.* (1985) have shown that Ins-1,3,4-P_3 is formed by removal of the 5-phosphate from L-*myo*-inositol-1,3,4,5-tetrakis-phosphate (Ins-1,3,4,5-P_4), which was shown to be produced during carbachol stimulation of rat brain cortex slices (Batty *et al.*, 1985) and vasopressin-stimulated hepatocytes (C. Hansen and J. R. Williamson, unpublished experiments.) Ins-1,3,4,5-P_4 itself is formed from Ins-1,4,5-P_3 by a soluble kinase present in a number of tissues, including brain and liver.

In order to establish that hormone-stimulated breakdown of polyphosphoinositides and formation of Ins-1,4,5-P_3 have a direct signaling role in the mobilization of intracellular Ca^{2+}, a number of criteria should be fulfilled (Williamson *et al.*, 1985). The most important of these are as follows: (1) hydrolysis of PIP_2 should not be a secondary event to the rise of cytosolic free Ca^{2+}; (2) the formation of Ins-1,4,5-P_3 should precede the rise of cytosolic free Ca^{2+}; (3) PIP_2 breakdown and Ins-1,4,5-P_3 formation should show a sensitivity to hormone concentration similar to that for the increase of the cytosolic free Ca^{2+}; (4) removal of the agonist or inhibition of Ins-1,4,5-P_3 formation should cause a fall of the cytosolic free Ca^{2+}; (5) Ins-1,4,5-P_3 at concentrations equivalent to those reached in the intact cell by hormone stimulation should cause Ca^{2+} release from a cell-free preparation or permeabilized cell. These criteria have not been rigorously tested in most tissues where Ins-1,4,5-P_3 is thought to be important in releasing Ca^{2+}, but they remain of importance, particularly for investigations of the possible role of Ins-1,4,5-P_3 in Ca^{2+} release from cells such as nerve and skeletal and cardiac muscle, which have well-developed voltage-dependent Ca^{2+} gates in the plasma membrane (Reuter, 1983). It is of interest to note that whereas the administration of Ins-1,4,5-P_3 to mechanically skinned frog skeletal muscle preparation caused a contraction apparently mediated by Ca^{2+} release (Vergara *et al.*, 1985), Ins-1,4,5-P_3 was without effect in causing Ca^{2+} release from saponin-permeabilized myocytes or isolated Ca^{2+}-loaded cardiac sarcoplasmic reticulum (Movsesian *et al.*, 1985).

A detailed discussion of the evidence supporting the second messenger role of Ins-1,4,5-P_3 in liver is presented elsewhere (Williamson *et al.*, 1985). Figure 2 presents the kinetic relationships among PIP_2 breakdown, IP_3 formation, the cytosolic free Ca^{2+} concentration, and phosphorylase *a* activity in isolated hepatocytes following administration of a low (1 nM) and a near-saturating (20 nM) concentration of vasopressin. These data are based on previously published studies using cells prelabeled with [^{32}P]-phosphate or [^{3}H]-inositol combined with more recent measurements of cytosolic free Ca^{2+} changes with Quin 2. With the high vasopressin concentration, about 50% of the total cell PIP_2 content is degraded with a half-time of about 6 sec (Fig. 2A), whereas IP_3 accumulation occurs on a slower time scale with a half-time of about 14 sec (Fig. 2B). As previously mentioned, however, it is now known that the accumulation of IP_3 after the first 10 sec of vasopressin stimulation is largely accounted for by the 1,3,4 isomer. Cytosolic free Ca^{2+} increased with a half-time of about 3 sec with the high vasopressin concentration (Fig. 2C), and the increase of phosphorylase was slightly slower (Fig. 2D). At 1 nM vasopressin, the rate of PIP_2 breakdown is considerably decreased, although the maximum extent of breakdown is about the same as at 20 nM. In contrast, both the rate and amount of IP_3 accumulation are decreased with the lower vasopressin concentration. Under these conditions, the rise of the cytosolic free Ca^{2+} is characterized by a delay of a few seconds with a slower rate of increase and a shift of the peak Ca^{2+} by about 8 sec. The changes of phosphorylase *a* activity showed similar effects.

It is apparent that comparisons of peak responses for the various parameters with different hormones concentrations are inappropriate because these occur at different times. However, when the initial rates of change of PIP_2 breakdown, IP_3 formation, and increase of cytosolic free Ca^{2+} are compared over a range of vasopressin concentrations, there is a very close correlation among these parameters (Thomas *et al.*, 1984). Furthermore, it is apparent from recent measurements of Ins-1,4,5-P_3 and Ins-1,3,4-P_3 in vasopressin-stimulated hepatocytes that the initial rate of change of total IP_3 is likely to correspond closely to measurements of the rate of Ins-1,4,5-P_3 formation (Burgess *et al.*, 1985).

Although studies such as those described above and others reviewed in detail elsewhere (Berridge and Irvine, 1984; Williamson *et al.*, 1985) strongly suggest a close coupling between polyphosphoinositide metabolism and Ca^{2+} mobilization, the postulation of a Ca^{2+}-signaling role for Ins-1,4,5-P_3 became much more convincing after the direct demonstration that addition of physiologically relevant concentrations of the putative messenger to permeabilized cells caused a release of Ca^{2+} from an intracellular vesicular calcium pool. Figure 3 illustrates this effect for saponin-permeabilized rat hepatocytes, where changes of Ca^{2+} in the medium were recorded with a Ca^{2+} electrode. The cells were incubated in a HEPES-buffered high-K^+ medium containing $MgATP^{2-}$, an ATP-regenerating system, together with antimycin A and oligomycin to deplete the mitochondrial Ca^{2+} stores and prevent mitochondrial Ca^{2+} uptake. Prior to the start of recordings, saponin was added to permeabilize the cells, and most of the endogenous Ca^{2+} in the medium was removed by an ATP-dependent sequestration into the endoplasmic reticulum. In trace

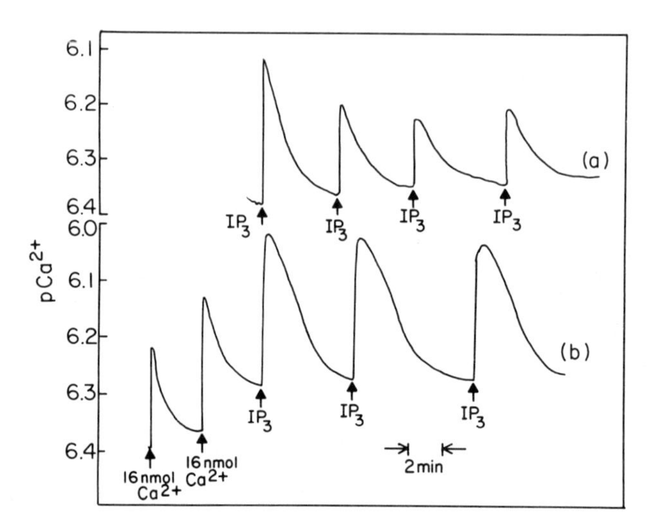

Figure 3. Effects of repeated additions of 1 μM of inositol-1,4,5-trisphosphate (IP_3) on Ca^{2+} release from saponin-permeabilized hepatocytes. The incubation medium contained antimycin A and oligomycin to prevent uptake of Ca^{2+} by the mitochondria, together with $MgATP^{2-}$ and an ATP-regenerating system. The medium free Ca^{2+} concentration was measured with a Ca^{2+} electrode.

a (Fig. 3), four successive additions of a maximal concentration of Ins-1,4,5-P_3 (1 μM) were made. Each addition was associated with a rapid release of Ca^{2+} followed by a slower rate of Ca^{2+} reuptake, which is associated with hydrolysis of the added Ins-1,4,5-P_3 (Joseph *et al.*, 1984a).

The first addition of Ins-1,4,5-P_3 caused a greater Ca^{2+} release than succeeding ones. One explanation for this phenomenon is that the Ca^{2+} initially released from the Ins-1,4,5-P_3-sensitive pool was sequestered partly in the calcium pool from which it originated and partly in other nonmitochondrial calcium storage sites. In order to test this possibility, small Ca^{2+} additions were made to the permeabilized cells to saturate the endoplasmic reticulum calcium pools, followed by successive additions of Ins-1,4,5-P_3 (trace b of Fig. 3). Each of these gave the same Ca^{2+} release, thus verifying the above explanation. It is apparent from these and other studies with saponin-permeabilized RIN m5F insulinoma cells (Prentki *et al.*, 1985) that the putative Ins-1,4,5-P_3 receptor does not become desensitized to Ins-1,4,5-P_3 and that the amount of Ca^{2+} released by saturating concentrations of Ins-1,4,5-P_3 is dependent on the amount of calcium in the Ins-1,4,5-P_3-sensitive calcium pool.

Figure 4 shows that in hepatocytes this pool represents only a fraction of the total calcium content of the nonmitochondrial calcium stores. After

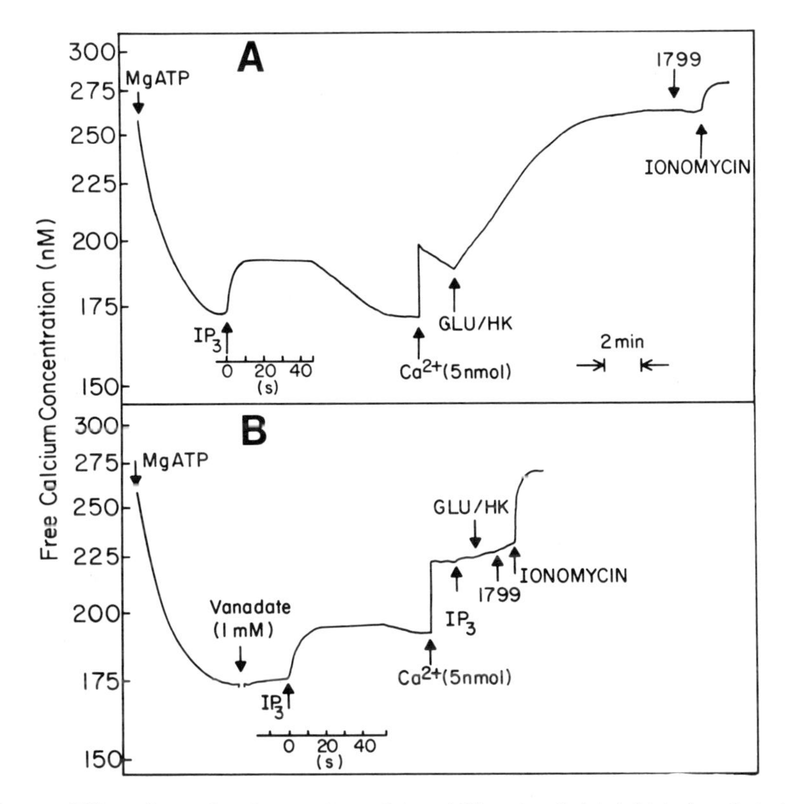

Figure 4. Effect of vanadate (comparison of A and B) on inositol-1,4,5-trisphosphate (IP_3)-mediated Ca^{2+} release from a fraction of the total endoplasmic reticulum calcium pool in saponin-permeabilized hepatocytes. The medium contained inhibitors of mitochondrial Ca^{2+} uptake, and addition of glucose and hexokinase (Glu/HK) was used to deplete the medium of ATP and cause Ca^{2+} release from intracellular Ca^{2+} stores. Addition of the nonfluorescent uncoupling agent 1799 before or after Glu/HK showed the absence of mitochondrial calcium. Residual vesicular Ca^{2+} was released by ionomycin. The medium free Ca^{2+} concentration was measured with Quin 2.

saponin was added to permeabilize the cells, endogenous Ca^{2+} (about 1 μM) was removed from the medium until a steady-state Ca^{2+} concentration of about 170 nM was attained. Figure 4A shows that the addition of a maximal concentration of Ins-1,4,5-P_3 (1 μM) caused a rapid release of part of the sequestered Ca^{2+}, which was complete in a few seconds. Reuptake of this Ca^{2+} occurred over the ensuing 3 min (note the change of time scale in the figure). The total nonmitochondrial calcium storage sites were not saturated at this point, as demonstrated by the gradual reuptake of Ca^{2+} after a small addition of Ca^{2+}. The subsequent addition of glucose and hexokinase in order to deplete ATP was accompanied by the gradual release of most of the previously sequestered Ca^{2+}. Finally, the addition of the mitochondrial uncoupling agent 1799 caused no further Ca^{2+} release, indicating the absence of sequestered mitochondrial calcium.

A similar experimental protocol was followed in Fig. 4B, but in this case vanadate was added to inhibit the endoplasmic reticulum Ca^{2+} ATPase. Under these conditions, addition of Ins-1,4,5-P_3 caused a release of Ca^{2+} similar to that in the absence of vanadate, indicating that uptake and release of Ca^{2+} must occur through separate and distinct mechanisms and that in the presence of saturating Ins-1,4,5-P_3, the Ca^{2+} uptake pathway is unable to compete against the Ins-1,4,5-P_3-stimulated release mechanism.

Previous studies using saponin-permeabilized hepatocytes showed that half-maximum effects of Ins-1,4,5-P_3 on Ca^{2+} release were attained with 0.1 μM Ins-1,4,5-P_3, and maximal effects were attained at about 0.5 μM (Joseph *et al.*, 1984a). As noted above, however, the characteristic response to a pulse addition of Ins-1,4,5-P_3 is a rapid Ca^{2+} release followed by reuptake as a consequence of Ins-1,4,5-P_3 degradation. This is different from the situation in hormone-stimulated cells, in which there is a sustained breakdown and resynthesis of PIP_2 during receptor occupancy and a continuous production of IP_3 (Fig. 2; Thomas *et al.*, 1983, 1984). One can therefore expect that under physiological conditions the concentration of intracellular Ins-1,4,5-P_3 will achieve a steady state at a level determined by its rates of production and degradation. In a permeabilized cell system, this situation can be modeled by the continuous infusion of Ins-1,4,5-P_3.

Figure 5 shows the results of an experiment using saponin-permeabilized RINm4F insulinoma cells, which respond to additions of Ins-1,4,5-P_3 in a fashion similar to hepatocytes but have a considerably larger Ins-1,4,5-P_3-sensitive calcium store (Prenki *et al.*, 1985). The addition of cells to medium containing $MgATP^{2-}$ and an ATP-regenerating system resulted in Ca^{2+} sequestration into the endoplasmic reticulum with a decrease to about 0.3 μM of the medium Ca^{2+} concentration. The infusion of Ins-1,4,5-P_3 caused the maintenance of different steady-state levels of free Ca^{2+} followed by a Ca^{2+} reuptake when the Ins-1,4,5-P_3 infusion was stopped. A similar sustained increase in the concentration of the medium Ca^{2+} was also observed in saponin-permeabilized hepatocytes on the infusion of Ins-1,4,5-P_3. These experiments indicate that Ca^{2+} cycling across the endoplasmic reticulum membrane by an Ins-1,4,5-P_3-sensitive efflux pathway and a Ca^{2+} ATPase influx mechanism is likely to occur in the intact cell, where steady-state Ins-1,4,5-P_3 concentrations

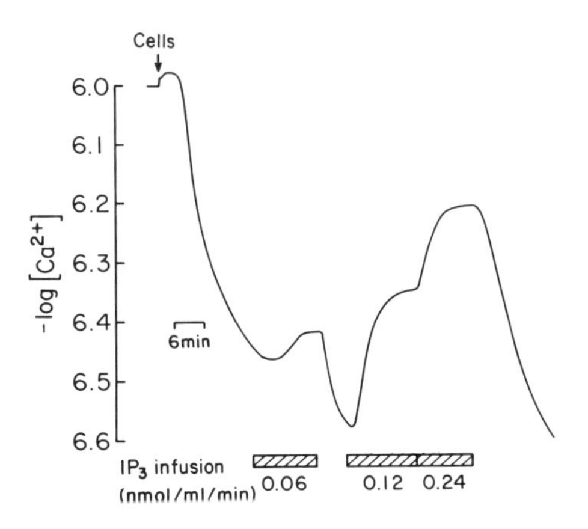

Figure 5. Effect of various rates of inositol-1,4,5-trisphosphate (IP_3) infusion on sustained Ca^{2+} release from saponin-permeabilized RIN m5F insulinoma cells incubated with MgATP, a creatine phosphate/creatine phosphokinase ATP-regenerating system, and antimycin A plus oligomycin to prevent mitrochondrial Ca^{2+} uptake. The data are reproduced with permission from Prentki *et al.* (1985).

give a graded response of net Ca^{2+} release. Thus, with both permeabilized liver cells and permeabilized RINm5F insulinoma cells, infusion of Ins-1,4,5-P_3 resulted in a sustained elevation of the medium free Ca^{2+}, indicating a lack of desensitization of the Ins-1,4,5-P_3 response. These experiments also suggest that Ins-1,4,5-P_3 contributes towards determining both the steady-state cytosolic free Ca^{2+} concentration and the degree of filling of the hormone-sensitive calcium pool by regulating the rate of Ca^{2+} cycling across the endoplasmic reticulum (Prentki *et al.*, 1985).

Relatively little is known about the mechanism of the Ins-1,4,5-P_3-induced Ca^{2+} release process. An important question is whether it is inhibited by an increase of the extravesicular Ca^{2+} concentration. This was investigated in an experiment in which saponin-permeabilized hepatocytes were allowed to accumulate $^{45}Ca^{2+}$ in medium containing $MgATP^{2-}$ and a mixture of mitochondrial inhibitors (ruthenium red, antimycin A, FCCP, and oligomycin). The medium free Ca^{2+} concentration was varied by the addition of different amounts of Ca^{2+} to 0.5 μM EGTA; 1 μM Ins-1,4,5-P_3 was added after a 25-min incubation period at each free Ca^{2+} concentration, and 40 sec later the cells were centrifuged through silicone oil to measure the vesicular $^{45}Ca^{2+}$ content. Figure 6 compares the amount of Ins-1,4,5-P_3-mediated Ca^{2+} release with the total $^{45}Ca^{2+}$ content of the permeabilized cells at each ambient Ca^{2+} concentration. The data show that the amount of Ca^{2+} released by Ins-1,4,5-P_3 is independent of the medium free Ca^{2+} concentration over the range from 1 μM to 100 μM. This finding does not agree with the results of Suematsu *et al.* (1984), who found that the amount of Ca^{2+} released from permeabilized bovine aorta smooth muscle cells by Ins-1,4,5-P_3 was reduced by external free Ca^{2+} concentrations in the range of 10 μM. The different

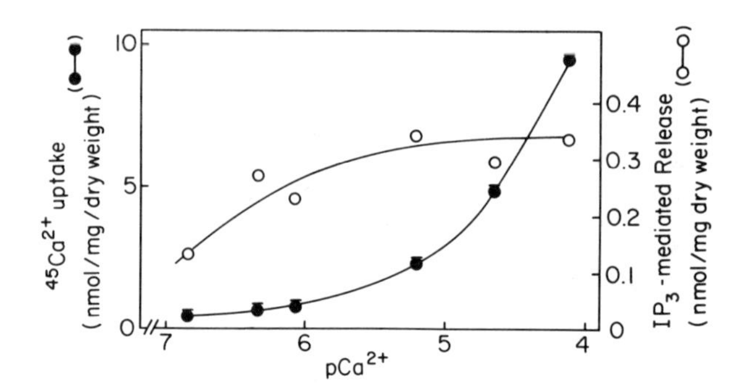

Figure 6. The influence of extracellular Ca^{2+} concentration on inositol-1,4,5-trisphosphate (IP_3)-mediated Ca^{2+} release from saponin-permeabilized hepatocytes. The permeabilized cells were incubated for 25 min in medium containing $^{45}Ca^{2+}$, $MgATP^{2-}$, inhibitors of mitochondrial Ca^{2+} uptake (ruthenium red, antimycin A, FCCP, and oligomycin), and the free Ca^{2+} concentration was varied using different amounts of calcium added to 0.5 mM EGTA. The loss of sequestered $^{45}Ca^{2+}$ and its appearance in the medium were measured 40 sec after addition of 1 μM of inositol-1,4,5-trisphosphate (IP_3) by centrifugation of the cells through silicone oil.

results may be accounted for by tissue differences. For instance, the driving force for Ca^{2+} efflux (i.e., the Ca^{2+} concentration gradient across the membrane of the Ins-1,4,5-P_3-sensitive calcium store) may be lower in smooth muscle than in liver because of a greater number of Ca^{2+} binding sites in the muscle sarcoplasmic reticulum.

The specificity of the IP_3 receptor has been investigated by a number of groups (see Berridge and Irvine, 1984), who have come to the conclusion that vicinal phosphates at the 4 and 5 positions of the inositol ring are required for Ca^{2+}-releasing activity and that a phosphate at the 1 position enhances sensitivity. A compound of particular interest is glycerophosphoinositol-4,5-bisphosphate ($GPIP_2$) (the product obtained by deacylation of PIP_2), which is an active agonist although about ten times less effective than Ins-1,4,5-P_3 in releasing Ca^{2+} from permeabilized hepatocytes (S. K. Joseph, unpublished experiments). Figure 7 shows a comparison between the effects of Ins-1,4,5-P_3 and $GPIP_2$ as Ca^{2+}-releasing agents on microsomes prepared from Syrian hamster insulinomas incubated in the presence of $MgATP^{2-}$ (Joseph *et al.*, 1984b). This preparation is free of mitochondria but is contaminated with plasma membranes and contains Ins-1,4,5-P_3 phosphatase activity. Consequently, addition of 0.5 μM Ins-1,4,5-P_3 causes a cycle of rapid Ca^{2+} release followed by reuptake as the added Ins-1,4,5-P_3 is degraded. Addition of $GPIP_2$ at a tenfold higher concentration caused a similar release of Ca^{2+}, but the

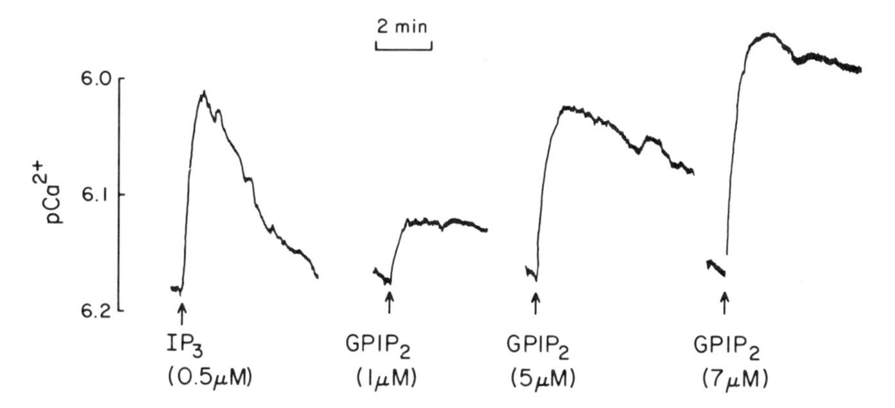

Figure 7. Comparison of the effects of inositol-1,4,5-trisphosphate (IP_3) and glycerophosphatidylinositol-4,5-bisphosphate ($GPIP_2$) on Ca^{2+} release from mitochondria-free microsomes prepared from Syrian hamster insulinoma cells.

reuptake of Ca^{2+} was very slow. This suggests that the hydrolysis of $GPIP_2$ by the contaminating Ins-1,4,5-P_3 phosphatase is much slower than that of Ins-1,4,5-P_3. The fact that graded, relatively stable steady-state levels of extravesicular free Ca^{2+} are maintained following additions of $GPIP_2$ (cf. Fig. 5 with Ins-1,4,5-P_3 infusions) provides evidence supporting the concept of regulation of the amount of calcium released from a functionally separate Ins-1,4,5-P_3-sensitive Ca^{2+} pool by the steady-state concentration of Ins-1,4,5-P_3. Ins-1,4,5-P_3 binding sites with a K_d of 5 nM have been identified in adrenal cortex microsomes (Baukal *et al.*, 1985) suggested that Ins-1,4,5-P_3 exerts its effect by binding to a specific receptor in specialized regions of the endoplasmic reticulum, thereby regulating the activity of a Ca^{2+} channel.

The spatial location of this calcium pool in mammalian cells has not been ascertained, but it is of interest that recent studies using iontophoresis of Ins-1,4,5-P_3 into frog (*Xenopus laevis*) eggs with recording of the elicited Ca^{2+} response by Ca^{2+}-selective microelectrodes have shown that although both shallow and deep injections of low concentrations of Ins-1,4,5-P_3 trigger local increases of intracellular free Ca^{2+}, only shallow injections will elicit a Ca^{2+}-triggered Ca^{2+} release like that which accompanies activation of physiological responses (Busa *et al.*, 1985). This major release of Ca^{2+}, which is much greater than that induced by Ins-1,4,5-P_3, is thought to be derived from a calcium pool adjacent to the egg cortex. The fact that injections of Ins-1,4,5-P_3 cause small pulses of Ca^{2+} to be released independently of their site of injection suggests that the

IP_3-sensitive calcium pool is distributed throughout the cell. Whether this is true for all cell types or whether the IP_3-sensitive calcium pool is located adjacent to the plasma membrane in some cell types requires investigation.

3. Hormone Effects on Ca^{2+} Fluxes

The major role of the Ins-1,4,5-P_3-mediated Ca^{2+} release mechanism is to provide an increase of the cytosolic free Ca^{2+} in hormone-stimulated cells. The previous discussion has summarized current knowledge concerning the site from which Ca^{2+} is released in the cell and the manner in which it is released. Further important aspects concern the role of the Ca^{2+} transport systems located in the plasma membrane and the mitochondria in the regulation of the cytosolic free Ca^{2+} during hormone stimulation and some quantitative considerations regarding how much Ca^{2+} is transported. The Ca^{2+} flux relationships between intra- and extracellular Ca^{2+} and between intracellular Ca^{2+} and Ca^{2+} sequestered in the endoplsamic reticulum and the mitochondria are depicted in Fig. 8. Figure 8 also shows the postulated transducing role of a GTP-binding protein in the receptor-mediated activation of phospholipase C and the separate signaling functions of Ins-1,4,5-P_3 and diacylglycerol.

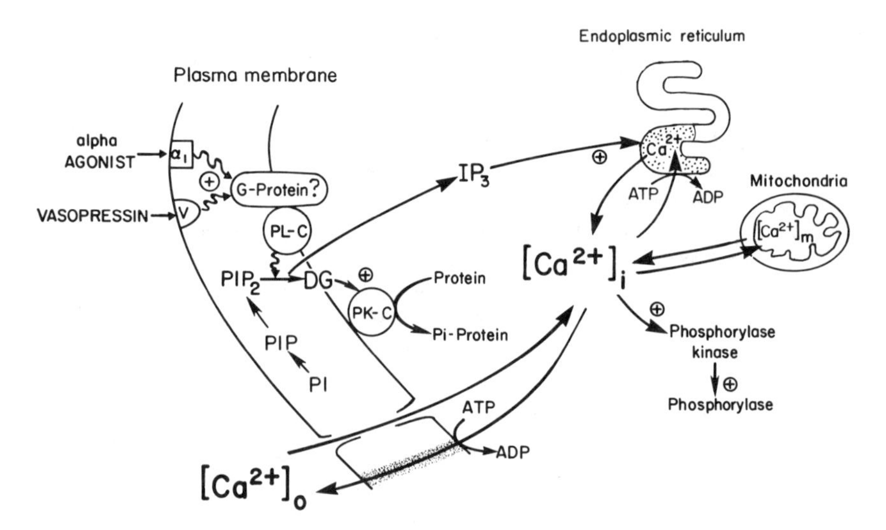

Figure 8. Schematic representation of the Ca^{2+} transport systems affecting cellular calcium homeostasis during hormonal stimulation.

The cytosolic free Ca^{2+} is depicted in Fig. 8 as being influenced by six Ca^{2+} fluxes: influx and efflux across the plasma, endoplasmic reticulum, and mitochondrial membranes (for reviews see Williamson *et al.*, 1981, 1985; Reinhart *et al.*, 1984a; Rasmussen and Barritt, 1984). In the resting cell with a constant total calcium content, the influx and efflux of Ca^{2+} across each of these membranes will be equal. When the calcium homeostasis of the cell is disturbed by the addition of hormone, the observed increase of the cytosolic free Ca^{2+} is determined by the following factors: (1) the amount of Ca^{2+} released from the IP_3-sensitive pool, (2) the activity coefficient of Ca^{2+} in the cytosol (the ratio of bound to free Ca^{2+}), and (3) Ca^{2+} flux changes across the plasma and mitochondrial membranes in response to the elevated Ca^{2+} concentration. These flux changes result in a net loss of Ca^{2+} from the cell during hormone stimulation even in the presence of physiological concentrations of extracellular Ca^{2+} (Blackmore *et al.*, 1982, 1983; Reinhart *et al.*, 1982) and a net increase of mitochondrial calcium (Shears and Kirk, 1984).

The kinetics of the changes of the cytosolic free Ca^{2+} as measured with Quin 2 in hepatocytes stimulated maximally with vasopressin are compared in Fig. 9 with the kinetics of net Ca^{2+} efflux from the cells. These data illustrate the effects obtained under conditions of low (approximately 10 μM) extracellular Ca^{2+} in order to exclude Ca^{2+} entry. The cytosolic free Ca^{2+} typically increased to a peak value in 15 sec and subsequently declined to prestimulation levels after about 2 min. The increase of cytosolic free Ca^{2+} can be attributed entirely to mobilization of intracellular calcium from the Ins-1,4,5-P_3-sensitive calcium pool. The decline of cytosolic free Ca^{2+} from the peak coincided approximately with the onset of net Ca^{2+} efflux from the cells as monitored by arsenazo III in the medium (Joseph and Williamson, 1983).

Interestingly, there is a delay of about 10 sec after the addition of vasopressin before net efflux of Ca^{2+} can be detected (Joseph and Williamson, 1983). This has been attributed to an initial, and possibly transient, inhibition of the plasma membrane Ca^{2+} ATPase (Lin *et al.*, 1983; Prpic *et al.*, 1984). The mechanism responsible for this inhibition has not been elucidated, but it persists during isolation of plasma membranes from hormone-treated livers. It may be caused by the decreased PIP_2 content of the plasma membranes, since this phospholipid has been shown to be an activator of purified erythrocyte Ca^{2+} ATPase reconstituted into phosphatidylcholine liposomes (Choquette *et al.*, 1984). Inhibition of the Ca^{2+} ATPase during the initial phase of intracellular Ca^{2+} mobilization will have the effect of facilitating the increase of the cytosolic free Ca^{2+} and augmenting the peak of the Ca^{2+} transient. Subsequently, either because of removal of the inhibitory influence or as a secondary consequence of

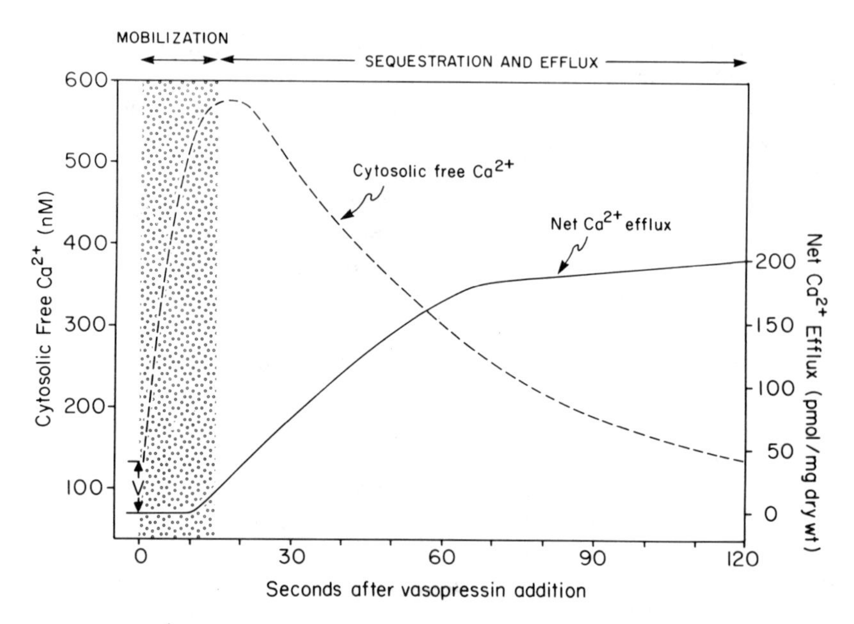

Figure 9. Comparison of the kinetics of the increase of cytosolic free Ca^{2+} and net Ca^{2+} efflux in hepatocytes stimulated with 10 μM vasopressin (V). The data for Ca^{2+} efflux from hepatocytes incubated with arsenazo III at low extracellular Ca^{2+} concentrations were replotted from the results of Joseph and Williamson (1983). The data for changes of the cytosolic free Ca^{2+} in hepatocytes incubated at low extracellular Ca^{2+} concentrations were replotted from the results of Joseph *et al.* (1985).

the elevated cytosolic free Ca^{2+}, calcium is pumped out of the cell even against the high Ca^{2+} concentration gradient prevailing at physiological extracellular Ca^{2+} concentrations. However, it is possible that not all of the calcium mobilized from the IP_3-sensitive calcium pool is lost from the cell, and a fraction may be sequestered in the mitochondria through stimulation of the Ca^{2+} uniporter by the elevated extramitochondrial Ca^{2+} concentration.

Direct proof of a hormone-induced redistribution of intracellular calcium from the endoplasmic reticulum to the mitochondria has been difficult to obtain experimentally (for reviews see Denton and McCormack, 1985; Williamson *et al.*, 1985). Nevertheless, four lines of evidence suggest an increased uptake of calcium by the mitochondria during stimulation of hepatocytes with Ca^{2+}-mobilizing hormones. (1) A small net increase of mitochondrial calcium has been observed in vasopressin-treated hepatocytes following rapid cell disruption (Shears and Kirk, 1984). (2) the activity of intramitochondrial Ca^{2+}-sensitive enzymes such as pyruvate dehydrogenase has been shown to be increased in mitochondria isolated from hormone-treated livers (Hems *et al.*, 1978; Assimacopoulos-Jeannet *et al.*, 1983; Sies *et al.*, 1983; McCormack, 1985). (3) New evidence has been presented indicating that mitochondria actively recycle Ca^{2+} at physiological levels of extramitochondrial Ca^{2+} (Nicchitta and Williamson, 1984). (4) Measurements of the calcium balance in the cell in the steady state after hormone stimulation indicate that net Ca^{2+} efflux is less than the amount of Ca^{2+} mobilized by Ins-1,4,5-P_3.

Studies with permeabilized hepatocytes have shown that the maximum amount of Ins-1,4,5-P_3-induced Ca^{2+} release is approximately 0.5 μmole/g cell dry weight (Joseph *et al.*, 1984a), whereas values for net Ca^{2+} efflux range from 0.2 to 0.4 μmole/g cell dry weight (Blackmore *et al.*, 1982, 1983; Reinhart *et al.*, 1982, 1984c; Joseph *et al.*, 1983). If there is a net gain of 0.2 μmole/g cell dry weight of calcium in the mitochondria, this would correspond to an increase of about 0.8 nmole/mg mitochondrial protein or about a 1 μM increase of free Ca^{2+} in the matrix space (Coll *et al.*, 1982). A further calculation of interest may be made by comparing the maximum amount of calcium released by Ins-1,4,5-P_3 (0.5 μmole/g cell dry weight) with the increment of peak cytosolic free Ca^{2+} induced by addition of a maximum hormone concentration to intact cells (0.5 μM). Assuming that this is distributed in 2 ml/g cell dry weight of cytosol, the ratio of bound to free Ca^{2+} in the cytosol would be on the order of 500. This high ratio accounts for the fact that a relatively large proportion (up to 20%) of the total cell calcium is mobilized during maximal hormone stimulation.

The influence of extracellular Ca^{2+} on the hormone-induced transient increase of cytosolic free Ca^{2+} can best be investigated by comparing the Quin 2 Ca^{2+} response obtained with cells incubated in the presence of 1.3 mM extracellular Ca^{2+} with the response to hormone obtained immediately after addition of a slight excess of EGTA. The results of such an experiment in hepatocytes after addition of vasopressin are shown in Fig. 10. Removal of extracellular Ca^{2+} had no immediate effect on the cytosolic free Ca^{2+} concentration prior to the addition of vasopressin or on the peak of the Ca^{2+} transient, but it caused the cytosolic free Ca^{2+} to return to its prestimulated value after a few minutes. In contrast, in the presence of extracellular Ca^{2+}, the cytosolic free Ca^{2+} remained elevated in the steady state after hormone addition, significantly above the control level. These data suggest that the primary function of extracellular Ca^{2+} in the hormone response is to cause a small but prolonged elevation of the cytosolic free Ca^{2+}.

Analysis of many experiments has shown that the hormone-induced increase of the steady-state cytosolic free Ca^{2+} is 70 ± 8 nM (Joseph *et*

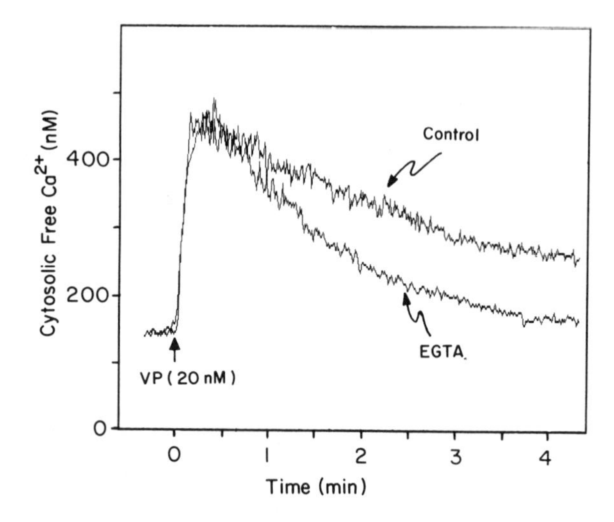

Figure 10. Effect of removal of extracellular Ca^{2+} on the vasopressin (VP)-induced cytosolic free Ca^{2+} in hepatocytes. Control cells were incubated in the presence of 1.3 mM Ca^{2+}. Calcium was removed from this medium for the "EGTA" trace by addition of 1.4 mM EGTA 30 sec before vasopressin.

al., 1985). This small sustained elevation of the cytosolic free Ca^{2+} has important effects on cell function, since it maintains the cell in an active state for prolonged periods. In liver this elevation accounts for sustained hormonal stimulations of various metabolic processes such as glycogenolysis and gluconeogenesis and increased oxygen consumption (Reinhart *et al.*, 1984b). Thus, as shown in Fig. 11, phosphorylase *a* activity in hepatocytes after vasopressin stimulation remains elevated despite the abrupt fall of the cytosolic free Ca^{2+} (see also Rasmussen and Barritt, 1984). One interpretation of this effect is that there is a simultaneous inhibition of protein phosphatase activity in conjunction with a transient stimulation of Ca^{2+}-dependent protein kinase activity. This may result in a slower turnover of covalently attached phosphate in target proteins such as phosphorylase, so that they are maintained in a more highly phosphorylated state despite a decrease of Ca^{2+} as the signal initiator of the functional response.

Two different experimental approaches have indicated that despite a net loss of calcium from the liver during hormonal stimulation, there is a hormone-mediated increase of Ca^{2+} influx into the cell. Reinhart *et al.* (1984c), using $^{45}Ca^{2+}$, have measured a twofold stimulation of Ca^{2+} exchange across the plasma membrane of perfused liver in the steady state following phenylephrine administration, and similar results have been ob-

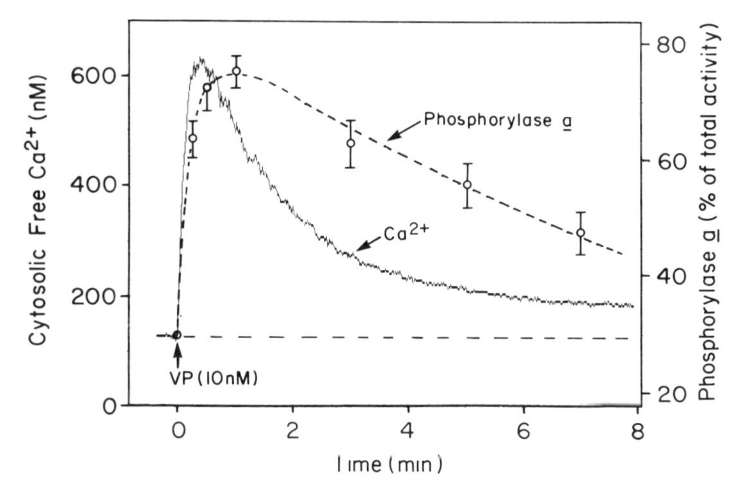

Figure 11. Comparison of changes of cytosolic free Ca^{2+} and phosphorylase *a* in hepatocytes incubated in the presence of 1.3 mM extracellular Ca^{2+}. See Joseph *et al.* (1985) for experimental details.

tained with hepatocytes using vasopressin, angiotensin II, or norepinephrine (Mauger *et al.*, 1984). A different approach of adding Ca^{2+} to hepatocytes incubated in the absence of extracellular Ca^{2+} with and without vasopressin and measuring changes of cytosolic free Ca^{2+} on adding Ca^{2+} after the termination of the hormone-induced transient of intracellular Ca^{2+} mobilization (Joseph *et al.*, 1985) also showed about a twofold hormone-induced increased permeability of the plasma membrane, with maximal effects at 2 mM extracellular Ca^{2+}. The maximum rate of increase of the cytosolic free Ca^{2+} was 6 nM/sec after Ca^{2+} was added to hepatocytes incubated for 8 min in Ca^{2+}-free medium in the presence of vasopressin (Fig. 12). Using the value derived earlier for the ratio of bound to free Ca^{2+} in the cytosol, this corresponds to an increase of total cytosolic calcium of about 6 nmole/g cell dry weight per second. This value may be compared with a rate of net Ca^{2+} efflux from the cell (see Fig. 9) of about 4 nmole/g cell dry weight per second and a rate of intracellular Ca^{2+} mobilization during the peak of the hormone-induced Ca^{2+} transient (see Fig. 9) of approximately 50 nmole/g cell dry weight per second. Although none of these estimations provides absolute values for unidirectional Ca^{2+} flux, they suggest that the hormone-induced changes of Ca^{2+} flux across the plasma membrane are about an order of magnitude lower than the Ins-1,4,5-P_3-induced mobilization of intracellular Ca^{2+}.

The onset of the increased permeability of the plasma membrane to

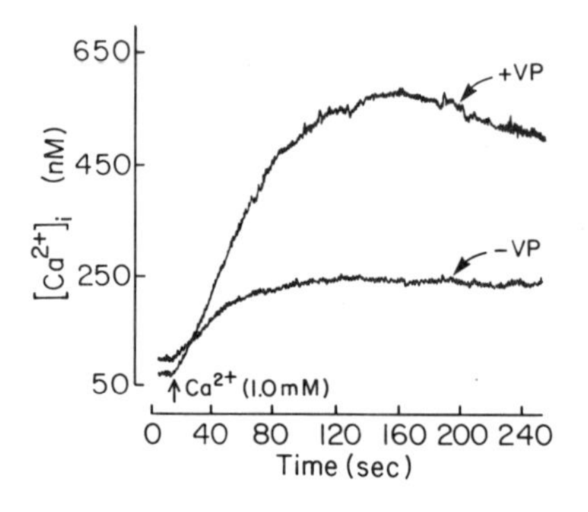

Figure 12. Changes of cytosolic free Ca^{2+} on addition of extracellular Ca^{2+} to hepatocytes incubated in Ca^{2+}-free medium. For hormone-treated cells, 10 nM vasopressin (+VP) was added at time zero, and 1 mM Ca^{2+} was added at 8 min. Control cells (−VP) were treated similarly except for the absence of vasopressin.

Ca^{2+} appears to occur rapidly after hormone addition (Mauger *et al.*, 1984; Reinhart *et al.*, 1984c; Joseph *et al.*, 1985) but is too small to have a significant effect on the peak rise of the Ca^{2+} transient, as illustrated in Fig. 10. However, in the steady state after the peak of the hormone-induced Ca^{2+} transient, the increased rate of Ca^{2+} entry compared with the resting state is associated with an increased steady-state cytosolic free Ca^{2+} (Fig. 11) and a correspondingly increased rate of Ca^{2+} efflux.

The mechanism responsible for the hormone-induced increase of Ca^{2+} entry into liver cells has not been elucidated. However, as shown in Fig. 13, when hepatocytes were incubated for 5 min with 0.5 mM of the Ca^{2+} antagonist diltiazem in the presence of 1.3 mM extracellular Ca^{2+} prior to addition of 10 nM vasopressin, the kinetics of the decline of phosphorylase *a* activity were considerably faster in diltiazem-treated

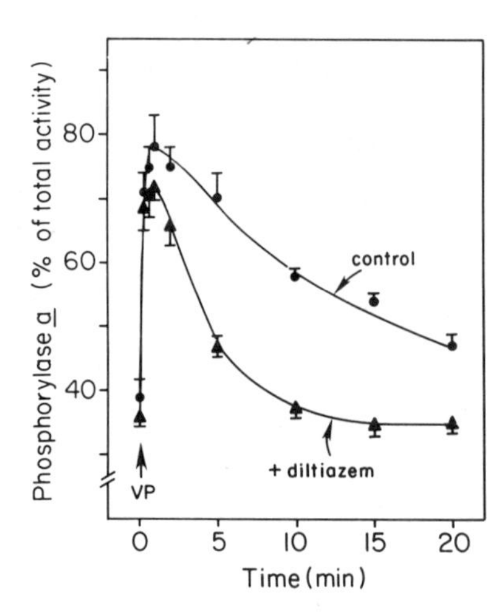

Figure 13. Effect of diltiazem on changes of phosphorylase *a* in hepatocytes induced by addition of 10 nM vasopressin (VP). Hepatocytes were incubated in the presence of 1.3 mM Ca^{2+}, and 0.5 mM diltiazem was added 5 min before vasopressin.

cells than in control cells, although the initial increase and peak of phosphorylase *a* activity were little affected. Qualitatively, this effect resembles the attenuation of the vasopressin-induced cytosolic free Ca^{2+} transient observed after addition of EGTA (Fig. 10), suggesting that the compound may interfere with the hormone-mediated changes in the Ca^{2+} permeability of the plasma membrane. Further studies by Joseph *et al.* (1985) have shown that exposing vasopressin-treated hepatocytes to diltiazem in low-Ca^{2+} medium markedly inhibited the reactivation of phosphorylase observed on the subsequent addition of Ca^{2+}, with half-maximal and maximal effects obtained at 0.12 and 0.5 mM diltiazem, respectively. These concentrations of diltiazem are two to three orders of magnitude greater than the effective concentrations required to block voltage-dependent Ca^{2+} channels in cardiac or smooth muscle (Schwartz *et al.*, 1981; Millard *et al.*, 1983); hence, its site of action may not be at a Ca^{2+} binding protein in the plasma membrane (e.g., associated with Ca^{2+}/Na^{+} or Ca^{2+}/H^{+} exchange). Alternatively, it may interfere with a signal transduction mechanism responsible for modulating plasma membrane Ca^{2+} influx.

It is important to emphasize that although there is a hormone-mediated increased exchange of Ca^{2+} across the plasma membrane, receptor occupancy is associated with a net loss of calcium from the cell through depletion of the Ins-1,4,5-P_3-sensitive calcium pool. Thus, as shown in Fig. 14A, the addition of a maximum concentration of vasopressin failed to elicit a second cytosolic free Ca^{2+} transient, even though the cytosolic free Ca^{2+} had fallen almost to its new steady-state value. Presumably, under these circumstances, the intracellular concentration of Ins-1,4,5-P_3 was sufficient to saturate the putative IP_3-binding sites associated with Ca^{2+} release. However, when cells were stimulated with a very low (0.2 nM) concentration of vasopressin, which produced a submaximal peak increase of cytosolic free Ca^{2+}, a subsequent addition of phenylephrine caused a second burst of Ca^{2+} release (Fig. 14B), indicating that the first addition of hormone caused only a partial depletion of the Ins-1,4,5-P_3-sensitive calcium pools.

Figure 14C shows that when a vasopressin antagonist is added after an initial addition of 0.2 nM vasopressin, the cytosolic free Ca^{2+} rapidly returns to control levels, and a subsequent addition of phenylephrine gives a larger response than in the absence of vasopressin antagonist (cf. Fig. 14B,C). However, the peak response to this phenylephrine addition was lower than the peak response to an initial maximal addition of hormone (maximum vasopressin and phenylephrine give identical peak increases of the Ca^{2+} transient). These results suggest that internal redistribution of Ca^{2+} on removal of the agonist, which presumably leads to a cessation

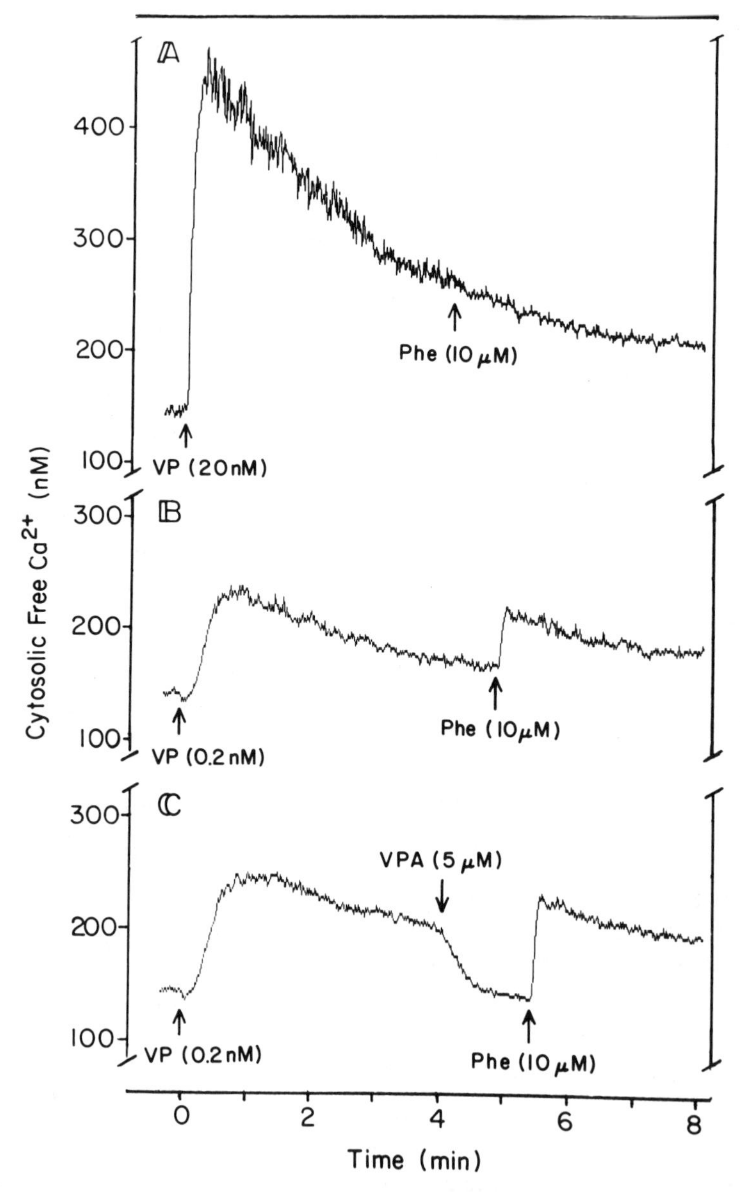

Figure 14. Apparent desensitization to heterologous hormone additions in hepatocytes by partial or complete emptying of the Ins-1,4,5-P_3-sensitive intracellular Ca^{2+} pool. In trace A, saturating concentrations of both vasopressin (VP) and phenylephrine (Phe) were added. In trace B, 10 μM phenylephrine was added after a submaximal (0.2 nM) concentration of vasopressin. In trace C, Vasopressin (0.2 nM) was displaced from its receptor by addition of vasopressin antagonist (VPA), followed by addition of phenylephrine.

of Ins-1,4,5-P_3 production, is not sufficient to refill the Ins-1,4,5-P_3-sensitive Ca^{2+} pool. A net increase of cellular calcium has been shown to occur after termination of hormonal stimulation of perfused rat liver (Reinhart *et al.*, 1982). However, repletion of the Ins-1,4,5-P_3-sensitive Ca^{2+} pool by extracellular Ca^{2+} is a relatively slow process. Thus, when the time interval between additions of vasopressin antagonist and phenylephrine is prolonged (cf. Fig. 14C), it requires 5–10 min for the size of the phenylephrine-induced peak of the Ca^{2+} transient to become equal to that of the control (data not shown). Preliminary results indicate that Ins-1,3,4,3,P_4 caused Ca^{2+} release from liver plasma membrane vesicles loaded with $^{45}Ca^{2+}$ (C. Hansen and J. R. Williamson, unpublished experiments) suggesting that Ins-1,3,4,5-P_4 may have a second messenger role in regulating Ca^{2+} entry into the cell.

4. Summary

On the basis of the data presented in this chapter and the associated discussion, we suggest the following model to account for hormone-induced changes of calcium homeostasis in liver. The rise of cytosolic free Ca^{2+} to its peak value, which occurs within 10 sec of the addition of hormone to hepatocytes, is caused entirely by the mobilization of calcium from intracellular nonmitochondrial storage sites. The subsequent decline of cytosolic free Ca^{2+}, which occurs over the ensuing 2–3 min, can be attributed partly to net loss of Ca^{2+} from the cell and partly to an increased Ca^{2+} sequestration by intracellular organelles, notably the mitochondria. Both of these effects are secondary to the elevated cytosolic free Ca^{2+}. In the absence of extracellular Ca^{2+} the cytosolic free Ca^{2+} declines to its prestimulated level, but in the presence of normal extracellular Ca^{2+} it remains slightly elevated because of a slow but sustained increased rate of Ca^{2+} entry, which balances an increased rate of Ca^{2+} efflux. A small sustained increase of the cytosolic free Ca^{2+} is sufficient to keep phosphorylase *a* levels elevated above control values. With saturating hormone additions, the Ins-1,4,5-P_3-sensitive calcium pool remains depleted during receptor occupancy, but with submaximal hormone concentrations, this pool is only partially depleted, and the remainder can be mobilized by the addition of a different Ca^{2+}-mobilizing hormone. After displacement of the hormone from the receptor, the Ins-1,4,5-P_3-sensitive calcium pool is refilled by a net influx of Ca^{2+} from the extracellular medium, possibly mediated by Ins-1,3,4,5-P_4.

ACKNOWLEDGMENTS. This work was supported by NIH grants AM-15120 and AA-05662.

References

Abdel-Latif, A. A., and Akhtar, R. A., 1982, Cations and the acetylcholine-stimulated ^{32}P-labeling of phosphoinositides in the rabbit iris, in: *Phospholipids in the Nervous System*, Vol. 1 (L. Horrocker, ed.), Raven Press, New York, pp. 251–264.

Assimacopoulos-Jeannet, F., McCormack, J. G., and Jeanrenaud, B., 1983, Effect of phenylephrine on pyruvate dehydrogenase activity in rat hepatocytes and its interaction with insulin and glucagon, *FEBS Lett.* **159:**83–88.

Baraban, J. M., Gould, R. J., Peroutka, S. J., and Snyder, S. H., 1985, Phorbol ester effects on neurotransmission: Interaction with neurotransmitters and calcium in smooth muscle, *Proc. Natl. Acad. Sci. U.S.A.* **82:**604–607.

Batty, I. R., Nahorski, S. R., and Irvine, R. F., 1985, Rapid formation of inositol 1,3,4,5-tetrakisphosphate following muscarinic receptor stimulation of rat cerebral cortical slices, *Biochem. J.* **323:**211–215.

Baukal, A. J., Guillemette, G., Rubin, R., Spat, A., and Catt, K. J., 1985, Binding sites for inositol trisphosphate in the bovine adrenal cortex, *Biochem. Biophys. Res. Commun.* **133:**532–538.

Berridge, M. J., 1983, Rapid accumulation of inositol trisphosphate reveals that agonists hydrolase polyphosphoinsitides instead of phosphatidyl inositol, *Biochem. J.* **212:**849–858.

Berridge, M. J., 1984, Inositol trisphosphate and diacylglycerol as second messengers, *Biochem. J.* **220:**345–360.

Berridge, M. J., and Irvine, R. F., 1984, Inositol trisphosphate, a novel second messenger in cellular signal transduction, *Nature* **312:**315–321.

Besterman, J. M., and Cuatrecasas, P., 1984, Phorbol esters rapidly stimulate amiloride-sensitive Na^+/H^+ exchange in a human leukemia cell line, *J. Cell Biol.* **99:**340–343.

Blackmore, P. F., Hughes, B. P., Shuman, E. A., and Exton, J. H., 1982, α-Adrenergic activation of phosphorylase in liver cells involves mobilization of intracellular calcium without influx of extracellular calcium, *J. Biol. Chem.* **257:**190–197.

Blackmore, P. F., Hughes, B. P., Charest, R., Shuman, E. A., and Exton, J. H., 1983, Time course of α-adrenergic and vasopressin actions on phosphorylase activation, calcium efflux, pyridine nucleotide reduction and respiration in hepatocytes, *J. Biol. Chem.* **258:**10488–10494.

Brown, J. E., Rubin, L. J., Ghalayini, A. J., Tarver, A. P., Irvine, R. F., Berridge, M. J., and Anderson, R. E., 1984, *Myo*-inositol polyphosphate may be a messenger for visual excitation in *Limulus* photoreceptors, *Nature* **311:**160–163.

Burgess, G. M., Godfrey, P. P., McKinney, J. S., Berridge, M. J., Irvine, R. F., and Putney, J. W., Jr., 1984, The second messenger linking receptor activation to internal Ca^{2+} release in liver, *Nature* **309:**63–66.

Burgess, G. M., McKinney, J. S., Irvine, R. F., and Putney, J. W., 1985, Inositol-1,4,5-trisphosphate and inositol 1,3,4-trisphosphate formation in Ca^{2+} mobilizing hormone-activated cells, *Biochem. J.* **323:**237–248.

Burns, C. P., and Rozengurt, E., 1983, Serum, platelet-derived growth factor vasopressin and phorbol esters increase intracellular pH in Swiss 3T3 cells, *Biochem. Biophys. Res. Commun.* **116:**931–938.

Busa, W. B., Ferguson, J. E., Joseph, S. K., Williamson, J. R., and Nuccittelli, R., 1985, Activation of frog (*Xenopus laevis*) eggs by inositol trisphosphate. I. Characterization of Ca^{2+} release from intracellular stores, *J. Cell. Biol.* **101:**677–682.

Castagna, M. Y., Takai, K., Kaibuchi, K., Sano, K., Kikkawa, J., and Nishizuka, Y., 1982,

Direct activation of calcium-activated, phospholipid-dependent protein kinase by tumor-promoting phorbol esters, *J. Biol. Chem.* **257**:7847–7851.

Choquette, D., Hakim, G., Filoteo, A. G., Plishker, G. A., Bostwick, J. R., and Penniston, J. T., 1984, Regulation of plasma membrane Ca^{2+} ATPases by lipids of the phosphatidylinositol cycle, *Biochem. Biophys. Res. Commun.* **125**:908–915.

Coll, K. E., Joseph, S. K., Corkey, B. E., and Williamson, J. R., 1982, Determination of the matrix free Ca^{2+} concentration and kinetics of Ca^{2+} efflux in liver and heart mitochondria, *J. Biol. Chem.* **257**:8696–8704.

Cooper, R. H., Coll, K. E., and Williamson, J. R., 1985, Differential effects of phorbol ester on phenylephrine and vasopressin-induced Ca^{2+} mobilization in isolated hepatocytes, *J. Biol. Chem.* **260**:3281–3288.

Creba, J. A., Downes, C. P., Hawkins, P. T., Brewster, G., Michell, R. H., and Kirk, C. J., 1983, Rapid breakdown of phosphatidylinositol 4-phosphate and phosphatidylinositol 4,5 bisphosphate in rat hepatocytes stimulated by vasopressin and other Ca^{2+}-mobilizing hormones, *Biochem. J.* **212**:733–747.

De Chaffoy de Courcelles, D., Roevens, P., and van Belle, H., 1984, 12-O-Tetradecanoylphorbol 13-acetate stimulates inositol lipid phosphorylation in intact human platelets, *FEBS Lett.* **173**:389–393.

Denton, R. M., and McCormack, J. G., 1985, Ca^{2+} transport by mammalian mitochondria and its role in hormone action, *Am. J. Physiol.* **249**:E543–E554.

Drust, D. S., and Martin, T. F. J., 1984, Thyrotropin-releasing hormone rapidly activates protein phosphorylation in GH_3 pituitary cells by a lipid-linked protein kinase C-mediated pathway, *J. Biol. Chem.* **259**:14520–14530.

Fein, A., Payne, R., Corson, D. W., Berridge, M. J., and Irvine, R. F., 1984, Photoreceptor excitation and adaptation by inositol 1,4,5-trisphosphate, *Nature* **311**:157–160.

Exton, J. H., 1980, Mechanisms involved in α-adrenergic phenomena: Role of calcium ions in actions of catecholamines in liver and other tissues, *Am. J. Physiol.* **238**:E3–E12.

Exton, J. H., 1981, Molecular mechanisms involved in α-adrenergic responses, *Mol. Cell. Endocrinol.* **23**:233–264.

Fujita, S., Irita, K., Takeshige, K., and Minakami, S., 1984, Diacylglycerol, 1-oleoyl-2-acetyl-glycerol, stimulates superoxide-generation from human neutrophils, *Biochem. Biophys. Res. Commun.* **120**:318–324.

Garrison, J. C., Johnsen, D. E., and Campanile, C. P., 1984, Evidence for the role of phosphorylase kinase, protein kinase C, and other Ca^{2+}-sensitive protein kinases in the response of hepatocytes to angiotensin II and vasopressin, *J. Biol. Chem.* **259**:3283–3292.

Gilman, A. G., 1984, G proteins and dual control of adenylate cyclase, *Cell* **36**:577–579.

Gomperts, B. D., 1983, Involvement of guanine nucleotide binding protein in the gating of Ca^{2+} by receptors, *Nature* **306**:64–66.

Grynkiewicz, G., Poenie, M., and Tsien, R. Y., 1985, A new generation of Ca^{2+} indicators with greatly improved fluorescence properties, *J. Biol. Chem.* **260**:3440–3450.

Halenda, S. P., and Feinstein, M. B., 1984, Phorbol myristate acetate stimulates formation of phosphatidyl inositol 4-phosphate and phosphatidyl inositol 4,5-bisphosphate in human platelets, *Biochem. Biophys. Res. Commun.* **124**:507–513.

Haslam, R. J., and Davidson, M. M. L., 1984, Receptor-induced diacylglycerol formation in permeabilized platelets; possible role for a GTP-binding protein, *J. Receptor Res.* **4**:605–629.

Hawthorne, J. N., 1983, Polyphosphoinositide metabolism in excitable membranes, *Biosci. Rep.* **3**:887–904.

Hems, D. A., and Whitton, P. D., 1980, Control of hepatic glycogenolysis, *Physiol. Rev.* **60**:1–50.

Hems, D. A., McCormack, J. G., and Denton, R. M., 1978, Activation of pyruvate dehydrogenase in the perfused rat liver by vasopressin, *Biochem. J.* **176:**627–629.

Irvine, R. F., Anggard, E. E., Letcher, A. J., and Downes, C. P., 1985, Metabolism of inositol 1,4,5-trisphosphate in rat parotid glands, *Biochem. J.* **229:**505–511.

Irvine, R. F., Dawson, R. M. C., and Freinkel, N., 1982, Stimulated phosphatidylinositol turnover: A brief appraisal, in: *Contemporary Metabolism*, Vol. 2 (N. Freinkel, ed.), Plenum Press, New York, pp. 301–342.

Irvine, R. F., Letcher, A. J., Lander, D. J., and Downes, C. P., 1984, Inositol trisphosphate in carbachol-stimulated rat parotid glands, *Biochem. J.* **223:**237–243.

Joseph, S. K., and Williams, R. J., 1985, Subcellular localization and some properties of the enzymes hydrolyzing inositol polyphosphates in rat liver, *FEBS Lett.* **180:**150–154.

Joseph, S. K., and Williamson, J. R., 1983, The origin, quantitation and kinetics of intracellular calcium mobilization by vasopressin and phenylephrine in hepatocytes, *J. Biol. Chem.* **258:**10425–10432.

Joseph, S. K., Thomas, A. P., Williams, R. J., Irvine, R. F., and Williamson, J. R., 1984a, *Myo*-inositol 1,4,5-trisphosphate: A second messenger for the hormonal mobilization of intracellular Ca^{2+} in liver, *J. Biol. Chem.* **259:**3077–3081.

Joseph, S. K., Williams, R. J., Corkey, B. E., Matschinsky, F. M., and Williamson, J. R., 1984b, The effect of inositol trisphosphate on Ca^{2+} fluxes in insulin-secreting tumor cells, *J. Biol. Chem.* **259:**12952–12955.

Joseph, S. K., Coll, K. E., Thomas, A. P., Rubin, R., and Williamson, J. R., 1985, The role of extracellular Ca^{2+} in the response of the hepatocyte to Ca^{2+}-dependent hormones, *J. Biol. Chem.* **260:**12508–12515.

Kaibuchi, K., Takai, Y., Sawamura, M., Hoshijima, M., Fujikura, T., and Nishizuka, Y., 1983, Synergistic functions of protein phosphorylation and calcium mobilization in platelet activation, *J. Biol. Chem.* **258:**6701–6704.

Kaibuchi, K., Takai, Y., and Nishizuka, Y., 1985, Protein kinase C and calcium ion in mitogenic response of macrophage-depleted human peripheral lymphocytes, *J. Biol. Chem.* **260:**1366–1369.

Kojima, I., Lippes, H., Kojima, K., and Rasmussen, H., 1983, Aldosterone secretion: Effect of phorbol ester and A23187, *Biochem. Biophys. Res. Commun.* **116:**555–562.

Kraus-Friedman, N., 1984, Hormonal regulation of hepatic gluconeogenesis, *Physiol. Rev.* **64:**170–259.

Labarca, R., Janowsky, A., Patel, J., and Paul, S. M., 1984, Phorbol esters inhibit agonist induced [^{3}H]-inositol-1-P accumulation in rat hippocampal slices, *Biochem. Biophys. Res. Commun.* **123:**703–709.

L'Allemain, G., Franchi, A., Cragoe, E., Jr., and Pouyssegur, J., 1984, Blockade of the Na^+/H^+ antiport abolishes growth factor-induced DNA synthesis in fibroblasts, *J. Biol. Chem.* **259:**4313–4319.

Lapetina, E. G., and Siegel, F. L., 1983, Shape change induced in human platelets by platelet-activation factor: Correlation with formation of phosphatidic acid and phosphorylation of a 40,000 dalton protein, *J. Biol. Chem.* **258:**7241–7244.

Lin, S.-H., Wallace, M. A., and Fain, J. N., 1983, Regulation of Ca^{2+}–Mg^{2+}-ATPase activity in hepatocyte plasma membranes by vasopressin and phenylephrine, *Endocrinology* **113:**2268–2275.

Lynch, C. J., Charest, R., Bocckino, S. B., Exton, J. H., and Blackmore, P. F., 1985, Inhibition of hepatic α_1-adrenergic effects and binding by phorbol myristate acetate, *J. Biol. Chem.* **260:**2844–2851.

Macara, I. G., 1985, Oncogenes, ions, and phospholipids, *Am. J. Physiol.* **248:**C3–C11.

Macara, I. G., Marinetti, G. V., and Balduzzi, P. C., 1984, Transforming protein of avian sarcoma virus UR2 is associated with phosphatidylinositol kinase activity: Possible role in tumorigenesis, *Proc. Natl. Acad. Sci. U.S.A.* **81:**2728–2732.

MacIntyre, D. E., McNicol, A., and Drummond, A. H., 1985, Tumour-promoting phorbol esters inhibit agonist-induced phosphatidate formation and Ca^{2+} flux in human platelets, *FEBS Lett.* **180:**160–164.

Mauger, J.-P., Poggioli, J., Guesdon, F., and Claret, M., 1984, Noradrenaline, vasopressin and angiotensin increases Ca^{2+} influx by opening a common pool of Ca^{2+} chhannels in isolated rat liver cells, *Biochem. J.* **221:**121–127.

McCormack, J. G., 1985, Studies on the activation of rat liver pyruvate dehydrogenase by adrenaline and glucagon, *Biochem. J.* **231:**597–608.

Michell, R. H., 1975, Inositol phospholipids and cell surface receptor function, *Biochim. Biophys. Acta* **415:**81–147.

Michell, R. H., 1979, Inositol phospholipids in membrane function, *Trends Biochem. Sci.* **4:**128–131.

Michell, R. H., 1982, Inositol lipid metabolism in dividing and differentiating cells, *Cell Calcium* **3:**429–440.

Michell, R. H., Kirk, C. J., Jones, L. M., Downes, C. P., and Creba, J. A., 1981, The stimulation of inositol lipid metabolism that accompanies calcium mobilization in stimulated cells: Defined characteristics and unanswered questions, *Trans. R. Soc. Lond.* [*Biol.*] **296:**123–137.

Millard, R. W., Grupp, G., Grupp, T. L., Disalvo, J., DePover, A., and Schwartz, A., 1983, Chronotropic, inotropic, and vasodilator actions of diltiazem, nifedipine, and verapamil, *Circ. Res.* **52**(Suppl. I):I29–I39.

Moolenaar, W. H., Tertoolen, L. G. J., and deLaat, S. W., 1984a, Growth factors immediately raise cytoplasmic free Ca^{2+} in human fibroblasts, *J. Biol. Chem.* **259:**8066–8069.

Moolenaar, W. H., Tertoolen, L. G. J., and deLaat, S. W., 1984b, Phorbol esters and diacylglycerol mimic growth factors in raising cytoplasmic pH, *Nature* **312:**371–374.

Movsesian, M. A., Thomas, A. P., Selak, M., and Williamson, J. R., 1985, Inositol trisphosphate does not release Ca^{2+} from permeabilized myocytes and cardiac sarcoplasmic reticulum, *FEBS Lett.* **185:**328–332.

Naccache, P. H., Molski, T. F. P., Borgeat, P., White, J. R., and Sha'afi, R. I., 1985, Phorbol esters inhibit FMet-Leu-Phe and leukotriene B_4 stimulated calcium mobilization and enzyme secretion in rabbit neutrophils, *J. Biol. Chem.* **260:**2125–2131.

Nicchitta, C. V., and Williamson, J. R., 1984, Spermine: A regulator of mitochondrial calcium cycling, *J. Biol. Chem.* **259:**12978–12983.

Nishizuka, Y., 1984a, Protein kinases in signal transduction, *Trends Biochem. Sci.* **9:**163–166.

Nishizuka, Y., 1984b, The role of protein kinase C in cell surface signal transduction and tumor promotion, *Nature* **308:**693–698.

Nishizuka, Y., Takai, Y., Kishimoto, U. K., and Kaibuchi, K., 1984, Phospholipid turnover in hormone action, *Recent Prog. Horm. Res.* **40:**301–341.

Orellana, S. A., Solski, P. A., and Brown, S. H., 1985, Phorbol ester inhibits phosphoinositide hydrolysis and calcium mobilization in cultured astrocytoma cells, *J. Biol. Chem.* **260:**5236–5239.

Oron, Y., Dascal, N., Nadler, E., and Lupu, M., 1985, Inositol 1,4,5-trisphosphate mimics muscarinic response in *Xenopus* oocytes, *Nature* **313:**141–143.

Pilkis, S. J., Park, C. R., and Claus, T. H., 1978, Hormonal control of hepatic gluconeogenesis, *Vitam. Horm.* **36:**383–460.

Prentki, M., Biden, T. J., Janjic, D., Irvine, R. F., Berridge, M. J., and Wollheim, C. B., 1984, Rapid mobilization of Ca^{2+} from rat insulinoma microsomes in inositol-1,4,5-trisphosphate, *Nature* **309:**562–564.

Prentki, M., Corkey, B. E., and Matschinsky, F. M., 1985, Inositol 1,4,5-trisphosphate and the endoplasmic reticulum Ca^{2+} cycle of a rat insulinoma cell line, *J. Biol. Chem.* **260:**9185–9190.

Prpic, V., Green, K. C., Blackmore, P. F., and Exton, J. H., 1984, Vasopressin, angiotensin II and α_1-adrenergic-induced inhibition of Ca^{2+} transport by rat liver plasma membrane vesicles, *J. Biol. Chem.* **250:**1382–1385.

Putney, J. W., Jr., McKinney, J. S., Aub, D. L., and Leslie, B. A., 1984, Phorbol ester-induced protein secretion in rat parotid gland: Relationship to the role of inositol lipid breakdown and protein kinase C activation in stimulus–secretion coupling, *Mol. Pharmacol.* **26:**261–266.

Rasmussen, H., and Barritt, P. Q., 1984, Calcium messenger system: An integrated view, *Physiol. Rev.* **64:**938–984.

Reinhart, P. H., Taylor, W. M., and Bygrave, F. L., 1982, Calcium ion fluxes induced by the action of α-adrenergic agonists in perfused rat liver, *Biochem. J.* **208:**619–630.

Reinhart, P. H., Taylor, W. M., and Bygrave, F. L., 1984a, The role of calcium ions in the mechanism of action of α-adrenergic agonists in rat liver, *Biochem. J.* **223:**1–13.

Reinhart, P. H., Taylor, W. M., and Bygrave, F. L., 1984b, The contribution of both extracellular and intracellular calcium to the action of α-adrenergic agonists in perfused rat liver, *Biochem. J.* **220:**35–42.

Reinhart, P. H., Taylor, W. M., and Bygrave, F. L., 1984c, The action of α-adrenergic agonists on plasma membrane calcium fluxes in perfused rat liver, *Biochem. J.* **220:**43–50.

Reuter, H., 1983, Calcium channel modulation by neurotransmitters, enzymes, and drugs, *Nature* **301:**569–574.

Rink, T. J., Sanchez, A., and Hallam, T. J., 1983, Diacylglycerol and phorbol ester stimulate secretion without raising cytoplasmic free calcium in human platelets, *Nature* **305:**317–319.

Rosoff, P. M., Stein, L. F., and Cantley, L. C., 1984, Phorbol esters induce differentiation in a pre-B-lymphocyte cell line by enhancing Na^+/H^+ exchange, *J. Biol. Chem.* **259:**7056–7060.

Schwartz, A., Grupp, G., Millard, R. W., Grupp, I. L., Lathrop, D. A., Matlib, M. A., Vaghy, P., and Valle, J. R., 1981, Calcium-channel blockers: Possible mechanisms of protective effects on the ischemic myocardium, in: *New Perspectives on Calcium Antagonists* (G. D. Weiss, ed.), Waverly Press, Baltimore, pp. 191–210.

Seyfred, M. A., Farrell, L. E., and Wells, W. W., 1984, Characterization of D-*myo*-inositol 1,4,5-trisphosphate phosphatase in rat liver plasma membranes, *J. Biol. Chem.* **259:**13204–13208.

Shears, S. B., and Kirk, C. J., 1984, Determination of mitochondrial calcium content in hepatocytes by a rapid cellular fractionation technique, *Biochem. J.* **219:**383–389.

Sies, H., Graf, P., and Crane, D., 1983, Decreased flux through pyruvate dehydrogenase during calcium ion movements induced by vasopressin, α-adrenergic agonists, and the ionophore A23187 in perfused rat liver, *Biochem. J.* **212:**271–278.

Storey, D. J., Shears, S. B., Kirk, C. J. and Michell, R. H., 1984, Stepwise enzymatic dephosphorylation of inositol 1,4,5-trisphosphate to inositol in liver, *Nature* **312:**374–376.

Streb, H., Irvine, R. F., Berridge, M. J., and Schulz, I., 1983, Release of Ca^{2+} from a non-mitochondrial intracellular store in pancreatic acinar cells by inositol 1,4,5-trisphosphate, *Nature* **306:**67–69.

Suematsu, E., Hirata, M., Hashimoto, T., and Kuriyama, H., 1984, Inositol 1,4,5-trisphosphate releases Ca^{2+} from intracellular store sites in skinned single cells of porcine coronary artery, *Biochem. Biophys. Res. Commun.* **120:**481–485.

Sugimoto, Y., Whitman, M., Cantley, L. C., and Erikson, R. L., 1984, Evidence that the Rous sarcoma virus transforming gene product phosphorylates phosphatidylinositol and diacylglycerol, *Proc. Natl. Acad. Sci. U.S.A.* **81:**2117–2121.

Taylor, M. V., Metcalfe, J. C., Hesketh, T. R., Smith, G. A., and Moore, J. P., 1984, Mitogens increase phosphorylation of phosphoinositides in thymocytes, *Nature* **312:**462–465.

Thomas, A. P., Marks, J. S., Coll, K. E., and Williamson, J. R., 1983, Quantitation and early kinetics of inositol lipid changes induced by vasopressin in isolated and cultured hepatocytes, *J. Biol. Chem.* **258:**5716–5725.

Thomas, A. P., Alexander, J., and Williamson, J. R., 1984, Relationship between inositol polyphosphate production and in the increase of cytosolic free Ca^{2+} induced by vasopressin in isolated hepatocytes, *J. Biol. Chem.* **259:**5574–5584.

Tsien, R. Y., 1983, Intracellular measurements of ion activities, *Annu. Rev. Biophys. Bioeng.* **12:**91–116.

Vergara, T., Tsien, R. Y., and Delay, M., 1985, Inositol 1,4,5-trisphosphate: A possible chemical link in excitation-contraction coupling, *Proc. Natl. Acad. Sci. U.S.A.* **82:**6352–6356.

Vincentini, L. M., DiVirgilio, F., Ambrosini, A., Pozzan, T., and Meldolesi, J., 1985, Tumor promoter phorbol 12-myristate, 13-acetate inhibits phosphoinositide hydrolysis and cytosolic Ca^{2+} rise induced by the activation of muscarinic receptors in PC12 cells, *Biochem. Biophys. Res. Commun.* **127:**310–317.

Watson, S. P., and Lapetina, E., 1985, 1,2-Diacylglycerol and phorbol ester inhibit agonist-induced products of inositolphosphate in human platelets. Possible implications for negative feedback regulation of inositol phospholipid hydrolysis, *Proc. Natl. Acad. Sci. U.S.A.* **82:**2623–2626.

Williamson, J. R., 1986, Role of inositol lipid breakdown in the generation of intracellular signals, *Hypertension*, in press.

Williamson, J. R., Cooper, R. H., and Hoek, J. B., 1981, Role of calcium in the hormonal regulation of liver metabolism, *Biochim. Biophys. Acta* **639:**243–295.

Williamson, J. R., Cooper, R. H., Joseph, S. K., and Thomas, A. P., 1985, Inositol trisphosphate and diacylglycerol as intracellular second messengers in liver, *Am. J. Physiol.* **248:**C203–C216.

Wilson, D. B., Bross, T. E., Hofmann, S. L., and Majerus, P. W., 1984, Hydrolysis of polyphosphoinositides by purified sheep seminal vesicle phospholipase C enzymes. *J. Biol. Chem.* **259:**11718–11724.

Whitaker, M., and Irvine, R. F., 1984, Inositol 1,4,5-trisphosphate microinjection activates sea urchin eggs, *Nature* **312:**636–639.

Zawalich, W., Brown, C., and Rasmussen, H., 1983, Insulin secretion: Combined effects of phorbol ester and A23187, *Biochem. Biophys. Res. Commun.* **117:**448–455.

12

Comparison of the Na^+ Pump and the Ouabain-Resistant K^+ Transport System with Other Metal Ion Transport ATPases

LEIGH ENGLISH, BENJAMIN WHITE, and LEWIS CANTLEY

1. Introduction

Digitalis has long been of medical importance and was mentioned in herbal treatment as early as 1250 (Schery, 1972). This drug entered into accepted medical practice in 1785 following the experimental observations of William Withering (Withering, 1785). Today it is commonly used as a cardiac drug to strengthen heart muscle contraction. In the 1950s the receptor for this drug was shown to be the plasma membrane Na^+–K^+ ATPase, which actively pumps Na^+ out of the cell and K^+ into the cell to maintain cytoplasmic ion concentrations (Skou, 1965). The ability of ouabain, a member of the digitalis family of cardiac glycosides, to inhibit the Na^+–K^+ ATPase has been widely used to characterize this important enzyme. Recently, a gene was cloned that, when transfected into green monkey fibroblasts, rescued cells from ouabain toxicity (Levenson, 1984). Characterization of the resistant cells indicated the presence of a new ouabain-resistant potassium transport system with characteristics similar to but distinct from the native Na^+–K^+ ATPase (English *et al.*, 1985a,b). In this chapter we discuss some of the structural and kinetic properties of

LEIGH ENGLISH, BENJAMIN WHITE, and LEWIS CANTLEY • Department of Biochemistry and Molecular Biology, Harvard University, Cambridge, Massachusetts 02138.

the Na^+–K^+ ATPase and a family of related cation transport systems. The nature of the transport system induced by the ouabain resistance gene is discussed in relation to this family of proteins.

2. ATP-Dependent Metal Ion Pumps

Four ATP-dependent metal ion pumps have been well characterized: the Na^+–K^+ ATPase (Cantley, 1981), the sarcoplasmic reticulum Ca^{2+} ATPase (de Meis and Vianna, 1979), the bacterial K^+ ATPase (Hesse *et al.*, 1984), and the gastric K^+–H^+ ATPase (Sachs *et al.*, 1976). These four enzymes have been shown to have many features in common. They all form a phosphorylated intermediate during ATP hydrolysis and have catalytic subunits of similar molecular weight. In addition, all four of these enzymes have sequence homology around the ATP-binding region. These features are discussed for comparison with the normal Na^+–K^+ pump and the K^+ transport system induced by the ouabain resistance gene.

The Na^+ pump is composed of two polypeptides, α and β, of approximate molecular weights 100,000 and 55,000, respectively, and both appear to span the bilayer. The α subunit contains the catalytic site and is labeled by affinity analogues of ouabain (Ruoho and Kyte, 1974; Rogers and Lazdunski, 1979). Two different forms of the Na^+ pump can be isolated from mammalian brain (Sweadner, 1979), canine cardiac muscle (Matsude *et al.*, 1984), and rat adipocytes (Lytton, 1985). These forms differ in apparent molecular weight of both α and β subunits and in functional properties. Notably, the form of the enzyme with a slightly larger α subunit and smaller β subunit binds ouabain with higher affinity. It is not known whether these forms result from expression of different genes or from post-translational modification.

At present no function can be attributed to the β subunit, although it cannot be separated from α without loss of catalytic activity. An affinity analogue of digitoxin has been shown to preferentially label the β subunit, indicating that this subunit may make up part of the cardiac glycoside binding site (Hall and Ruoho, 1980). This research demonstrates that the β subunit is in close proximity to the α subunit. The β subunit is embedded in the hydrophobic lipid bilayer, as indicated by labeling with hydrophobic affinity labels (Montecucco *et al.*, 1981), and a segment of the subunit is extracellular as indicated by the covalently bound carbohydrate. The precise site of the carbohydrate on the β subunit is not known. The function of the β subunit is not at all certain; however, its stoichiometric presence in all active preparations of the enzyme indicates an essential structural or regulatory role.

The Ca^{2+} ATPase of the sarcoplasmic reticulum is a single polypeptide chain nearly identical in size to the α subunit of Na^+–K^+ ATPase (~100 kD; MacLennan *et al.*, 1971). The functional mechanisms of these two enzymes are strikingly similar (Cantley, 1981), and sequence homology suggests that they evolved from a common gene (Farley *et al.*, 1984). However, the Ca^{2+} ATPase is unaffected by cardiac glycosides, and no subunit analogous to the Na^+–K^+ ATPase β subunit is necessary for catalytic activity. The fact that the Ca^{2+} ATPase functions without a β subunit suggests alternative roles for this subunit such as specific localization of the Na^+ pump in the plasma membrane or some as yet undetermined regulatory function.

The gastric H^+–K^+ ATPase is also a membrane-bound enzyme responsible for the coupled regulation of H^+ and K^+ in specialized cells of the stomach. This enzyme transports K^+ into the cell in exchange for H^+. Like the Na^+ pump and the Ca^{2+} ATPase, the H^+–K^+ ATPase has a catalytic subunit of approximately 100 kD and during enzyme turnover forms a β-aspartyl phosphorylated intermediate (Saccomani *et al.*, 1983). Two other proteins of about 100 kD copurify with the H^+–K^+ ATPase. One, a glycoprotein, contains no catalytic site and like the β subunit of the Na^+ pump is relatively insensitive to trypsin digestion. Like the Ca^{2+} ATPase, this enzyme is insensitive to ouabain. A similar enzyme appears to function in some insect cells, coregulating both cellular pH and K^+ (English and Cantley, 1984, 1985).

The bacterial K^+ ATPase, like the other ATPases mentioned, is a membrane-bound ATPase found in the inner bacterial membrane of *E. coli*. This is one of the systems used by the bacteria to regulate cellular K^+. Like the Na^+ pump, H^+–K^+ ATPase, and the Ca^{2+} ATPase, the catalytic subunit (A 90-kD polypeptide) is transiently phosphorylated at an aspartyl residue as an intermediate in turnover (Hesse *et al.*, 1984). This enzyme also has two other subunits of 47 and 22 kD (Laimino *et al.*, 1978), but, as with the β subunit of the Na^+ pump, the function of these subunits is not known. The primary sequences of all three subunits have been determined, and the 90-kD catalytic subunit has significant sequence homology with the Ca^{2+} ATPase, indicating evolutionary conservation of this type of ATPase (Hesse *et al.*, 1984).

3. Sodium Pump Structure

Three-dimensional crystals do not exist for any of these ATPases; however, some information about the tertiary structure of the Na^+ pump and the Ca^{2+} ATPase has been deduced from analysis of polypeptide

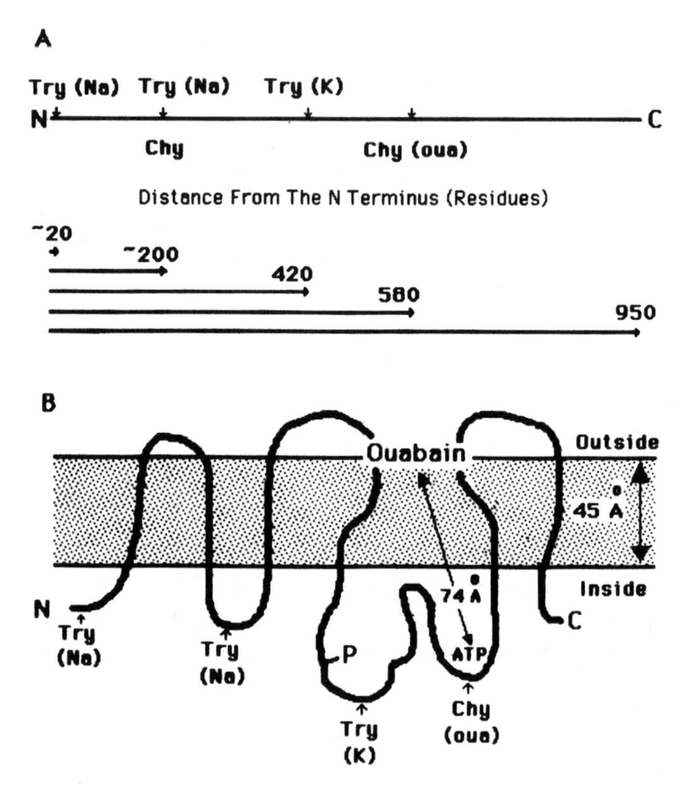

Figure 1. A: Proteolytic cleavage sites on the α subunit of the Na^+ pump. Proteolytic sites are labeled with arrows. Abbreviations: Try, trypsin; Chy, chymotrypsin; oua, ouabain. Ion requirements for each of the proteolytic sites are given in parentheses. B: Schematic diagram of the α subunit of the Na^+ pump in the membrane. N and C refer to the N-terminal and C-terminal ends of the peptide; P refers to the phosphorylated active site. ATP analogues bind near the chymotryptic sites generated in the presence of ouabain [C(oua)].

fragments of these enzymes (Castro and Farley, 1979; Thorley-Lawson *et al.*, 1975). In particular, tryptic and chymotryptic cuts of the α subunit of the Na^+ pump provide considerable insight into the mechanism of ATPase activity of this enzyme. All of the proteolytic cuts are on the cytoplasmic side (Chin and Forgac, 1984). The size of these cuts indicate that most of the enzyme is located in the bilayer or cytoplasm. The major proteolytic fragments remain associated with the membrane. These trypsin-insensitive portions of the peptide label with hydrophobic reagents and are sufficient in length to cross the bilayer several times.

A model for folding the catalytic subunit of the Na^+ pump across the plasma membrane is proposed in Fig. 1B. This model is based on the previously described experiments and on site-specific labeling studies. The structure proposed has four transmembrane segments on the amino-terminal half of the molecule and two transmembrane segments on the carboxy-terminal half, with a central hydrophilic domain containing the ATP hydrolysis site located on the cytoplasmic surface. This structure is consistent with the location of hydrophobic transmembrane stretches of the bacterial K^+ ATPase predicted from the complete sequence of the catalytic subunit (Hesse *et al.*, 1984).

Binding of Na^+ or K^+ to the enzyme affects the protein conformation and the sites of preferential proteolysis by trypsin. The primary tryptic cleavage sites in the presence of Na^+ [designated T(Na)] are near the amino terminus, whereas in the presence of K^+ [designated T(K)], trypsin hydrolyzes at a series of sites near the center of the polypeptide chain. Binding ouabain to the enzyme stabilizes a conformation similar to that induced by K^+ as judged by tryptic digestion. However, new sites for chymotryptic cleavage become exposed when ouabain is added to the enzyme. These sites [designated C(oua)] are not exposed in the presence of Na^+ alone or K^+ alone (Castro and Farley, 1979). The tryptic and chymotryptic proteolysis sites are illustrated in Fig. 1A.

Ouabain binds to the Na^+–K^+ ATPase on a site exposed to the outside of the cell. Ouabain affinity analogues label regions of the α subunit from both the amino-terminal half and the carboxy-terminal half. The β subunit is also labeled by one affinity probe (Hall and Ruoho, 1980). Based on the interaction of the Na^+ pump with fluorescence-5′-isothiocyanate (FITC), a reagent that appears to bind at the cytoplasmic ATP hydrolysis site, Carilli *et al.* (1982) were able to demonstrate that this site is approximately 74 Å from the ouabain binding site. Interestingly, the FITC site is near the chymotryptic cleavage site that is exposed when ouabain binds to the enzyme (see Fig. 1A), suggesting that a major conformational change occurs on ouabain binding. The sequence of the FITC binding region and the peptides surrounding the aspartyl active site are also known for the Ca^{2+} ATPase and have remarkable sequence homology with the same region on the Na^+–K^+ ATPase (Allen *et al.*, 1980; Mitchinson *et al.*, 1982; Bastide *et al.*, 1973; Farley *et al.*, 1984). The peptide sequences surrounding the aspartyl active site of the K^+ ATPase and the FITC binding region of the H^+–K^+ ATPase share sequence homology with both the Na^+–K^+ ATPase and the Ca^{2+} ATPase (Hesse *et al.*, 1984; Farley and Faller, 1985) (Fig. 2).

The unique proteolytic cuts of the Na^+ pump α subunit are interesting in view of the Na^+- and K^+-dependent mechanisms of ATP hydrolysis. In the presence of Na^+ the enzyme is able to reversibly bind ATP and generate the phosphorylated aspartyl intermediate; K^+ subsequently

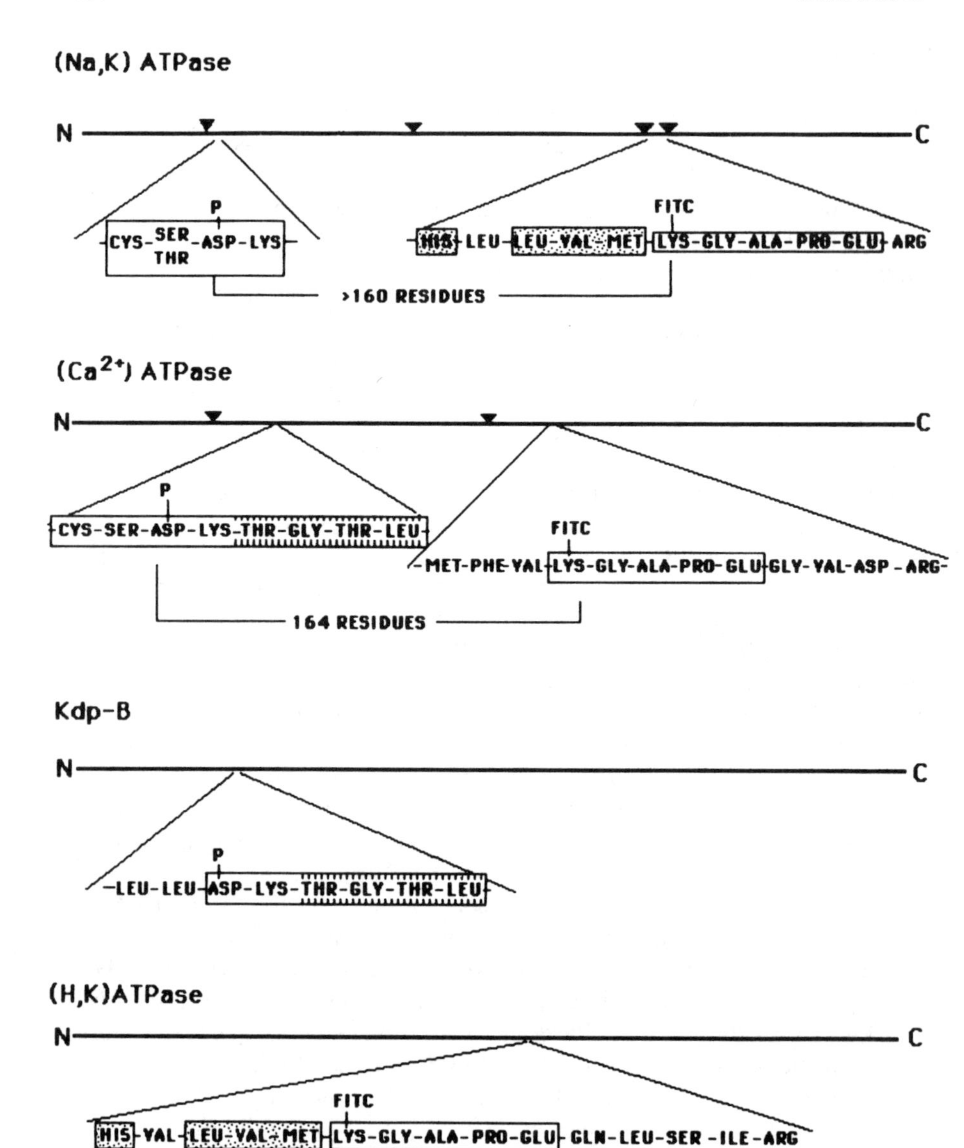

Figure 2. Comparison of the primary sequence around the active site of the Na^+–K^+ ATPase, Ca^{2+} ATPase, *E. coli* K^+ ATPase (Kdp ATPase), and the gastric H^+–K^+ ATPase. Sequence information for the Na^+–K^+ ATPase, Ca^{2+} ATPase, K^+ ATPase, and H^+–K^+ ATPase was taken from Farley *et al.* (1984), Mitchenson *et al.* (1982), Hesse *et al.* (1984), and Farley and Faller (1985), respectively. Homologous residues are framed. Proteolytic sites are indicated by (▼).

causes the enzyme to break the covalent aspartyl phosphate bond by inducing a conformational change in the enzyme. The K^+-dependent form of the enzyme is capable of reacting with inorganic phosphate to form an aspartyl-phosphate anhydride bond by a mechanism similar to the reverse of the final step in hydrolysis (Cantley, 1981). The conformation of the Na^+ pump in the presence of Na^+ has been named E_1, whereas the K^+-stabilized conformation is named E_2. Ouabain binds to the E_2 conformation and stabilizes the enzyme in a state in which the phosphoaspartate bond is both thermodynamically and kinetically more stable.

The Na^+ pump is able to hydrolyze substrates other than ATP. For example, paranitrophenylphosphate (pNPP) is hydrolyzed in a K^+-dependent reaction that does not require Na^+ and occurs on the E_2 state of the enzyme (Robinson 1976). The hydrolysis of pNPP by the Na^+ pump is inhibited by Na^+ because the enzyme conformation is altered to the E_1 form. This inhibition can be relieved by adding micromolar ATP, which phosphorylates the enzyme and thereby shifts it to the E_2 state. Apparently pNPP is hydrolyzed by the E_2 state after dephosphorylation and prior to the relaxation back to the E_1 form (Chipman and Lev, 1983). Like ATP hydrolysis, pNPP hydrolysis by the Na^+ pump is completely inhibited by ouabain.

4. Ouabain-Resistant K^+ Transport System

Various attempts have been made to obtain the gene for the mammalian Na^+ pump. Most recently, by using ouabain selection pressure to force mutagenized cells to enhance their production of the Na^+ pump, a collection of ouabain-resistant mutants was obtained from mouse fibroblasts (Levenson *et al.*, 1984). When a 6.5-kb fragment of the genome of these cells was transfected into otherwise ouabain-sensitive CV 1 green monkey fibroblasts, the resulting cells were also ouabain resistant. Our investigation of these cells demonstrated that the transfected gene is able to induce ouabain resistance in these cells by the enhancement of a K^+ transport system that is able to maintain high cellular K^+ in the presence of ouabain. The cellular Na^+ level (60 mM) is not as low as in normal cells but is still lower than extracellular Na^+ (140 mM) (English *et al.*, 1985a,b).

We have been able to show that the ouabain-resistant K^+ transport system induced by the 6.5-kb genomic DNA is actually very similar to the Na^+ pump. Although plasma membranes from the transfected cells have little Na^+-stimulated ATPase activity (English *et al.*, 1985a), a K^+-dependent and ouabain-resistant pNPPase activity with properties similar

Table I. Potassium pNPPase Activity[a]

	Oua^R6	Oua^R6 oua	Na^+-K^+ ATPase
V_{max}[b]	2.1	2.7	4700
$K_{0.5}$ K^+ (mM)	1.4	0.9	1.0
Ouabain 1 mM[b]	0.9 (59%)	2.4 (12%)	0 (100%)
Vanadate 100 μM[b]	0.0 (100%)	0.0 (100%)	0.0 (100%)
NaCl 100 mM[b]	0.2 (92%)	0.2 (92%)	235 (95%)
$K_{0.5}$ Na^+ (mM)	40	40	40
ATP 10 μM + NaCl 100 mM[b]	2.1 (0%)	2.7 (0%)	4700 (0%)

[a] Potassium pNPPase activity in plasma membranes isolated from the African green monkey kidney fibroblasts (CV1 cells) transfected with the ouabain resistance gene (Oua^R6) and from the same cells induced to express the ouabain-resistant K^+ transport system by growing the cells in 10^{-5} M ouabain for at least 24 hr (Oua^R6 oua). Values are compared with purified dog kidney Na^+-K^+ ATPase. Membranes were isolated according to the method of Brunette and Till (1971). Assay included 2.5 mM *p*-nitrophenylphosphate, 3 mM $MgCl_2$, 1 mM dithiothreitol, 25 mM Tris HCl, pH 7.5.
[b] Specific activity expressed as nmole/mg per min.

to the Na^+-K^+ ATPase is apparent. Like the K^+ pNPPase of purified dog kidney Na^+-K^+ ATPase, K^+-dependent pNPPase activities in these membranes is inhibited by vanadate and Na^+. The Na^+ inhibition is relieved by micromolar concentrations of ATP (Table I). The K_m for K^+ activation and K_i for Na^+ inhibition are also similar to those found for dog kidney enzyme and for the native monkey fibroblasts Na^+-K^+ ATPase. Thus, the only major difference between the enzyme induced by the ouabain resistance gene and the normal Na^+-K^+ pump appears to be the ability of ouabain to inhibit activity.

Further evidence for similarity between the native Na^+-K^+ pump and the enzyme induced by the ouabain resistance gene is provided by active site labeling. A 100-kD peptide with electrophoretic mobility similar to that of dog kidney Na^+-K^+ ATPase α subunit can be phosphorylated when membranes from cells expressing the ouabain resistance gene are incubated with [γ-^{32}P]-ATP. This phosphorylation, like that of the dog kidney enzyme, is enhanced by Na^+ and decreased by K^+. However, when dog kidney enzyme is incubated in ouabain plus unlabeled inorganic phosphate prior to the brief labeling with [γ-^{32}P]-ATP, the active site phosphorylation is completely blocked. Preincubation with ouabain plus inorganic phosphate has no effect on [γ-^{32}P]-ATP phosphorylation of the 100-kD peptide from cells transfected and expressing the ouabain resistance gene. Thus, like the K^+ pNPPase activity, the active site phosphorylation of the enzyme induced by the ouabain resistance gene is unaffected by high concentrations (1 mM) of ouabain. These data suggest that the ouabain-resistant cells express an ATPase similar to the Na^+ pump. It has a catalytic subunit of approximately 100 kD and can be phosphorylated by ATP in the presence of Na^+.

These data illustrate that the induced K^+ transport system is similar to the Na^+–K^+ ATPase with the exception that it is now ouabain insensitive. This is particularly interesting in light of the fact that the size of the message encoded by the ouabain resistance gene is 1.2 kb, indicating that the protein must be less than 40 kD (English *et al.*, 1985a). None of the enzymes in the family of ATP-dependent metal ion pumps characterized thus far has a catalytic subunit this small. This leads us to speculate that the message may actually be coding for the β subunit of the Na^+ pump and that this subunit is specifically responsible for regulating the ion and ouabain affinity of the Na^+ pump. Although this may not be true, and other proteins may be affecting the ouabain binding site, it is clear that the transfection and expression of the ouabain resistance gene results in appearance of a 100-kD protein with properties similar to those of the normal Na^+ pump but with no sensitivity to ouabain. The nature of the protein encoded by the ouabain resistance gene will be informative in our understanding of the structure and mechanism of this important transport system.

ACKNOWLEDGMENT. Supported by NIH grants GM 36133 and GM 35912.

References

Allen, G., Trinnaman, B. J., and Green, N. M., 1980, The primary structure of the calcium ion-transporting adenosine triphosphatase protein of rabbit skeletal sarcoplasmic reticulum, *Biochem. J.* **187**:591–616.

Avruch, J., and Fairbanks, G., 1972, Demonstration of a phosphopeptide intermediate in Mg^{2+}-dependent Na^+- and K^+-stimulated adenosine triphosphatase reaction of the erythrocyte membrane, *Proc. Natl. Acad. Sci. U.S.A.* **69**:1216–1220.

Bastide, F., Meissner, G., Fleischer, S., and Post, R. L., 1973, Similarity of the active site of phosphorylation of the adenosine triphosphatase for transport of sodium and potassium ions in kidney to that for transport of calcium ions in the sarcoplasmic reticulum of muscle, *J. Biol. Chem.* **248**:8385–8391.

Brunette, D. M. and Till, J. E., 1971, A rapid method for the isolation of L cell surface membranes using an aqueous two-phase polymer system, *J. Membr. Biol.* **5**:215–224.

Cantley, L. C., 1981, Structure and mechanism of the (Na^+, K^+) ATPase, *Curr. Top. Bioenerg.* **11**:201–237.

Carilli, C. T., Farley, R. A., Perlman, D. M., and Cantley, L. C., 1982, The active site structure of Na^+- and K^+-stimulated ATPase, *J. Biol. Chem.* **257**:5601–5606.

Castro, J., and Farley, R. A., 1979, Proteolytic fragmentation of the catalytic subunit of the sodium and potassium adenosine triphosphatase, *J. Biol. Chem.* **254**:2221–2228.

Chin, G., and Forgac, M., 1984, Purification and proteolysis of vesicles containing inside-out and rightside-out oriented reconstituted (Na^+,K^+)-ATPase, *J. Biol. Chem.* **259:**5255–5263.

Chipman, D. M., and Lev, A., 1983, Modification of the conformational equilibria in the sodium and potassium dependent adenosinetriphosphatase with glutaraldehyde, *Biochemistry* **22:**4450–4459.

de Meis, L., and Vianna, A. L., 1979, Energy interconversion by the Ca^{2+}-dependent ATPase of the sarcoplasmic reticulum, *Annu. Rev. Biochem.* **48:**275–292.

English, L. H., and Cantley, L. C., 1984, Characterization of monovalent ion transport in an insect cell line (*Manduca sexta* embryonic cell line CHE), *J. Cell. Physiol.* **121:**125–132.

English, L. H. and Cantley, L. C., 1985, Delta endotoxin inhibits K^+-uptake, lowers cytoplasmic pH, and inhibits a K^+ATPase in the *Manduca sexta* CHE cell, *J. Membr. Biol.* **85:**199–204.

English, L. H., Epstein, J., Cantley, L., Housman, D., and Levenson, R., 1985a, Expression of an ouabain resistance gene in transfected cells, *J. Biol. Chem.* **260:**1114–1119.

English, L. H., Epstein, J., Cantley, L., Housman, D., and Levenson, R., 1985b, Ouabain treatment induces an amiloride-sensitive K^+ transport system in cells transfected with the ouabain resistance gene, in: *The Sodium Pump* (I. M. Glynn and J. C. Ellory, eds.), The Company of Biologists Ltd., London, pp. 193–196.

Farley, R. A., and Faller, L. D., 1985, The amino acid sequence of a fluorescein-labeled peptide from the ATP-binding site of gastric H,K-ATPase, *Fed. Proc.* **44:**2527.

Farley, R. A., Goldman, D., and Bayley, H., 1980, Identification of the regions of the catalytic subunit of (Na–K)-ATPase embedded within the cell membrane, *J. Biol. Chem.* **255:**860–864.

Farley, R. A., Tran, C. M., Carilli, C. T., Hawke, D., and Shively, J. E., 1984, The amino acid sequence of a fluorescein-labeled peptide from the active site of (Na,K)-ATPase, *J. Biol. Chem.* **259:**9532–9535.

Hall, C., and Ruoho, A., 1980, Ouabain-binding-site photoaffinity probes that label both subunits of Na^+, K^+-ATPase, *Proc. Natl. Acad. Sci. U.S.A.* **77:**4529–4533.

Hesse, J. E., Wieczorek, L., Altendorf, K., Reicin, A. S., Dorus, E., and Epstein, W., 1984, Sequence homology between two membrane transport ATPases, the Kdp-ATPase of *Escherichia coli* and the Ca^{2+}-ATPase of sarcoplasm reticulum, *Proc. Natl. Acad. Sci. Sci. U.S.A.* **81:**4746–4750.

Jorgensen, P. L., 1982, Mechanism of the Na^+, K^+ pump: Protein structure and conformations of the pure (Na^+, K^+)ATPase, *Biochim. Biophys. Acta* **694:**27–68.

Laimins, L. A., Rhoads, D. B., Altendorf, K., and Epstein, W., 1978, Identification of the structural proteins of an ATP-driven potassium transport system in *Escherichia coli, Proc. Natl. Acad. Sci. U.S.A.* **75:**3216–3219.

Levenson, R., Racaniello, V., Albritton, L., and Housman, D., 1984, Molecular cloning of the mouse ouabain-resistance gene, *Proc. Natl. Acad. Sci. U.S.A.* **81:**1489–1493.

Lytton, J., Lin, J. C., and Guidotti, G., 1985, Identification of two molecular forms of (Na^+K^+)-ATPase in rat adipocytes, *J. Biol. Chem.* **260:**1177–1184.

MacLennan, D. H., Seeman, P., Iles, G. H., and Yip, C. C., 1971, Membrane formation by the adenosine triphosphatase of sarcoplasmic reticulum, *J. Biol. Chem.* **246:**2702–2710.

MacLennan, D. H., Reithmeier, R. A. F., Shoshan, V., Campbell, K. P., and LeBel, D., 1980, Ion pathways in protein of the sarcoplasmic reticulum, *Ann. N.Y. Acad. Sci.* **358:**138–148.

Mitchinson, C., Wilderspin, A. F., Trinnaman, B. J., and Green, N. M., 1982, Identification of a labelled peptide after stoichiometric reaction of fluorescein isothiocyanate with

Ca^{2+}-dependent adenosine triphosphatase of sarcoplasmic reticulum, *FEBS Lett.* **146:**87–92.

Montecucco, C., Bisson, R., Gach, C., and Johansson, A., 1981, Labelling of the hydrophobic domain of the Na^+, K^+-ATPase, *FEBS Lett.* **128:**17–21.

Robinson, J. D., 1976, Substrate sites of the ($Na^+ + K^+$)-dependent ATPase, *Biochim. Biophys. Acta* **429:**1006–1019.

Rogers, T. B., and Lazdunski, M., 1979, Photoaffinity labeling of the digitalis receptor in the (sodium + potassium)-activated adenosine triphosphatase, *Biochemistry* **18:**135–140.

Ruoho, A., and Kyte, J., 1974, Photoaffinity labeling of the ouabain-binding site on ($Na^+ + K^+$) adenosinetriphosphatase, *Proc. Natl. Acad. Sci. U.S.A.* **71:**2352–2356.

Sachs, G., Chang, H. H., Rabon, E., Shackman, R., Lewin, M., and Saccomani, G., 1976, A nonelectrogenic H^+ pump in plasma membranes of hog stomach, *J. Biol. Chem.* **251:**7690–7698.

Schery, R. W., 1972, *Plants For Man,* Prentice-Hall, Englewood Cliffs, NJ, p. 312.

Skou, J. C., 1965, Enzymatic basis for active transport of Na^+ and K^+ across cell membrane, *Physiol. Rev.* **45:**596–617.

Sweadner, K. J., 1979, Two molecular forms of ($Na^+ + K^+$)-stimulated ATPase in Brain, *J. Biol. Chem.* **254:**6060–6067.

Thorley-Lawson, D. A., and Green, N. M., 1975, Separation and characterization of tryptic fragments from the adenosine triphosphatase of sarcoplasmic reticulum, *Eur. J. Biochem.* **59:**193–200.

Withering, W., 1785, An account of the foxglove and some of its medical uses: with practical remarks on dropsy and other diseases, Swiney, Birmingham.

13

Current Concepts of Tumor Promotion by Phorbol Esters and Related Compounds

CATHERINE A. O'BRIAN, ROB M. LISKAMP, JOHN P. ARCOLEO, W.-L. WENDY HSIAO, GERARD M. HOUSEY, and I. BERNARD WEINSTEIN

1. The Two-Stage Model of Carcinogenesis

Studies of carcinogenesis on mouse skin have shown that the mechanism of carcinogenesis clearly involves at least two qualitatively distinct stages (Weinstein, 1981a; Weinstein *et al.*, 1982, and references therein). Initiation, the first stage in the two-stage model, is rapid and apparently irreversible. The observed molecular actions of initiators appear to be rapid and irreversible, since several initiators of carcinogenesis or their metabolites are electrophiles that covalently modify DNA *in vivo*. The second stage, promotion, is a slow and often reversible process during which initiated cells become neoplastic. Promotion is also distinguished from initiation at the molecular level, since certain tumor promoters do not bind covalently to cellular DNA but do bind reversibly to cell membrane receptors. The two-stage mouse skin carcinogenesis model has been used as a paradigm for studies on multistage carcinogenesis in several other systems. Evidence that hepatocellular cancer, bladder cancer, colon cancer, and breast cancer also proceed via processes analogous to initiation and promotion has been reviewed elsewhere (Slaga *et al.*, 1978; Weinstein

---

CATHERINE A. O'BRIAN, ROB M. LISKAMP, JOHN P. ARCOLEO, W.-L. WENDY HSIAO, GERARD M. HOUSEY, and I. BERNARD WEINSTEIN • Division of Environmental Sciences, Department of Human Genetics, and Cancer Center/Institute of Cancer Research, Columbia University, New York, New York 10032.

et al., 1981b). The concept of promotion appears to be especially relevant to an understanding of the causation of human breast cancer (Weinstein *et al.*, 1981b).

2. *Cellular Effects of Phorbol Esters and Membrane-Associated Receptors*

In recent years, many cellular and biochemical markers for the action of tumor promoters have been identified in cell culture systems and on mouse skin (Slaga *et al.*, 1978; Weinstein *et al.*, 1979, 1980). We have classified the numerous effects of tumor promoters into three categories: (1) mimicry of transformation; (2) modulation of differentiation; and (3) membrane effects. Some of these membrane effects are listed in Table I. Since tumor promoters induce in normal cells the expression of several phenotypic traits characteristic of tumor cells and also enhance the expression of some of these traits in transformed cells (Slaga *et al.*, 1978; Weinstein *et al.*, 1979, 1980), these promoters are excellent pharmacological tools for studying cellular mechanisms that control the expression of these genes. Tumor promoters can also inhibit or induce differentiation in numerous cell systems (Slaga *et al.*, 1978; Weinstein *et al.*, 1979, 1980), suggesting that a major biological activity of tumor promoters is the modulation of differentiation. The ability of tumor promoters to inhibit ter-

Table I. Effects of TPA on Cell Surfaces and Membranes in Cell Culture[a]

Altered Na^+–K^+ ATPase
Increased uptake 2-DG, ^{32}P
Increased membrane lipid "fluidity"
Increased phospholipid turnover
Increased release arachidonic acid, prostaglandins
Altered morphology and cell–cell orientation
Altered cell adhesion
Increased pinocytosis
Altered fucose glycopeptides
Decreased LETS protein
"Uncoupling" of β-adrenergic receptors
Inhibition of binding of EGF to receptors
Decrease in acetylcholine receptors
Synergistic interaction with growth factors
Inhibition of metabolic cooperation

[a] For specific references see Weinstein (1981a).

minal differentiation may be central to their action as tumor promoters (Slaga *et al.*, 1978; Weinstein *et al.*, 1980).

High-affinity receptors for tumor-promoting phorbol esters have been observed in the membranes of numerous cell types. These receptors also bind tumor-promoting polyacetates (such as aplysiatoxin) and indole alkaloids (such as teleocidin) with high affinity. However, these receptors do not specifically bind nonpromoting phorbol esters, polyacetates, or indole alkaloids (Horowitz *et al.*, 1983). The high affinity binding of chemically diverse tumor promoters (but not nonpromoting structural analogues) to the same membrane receptor provides evidence that the receptor may actually be the primary target of action of these tumor promoters. The importance of membrane receptors in mediating the biological action of tumor promoters is strongly suggested by the effects of tumor promoters on cells that are mediated directly at the level of cell membranes. Several effects that tumor promoters have on cells occur within seconds or minutes of exposure to these compounds and are not precluded by inhibitors of RNA or protein synthesis (Weinstein *et al.*, 1981a).

3. *Effects of Tumor Promoters on Protein Kinase C*

There is considerable evidence that protein kinase C (PKC), the Ca^{2+}- and phospholipid-dependent protein kinase, is the membrane-associated tumor promoter receptor. Protein kinase C and phorbol ester receptor have been copurified to apparent homogeneity from both bovine brain and rat brain (Kikkawa *et al.*, 1982; Parker *et al.*, 1984). In addition, PKC can be activated by tumor-promoting phorbol esters *in vivo*, and the relative potencies of a series of phorbol esters in tumor promotion tend to correlate with their relative potencies in PKC activation (Castagna *et al.*, 1982.

We isolated PKC from rat brain (O'Brian *et al.*, 1984) as well as from bovine brain (Arcoleo and Weinstein, 1985), and we examined the regulation of the activity of each enzyme using Mg-[γ^{32}P]-ATP as a phosphodonor substrate and histone HI as a phosphoacceptor substrate. Reaction mixtures were buffered with 20 mM Tris HCl at pH 7.5 and 30°C, and reactions were terminated within a linear time course (2–15 min). We found with each enzyme preparation that PKC activity was stimulated ten- to 30-fold by 1 mM Ca^{2+} plus phospholipid (10–100 μg/ml phosphatidylserine or phosphatidylinositol) and to a similar extent when 200 nM

TPA* and 1 mM EGTA were added in lieu of Ca^{2+}. Neither TPA nor Ca^{2+} significantly stimulated PKC in the absence of phospholipid. Protein kinase C activation by 1 mM Ca^{2+} plus phospholipid was not enhanced by 200 nM TPA, and the activation by 200 nM TPA plus phospholipid was not enhanced by the replacement of 1 mM EGTA with 1 mM Ca^{2+} (O'Brian *et al.,* 1984; Arcoleo and Weinstein, 1985). Other investigators have also reported maximal stimulation of PKC activity by tumor-promoting phorbol esters and phospholipids in reaction mixtures containing EGTA but no added calcium (Donnelly and Jensen, 1983; Couturier *et al.,* 1984).

We have found that tumor-promoting phorbol esters other than TPA (PDBu, HHPA, and mezerein) also activate PKC in the absence of added Ca^{2+}, whereas under similar conditions biologically inactive phorbol compounds (phorbol, 4-α-PDD) do not stimulate PKC (Arcoleo and Weinstein, 1985). In view of our observation that tumor-promoting phorbol esters could activate PKC in the absence of added Ca^{2+}, it was of interest to test related tumor promoters from different chemical classes for similar effects. When tested at 100 nM in the absence of added Ca^{2+}, the indole alkaloid tumor promoters teleocidin B and lyngbyatoxin A, as well as the polyacetate tumor promoters aplysiatoxin and debromoaplysiatoxin, markedly stimulated PKC activity. Anhydrodebromoaplysiatoxin, which is not a tumor promoter, had little effect on PKC activity (Arcoleo and Weinstein, 1985).

We determined the concentration dependence for activation of PKC by various compounds in the absence of added Ca^{2+}; HHPA and TPA were the most potent compounds tested, producing 50% maximal activation at about 1.0 and 2.5 nM, respectively. we found that aplysiatoxin and debromoaplysiatoxin were equipotent in stimulating PKC, with half-maximal stimulation occurring at about 12 nM. Teleocidin B was also quite potent, producing half-maximal activation at 18 nM (Arcoleo and Weinstein, 1985).

Our results indicate that tumor-promoting indole alkaloids, polyacetates, and phorbol esters activate PKC by apparently similar mechanisms. The results are consistent with our previously postulated model in which the hydrophilic regions of these amphiphilic compounds share structural features that allow them to interact in a similar manner with a common protein receptor, whereas the lipophilic regions of these compounds in-

* Abbreviations used: HHPA, 12-O-hexadecanoyl-16-hydroxylphorbol-13-acetate; PS, phosphatidylserine; PI, phosphatidylinositol; PMSF, phenylmethylsulfonyl fluoride; PIPES, piperazine-N,N'-bis[2-ethanesulfonic acid]; α-PDD, 4-α-phorbol-12,13-didecanoate; PDBu, phorbol-12,13-dibutyrate; PKC, calcium- and phospholipid-dependent protein kinase; TPA, 12-O-tetradecanoylphorbol-13-acetate.

teract with the lipid associated with this protein (Weinstein *et al.*, 1984). We believe that an elucidation of the precise structural requirements for tumor promoters may be obtained from further structure–activity studies. A structural understanding of tumor promoters could lead to the development of a potent PKC inhibitor through the exploitation of determinants necessary for high-affinity binding to PKC.

It has been proposed that diacylglyceride is the endogenous activator of PKC and that diacylglyceride is structurally related to TPA (Nishizuka, 1984). However, TPA is approximately four orders of magnitude more potent than diacylglycerol with respect to *in vitro* stimulation of PKC (Castagna *et al.*, 1982). The tremendous quantitative disparity between the potencies of TPA and diacylglycerol as PKC activators argues against the hypothesis that they activate PKC by essentially the same mechanism. Our structural models of tumor promoters (Weinstein *et al.*, 1984) suggest that TPA, teleocidin, and aplysiatoxin have a similar and very specific hydrophilic domain, which is not present in diacylglycerol, and that this domain probably explains the biological potency and structural specificity of these compounds. We encourage the search for lipid metabolites that are more potent that diacylglycerol in the activation of PKC rather than the assumption that diacylglycerol is the only biologically relevant lipid metabolite that activates PKC. The possibility of endogenous TPA-like compounds is of further interest, since TPA plus phospholipid activates PKC *in vitro* in the absence of added calcium. It is important to determine whether *in vivo* PKC is regulated by intracellular levels of Ca^{2+}, or whether it is, in part, under a Ca^{2+}-independent regulation by endogenous TPA-like compounds.

4. *Inhibition of Protein Kinase C by Tamoxifen*

Tamoxifen, a synthetic nonsteroidal antiestrogen, is used to treat human breast cancer (Mouridsen *et al.*, 1978). The mechanism by which tamoxifen exerts its antitumor effect is not yet understood (Sutherland and Murphy, 1982). Tamoxifen binds with high affinity to the estrogen receptor, and certain biological effects of tamoxifen can be reversed by estrogen (Lippman *et al.*, 1976; Coezy *et al.*, 1982; Reddel *et al.*, 1983). Since some of the biological effects of tamoxifen cannot be reversed by estrogen (Reddel *et al.*, 1983; Sutherland *et al.*, 1983a,b), it appears that not all of its biological effects are mediated by the estrogen receptor. It has recently been reported that tamoxifen inhibits the activation of bovine brain cAMP phosphodiesterase by calmodulin and that the inhibition can be overcome by increasing the concentrations of calmodulin (Lam, 1984).

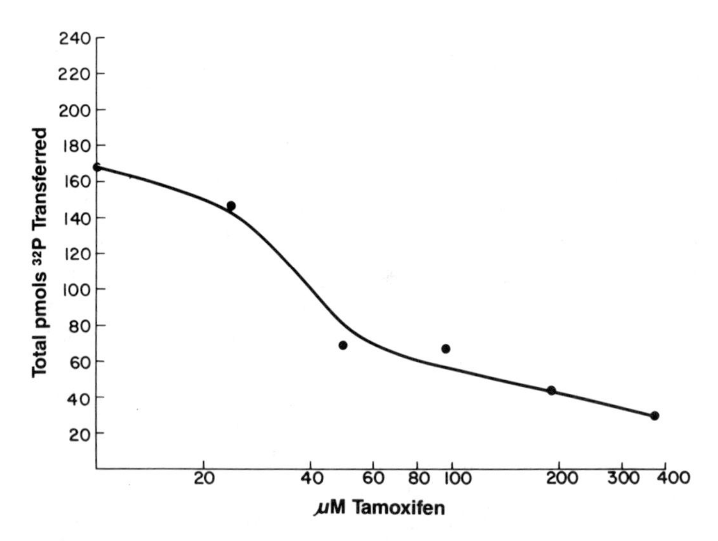

Figure 1. Inhibition of PKC by tamoxifen in the presence of added TPA and phospholipid. Rat brain PKC was assayed in the presence of TPA and phospholipid. Reactions were initiated by the addition of enzyme. All reaction mixtures contained 4% DMSO. The extent of histone phosphorylation observed in PKC assays done in the presence and absence of 4% DMSO agreed within 10%. Ordinate represents the picomoles of ^{32}P transferred from $[\gamma^{32}P]$-ATP to histone III-S in a TPA- and phospholipid-dependent manner. Each point represents the average of triplicate assays, which agreed within 10%.

In view of the latter results, we tested whether tamoxifen affected the activity of PKC, another Ca^{2+}-interacting enzyme. We found that tamoxifen inhibits the Ca^{2+}- and phospholipid-dependent activity of rat brain PKC with an IC_{50} of approximately 100 μM when the phospholipid concentration is 10 μg/ml. The TPA- and phospholipid-stimulated activity of PKC is also inhibited by tamoxifen, as illustrated in Fig. 1. The IC_{50} is approximately 40 μM when the concentration of TPA is 200 nM and that of phospholipid is 10 μg/ml. In addition, we found that tamoxifen inhibits teleocidin- and phospholipid-dependent PKC activity (O'Brian *et al.,* 1985).

Protamine sulfate is a PKC substrate that is phosphorylated by PKC in a Ca^{2+}- and phospholipid-independent reaction (Kikkawa *et al.,* 1982). We found no measureable effect on PKC-catalyzed protamine sulfate phosphorylation by tamoxifen whether or not Ca^{2+} and phospholipid were present (O'Brian *et al.,* 1985). This result provides evidence that tamoxifen does not interact directly with the active site of PKC.

We observed that the potency of tamoxifen as a PKC inhibitor is decreased as the concentration of phospholipid is increased whether PKC is activated by TPA plus phospholipid, Ca^{2+} plus phospholipid, or teleocidin plus phospholipid (O'Brian *et al.,* 1985). The observation that increasing phospholipid concentrations overcomes the inhibition of PKC by tamoxifen suggests that this tamoxifen-mediated inhibition is competitive with phospholipid and lends further support to our evidence that tamoxifen is not an active-site inhibitor of PKC.

The phenothiazine drug chlorpromazine, dibucaine, imipramine, phentolamine, tetracaine, and verapamil inhibit PKC (Mori *et al.,* 1980). These drugs do not appear to interact with the active site of PKC, and their inhibitory actions are reduced in the presence of increasing concentrations of phospholipid, suggesting that they inhibit PKC by competing with lipids (Mori *et al.*, 1980). As mentioned above, we have found that, like these phospholipid-interacting drugs, tamoxifen does not appear

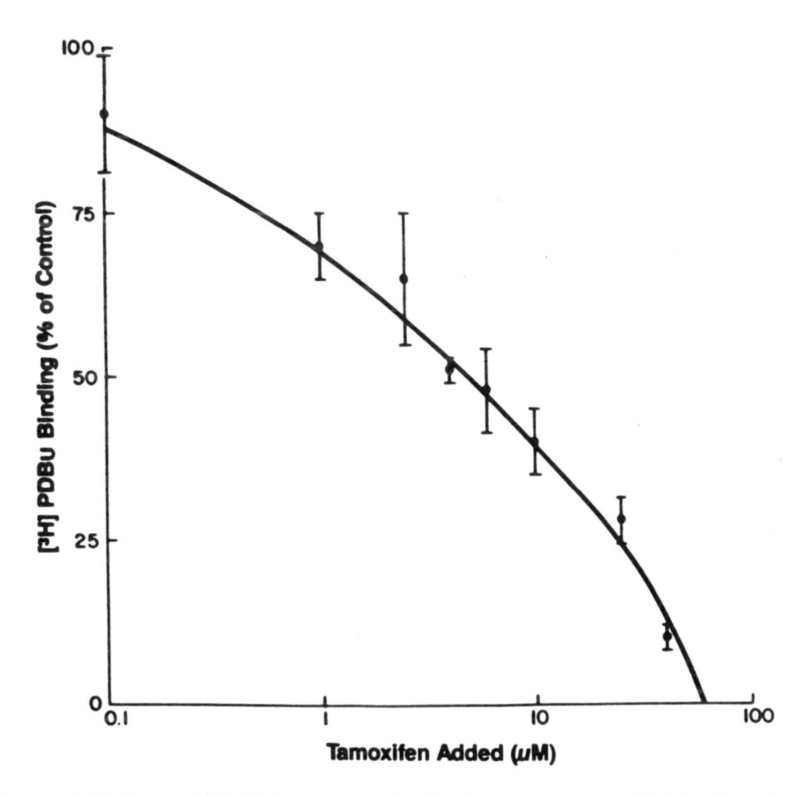

Figure 2. Inhibition of [^{3}H]-PDBu receptor binding by tamoxifen in C3H 10T1/2 cells. Inhibition of [^{3}H]-PDBu binding by tamoxifen was assayed as previously described (O'Brian *et al.*, 1985). Points are averages of triplicate assays and are expressed as percentage of the control value (i.e., in the absence of tamoxifen).

to interact with the active site of PKC and that the inhibition of PKC by tamoxifen is also overcome by increasing concentrations of phospholipid. These observations suggest that tamoxifen and the phenothiazines may inhibit PKC by a similar mechanism.

In order to determine the potency of the interaction between tamoxifen and PKC in the lipid environment of intact cells, we studied the inhibition of[^{3}H]-PDBu binding by tamoxifen in mouse fibroblast C3H 10T1/2 cells. An IC_{50} of 5 μM was found in this system (Fig. 2). Interestingly, it has been shown that this concentration of tamoxifen inhibits cell growth in MCF-7 human breast cells (Reddel *et al.*, 1983), suggesting that the inhibition of PKC by tamoxifen may mediate certain antiproliferative activities of the drug.

5. *Substrates for Protein Kinase C*

The strength of the evidence that PKC may play a critical role in carcinogenesis has stimulated research directed at the determination of PKC substrates. The EGF receptor (Hunter *et al.*, 1984), guanylate cyclase (Zwiller, *et al.*, 1984), glycogen synthase (Ahmad *et al.*, 1984), histone H1 (O'Brian *et al.*, 1984), high-mobility-group proteins 1, 2, and 17 (Ramachandran *et al.*, 1984), and myosin light-chain kinase (Nishikawa *et al.*, 1984) are all phosphorylated by PKC *in vitro*. The importance of the observed phosphorylations of histone H1 and the high-mobility-group proteins is not clear, since it is not known whether PKC activity exists in cell nuclei. However, the fact that *in vitro* PKC phosphorylation alters the activities of the EGF receptor (Hunter *et al.*, 1984), guanylate cyclase (Zwiller *et al.*, 1984), and glycogen synthase (Ahmad *et al.*, 1984) suggests that these phosphorylations may be biologically significant.

We have synthesized a peptide that is comprised of the sequence of the major site of PKC-catalyzed phosphorylation on histone H1 (Arg-Arg-Lys-Ala-Ser-Gly-Pro-Pro-Val). We have found that this peptide is phosphorylated by PKC in a Ca^{2+} and phospholipid-dependent manner as well as in a TPA- and phospholipid-dependent manner with a K_{mapp} of 130 μM, indicating that the peptide contains features characteristic of PKC phosphoacceptor substrates (O'Brian *et al.*, 1984). The sequence of this peptide bears some similarity to those of other known sites of PKC-catalyzed phosphorylation located on the EGF receptor and on high-mobility-group proteins. Both of these sequences, like the histone H1 sequence, contain a cluster of basic residues preceding the phosphoacceptor residue (Hunter *et al.*, 1984; Ramachandran *et al.*, 1984), suggesting that PKC may recognize serines and threonines that follow multiple basic residues,

Table II. Sites of Protein Kinase C Phosphorylation

Source	Sequence[a]
Histone H1	Arg-Arg-Lys-Ala-Ser*-Gly-Pro-Pro-Val
HM 17	Gln-Arg-Arg-Ser*-Ala-Arg-Lys-Ser-Ala
EGF-Receptor	Val-Arg-Lys-Arg-Thr*-Leu-Arg-Arg-Leu-Leu

[a] Asterisks indicate sites of PKC-catalyzed phosphorylation.

as shown in Table II. However, the relative importance of different recognition groups on these PKC phosphoacceptor substrates has not yet been studied directly.

6. *Compartmentalization of Protein Kinase C*

In order to study the compartmentalization of PKC, we have synthesized a series of fluorescent derivatives of TPA in which a dansyl group is linked to the 12 carboxylic acid residue. We have found that dansyl-TPA has potent biological activity, but dansyl-TPA-20-acetate and dansyl-13-desacetate are relatively weak biologically (R. M. Liskamp *et al.*, 1985). For instance, dansyl-TPA was extremely potent as an inhibitor of [^{3}H]-PDBu–receptor binding, with an IC_{50} of 2 nM. The IC_{50} for TPA in this assay is 4 nM. Dansyl-TPA-20-acetate was much less potent, having an IC_{50} of 30 nM, and dansyl-TPA-13-desacetate was extremely weak, with an IC_{50} of about 1000 nM. We also found that dansyl-TPA is equipotent with TPA in stimulating PKC activity in the presence of phospholipids: TPA induces characteristic changes in the morphology of both chick embryo and C3H 10T1/2 cells; within 2 hr of exposure of C3H 10T1/2 cells to dansyl-TPA (100 nM), the cells became spherical and highly refractile. Cells treated with either of the other two dansyl derivatives, at concentrations as high as 500 nM, did not show these morphological changes.

When C3H 10T1/2 cells were incubated with 100 nM dansyl-TPA for 1 hr at 37°C, the entire cytoplasm, including cytoplasmic processes, showed an intense and diffuse fluorescence with occasional intracytoplasmic granular fluorescent spots. However, the nucleus was virtually devoid of fluorescence. Several types of studies were carried out to demonstrate that the fluorescence seen with dansyl-TPA reflected specific uptake and cellular binding of the compound rather than nonspecific uptake of a hydrophobic compound. We found that the binding seen with 100 nM dansyl-TPA was almost completely suppressed when the cells were simultaneously incubated with 100 nM dansyl-TPA plus 40 μM PDBu or 40 μM TPA. Additional evidence of specific binding was obtained by employing the other two dansyl-TPA derivatives. When studied in parallel, cells incubated with 100 nM dansyl-TPA again showed intense cytoplasmic fluorescence, whereas cells incubated with 100 nM dansyl-TPA-20-acetate showed only weak fluorescence, and cells incubated with 100 nM dansyl-TPA-13-desacetate showed very faint fluorescence (R. M. Liskamp *et al.*, 1985).

Our fluorescence microscopy studies indicate that dansyl-TPA is taken up extensively by C3H 10T1/2 cells and is distributed diffusely within the cytoplasm, but not the nucleus, of these cells. These findings suggest that at least some of the pleiotropic effects of the phorbol ester tumor promoters might reflect direct effects within the cytoplasm of target cells.

7. *Tumor Promotion, Growth Factors, and Oncogenes*

The importance of PKC in the process of carcinogenesis is also suggested by recent studies of growth factor action: PKC activation appears to be elicited by certain growth factors and hormones. Many growth factors and hormones stimulate phosphatidylinositol turnover and thus cause the production of diacyglycerol, an activator of PKC (Berridge and Irvine, 1984).

We have been interested in whether tumor promoters can act synergistically with oncogenes. In examining possible synergistic interactions between tumor promoters and a cloned oncogene, we have used C3H 10T1/2 cells as recipients. These cells are particularly well suited to studies of the action of phorbol ester tumor promoters, since they contain an abundance of high-affinity receptors for these and related compounds and display striking changes in cell morphology and membrane-related properties in response to these agents (Horowitz *et al.*, 1983).

We first assessed the general competence of C3H 10T1/2 cells for DNA-mediated transfection. For this purpose we chose the dominant drug resistance marker GPT linked to the early region of SV40 to enhance its transcription, and our results indicated that the C3H 10T1/2 cells were reasonably competent for transfection (Hsiao *et al.*, 1984). Therefore, we also assessed the ability of the mutated human bladder cancer c-*ras*^{H} oncogene to induce foci of transformation on C3H 10T1/2 cells in the DNA transfection studies using the recombinant plasmid pT24. Parallel studies were done with NIH 3T3 cells as recipients. The oncogene yielded many transformed foci on NIH 3T3 cells. On the other hand, we obtained only a few transformed foci on C3H 10T1/2 cells (Hsiao *et al.*, 1984).

It was of interest to determine whether the low frequency of oncogene-induced transformation obtained with C3H 10T1/2 cells could be enhanced by the tumor promoter 12-O-tetradecanoylphorbol-13-acetate (TPA). We also ran parallel studies with the GPT marker plus TPA to determine the specificity of the effect. The exposure of C3H 10T1/2 cells to TPA (100 ng/ml) during transfection with pSV2-*gpt* and selection in mycophenolic acid inhibited the number of Gpt^+ colonies obtained. On the other hand, TPA caused a five- to tenfold increase in the number of transformed foci obtained in C3H 10T1/2 cells transfected with the T24 oncogene. We found a similar effect using a rat fibroblast cell line as recipients.

Our results suggest that the initiating carcinogen might function by activating a cellular protooncogene during multistage carcinogenesis, whereas tumor promoters might enhance the outgrowth of the altered cells or the expression of other cellular genes that complement the function of the activated oncogene. A major unanswered question is how the effects of tumor promoters, acting presumably through PKC and protein phosphorylation, lead to alterations in gene expression and enhancement of cell transformation.

8. Applied Aspects in Therapeutics

We consider the design of specific PKC antagonists an important approach for the development of antiproliferative drugs. Specific PKC antagonists may be useful drugs in the control of abnormal proliferative disease states such as psoriasis, particularly since PKC appears to be an important mediator of growth factor action. In addition, PKC antagonists may be useful in the treatment of cancer. The current understanding of PKC–tumor promoter interactions suggests that a PKC antagonist may block tumor promotion and may even inhibit the growth of established tumors. Site-directed PKC antagonists may be able to modulate the function of specific membrane receptors and channels by inhibiting their phosphorylation in intact cell by PKC.

The studies of PKC and tumor promotion reviewed here indicate the importance of the cell membrane in PKC activation and presumably in tumor promotion. As the regulation of PKC activity by lipid metabolites is further elucidated, the role of cell membrane composition in carcinogenesis may become clearer. This may result in a more coherent understanding of dietary lipid and cancer risk and could point towards dietary modification or the development of membrane-active drugs for use in cancer prevention.

ACKNOWLEDGMENTS. We acknowledge the excellent secretarial assistance of Ms. Lucy Morales and Ms. Evelyn Emeric. This research was supported by National Cancer Institute Grant CA-02656.

References

Ahmad, Z., Lee, F. T., DePaoli-Roach, A., and Roach, P. J., 1984, Phosphorylation of glycogen synthase by the Ca^{2+} and phospholipid-activated protein kinase (protein kinase C), *J. Biol. Chem.* **259:**8743–8747.

Arcoleo, J. P., and Weinstein, I. B., 1985, Activation of protein kinase C by tumor promoting phorbol esters, teleocidin, and aplysiatoxin in the absence of added calcium, *Carcinogenesis* **6:**213–217.

Berridge, M. J., and Irvine, R. F., 1984, Inositol trisphosphate, a novel second messenger in cellular signal transduction, *Nature* **312:**315–321.

Castagna, M., Takai, Y., Kaibuchi, K., Sano, K., Kikkawa, U., and Nishizuka, Y., 1982, Direct activation of calcium-activated, phospholipid-dependent protein kinase by tumor-promoting phorbol esters, *J. Biol. Chem* **257:**7847–7851.

Coezy, E., Borgna, J. L., and Rochefort, H., 1982, Correlation between binding to estrogen receptor and inhibition of cell growth, *Cancer Res.* **42:**317–323.

Couturier, A., Bazgar, S., and Castagna, M., 1984, Further characterization of tumor promoter-mediated activation of protein kinase C, *Biochem. Biophys. Res. Commun.* **121:**448–455.

Dicker, P., and Rozengurt, E., 1978, Stimulation of DNA synthesis by tumor promoter and pure mitogenic factors, *Nature* **276:**723–726.

Donnelly, T. E., and Jensen, R., 1983, Effect of fluphenazine on the stimulation of calcium-sensitive, phospholipid-dependent protein kinase by TPA, *Life Sci.* **33:**2247–2253.

Horowitz, A. D., Fujiki, H., Weinstein, I. B., Jeffrey, A., Okin, E., Moore, R. E., and Sugimura, T., 1983, Comparative effects of aplysiatoxin, debromoaplysiatoxin, and teleocidin on receptor binding and phospholipid metabolism, *Cancer Res.* **43:**1529–1535.

Hsiao, W.-L. W., Gattoni-Celli, S., and Weinstein, I. B., 1984, Oncogene-induced transformation of C3H 10T1/2 cells is enhanced by tumor promoters, *Science* **226:**552–555.

Hunter, T., Ling, N., and Cooper, J. A., 1984, Protein kinase C phosphorylation of the EGF receptor at a threonine residue close to the cytoplasmic face of the plasma membrane, *Nature* **311:**480–483.

Kikkawa, U., Takai, Y., Minakuchi, R., Inohara, S., and Nishizuka, Y., 1982, Calcium-activated, phospholipid-dependent protein kinase from rat brain, *J. Biol. Chem.* **257:**13341–13348.

Lam, H.-Y. P., 1984, Tamoxifen is a calmodulin antagonist in the activation of a cA,P phosphodiesterase, *Biochem. Biophys. Res. Commun.* **118:**27–32.

Lippman, M. E., Bolan, G., and Huff, K. K., 1976, The effects of estrogens and antiestrogens on hormone responsive human breast cancer in long term tissue culture, *Cancer Res.* **36:**4595–4601.

Liskamp, R. M. J., Brothman, A. R., Areoleo, J. P., Miller, O. J., and Weinstein, I. B.,

1985, Cellular uptake and localization of fluorescent derivatives of phorbol ester tumor promoters, *Biochem. Biophys. Res. Commun.* **131**:920–927.

Mori, T., Takai, Y., Minakuchi, R., Yu, B., and Nishizuka, Y., 1980, Inhibitory action of chlorpromazine, dibucaine, and other phospholipid-interacting drugs on calcium-activated, phospholipid-dependent protein kinase, *J. Biol. Chem.* **255**:8378–8380.

Mouridsen, H., Palshof, T., Patterson, J., and Battersby, L., 1978, Tamoxifen in advanced breast cancer, *Cancer Treat. Rev.* **5**:131–141.

Nishikawa, M., Sellers, J. R., Adelstein, R. S., and Hidaka, H., 1984, Protein kinase C modulates *in vitro* phosphorylation of the smooth muscle heavy meromyosin by myosin light chain kinase, *J. Biol. Chem.* **259**:8808–8814.

Nishizuka, Y., 1984, The role of protein kinase C in cell surface signal transduction and tumor promotion, *Nature* **308**:693–698.

O'Brian, C. A., Lawrence, D. S., Kaiser, E. T., and Weinstein, I. B., 1984, Protein kinase C phosphorylates the synthetic peptide Arg-Arg-Lys-Ala-Ser-Gly-Pro-Pro-Val in the presence of phospholipid plus either Ca^{2+} or a phorbol ester tumor promoter, *Biochem. Biophys. Res. Commun.* **124**:296–302.

O'Brian, C. A., Liskamp, R. M., Solomon, D. H., and Weinstein, I. B., 1985, Inhibition of protein kinase C by tamoxifen, *Cancer Res.* **45**:2462–2465.

Parker, P. J., Stabel, S., and Waterfield, M. D., 1984, Purification to homogeneity of protein kinase C from bovine brain—identity with the phorbol ester receptor, *EMBO* **3**:953–959.

Ramachandran, C., Yau, P., Bradburg, E. M., Shyamala, G., Yasuda, H., and Walsh, D. A., 1984, Phosphorylation of high-mobility group proteins by the calcium-phospholipid-dependent protein kinase and the cyclic AMP dependent protein kinase, *J. Biol. Chem.* **259**:13495–13503.

Reddel, R. R., Murphy, L. C., and Sutherland, R. L., 1983, Effects of biologically active metabolites of tamoxifen on the proliferation kinetics of MCF-7 human breast cancer cells *in vitro, Cancer Res.* **43**:4618–4624.

Slaga, T. J., Sivak, A., and Boutwell, R. K., 1978, *Mechanisms of Tumor Promotion and Cocarcinogenesis*, Vol. 2, Raven Press, New York.

Sutherland, R. L., and Murphy, L. C., 1982, Mechanisms of oestrogen antagonism by non-steroidal antiestrogens, *Mol. Cell. Endocrinol.* **25**:5–23.

Sutherland, R. L., Green, M. D., Hall, R. E., Reddel, R. R., and Taylor, I. W., 1983a, Tamoxifen induces accumulation of MCF-7 human mammary carcinoma cells in the G0/G1 phase of the cell cycle, *Eur. J. Cancer Clin. Oncol.* **19**:615–621.

Sutherland, R. L., Hall, R. E., and Taylor, I. W., 1983b, Cell proliferation kinetics of MCF-7 human mammary carcinoma cells in culture and effects of tamoxifen on exponentially growing and plateau phase cells *Cancer Res.* **43**:3998–4006.

Weinstein, I. B., 1981a, Current concepts and controversies in chemical carcinogenesis, *J. Supramol. Struct. Cell. Biochem.* **17**:99–120.

Weinstein, I. B., 1981b, Studies on the mechanism of action of tumor promoters and their relevance to mammary carcinogenesis, in: *Cell Biology of Breast Cancer* (C. M. McGrath, M. J. Brennan, and M. A. Rich, eds.), Academic Press, New York, pp. 425–450.

Weinstein, I. B., Lee, L. S., Fisher, P. B., Mufson, R. A., and Yamasaki, H., 1979, Cellular and biochemical events associated with the action of tumor promoters, in: *Naturally Occurring Carcinogens—Mutagens and Modulators of Carcinogenesis* (E. C. Miller, ed.), Japan Science Press, Tokyo, pp. 301–303.

Weinstein, I. B., Mufson, R. A., Lee, L. S., Laskin, P. B., Horowitz, A. D., and Ivanovic, V., 1980, Membrane and other biochemical effects of the phorbol esters and their relevance to tumor promotion, in: *Carcinogenesis: Fundamental Mechanisms and Environmental Effects* (B. Pullman, P. O. P. Ts'O, and H. Gelboin, eds.), R. Reidel, Amsterdam, pp. 543–563.

Weinstein, I. B., Horowitz, A. D., Fisher, P., Ivanovic, V., Gattoni-Celli, S., and Kirschmeier, P., 1982, Mechanisms of multi-stage carcinogenesis and their relevance to tumor cell heterogeneity, in: *Tumor Cell Heterogeneity* (A. H. Owens, D. S. Coffey, and S. B. Baylin, eds.), Academic Press, New York, pp. 261–283.

Weinstein, I. B., Arcoleo, J. P., Backer, J., Jeffrey, A. M., Hsiao, W., Gattoni-Celli, S., and Kirschmeier, P., 1984, Molecular mechanisms of tumor promotion and multistage carcinogenesis, in: *Cellular Interactions by Environmental Tumor Promoters* (H. Fujiki, ed.), Japan Science Press, Toyko, pp. 59–74.

Zwiller, J., Revel, M. O., and Melviya, A. N., 1984, Protein kinase C catalyzes the phosphorylation of guanylate cyclase *in vitro*, *J. Biol. Chem.* **260:**1350–1353.

V

CELL POLARITY AND MEMBRANE TRANSPORT PROCESSES

14

Intracellular Protein Topogenesis

GÜNTER BLOBEL, PETER WALTER, and REID GILMORE

A cell contains millions of protein molecules, which are continually being synthesized and degraded. At homeostasis, a given species of protein is represented by a characteristic number of molecules that is kept constant within a narrow range. Very little is known about the accounting procedures of the cell, i.e., how it balances and controls biosynthesis and biodegradation.

An important aspect of biosynthesis (Blobel, 1980) as well as biodegradation (Blobel, 1979) is the intracellular topology of proteins. Many protein species spend their entire lives in the compartment in which they are synthesized, whereas others have to be translocated across the hydrophobic barrier of one or sometimes two distinct cellular membranes to reach the intracellular compartment or extracellular site where they exert their function. Numerous protein species have to be integrated asymmetrically into distinct cellular membranes. For many proteins this requires partial translocation, i.e., the selective transfer of one or several distinct hydrophilic or charged segments of the polypeptide chain across the hydrophobic barrier of one or two intracellular membranes. Following complete or partial translocation across a translocation-competent membrane(s), subpopulations may undergo further posttranslocational traffic (Palade, 1975). Soluble or membrane proteins may be shipped in bulk or by receptor-mediated processes from a translocation-competent donor compartment to a translocation-incompetent receiver compartment. This posttranslocational traffic may be unidirectional (in which case the protein becomes a permanent resident of a particular cellular compartment or membrane) or may follow a cyclical pattern between distinct cellular membranes (e.g., the recycling of receptors).

GÜNTER BLOBEL, PETER WALTER, and REID GILMORE • Laboratory of Cell Biology, The Rockefeller University, New York, New York 10021.

The collective term topogenesis was introduced (Blobel, 1980) to encompass partial or complete protein translocation across membranes as well as subsequent posttranslocational protein traffic. Proposals have been made for the mechanisms involved in protein topogenesis (Blobel, 1980). The essence of these proposals is that the information for intracellular protein topogenesis resides in discrete topogenic sequences that constitute permanent or transient parts of the polypeptide chain. The repertoire of distinct topogenic sequences was predicted to be relatively small because many different proteins would be topologically equivalent, i.e., targeted to the same intracellular address.

Four types of topogenic sequences were distinguished (Blobel, 1980). (1) Signal sequences initiate translocation of proteins across specific membranes and are decoded by protein translocators that effect unidirectional translocation of proteins across specific cellular membranes by virtue of their signal-sequence-specific recognition and their location in distinct cellular membranes. (2) Stop-transfer sequences interrupt the translocation process that was previously initiated by a signal sequence and yield asymmetric integration of proteins into translocation-competent membranes by excluding a distinct segment of the polypeptide chain from translocation. (3) Sorting sequences act as determinants for posttranslocational traffic of protein subpopulations, originating in translocation-competent membranes (and compartments) and leading to translocation-incompetent receiver membranes (and compartments). (4) Insertion sequences interact with the lipid bilayer directly and thereby anchor a protein to the hydrophobic core of the lipid bilayer.

Translocation is understood here as the transport of an entire polypeptide chain across one (or two) membrane(s), proceeding unidirectionally from the protein biosynthetic compartment. Hypothetical models for intracellular protein translocation must deal with two essential tenets that appear to underlie the observed phenomenology of this process. First, the permeability barrier of the membrane appears to be reversibly modified for the passage of each translocated polypeptide chain while being maintained for other solutes. Second, the species of protein to be translocated and the type of membrane across which a given protein is translocated are highly specific. Both of these tenets can be readily satisfied by postulating that protein translocation is a receptor-mediated process (Blobel, 1980) in which specificity is achieved by signal sequences in the proteins to be translocated and by signal-sequence-specific translocation systems that are restricted in their location to distinct cellular membranes.

There are several translocation-competent intracellular membrane systems in eucaryotic cells (Blobel, 1980). The system of the endoplasmic reticulum (ER) is presently the best understood, and it is therefore the only one discussed here. Three distinct classes of proteins are known to utilize this translocation system: secretory (Palade, 1975) and lysosomal (Erickson *et al.,* 1981) proteins are translocated across the ER membrane, and certain classes of integral membrane proteins (Lingappa *et al.,* 1978) are integrated into it.

It has been shown that each polypeptide in these three groups of proteins studied does in fact contain a signal sequence. This signal sequence is usually located at the amino terminus of the polypeptide. It comprises 15–30 (mostly hydrophobic) residues and is usually cleaved during or after translocation. However, the signal sequence is not always located at the amino terminus and is not always cleaved; i.e., it may remain part of the mature protein. Although there is no homology in the primary structure, it appears that there is a consensus tertiary structure consisting of a hydrophobic α helix (or β-pleated sheet) followed by a β turn.

Until recently, the machinery decoding the ER-targeted signal sequence remained more or less fictitious. Only after the development of a cell-free translocation system that appeared to faithfully reproduce translocation of newly synthesized secretory protein (Blobel and Dobberstein, 1975a,b) did it become possible to identify and characterize the components of this machinery.

So far, two components have been purified from canine pancreas and shown to be required for translocation. One of these is signal recognition particle (SRP), an 11 S cytoplasmic ribonucleoprotein. The SRP consists of six nonidentical polypeptide chains with apparent molecular masses of 72, 68, 54, 19, 14 and 9 kD, respectively (Walter and Blobel, 1980), and one molecule of 7 S RNA of about 300 nucleotides (Walter and Blobel, 1982). When separated, these components could be reconstituted into an active particle (Walter and Blobel, 1983a,b). The second isolated component of the protein translocation machinery is SRP receptor (Gilmore *et al.,* 1982) (also termed "docking protein"; Meyer *et al.,* 1982). The SRP receptor is an integral membrane protein of the ER of 72 kD that has been purified by SRP affinity chromatography from detergent extracts of microsomal vesicles (Gilmore *et al.,* 1982). The SRP receptor contains a 60-kD cytoplasmic domain that can be severed from the membrane by a variety of proteases and can be added back to proteolysed membranes to reconstitute their translocation activity (Walter *et al.,* 1979; Meyer and Dobberstein, 1980). Both SRP and SRP receptor require a free sulfhydryl group(s) for their activity.

The function of these two components in the protein translocation process was deduced using *in vitro* assembled translating polysomes that were programmed with mRNAs coding for specific secretory proteins.

Thus, it became possible to follow directly the fate of a nascent secretory protein and thereby to analyze the discrete steps of its recognition and its subsequent translocation across the membrane of microsomal vesicles.

It was shown that free (soluble) SRP can exist in equilibrium with a membrane-bound form (bound presumably to SRP receptor) as well as with a ribosome-bound form (Walter and Blobel, 1983). On translation of an mRNA coding for a signal sequence addressed to the ER translocation system, there was an enhancement by three or four orders of magnitude of the apparent affinity of SRP for the translating ribosome (Walter and Blobel 1981). Thus, SRP could be shown to recognize the signal sequences of nascent secretory (Walter and Blobel, 1981; Stoffel *et al.*, 1981; Mueller *et al.*, 1982), lysosomal (Erickson *et al.*, 1983), and membrane (Anderson *et al.*, 1982) proteins.

Interestingly, concomitant with this increase in binding affinity, SRP specifically arrested the elongation of the initiated polypeptide chain at a discrete site (just after the signal peptide had emerged from the ribosome) and thereby prevented the completion of presecretory protein (Walter and Blobel 1981; Meyer *et al.*, 1982). The mechanism by which SRP interacts with the ribosome and the signal sequence to cause recognition and elongation arrest is unknown. However, the detection of the signal sequence by SRP clearly involves the recognition of the nascent polypeptide chain *per se* as distinguished from the recognition of the mRNA template, since a perturbation of the nascent polypeptide through the incorporation of certain amino acid analogues rendered signal sequences nonrecognizable (Hortin and Boime, 1980; Walter and Blobel, 1981).

The role of the SRP arrest *in vivo* remains unknown. However, it has been speculated that it might be physiologically important to prevent the complete synthesis of secretory and lysosomal proteins in the cytoplasmic compartment (Walter and Blobel, 1981). The SRP-mediated elongation arrest could also be of regulatory significance. Modulation of the arrest-releasing activity, either by other as yet undefined components or by direct modification of SRP and/or SRP receptor, could provide the cell with an on/off switch for translocation-coupled protein synthesis and thereby provide a mechanism for a fast and regulatable response to a variety of physiological stimuli. Recently, Anderson *et al.* (1983) demonstrated that the membrane integration of two integral membrane proteins containing uncleaved signal sequences is SRP dependent. However, they were unable to demonstrate an SRP-dependent elongation arrest for these proteins. Thus, the possibility exists that for certain proteins a tight (and thus detectable) SRP arrest is not obligatory for correct targeting and integration.

On interaction of the SRP-arrested ribosomes with microsomal membranes, the elongation arrest was released (Walter and Blobel, 1981). This arrest-releasing activity of the microsomal membrane fraction was localized to the SRP receptor (Gilmore *et al.*, 1982). Since the SRP receptor was purified by virtue of its affinity to SRP itself, it is likely that it is also the direct interaction between the SRP on the arrested ribosome and the SRP receptor that is responsible for the release of the arrest.

Preliminary quantitation indicated that in pancreatic cells both SRP and SRP receptor are present only in substoichiometric amounts with respect to membrane-bound ribosomes (Gilmore *et al.*, 1982). This suggested that the ribosome–SRP–SRP receptor interaction might be transient, merely targeting the SRP-arrested ribosome to a specific translocation-competent site on the ER membrane. This conjecture is supported by recent data that demonstrate that purified SRP receptor in detergent solution, concomitant with the release of the elongation arrest, causes SRP to lose high affinity for the signal-bearing ribosome (Gilmore and Blobel, 1983). Since the SRP receptor itself showed no measurable affinity to ribosomes, it appears that both SRP and SRP receptor are free to be recycled once the correct targeting of the ribosome to the ER membrane has been accomplished. It is for this reason that we prefer to use the originally proposed term "SRP receptor" (Walter and Blobel, 1981) rather than the term "docking protein" (Meyer *et al.*, 1982). The latter term implies the anchoring of the translating/translocating ribosome to the membrane rather than the transient information processing that SRP receptor provides on interaction with its ligand.

The steps following the initial targeting event are presently only poorly understood. Once targeting has occurred, the ribosome–SRP–SRP receptor interaction might be replaced by a direct interaction of the ribosome with other integral membrane proteins, leading to the formation of a functional ribosome–membrane junction. Two integral membrane proteins that might function as ribosome receptors, ribophorin I and II, have been described (Kreibich *et al.*, 1978; Marcantonio *et al.*, 1982) and appear to be involved in ribosome binding to the ER membrane. However, their direct involvement in protein translocation has not yet been demonstrated and has been questioned by other investigators (Bielinska *et al.*, 1979).

On termination of protein synthesis, the completed polypeptide is released into the lumen of the ER. The permeability barrier of the membrane is restored, and the ribosome is released from the membrane. It is then free to enter the soluble pool and complete the "ribosome cycle."

In the case of integral membrane proteins, the translocation process is aborted prior to completion of the protein, resulting in its asymmetric integration into the membrane. The information causing this arrest of translocation (a stop-transfer sequence) is probably contained in the nascent chain and interpreted by the translocation machinery. Some translocated proteins (such as the extracellular domain of the influenza virus hemagglutinin) contain long stretches of hydrophobic amino acids. Thus, it is not merely hydrophobicity that causes integral membrane proteins to become "stuck" in the lipid bilayer. It is not known whether a ribosome remains membrane bound or detaches from the membrane while synthesizing the cytoplasmic domain of a membrane protein.

Thus, the combination of two topogenic sequences, a signal sequence and a stop-transfer sequence, yields asymmetrically integrated membrane proteins. Various programs of topogenic sequences have been proposed (Blobel, 1980) to yield the myriad of asymmetric topologies that exist for integral membrane proteins.

References

Anderson, D. J., Walter, P., and Blobel, G., 1982, Signal recognition protein is required for the integration of acetylcholine receptor δ subunit, a transmembrane glycoprotein, into the endoplasmic reticulum membrane, *J. Cell Biol.* **93:**501–506.

Anderson, D. J., Mostov, K. E., and Blobel, G., 1983, Mechanisms of integration for *de novo* synthesized polypeptides into membranes. Signal recognition particle is required for the integration into microsomal membranes of calcium ATPase and of lens MP26 but not of cytochrome b_5, *Proc. Natl. Acad. Sci. U.S.A.* **80:**7249–7253.

Bielinska, M., Rogers, G., Rucinsky, T., and Boime, I., 1979, Processing *in vitro* of placental polypeptide hormones by smooth microsomes, *Proc. Natl. Acad. Sci. U.S.A.* **76:**6152–6156.

Blobel, G., 1979, Extralysosomal compartments for the turnover of intracellular macromolecules, in: *Limited Proteolysis in Microorganisms* (G. N. Cohen, H. Holzer, eds.), U.S. Department of Health, Education and Welfare, Washington, pp. 167–169.

Blobel, G., 1980, Intracellular protein topogenesis, *Proc. Natl. Acad. Sci. U.S.A.* **77:**1496–1500.

Blobel, G., and Dobberstein, B., 1975a, Transfer of proteins across membranes I. Presence of proteolytically processed and unprocessed nascent immunoglobulin light chains on membrane-bound ribosomes of murine myeloma, *J. Cell Biol.* **67:**835–851.

Blobel, G., and Dobberstein, B., 1975b, Transfer of proteins across membranes II. Reconstitution of functional rough microsomes from heterologous components, *J. Cell Biol.* **67:**852–862.

Erickson, A. H., Conner, G. E., and Blobel, G., 1981, Biosynthesis of a lysosomal enzyme. Partial structure of two transient and functionally distinct NH_2-terminal sequences in cathepsin D, *J. Biol. Chem.* **256:**11224–11231.

Erickson, A. H., Walter, P., and Blobel, G., 1983, Translocation of a lysosomal enzyme across the microsomal membrane requires signal recognition particle, *Biochem. Biophys. Res. Commun.* **115:**275–280.

Gilmore, R., and Blobel, G., 1983, Transient involvement of signal recognition particle and its receptor in the microsomal membrane prior to protein translocation, *Cell* **35:**677–685.

Gilmore, R., Walter, P., and Blobel, G., 1982, Protein translocation across the endoplasmic reticulum II. Isolation and characterization of the signal recognition particle receptor, *J. Cell Biol.* **95:**470–477.

Hortin, G., and Boime, I., 1980, Inhibition of preprotein processing in ascites tumor lysates by incorporation of a leucine analogue, *Proc. Natl. Acad. Sci. U.S.A.* **77:**1356–1360.

Kreibich, G., Ulrich, B. L., and Sabatini, D. D., 1978, Proteins of rough microsomal membranes related to ribosome binding I. Identification of ribophorins I and II, membrane proteins characteristic of rough microsomes, *J. Cell Biol.* **77:**464–487.

Lingappa, V. R., Katz, F. N., Lodish, H. F., and Blobel, G., 1978, A signal sequence for the insertion of a transmembrane glycoprotein. Similarities to the signals of secretory proteins in primary structure and function, *J. Biol. Chem.* **253:**8667–8670.

Marcantonio, E. E., Grebenau, R. C., Sabatini, D. D., and Kreibich, G., 1982, Identification of ribophorins in rough microsomal membranes from different organs of several species, *Eur. J. Biochem.* **124:**217–222.

Meyer, D. I., and Dobberstein, B., 1980a, A membrane component essential for vectorial translocation of nascent proteins across the endoplasmic reticulum: Requirements for its extraction and reassociation with the membrane, *J. Cell Biol.* **87:**498–502.

Meyer, D. I., and Dobberstein, B., 1980b, Identification and characterization of a membrane component essential for the translocation of nascent secretory proteins across the membrane of the endoplasmic reticulum, *J. Cell Biol.* **87:**503–508.

Meyer, D. I., Krause, E., and Dobberstein, B., 1982, Secretory protein translocation across membranes—the role of the 'docking protein,' *Nature* **297:**647–650.

Mueller, M., Ibrahimi, I., Chang, C. N., Walter, P., and Blobel, G., 1982, A bacterial secretory protein requires signal recognition particle for translocation across mammalian endoplasmic reticulum, *J. Biol. Chem.* **257:**11860–11863.

Palade, G., 1975, Intracellular aspects of the process of protein secretion, *Science* **189:**347–358.

Stoffel, W., Blobel, G., and Walter, P., 1981, Synthesis *in vitro* and translocation of apolipoprotein AI across microsomal vesicles, *Eur. J. Biochem.* **120:**519–522.

Walter, P., and Blobel, G., 1980, Purification of a membrane-associated protein complex required for protein translocation across the endoplasmic reticulum, *Proc. Natl. Acad. Sci. U.S.A.* **77:**7112–7116.

Walter, P., and Blobel, G., 1981, Translocation of proteins across the endoplasmic reticulum. III. Signal recognition protein (SRP) causes signal sequence-dependent and site-specific arrest of chain elongation that is released by microsomal membranes, *J. Cell Biol.* **91:**557–561.

Walter, P., and Blobel, G., 1982, Signal recognition particle contains a 7 S RNA essential for protein translocation across the endoplasmic reticulum, *Nature* **299:**691–698.

Walter, P., and Blobel, G., 1983a, Disassembly and constitution of signal recognition particle, *Cell* **34:**525–533.

Walter, P., and Blobel, G., 1983b, Subcellular distribution of signal recognition particle and 7SL-RNA determined with polypeptide-specific antibodies and complementary DNA probe, *J. Cell Biol.* **97:**1693–1699.

Walter, P., Jackson, R. C., Marcus, M. M., Lingappa, V. R., and Blobel, G., 1979, Tryptic dissection and reconstitution of translocation activity for nascent presecretory proteins across microsomal membranes, *Proc. Natl. Acad. Sci. U.S.A.* **76:**1795–1799.

Walter, P., Ibrahimi, I., and Blobel, G., 1981, Translocation of proteins across the endoplasmic reticulum. I. Signal recognition protein (SRP) binds to *in-vitro*-assembled polysomes synthesizing secretory protein, *J. Cell Biol.* **91:**545–550.

15

Analysis of Epithelial Cell Surface Polarity Development with Monoclonal Antibodies

GEORGE K. OJAKIAN

1. Introduction

The cells of transporting epithelia are organized into either sheets or tubules so that they provide a barrier between two compartments, the mucosal and serosal, which are essential for the proper physiological functioning of a variety of organs and tissues. The plasma membrane of epithelial cells is divided into two unique domains: the apical membrane, which borders the mucosal or luminal side, and the basolateral membrane, which contacts the basal lamina on the serosal side of the epithelium (Berridge and Oschman, 1972). At the boundary between the apical and basolateral domains is a differentiated region of the plasma membrane termed the junctional complex (Farquhar and Palade, 1963), and one of these membrane specializations, the tight junction, seals the lateral space, preventing the transepithelial movement of ions and larger molecules (Farquhar and Palade, 1963; Staehelin, 1974).

Biochemical studies on purified fractions of apical and basolateral membranes have demonstrated that epithelial cells are polarized, with each cell surface domain being characterized by a unique set of enzymes, proteins, and lipids (Fujita *et al.*, 1973; Kawai *et al.*, 1974; Semenza, 1976; Murer and Kinne, 1980; Rodriguez-Boulan, 1983). These results are supported by immunocytochemical and autoradiographic evidence that leucine aminopeptidase is localized to the apical cell surface (Reggio *et*

GEORGE K. OJAKIAN • Department of Anatomy and Cell Biology, State University of New York, Downstate Medical Center, Brooklyn, New York 11203.

al., 1982), whereas Na^+–K^+ is a basolateral enzyme (Kyte, 1976; Ernst and Mills, 1977; Reggio *et al.*, 1982). Ultrastructural studies have demonstrated that the asymmetric distribution of epithelial cell surface components is maintained by tight junctions (Pisam and Ripoche, 1976; Ziomek *et al.*, 1980). Physiologically, polarized epithelial cells are capable of transporting ions, sugars, amino acids, and proteins vectorially against transepithelial concentration gradients (see Berridge and Oschman, 1972; Murer and Kinne, 1980).

We have been investigating the development and maintenance of epithelial cell surface polarity in Madin–Darby canine kidney (MDCK) cells. This epithelial cell line was derived from adult dog kidney and has retained many of the differentiated properties normally associated with renal transporting epithelium, making it an excellent model for studying the mechanisms of cell polarity development. Structurally, MDCK cells have apical microvilli and basolateral membranes with numerous lateral interdigitations (Leighton *et al.*, 1970; Misfeldt *et al.*, 1976; Cereijido *et al.*, 1978). The MDCK cell surface is polarized, having asymmetric distributions of both membrane lipids (Van Meer and Simons, 1982) and proteins (Richardson and Simmons, 1979; Louvard, 1980; Reggio *et al.*, 1982; Herzlinger and Ojakian, 1984). The MDCK cells have retained a cell surface antigen unique to the distal nephron (Herzlinger *et al.*, 1982), respond physiologically to a variety of hormones (Rindler *et al.*, 1979), can support the vectorial transport of sodium and water (Leighton *et al.*, 1970; Misfeldt *et al.*, 1976; Cereijido *et al.*, 1978), and possess functional tight junctions (Misfeldt *et al.*, 1976; Cereijido *et al.*, 1978; Rabito *et al.*, 1978; Cramer *et al.*, 1980; Ojakian, 1981).

2. *Monoclonal Antibodies against the MDCK Cell Surface*

We have utilized the hybridoma technique of Köhler and Milstein (1975) to produce monoclonal antibodies against MDCK cell surface antigens. These antibodies are then used as highly specific probes to study the distribution of epithelial cell surface proteins during the development of tight junctions and to map the distribution of the proteins along the nephron.

Mice were immunized with MDCK cells and hybridomas were produced according to the fusion protocol of Gefter *et al.* (1977). Hybridoma supernatants were screened for monoclonal antibodies specific to MDCK cells by radioimmunoassay, and comparative binding studies were done in parallel to insure that clones secreting epithelial-specific antibodies were isolated (Herzlinger *et al.*, 1982). We have done extensive work with monoclonal antibodies from nine of these hybridoma lines, and these results are discussed in detail.

To identify the cell surface antigens recognized by these antibodies, MDCK cells were metabolically labeled with either [^{35}S]-methionine or [^{3}H]-glucosamine, and membrane proteins were extracted with detergent, immunoprecipitated, and analyzed by SDS polyacrylamide gel electrophoresis (SDS-PAGE). It has been determined that six of these monoclonal antibodies recognize a 25-kD glycoprotein and that the others are against glycoproteins of either 35 kD, 50 kD, or 60 kD (Herzlinger and Ojakian, 1984). The distributions of these cell surface glycoproteins have been mapped along the renal nephron, and these results are represented below.

3. *Cell Surface Distribution of Membrane Glycoproteins*

The antibodies secreted by hybridoma clones were screened for immunoreactivity on subconfluent MDCK cells to allow the antibodies access to the entire cell surface (Herzlinger *et al.*, 1982). When immunofluorescence microscopy was done to determine the distribution of the 35-kD, 50-kD, and 60-kD glycoproteins, it was clear not only that every cell expressed these proteins but that they appeared to be distributed over the entire cell surface (Herzlinger and Ojakian, 1984). These results were confirmed at higher resolution by electron microscopic immunocytochemistry. Using antibodies coupled to horseradish peroxidase (HRP), we were able to demonstrate that these glycoproteins appeared to be uniformly distributed over both the apical and basal cell surfaces (Fig. 1).

It was only possible to obtain a qualitative estimate of the amount of MDCK cell surface glycoproteins with HRP-labeled antibodies because of the diffuse nature of the electron-dense reaction product. Quantitative measurements of the distribution of the 50-kD and 60-kD glycoproteins were made using colloidal-gold labeled protein A (Langone, 1982; Muhlpfordt, 1982). With this procedure we were able to localize the 50-kD (data not shown) and 60-kD glycoproteins (Fig. 2) on both the apical and basal surfaces of subconfluent MDCK cells at the ultrastructural level. Visual examination of the electron micrographs suggested that there were equal amounts of those glycoproteins on both the apical and basal surfaces. Quantitative measurements of the colloidal gold particles bound to the MDCK cell surface demonstrated that the 50-kD (data not shown) and 60-kD glycoproteins (Table I) were uniformly distributed over both the apical and basal plasma membranes.

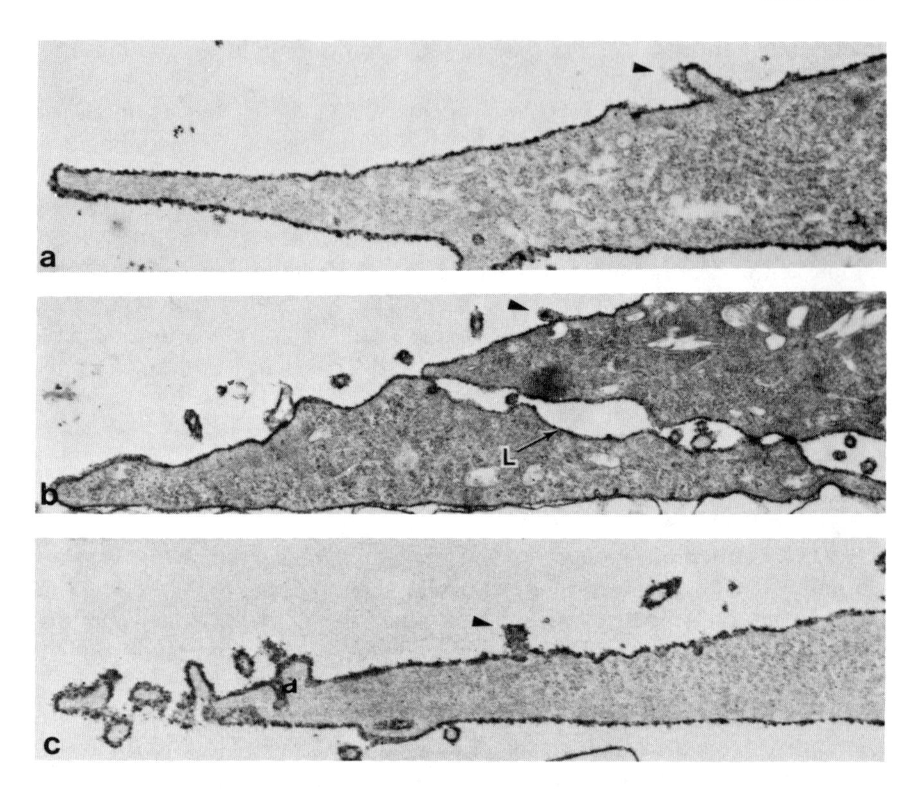

Figure 1. Ultrastructural distribution of cell surface glycoproteins on subconfluent MDCK cells. The 35-kD (a), 50-kD (b), and 60-kD (c) glycoproteins have been localized by incubation in monoclonal antibody followed by goat antimouse HRP (GAM-HRP). All micrographs are oriented with the apical surface at the top (microvilli are designated by arrowheads) and the basal surface at the bottom. Electron-dense reaction product appears to be uniformly distributed over both apical and basal cell surfaces (a–c) and was localized on the lateral membranes (L) of cells in contact (b). (a) 13,800 ×, (b) 12,100 ×, (c) 15,600 ×. (Reproduced from Herzlinger and Ojakian, 1984, by copyright permission of the Rockefeller University Press.)

The 35-kD, 50-kD, and 60-kD glycoproteins were localized on MDCK cells that had grown into confluent monolayers. Under these conditions, the monolayers have functional tight junctions (Misfeldt *et al.*, 1976; Cereijido *et al.*, 1978), and this laboratory typically records transepithelial electrical resistances of 150–300 Ω cm^2 (Cramer *et al.*, 1980; Ojakian, 1981). When the 50-kD glycoprotein was localized on confluent MDCK cells with HRP-labeled antibodies, this membrane protein had a polarized distribution localized primarily to the basolateral membrane (Fig. 3). Sim-

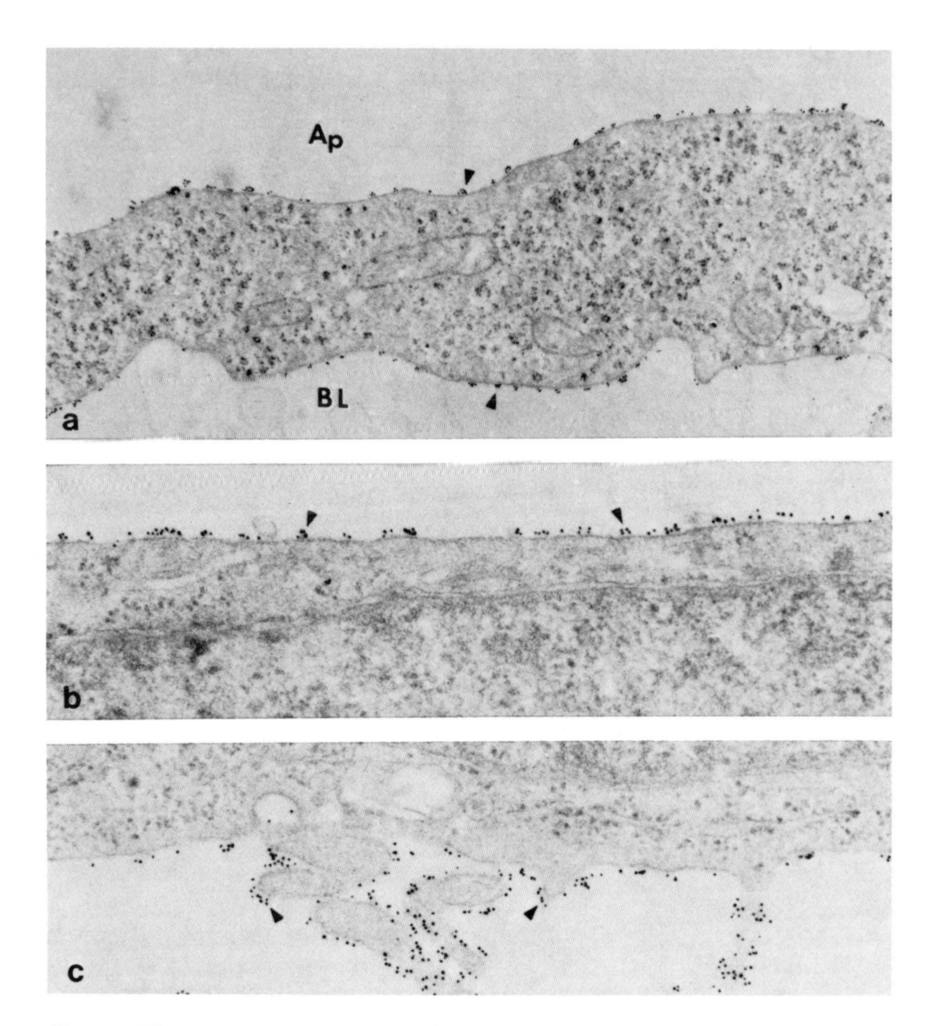

Figure 2. Ultrastructural localization of the 60-kD glycoprotein with colloidal gold. Subconfluent MDCK cells grown on micropore filters were fixed for electron microscopy and then incubated sequentially in monoclonal antibody G12, rabbit antimouse IgG, and 9-nm-diameter colloidal gold-labeled protein A for 2 hr each at 4°C to localize the 60-kD glycoprotein. (a) Colloidal gold label (arrowheads) can be visualized on both the apical (Ap) and basolateral (BL) cell surface of a subconfluent MDCK cell. The 60-kD glycoprotein localized by colloidal gold (arrowheads) can be clearly seen at higher magnification of the apical (b) and basolateral (c) cell surfaces. The quantitative measurements for the density of the 60-kD glycoprotein are presented in Table I. (a) 6,900 ×, (b,c) 17,300 ×.

Table I. Quantitative Measurement of the 60-kD Glycoprotein Density on Subconfluent MDCK Cell Surfaces[a]

Plasma membrane domain	Surface area (μm/cell)	Gold particles (number/cell)	Gold particles (number/μm)	$A_{P/BL}$
Apical	1.70	177.3	104.3 ± 6.5[b]	1.025
Basolateral	1.74	176.0	101.1 ± 4.8	

[a] Subconfluent MDCK cells grown on micropore filters were fixed with paraformaldehyde, and the 60-kD glycoprotein was localized on the plasma membrane by incubating in monoclonal antibody, RAM, and protein A–colloidal gold as described in Fig. 2. $N = 7$.
[b] Standard error of the mean.

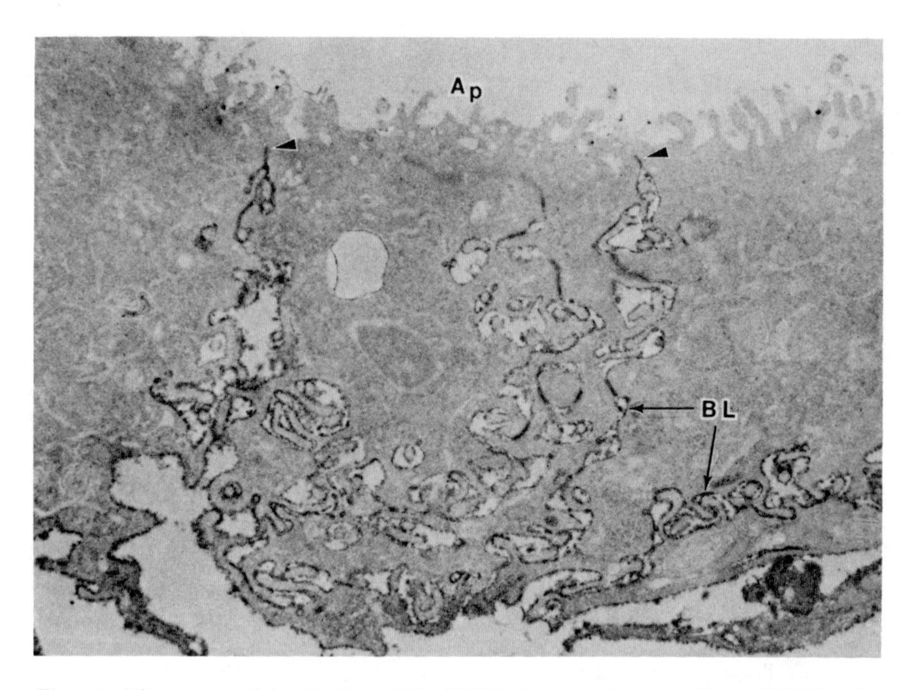

Figure 3. Ultrastructural localization of the 50-kD glycoprotein on confluent MDCK cells. Monolayers were fixed, scraped from the culture dish, and incubated in monoclonal antibody H6 followed by GAM-HRP. The distribution of electron-dense reaction product demonstrates that the 50-kD glycoprotein is localized to the basolateral membrane (BL) up to the tight junctions (arrowheads) and was not visualized on the apical cell surface (Ap). 14,800 ×. (Reproduced from Herzlinger and Ojakian, 1984, by copyright permission of the Rockefeller University Press.)

ilar distributions were also observed for the 35-kD and 60-kD glycoproteins (Herzlinger and Ojakian, 1984). These results provide additional evidence that MDCK are polarized cells having asymmetric distributions of plasma membrane glycoproteins similar to those previously reported for leucine aminopeptidase and Na^{+}–K^{+} ATPase (Louvard, 1980; Lamb *et al.*, 1981; Reggio *et al.*, 1982). They also support the hypothesis that tight junctions are necessary for the maintenance of epithelial cell surface polarity (Pisam and Ripoche, 1976; Meldolesi *et al.*, 1978; Ziomek *et al.*, 1980).

The random glycoprotein distributions observed in subconfluent cells are not unexpected since, in the absence of tight junctions, the fluidity of plasma membranes would allow the lateral movement of proteins (Singer and Nicolson, 1972). Rodriguez-Boulan and Sabatini (1978) have demonstrated that MDCK cells can support the polarized budding of lipid envelope viruses, and the envelope proteins of these viruses appear to be directly inserted into the proper membrane domain (Rodriguez-Boulan and Pendergast, 1980; Matlin and Simons, 1984; Rindler *et al.*, 1985). The observation that subconfluent MDCK cells can also support polarized viral budding (Rodriguez-Boulan *et al.*, 1983) appears to contradict our data; however, this study did not demonstrate that envelope proteins are inserted in a polarized manner prior to viral assembly. An ideal experimental system for testing this hypothesis would be to utilize temperature-sensitive envelope virus mutants, which allow the coordinated insertion of proteins into the epithelial plasma membrane within a well-defined time course (Rodriguez-Boulan *et al.*, 1984; Rindler *et al.*, 1985).

4. Tight Junctions and the Development of Cell Surface Polarity

Since the 35-kD, 50-kD, and 60-kD glycoproteins had a polarized cell surface distribution in confluent MDCK cells but a nonpolarized one in subconfluent cells, we decided to study glycoprotein distribution during cell growth into confluent monolayers in order to determine the role of tight junctions in cell polarity development. This was accomplished by measuring the transepithelial electrical resistance of MDCK plated on micropore filters at precise time intervals to monitor tight junction formation as the cells grew into confluent monolayers. These measurements were correlated with the distribution of the 35-kD, 50-kD and 60-kD glycoproteins at the same time intervals (Herzlinger and Ojakian, 1984). Our data demonstrate that tight junctions are formed within 48 hr after plating (Fig. 4) and that cell surface polarity can be observed by immunofluorescence microscopy at the same time (Fig. 5). It is important to note

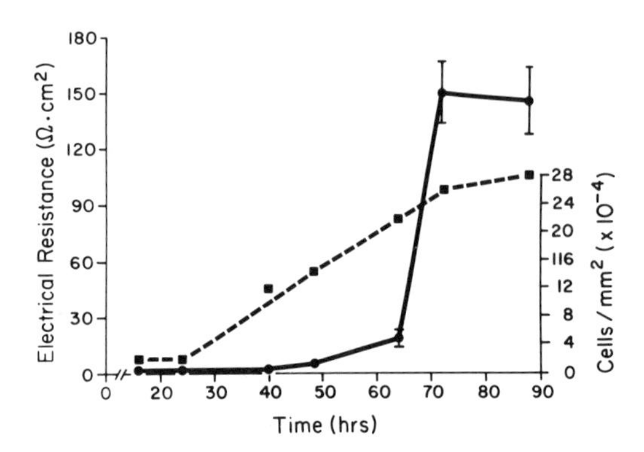

Figure 4. Development of MDCK tight junctions. The MDCK cells were plated on micropore filters at 1.5×10^5 cells/well, and the transepithelial electrical resistance (Ω cm^2) was measured at the times indicated as the cells grew into a confluent monolayer (solid line). Electrical resistances for four filters were recorded at each time, and the data are represented as the mean ± standard error. Density measurements of cells grown on micropore filters were taken at the same intervals (dashed line). (Reproduced from Herzlinger and Ojakian, 1984, by copyright permission of the Rockefeller University Press.)

that polarized distributions of glycoproteins were detected prior to the maximal transepithelial electrical resistance, demonstrating that the development of cell polarity is tightly coupled to tight junction formation and that complete junctional assembly is not required for the establishment of asymmetric distributions of plasma membrane glycoproteins.

We have also observed that the development of cell surface polarity appears to be a continuous process, with the levels of glycoproteins detected on the basolateral membrane by immunofluorescence increasing after the formation of tight junctions (Herzlinger and Ojakian, 1984). Quantitative immunoelectron microscopic studies are now in progress to determine whether these increases result from an increased density of glycoprotein molecules, an increased basolateral membrane area, or a combination of both.

Our observations on the coupling of tight junction assembly to the establishment of asymmetric distributions of cell surface glycoproteins confirm and extend those of Rabito and Tchao (1980), who demonstrated the redistribution of Na^+–K^+ ATPase during tight junction formation in MDCK cells. More recently, Rabito *et al.* (1984) have provided evidence that the polarized distribution of the apical cell surface enzymes alkaline phosphate and glutamyl transpeptidase is established during the development of tight junctions in the pig kidney epithelial cell line LLC-PK1. Furthermore, both of these studies from Rabito's laboratory also demonstrate that although polarized distributions of cell surface enzymes become established during the assembly of tight junctions, these proteins continue to be inserted into the plasma membrane after junctional formation is completed.

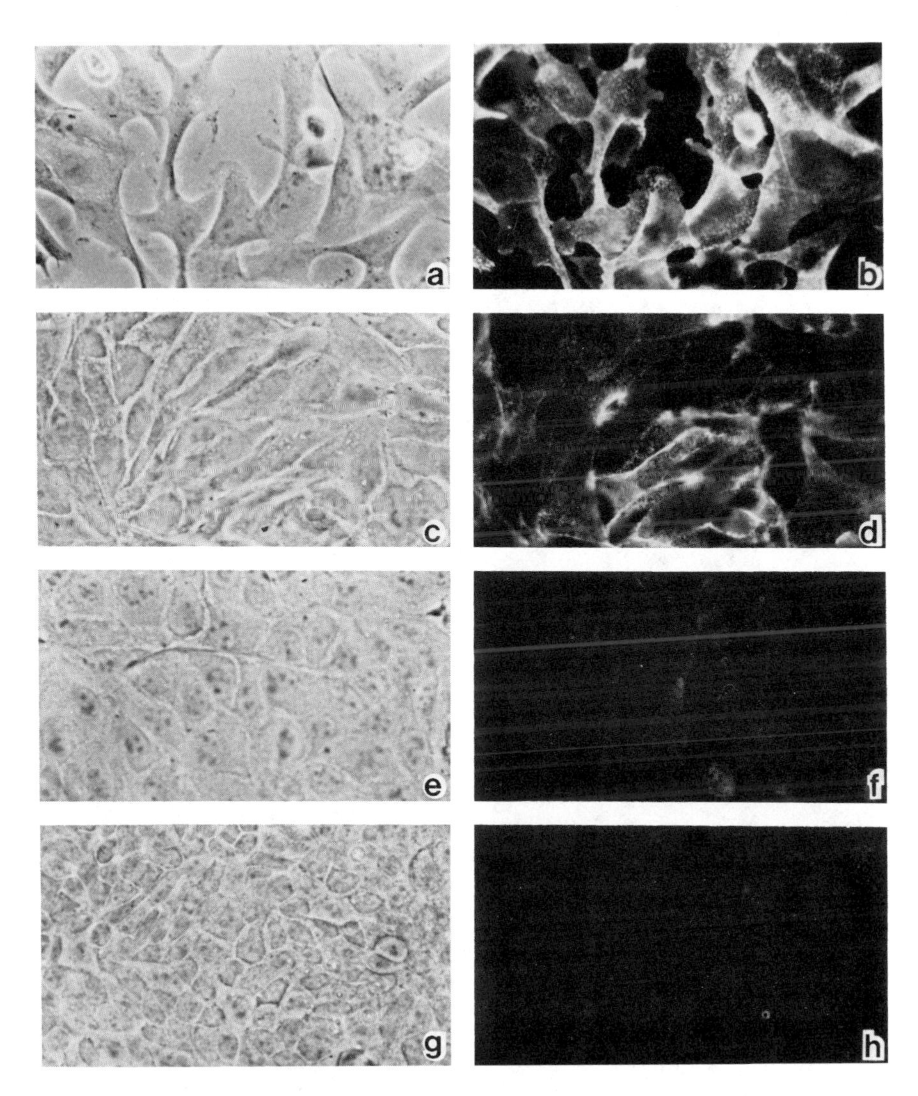

Figure 5. Development of MDCK cell surface polarity. Corresponding phase-contrast (a,c,e,g) and immunofluorescence (b,d,f,h) micrographs of developing MDCK monolayers stained with monoclonal antibody G12 to localize the 60-kD glycoprotein. When examined 24 hr after plating (a,b), intense apical cell surface staining was observed (b) on the sub-confluent cells (a). By 40 hr (c,d), many of the cells still had the 60-kD glycoprotein on their apical cell surface (d); 48 hr after plating (e,f), only a small amount of apical staining was observed (f), and these levels remained unchanged at 88 hr (g,h). 140×. (Reproduced from Herzlinger and Ojakian, 1984, by copyright permission of the Rockefeller University Press.)

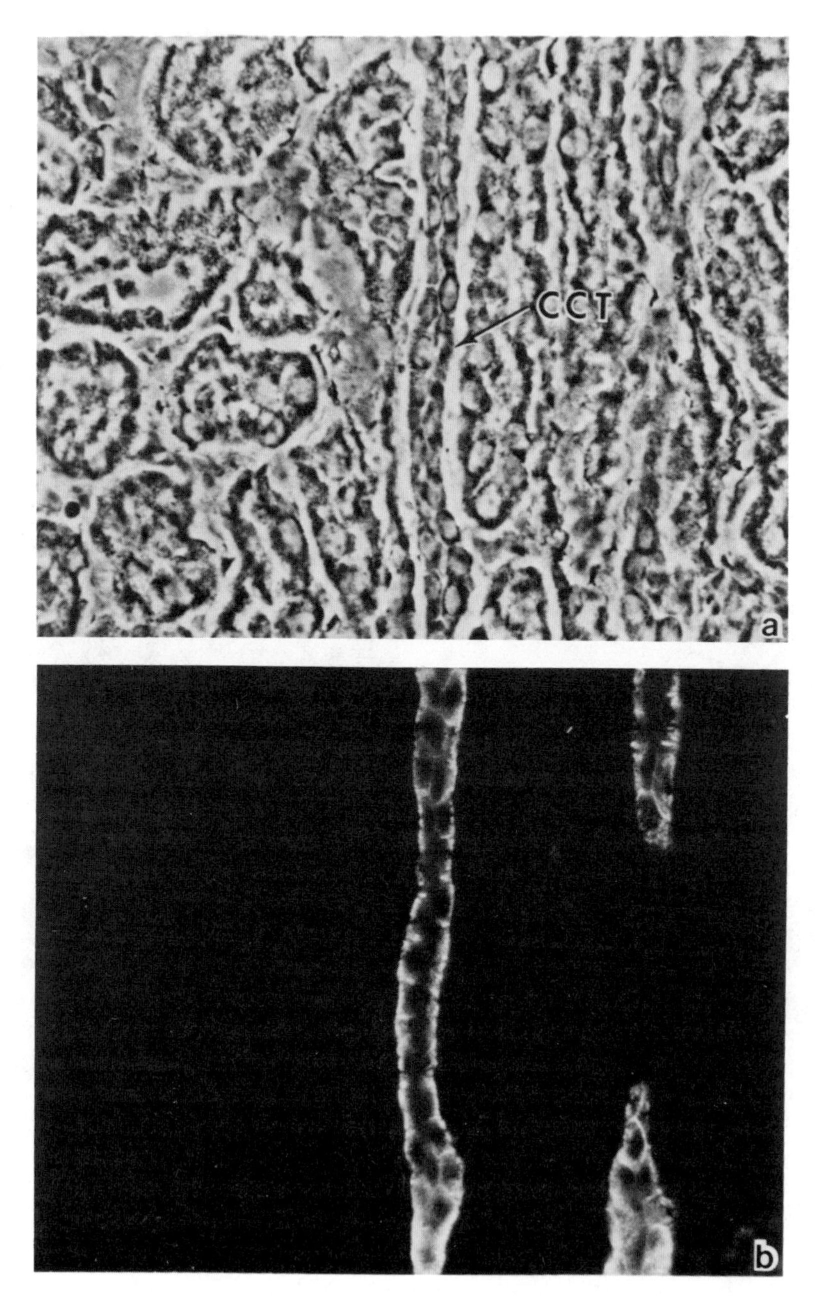

Figure 6. Distribution of the 25-kD glycoprotein in rabbit kidney cortex. Corresponding phase (a) and immunofluorescence (b) micrographs demonstrate that the 25-kD glycoprotein

The sorting mechanisms involved in establishing cell surface polarity have not yet been elucidated despite the large body of work done in this field. We have suggested previously (Herzlinger and Ojakian, 1984) that once tight junctions have been assembled, membrane proteins can be targeted to the proper domain, as has been demonstrated with envelope viruses (Rodriguez-Boulan and Pendergast, 1980; Rindler *et al.*, 1982; Roth *et al.*, 1983; Matlin and Simmons, 1984; Rindler *et al.*, 1985). When proteins are trapped in the wrong membrane domain by the assembling tight junctions, these proteins can then be removed from the cell surface by endocytosis and either degraded in lysosomes or recycled to the proper membrane domain (Matlin *et al.*, 1983; Personen and Simons, 1983).

5. *Identification of Nephron-Segment-Specific Cell Surface Proteins*

The renal nephron is a long epithelial tubule composed of a variety of morphologically and physiologically different cell types. Some of these functional differences reflect variations in hormonal sensitivity, with different nephron segments being responsive to different hormones (Morel, 1981). Since MDCK cells were derived from adult dog kidney, we decided to map the distribution of MDCK glycoproteins along the nephron to determine the segment of MDCK origin. Furthermore, by determining the cell surface distribution in the kidney, we would have probes available for studying the dynamics of epithelial cell surface polarity development using monoclonal antibodies against membrane proteins from different segments of the nephron. In some of our previous work, we have demonstrated that MDCK cells express a cell surface antigen that is localized primarily to epithelial cells of the thick ascending limb and distal convoluted tubule (Herzlinger *et al.*, 1982). We subsequently identified this antigen as the 35-kD glycoprotein discussed above (Herzlinger and Ojakian, 1984) and have demonstrated by affinity chromatography and SDS-PAGE that the monoclonal antibody to this glycoprotein recognizes a protein of identical molecular weight in the kidney (D. A. Herzlinger and G. K. Ojakian, unpublished data). Immunoelectron microscopy has localized the 35-kD glycoprotein to the basolateral membrane (Ojakian and

---

is localized to all epithelial cells of the cortical collecting tubule (CCT). Careful examination of the immunofluorescence staining (b) suggests that this glycoprotein is distributed on the basolateral membrane infoldings. 450×.

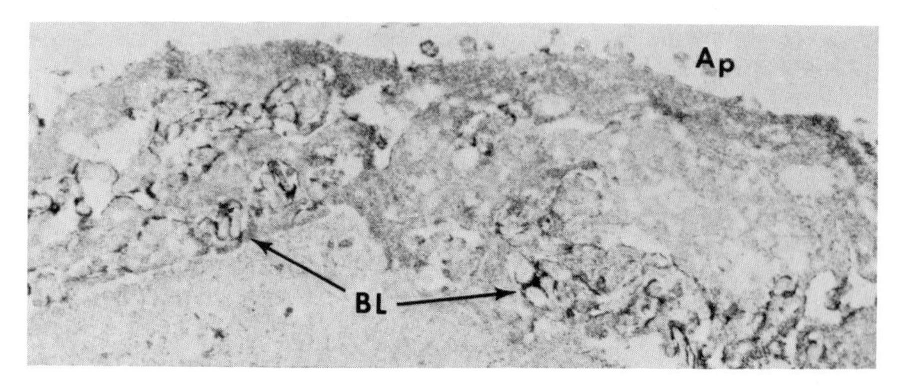

Figure 7. Ultrastructural localization of the 25-kD glycoprotein on the cortical collecting tubule. Dog kidney cortex was fixed and processed for electron microscopic immunocytochemistry with monoclonal antibody El and GAM-HRP according to the procedures of Sisson and Vernier (1980). The 25-kD glycoprotein has a polarized cell surface distribution on the collecting tubule, being localized to the basal–lateral membrane infoldings (BL) and not on the apical plasma membrane (Ap). 8800×.

Herzlinger, 1984), a distribution that is identical to that on MDCK cells (Herzlinger and Ojakian, 1984).

We have also obtained six monoclonal antibodies against a 25-kD glycoprotein on MDCK cells, determined that the antibodies obtained are against at least three different epitopes on the molecule, and localized the antigen to the epithelial cells of the cortical thick ascending limb, distal convoluted tubule, and both the cortical and medullary collecting tubule (Ojakian and Herzlinger, 1984). Immunofluorescence microscopy demonstrates that the 25-kD glycoprotein appears to be localized to the basolateral membrane of the collecting tubule (Fig. 6), and the polarized distribution of the 25-kD glycoprotein in this segment of the nephron has been confirmed by immunoelectron microscopy (Fig. 7).

When the 25-kD glycoprotein was localized on the distal convoluted tubule, a surprising result was obtained. Immunoelectron microscopy of dog kidney cortex demonstrated that this glycoprotein was uniformly distributed over both the apical and basolateral cell surfaces (Fig. 8). The physiological processes that are involved in the variety of vectorial epithelial transport functions (Berridge and Oschman, 1972; Murer and Kinne, 1980) usually require that cell surface enzymes have polarized distributions so that ions, sugars, amino acids, and other nutrients can be transported against concentration gradients. Since we have not yet ascertained the function of the 25-kD glycoprotein, we cannot speculate on the significance of its unpolarized distribution in the distal convoluted tubule. It is important to point out, however, that at least two other epithelial cell surface proteins also have been localized on both the apical and basolateral domains of epithelial cells: secretory component, the IgA transport protein (Solari and Kraehenbuhl, 1984), and the IgG receptor of the neonatal intestinal epithelium (Rodewald, 1980). Although both of these receptors are involved in transepithelial ligand transport, we do not presently have any evidence that the 25-kD glycoprotein performs a similar function in kidney epithelia.

6. *High-Resistance MDCK Cells as a Model for the Collecting Tubule*

When the cell surface distribution of the 25-kD glycoprotein was studied on MDCK cells by immunofluorescence microscopy, it was localized on the apical plasma membrane of both subconfluent and confluent cells (data not shown). Ultrastructural studies have confirmed this observation and have determined that the 25-kD glycoprotein is also localized to the basolateral membrane in quantities similar to that of the apical membrane (Fig. 9). Since confluent MDCK cells have polarized distributions of the 35-kD, 50-kD, and 60-kD glycoproteins (Herzlinger and Ojakian, 1984) as well as other membrane components (Richardson and Simmons, 1979; Louvard, 1980; Cereijido *et al.*, 1980; Reggio *et al.*, 1982), it is reasonable to assume that the cellular processes involved in the sorting out of membrane proteins (Sabatini *et al.*, 1982; Rindler *et al.*, 1982, 1985; Rodriguez-Boulan *et al.*, 1984) are functioning properly in MDCK cells. In addition, as discussed above, there are epithelial cell surface proteins that do not have a polarized distribution. In this respect, the 25-kD glycoprotein has a distribution similar to that of the distal convoluted tubule (Fig. 9), the nephron segment that is the likely origin of the MDCK cell line (Rindler *et al.*, 1979; Herzlinger *et al.*, 1982).

To further study the properties of membrane proteins involved in the establishment of epithelial cell surface polarity, we have been utilizing MDCK clonal cell lines. One of these lines (LR-MDCK) has a low transepithelial electrical resistance (100 Ω cm^2) that is similar to that of the parental MDCK cells (Misfeldt *et al.*, 1976; Cereijido *et al.*, 1978). Like MDCK cells, confluent LR-MDCK have a basolateral distribution of the 35-kD, 50-kD, and 60-kD glycoproteins (data not shown) but a nonpolarized distribution of the 25-kD glycoprotein (Fig. 10).

We have also been studying a clonal line of MDCK that has a transepithelial electrical resistance of 600–1000 Ω cm^2 and a transepithelial potential difference of 8–20 mV, with the apical side negative with respect

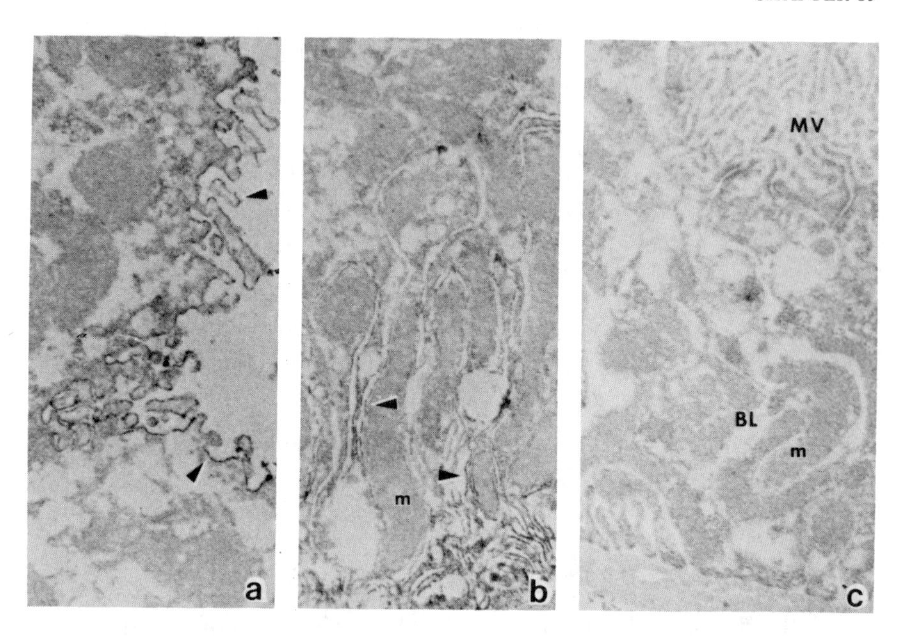

Figure 8. Ultrastructural localization of the 25-kD glycoprotein on the distal convoluted tubule. Electron-dense reaction product outlines both the apical (a) and basolateral (b) cell surfaces. As an internal control, an epithelial cell of the proximal convoluted tubule from the same thin section is presented in an adjacent panel (c). Since the proximal tubule does not express the 25-kD glycoprotein, the apical microvillar (mv) and the basolateral (BL) membranes are not delineated by the electron-dense HRP reaction product (c). Mitochondria (m) next to the basolateral membrane infoldings are also unstained in both the distal (b) and proximal (c) tubule cells. 14,000×.

to the basolateral (Table II). This high-resistance MDCK cell line (HR-MDCK) is physiologically similar to the strain 1 MDCK line studied by Richardson *et al.* (1981). The HR-MDCK have polarized distributions of not only the 35-kD, 50-kD, and 60-kD glycoproteins but also the 25-kD glycoprotein (Fig. 10). Immunofluorescence microscopy of HR-MDCK demonstrates that the levels of the 25-kD glycoprotein are extremely low on the apical cell surface (Fig. 10d) but are high on the basolateral membrane (Fig. 10e).

At the present time, we do not know why the 25-kD glycoprotein has a nonpolarized distribution in MDCK and LR-MDCK and a polarized one in HR-MDCK. Since other membrane glycoproteins have polarized distributions in the three MDCK cell lines, it is reasonable to assume that

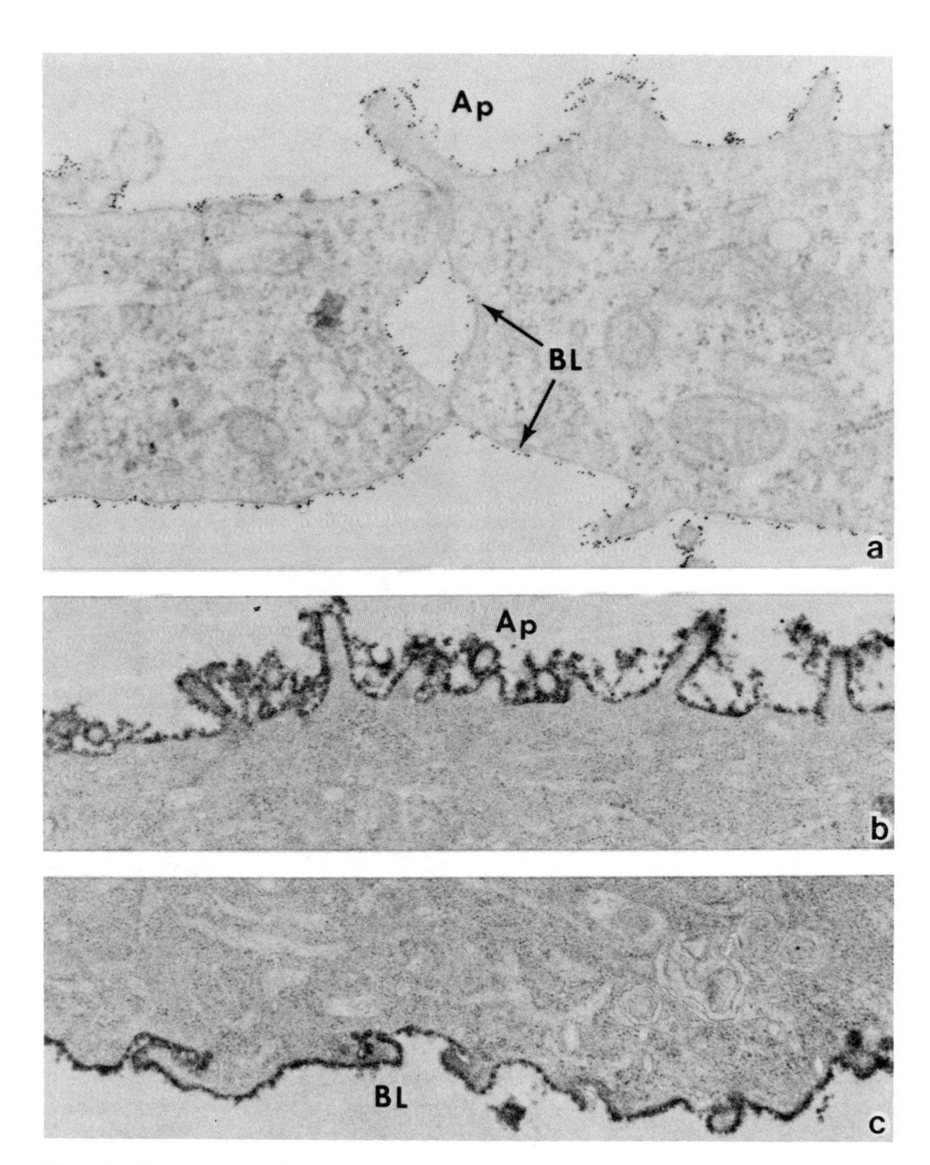

Figure 9. Ultrastructural distribution of the 25-kD glycoprotein on MDCK cells. Monoclonal antibody El and colloidal gold–protein A were used to localize the 25-kD glycoprotein on growing MDCK cells (a). In regions of the developing monolayer where the cells are in contact, this glycoprotein is distributed on both the apical and basolateral membranes (a). With GAM-HRP it was determined that the nonpolarized distribution of the 25-kD glycoprotein was maintained on the MDCK cells after growth into a confluent monolayer (b,c). In unstained thin sections, considerable staining could be detected on the apical cell surface (b), and the amount of apical electron-dense reaction product is comparable to that on the basolateral membrane (c). In contrast, compare these results to the polarized cell surface distribution of the 50-kD glycoprotein (Fig. 3). 17,500×.

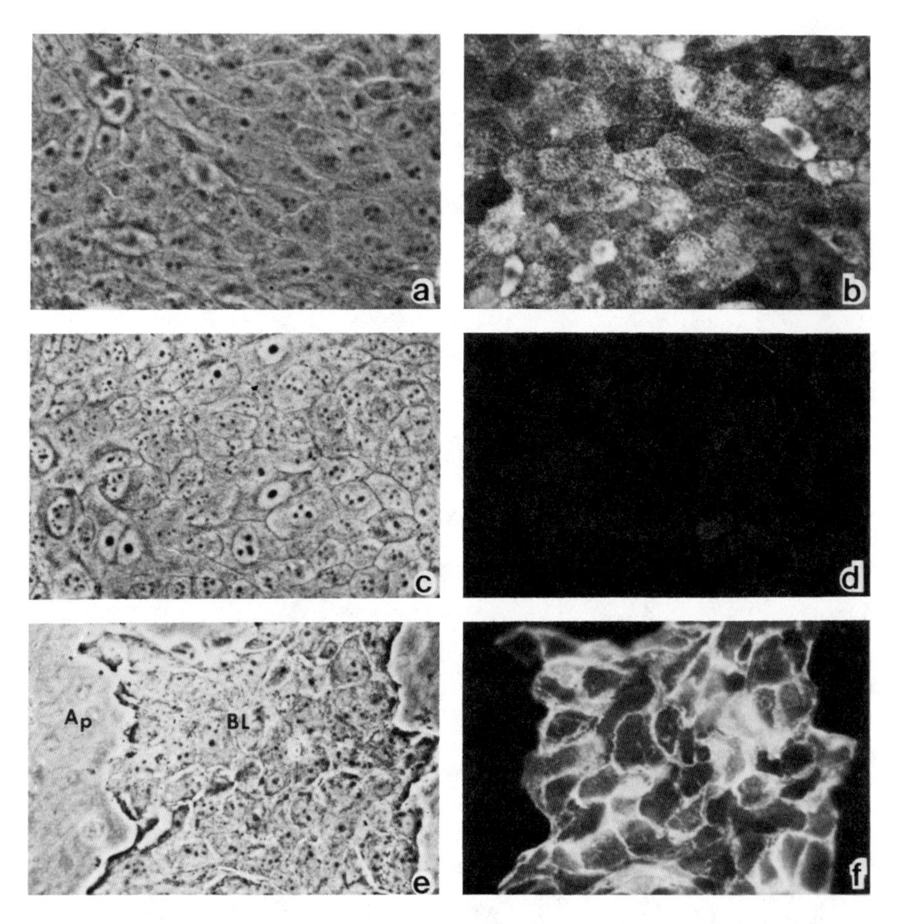

Figure 10. Cell surface distributions of the 25-kD glycoprotein on MDCK clonal cell lines. Phase-contrast micrographs of confluent LR-MDCK (a) and HR-MDCK (c,e) monolayers are presented. In adjacent micrographs, immunofluorescence staining of the identical area (b,d,f) with monoclonal antibody El demonstrates that the 25-kD glycoprotein is present on the apical surface of LR-MDCK (b) but not on HR-MDCK (d). In a region of the HR-MDCK monolayer with an experimentally induced tear (e), the 25-kD glycoprotein was localized to the basolateral membranes (BL) but could not be detected on the apical (Ap) cell surface (f). 210×.

the 25-kD glycoprotein has properties that differ in these cells and that elucidating these potential differences will give us valuable information on the signals involved in intracellular sorting of membrane proteins (Blobel *et al.*, 1979; Sabatini *et al.*, 1982).

To approach this problem we have initiated studies on the compar-

Table II. Comparative Summary of Differences between the MDCK and HR-MDCK Cell Lines

Properties	MDCK	HR-MDCK
Transepithelial electrical resistance	150–200 Ω cm^2	1000 Ω cm^2
Transepithelial potential difference	−1 mV	−10 mV
25-kD glycoprotein distribution	AP and BL	BL
Nephron segment identity	Distal tubule	Collecting tubule

ative biochemistry of the 25-kD glycoprotein in the MDCK cell lines and kidney membranes. The MDCK, LR-MDCK, and HR-MDCK were all grown on [^{35}S] methionine, and the radiolabeled glycoproteins were immunoprecipitated and analyzed by SDS-PAGE. Our results demonstrate that monoclonal antibodies to the 25-kD glycoproteins recognize a protein of identical molecular weight in all three cell lines (data not shown).

In addition, preliminary experiments using isoelectric focusing have demonstrated that the isoelectric points of the 25-kD glycoproteins are identical in MDCK and HR-MDCK cells. At the present time, we do not have any further information as to whether the differences in 25-kD glycoprotein sorting could be caused by differences in glycosylation, amino acid composition, or secondary modifications such as phosphorylation. All of these possibilities are currently under investigation as we work toward our long-range goal that comparative studies between the nonpolarized and polarized forms of the 25-kD glycoprotein will provide us with information on the sorting-out processes involved in the biogenesis of epithelial cell surfaces. Finally, the high-resistance lines of MDCK have many of the characteristics of the collecting tubule (see Table II; also Richardson *et al.*, 1981; Valentich, 1981) and should prove to be extremely useful tools for studying renal epithelial cell physiology in well-defined cell culture systems as advocated by Handler and his colleagues (1980).

References

Berridge, M. J., and Oschman, J. L., 1972, *Transporting Epithelia*, Academic Press, New York.

Blobel, G., Walter, P., Chung, C. N., Goldman, B. M., Erickson, A. H., and Lingappa, V.

R., 1979, Translocation of proteins across membranes: The signal hypothesis and beyond, *Symp. Soc. Exp. Biol.* **33**:9–36.

Cereijido, M., Robbins, E. S., Dolan, W. J., Rotunno, C. A., and Sabatini, D. D., 1978, Polarized monolayers formed by epithelial cells on a permeable and translucent support, *J. Cell Biol.* **77**:853–880.

Cereijido, M., Ehrenfeld, J., Meza, I., and Martinez-Palomo, A., 1980, Structural and functional membrane polarity in cultured monolayers of MDCK cells, *J. Membr. Biol.* **52**:147–159.

Cramer, E. B., Milks, L. C., and Ojakian, G. K., 1980, Transepithelial migration of human neutrophils: An *in vitro* model system, *Proc. Natl. Acad. Sci. U.S.A.* **77**:4069–4073.

Ernst, S. A., and Mills, J. W., 1977, Basolateral plasma membrane localization of ouabain-sensitive sodium transport sites in the secretory epithelium of the avian salt gland, *J. Cell. Biol.* **75**:74–94.

Farquhar, M. G., and Palade, G. E., 1963, Junctional complexes in various epithelia, *J. Cell Biol.* **17**:375–412.

Fujita, M., Kawai, K., Asano, K., and Nakao, M., 1973, Protein components of two different regions of an intestinal epithelial cell membrane, *Biochim. Biophys. Acta* **307**:141–151.

Gefter, M. L., Margulies, D. H., and Scharff, M. D., 1977, A simple method for polyethylene glycol-promoted hybridization of mouse myeloma cells, *Somat. Cell Genet.* **3**:231–236.

Handler, J. S., Perkins, F. M., and Johnson, J. P., 1980, Studies of renal cell function using cell culture techniques, *Am. J. Physiol.* **238**:F1–F9.

Herzlinger, D. A., and Ojakian, G. K., 1984, Studies on the development and maintenance of epithelial cell surface polarity with monoclonal antibodies, *J. Cell Biol.* **98**:1777–1787.

Herzlinger, D. A., Easton, T. G., and Ojakian, G. K., 1982, The MDCK epithelial cell line expresses a cell surface antigen of the kidney distal tubule, *J. Cell Biol.* **93**:269–277.

Hoi Sang, U., Saier, M. H., Jr., and Ellisman, M. H., 1979, Tight junction formation is closely linked to the polar redistribution of intramembranous particles in aggregating MDCK epithelia, *Exp. Cell Res.* **122**:384–391.

Kawai, K., Fujita, M., and Nakao, M., 1974, Lipid components of two different regions of an intestinal epithelial cell membrane of mouse, *Biochim. Biophys. Acta* **369**:222–233.

Köhler, G., and Milstein, C., 1975, Continuous cultures of fused cells secreting antibody of predefined specificity, *Nature* **256**:495–497.

Kyte, J., 19176, Immunoferritin determination of the distribution of ($Na^+ + K^+$) ATPase over the plasma membranes of renal convoluted tubules. I. Distal segment, *J. Cell Biol.* **68**:287–303.

Lamb, J. F., Ogden, P., and Simmons, N. L., 1981, Autoradiographic localization of [^{3}H]ouabain bound to cultured epithelial cell monolayers of MDCK cells, *Biochim. Biophys. Acta* **644**:333–340.

Langone, J. J., 1982, Use of labeled protein A in quantitative immunochemical analysis of antigens and antibodies, *J. Immunol. Methods* **51**:3–22.

Leighton, J., Estes, L. W., Mansukhani, S., and Brada, Z., 1970, A cell line derived from dog kidney (MDCK) exhibiting qualities of papillary adenocarcinoma and of renal tubular epithelium, *Cancer* **26**:1022–1028.

Louvard, D., 1980, Apical membrane aminopeptidase appears at sites of cell–cell contact in cultured kidney epithelial cells, *Proc. Natl. Acad. Sci. U.S.A.* **77**:4132–4136.

Matlin, K., and Simons, K., 1984, Sorting of an apical plasma membrane glycoprotein occurs before it reaches the cell surface in cultured epithelial cells, *J. Cell Biol.* **99**:2131–2139.

Matlin, K. S., Bainton, D. F., Personen, M., Louvard, D., Gentry, N., and Simons, K., 1983, Transepithelial transport of a viral membrane glycoprotein implanted into the apical plasma membrane of Madin–Darby canine kidney cells. I. Morphological evidence, *J. Cell Biol.* **97**:627–637.

Meldolesi, J., Castiglioni, G., Parma, R., Nassivera, N., and DeCamilli, P., 1978, Ca^{++}-dependent disassembly and reassembly of occluding junctions in guinea pig pancreatic acinar cells: Effect of drugs, *J. Cell Biol.* **79**:156–172.

Misfeldt, D. S., Hamamoto, S. T., and Pitelka, D. R., 1976, Transepithelial transport in cell culture, *Proc. Nat. Acad. Sci. U.S.A.* **73**:1212–1216.

Morel, F., 1981, Sites of hormone action in the mammalian nephron, *Am. J. Physiol.* **240**:F159–F164.

Mühlpfordt, H., 1982, The preparation of colloidal gold particles using tannic acid as an additional reducing agent, *Experientia* **38**:1127–1128.

Murer, H., and Kinne, R., 1980, The use of isolated membrane vesicles to study epithelial transport processes, *J. Membr. Biol.* **55**:81–95.

Ojakian, G. K., 1981, Tumor promoter-induced changes in the permeability of epithelial cell tight junctions, *Cell* **23**:95–103.

Ojakian, G. K., and Herzlinger, D. A., 1984, Analysis of epithelial cell surface polarity with monoclonal antibodies, *Fed. Proc.* **43**:2208–2216.

Pesonen, M., and Simons, K., 1983, Transepithelial transport of a viral membrane glycoprotein implanted into the apical plasma membrane of Madin–Darby canine kidney cells. II. Immunological evidence, *J. Cell Biol.* **97**:638–643.

Pisam, M., and Ripoche, P., 1976, Redistribution of surface macromolecules in dissociated epithelial cells, *J. Cell Biol.* **71**:907–920.

Rabito, C. A., and Tchao, R., 1980, [^{3}H]ouabain binding during the monolayer organization and cell cycle in MDCK cells, *Am. J. Physiol.* **238**:C43–C48.

Rabito, C. A., Tchao, T., Valentich, J., and Leighton, J., 1978, Distribution and characteristics of the occluding junctions in a monolayer of a cell line (MDCK) derived from canine kidney, *J. Membr. Biol.* **43**:351–365.

Rabito, C. A. Kreisberg, J. I., and Wight, D., 1984, Alkaline phosphatase and γ-glutamyl transpeptidase as polarization markers during the organization of LLC-PK_1 cells into an epithelial membrane, *J. Biol. Chem.* **259**:574–582.

Reggio, H., Courdrier, E., and Louvard, D., 1982, Surface and cytoplasmic domains in polarized epithelial cells, in: *Progress in Clinical and Biological Research Vol. 91* (J. F. Hoffman, G. H. Giebisch, L. Doris, eds.) Alan R. Liss, New York, pp. 89–105.

Richardson, J. C., and Simmons, N. L., 1979, Demonstration of protein asymmetries in the plasma membrane of cultured renal (MDCK) epithelial cells by lactoperoxidase-mediated iodination, *FEBS Lett.* **105**:201–204.

Richardson, J. C. W., Scalera, V., and Simmons, N. L., 1981, Identification of two strains of MDCK cells which resemble separate nephron tubule segments, *Biochim. Biophys. Acta* **673**:26–36.

Rindler, M. J., Chuman, L. M., Shaffer, L., and Saier, M. H., Jr., 1979, Retention of differentiated properties in an established dog kidney epithelial cell line (MDCK), *J. Cell Biol.* **81**:635–648.

Rindler, M. J., Ivanov, I. E., Rodriguez-Boulan, E., and Sabatini, D. D., 1982, Biogenesis of epithelial cell plasma membranes, *Ciba Found. Symp.* **92**:184–208.

Rindler, M. J., Ivanov, I. E., Plesken, H., and Sabatini, D. D., 1985, Polarized delivery of viral glycoproteins to the apical and basolateral plasma membrane of Madin–Darby canine kidney cells infected with temperature-sensitive viruses, *J. Cell Biol.* **100**:136–151.

Rodewald, R., 1980, Distribution of immunoglobulin G receptors in the small intestine of the young rat, *J. Cell Biol.* **85**:18–32.

Rodriguez-Boulan, E., 1983, Membrane biogenesis, enveloped RNA viruses, and epithelial polarity, in: *Modern Cell Biology*, Vol. 1 (B. Satir, ed.) Alan R. Liss, New York, pp. 119–170.

Rodriguez-Boulan, E., and Pendergast, M., 1980, Polarized distribution of viral envelope proteins in the plasma membrane of infected epithelial cells, *Cell* **20:**45–54.

Rodriguez-Boulan, E., and Sabatini, D. D., 1978, Asymmetric budding of viruses in epithelial monolayers: A model system for study of epithelial polarity, *Proc. Natl. Acad. Sci. U.S.A.* **75:**5071–5075.

Rodriguez-Boulan, E., Paskiet, K. T., Salas, P. J. I., and Bard, E., 1984, Intracellular transport of influenza virus hemagglutinin to the apical surface of Madin–Darby canine kidney cells, *J. Cell Biol.* **98:**308–319.

Roth, M. G., Compans, R. W., Giusti, L., Davis, A. R., Nayak, D. P., Gething, M. J., and Sambrook, J., 1983, Influenza virus hemagglutinin expression is polarized in cells infected with recombinant SV40 viruses carrying cloned hemagglutinin DNA, *Cell* **33:**435–443.

Semenza, G., 1979, Small intestinal disaccharidases: Their properties and role as sugar translocators across natural and artificial membranes, in: *The Enzymes of Biological Membranes*, Vol. 3 (A. Martonosi, ed.), Plenum Press, New York, pp. 349–382.

Singer, S. J., and Nicolson, G. L., 1972, The fluid mosaic model of the structure of cell membranes, *Science* **175:**720–731.

Sisson, S. P., and Vernier, R. L., 1980, Methods for immunoelectron microscopy: Localization of antigens in rat kidney, *J. Histochem. Cytochem.* **28:**441–452.

Solari, R., and Kraehenbuhl, J.-P., 1984, Biosynthesis of the IgA antibody receptor: A model for the transepithelial sorting of a membrane glycoprotein, *Cell* **36:**61–71.

Staehelin, L. A., 1974, Structure and function of intercellular junctions, *Int. Rev. Cytol.* **39:**191–283.

Valentich, J. D., 1981, Morphological similarities between the dog kidney cell line MDCK and the mammalian cortical collecting tubule, *Ann. N.Y. Acad. Sci.* **372:**384–405.

Van Meer, G., and Simons, K., 1982, Viruses budding from either the apical or the basolateral plasma membrane domain of MDCK cell have unique phospholipid compositions, *EMBO J.* **1:**847–852.

Ziomek, C. A., Schulman, S., and Edidin, M., 1980, Redistribution of membrane proteins in isolated mouse intestinal epithelial cells, *J. Cell Biol.* **86:**849–857.

16

Frequency and Time Domain Analysis of Epithelial Transport Regulation

SIMON A. LEWIS and JOHN W. HANRAHAN

1. Introduction

The prime responsibility of most epithelia is to maintain plasma electrolyte and nonelectrolyte composition. Once a perturbation in plasma composition is sensed by osmo-, chemo-, or pressure receptors, a hormonal or neural signal can then initiate a series of intracellular events that ultimately modify the rate and sometimes even the direction of epithelial transport. To perform this function, an epithelium must be constructed so that it can actively and on demand absorb, secrete, or restrict the movements of substances between the plasma and external environment.

An epithelium is a planar array of cells joined at its apical–lateral interface by a continuous hoop of lipoprotein called the tight junction. There are two pathways across an epithelium through which a substance can pass. The first of these requires passage through the lateral intercellular spaces and tight junctions and occurs only in the presence of a transepithelial driving force (i.e., electrical and/or chemical gradient). The second pathway is transcellular and requires the presence of specific transport proteins in both the apical and basolateral membranes. Regulation of the rate direction of epithelial transport may involve a modification of either or both membrane transport systems. Because large changes in epithelial transport might alter the composition and volume of the cytoplasm, epithelial cells must also possess regulatory systems that are designed to safeguard their own viability.

Three obvious mechanisms by which cells can modulate transport

---

SIMON A. LEWIS and JOHN W. HANRAHAN • Department of Physiology, Yale Medical School, New Haven, Connecticut 06510. Supported by NIH grant AM33243.

are (1) the chemical activation/deactivation of preexisting transport units in the membrane (e.g., by phosphorylation or methylation), (2) the insertion/withdrawal of cytoplasmic vesicles containing transport units, and, finally, (3) the modulation/modification of already active transport units by such factors as membrane voltage or cell ionic composition.

There are many examples of extrinsic and intrinsic regulation of epithelial transport. The best-known instance of extrinsic regulation is the aldosterone-induced increase in Na^+ transport across the distal tubule and cortical collecting duct of the kidney, the mammalian colon, and amphibian and mammalian urinary bladders. Under conditions such as Na^+ depletion and hemorrhage, aldosterone is released from the adrenal cortex into the bloodstream, where it is carried to those tissues containing mineralocorticoid receptors in their cytoplasm and nucleus. Following mRNA transcription and subsequent protein synthesis, target epithelia show an increase in transepithelial Na^+ transport, which occurs approximately 45 min after aldosterone challenge. As originally proposed for toad bladder by Handler *et al.* (1972) and later demonstrated for a number of other epithelia, one action of aldosterone is to increase the Na^+ permeability of the apical (lumen-facing) membrane. This increase in apical permeability could result from the activation of quiescent channels, insertion of new channels from a cytoplasmic store, or modification of preexisting channels.

Only in recent years has it been established that there is also intrinsic regulation of transport. Perhaps the first observation was for amiloride-sensitive Na^+ transport in tight epithelia, about which it was reported that the inhibition of basolateral membrane Na^+–K^+ ATPase by ouabain resulted in a decrease in apical membrane permeability to Na^+. This negative feedback system would provide a failsafe for the cell by reducing Na^+ influx, thereby avoiding excessive swelling and consequent cell lysis (Lewis *et al.*, 1976). More recently, an intrinsic regulatory system has been proposed for the small intestine that involves basolateral K^+ conductance (Hudson and Schultz, 1984). In the intestine, sugars and amino acids enter the cells from the lumen coupled to the inward movement of Na^+. Thus, when substrates such as glucose are first added to the mucosal solution, we would expect increased Na^+ entry to depolarize the membrane and elevate intracellular Na^+ activity. These changes do occur, but only transiently; intestinal cells have the same Na^+ activity and membrane potential whether glucose is in the luminal solution or not, under steady-state conditions. These and other observations strongly suggest that (1) basolateral membrane K^+ permeability increases during exposure to glucose (accounting for the repolarization) and (2) there are changes in the kinetic properties of the Na^+–K^+ ATPase that may account for the relatively constant cell Na^+ activity. However, as with external regulation, these effects could be explained by chemical activation, insertion, or modulation of transporters.

This chapter discusses how frequency domain analysis can be used to distinguish between the various mechanisms of epithelial transport regulation. Using examples, we describe evidence that all three mechanisms may be used to control different transport systems within a single epithelium.

2. *Differentiating between Possible Mechanisms of Transport Regulation*

In order to determine which of the three mechanisms is being used by the cell to regulate transport, we obviously must measure a property that is unique to each mechanism. To establish activation, we should measure changes in the number of transporting units. For vesicular insertion, we must be able to measure an increase in membrane area. To determine whether modulation is occurring, we must study changes in the properties of individual transporters.

This chapter focuses on electrophysiological techniques. We are therefore confined to transport processes that are conductive. Two electrical methods that are currently available for distinguishing between activation and modulation are fluctuation analysis and single-channel recording. However, neither of these methods can differentiate between activation and vesicular insertion, which requires the measurement of membrane surface area by impedance analysis.

The purpose of this section is to provide an overview of the advantages, disadvantages, and potential pitfalls of impedance analysis, fluctuation analysis, and patch clamping rather than to describe in detail the theory and methodology for each technique.

2.1. *Impedance Analysis*

What is impedance analysis, and what unique information does it offer? Impedance is defined as the ratio of the voltage response of an electrical circuit to an applied current; the current waveform is usually a sinusoid or contains sinusoidal components. Tissue impedance varies as a function of frequency because of the capacitance of the lipid bilayer, which has low impedance at high frequencies and has high impedance at low frequencies. Two parameters are measured at each frequency: the ratio of the magnitude of the sinusoidal voltage to the current and the amount of time (or phase angle) that the voltage lags the current. Plotting log magnitude and phase angle versus log frequency provides Bode plots, which can then be fitted to various circuit models consisting of resistors and capacitors. The equivalent circuit must be assumed *a priori,* because the Bode plots are not unique; i.e., more than one electrical network could give the same Bode plot. This method has been applied to many biological membrane systems since the early 1940s (see Cole, 1972).

If the input current is a square wave, the output voltage response is analyzed in the time domain and yields one or more relaxation time constants (e.g., Lewis and de Moura, 1984). However, in some epithelia it is difficult or impossible to resolve the magnitudes and time constants adequately by curve fitting to these exponentials. Also, transient analysis weights the lower frequencies, which are less useful in distinguishing between detailed models (e.g., the distributed model; Clausen *et al.,* 1979). Alternatively, a sine wave or burst of sine waves at a number of different frequencies can be used. The output or response waveform will be a periodic function shifted in time, which can be analyzed for magnitude and phase. This method provides high resolution, but it is achieved at the expense of speed. The best approach is to use a current composed of many frequencies (e.g., white noise or a pseudorandom noise). The voltage response to this type of signal can be analyzed quickly and accurately in the frequency domain (see Clausen *et al.,* 1979; Clausen and Fernandez, 1981; Lim *et al.,* 1984).

A crucial step in interpreting data is to design an electrical equivalent circuit that adequately represents the important morphological characteristics of the epithelium. The traditional "lumped" model for epithelia consists of a parallel resistor and capacitor network (RC) to represent the apical membrane, another RC for the basolateral membrane, and a third resistor in parallel with the two RCs to denote the tight junction and lateral intercellular spaces. In order to estimate the values for each resistor and capacitor, all the data must be fitted to the model using a computer-based curve-fitting routine. Consistent deviations between data and theory would suggest that a particular circuit is inappropriate as a model for the epithelium. Examples of this approach have been described by Clausen *et al.* (1979, 1983).

One important piece of information that impedance analysis provides is the membrane area, because the capacitances obtained by curve fitting the data are directly proportional to actual membrane areas, including any amplification caused by membrane infolding. The relationship between area and capacitance is $1\ cm^2 \simeq 1\ \mu F$. By estimating capacitance, impedance analysis can be used to find out whether regulation involves fusion of cytoplasmic vesicles (containing within their bilayer the transport proteins) with the plasma membrane of the cell; a change in membrane area coincident with a change in transport rate would suggest that vesicle insertion/withdrawal is involved.

2.2. *Noise (Fluctuation) Analysis*

Fluctuation analysis is a method that allows calculation of the total number of active channels in a given area of membrane (channel density) and, with certain assumptions, the conductance of single channels and the overall rate at which the channels open and close. In addition, fluctuation between blocked (closed or nonconducting) and unblocked (open or conducting) states can be used to calculate association and dissociation rate constants for reversible blockers.

Fluctuation analysis relies on the fact that a population of identical channels randomly opening and closing will demonstrate a mean number of open channels over time that is proportional to the macroscopic current. However, at any point in time, the absolute number of open channels (or current) varies around the mean. To obtain information about channel density, one voltage clamps the membrane and records the mean current across the membrane as well as the fluctuations of this current about the mean value observed after both high- and low-pass filtering. The digitized current fluctuations are fast Fourier transformed into a sum of sinusoids of various frequencies, weighted according to their relative amplitudes in the original current record. This is depicted as the power spectral density (PSD) in a plot of log power (in units of amp^2 sec) versus log frequency (in hertz).

For a population of identical, independent channels that have only two kinetic states (open and closed), this spectrum has a characteristic shape called a Lorentzian, with a plateau at low frequencies (S_o) and a slope of -2 at high frequencies. The frequency at which the PSD decreases to one-half the plateau value is called the corner frequency (f_c) and is proportional to the sum of the opening and closing rate constants. The plateau value is a function of these constants, the number of channels (M), and the single-channel current (i). Mean current (I) is the product of the single-channel current, the number of channels, and the probability of each channel being open (P_o). It is obvious that one cannot solve for the four unknowns (i.e., the two rate constants, M, and i) with only three equations (for f_c, S_o, and i). Therefore, either one of the unknowns must be determined separately or the system must be perturbed.

A very useful approach was introduced by Lindemann and Van Driessche (1977) to study apical Na^+ channels in frog skin. They were unable to observe spontaneous fluctuations in the mean (macroscopic) current, suggesting that either the Na^+ channels were always open or the spontaneous transition rates were outside the measurable frequency range. They reasoned that adding a reversible blocker such as amiloride at submaximal doses might induce fluctuations. Indeed, when low doses of amiloride were applied, a characteristic Lorentzian-shaped power spectrum was obtained. By assuming the opening (dissociation) rate constant to be independent of blocker concentration while the closing rate constant is in possession of a concentration term (i.e., pseudo-first-order kinetics) and by measuring the corner frequency at various amiloride concentrations, Lindemann and Van Driessche calculated the "on" and "off" rate constants for amiloride with the channel. Because the plateau of the Lorentzian and the mean macroscopic current contain information about the number of channels and current flow through a single channel, combining these two values with the rate constants enabled them to calculate M and i.

The application of fluctuation analysis to the study of epithelial tissues requires some caution, as stressed by Van Driessche and Gogelein (1980), Lindemann and De Felice (1981), and Van Driessche and Gullentops (1982). One general problem results from the large area of epithelial membranes. A large surface area contains more channels and thus yields a higher plateau value, but it also reduces the usable frequency range because of the larger capacitative currents induced by noise inherent in the voltage-clamp amplifier. Also, free solution resistance must be kept to a minimum to assure an adequate voltage clamp of the membrane of interest, and the resistance of the membrane that contains fluctuating channels should be much higher than that of the other cell border. Interpreting the PSD is more difficult when blocker-induced noise is superimposed on spontaneous fluctuations to derive rate constants, channel density, and single-channel currents or when two blockers (e.g., Na^+ and amiloride) compete for the open channel (Li *et al.*, 1982). Perhaps the most important caveat of noise analysis is that many different kinetic schemes for channel fluctuations will result in Lorentzian spectra, thus making calculated values model dependent.

Despite these reservations, the method of fluctuation analysis has been validated in simpler bilayer and excitable systems. It immediately tells us whether changes in macroscopic current are caused by alteration in channel density (number of channels). Combining impedance with fluctuation analysis allows us to discriminate between channel activation and insertion of new channels into the membrane.

2.3. Patch Clamp

This technique has become increasingly popular in epithelial research because it provides a very direct method for studying single ion channels in native membrane. In principle, it can tell us everything fluctuation analysis can, but with much less computational effort and fewer assumptions regarding single-channel properties (see above). The procedures involved, and the equipment required to perform patch clamping, have been described in the classic paper by Hamill *et al.* (1981). In brief, a glass micropipette having a fire-polished tip and containing a salt solution is connected to a current-to-voltage amplifier. The micropipette tip is pressed on the surface of a cell, and when negative pressure is applied, a bleb of membrane bows into the pipette, forming a high-resistance seal (*ca.* $10 \times 10^9\ \Omega$) with the glass.

In a single-channel recording, one sees discrete transitions caused by "shots" of current through the open channel when a transmembrane driving force is applied. Typically, the first step is to determine channel selectivity by excising the patch of membrane and using dilution potential measurements to determine whether the channel is anion or cation selective. Biionic substitutions can then be used to determine the selectivity among different cations or anions. Further experiments may include studies of blockers, voltage sensitivity, modulation (e.g., activation by Ca^{2+}), and detailed analyses of kinetic models (such as how many open and closed states there are and how they are related). Histograms of open and closed times yield information about the number of kinetic states the channel can assume and, for very simple schemes, the transition rates between them. Detailed information regarding the sensitivity of each state to voltage, blockers, activators, etc. can thus be obtained.

Some problems encountered by the patchologist are listed below.

1. In many instances the membrane must be enzymatically treated before a high-resistance seal can be obtained. This raises the question of channel integrity after enzyme exposure.
2. Channel kinetics are most easily studied when there is only one channel in the membrane patch. Unfortunately, channels are apparently clustered in many cells.
3. Excising patches can lead to alterations in channel properties. Examples from the literature on excitable membranes include Ca^{2+} channel "run-down" and activation curves shifting along the voltage axis. Such changes might result from a loss of one or more intracellular regulatory factors.
4. Determining rate constants from duration histograms requires a

kinetic model when there is more than one open and one closed state.

5. Whereas ensemble fluctuation analysis is ideal for studying populations of channels and determining channel density, single-channel recording is not generally suited for answering the question "How many?"
6. Access to the basolateral membrane is difficult in most epithelia because of the basement membrane.
7. One often must grow cells in culture to obtain high-resistance seals. The assumption made is that channel properties are similar in cultured cells.

The main advantage of patch clamping is that it allows the study of the properties and intrinsic regulation of individual channels. With this information we can return to the intact epithelium and interpret the regulation of macroscopic permeability.

3. *Results Obtained from Frequency Domain Analysis*

In the remainder of this paper we illustrate how each of the methods outlined above has been used to study epithelial transport regulation.

3.1. *Membrane Area*

Impedance analysis has been performed on a number of amphibian and mammalian epithelia. The values for apical capacitance (proportional to apical membrane area) and basolateral capacitance (when measured) are summarized in Table I. Although most of the capacitance values were derived by fitting data to the simplest circuit model (see Section 2.1), in four cases the circuit had to be modified because of consistent discrepancies between theory and data. A simple "lumped" circuit considers the epithelium as a flat sheet of cells so that invaginations that would offer a significant resistance to transepithelial current flow are ignored. However, in two of the epithelia (rabbit urinary bladder and *Necturus* gallbladder), the resistance of the lateral intercellular space (LIS) is >10% that of the basolateral membrane. As a consequence, the basolateral membrane must be considered a distributed resistance. By using such a model, it was possible to eliminate the discrepancy between impedance data and theory at high frequencies (see Clausen and Wills, 1981; Lim *et al.*, 1984). For the rabbit descending colon, the existence of crypts (some 20 μm in diameter and greater than 200 μm in length) also required a distributed

Table I. Epithelial Circuit Parameters Estimated Using Impedance Analysis[a]

Tissue	R_A (ohm cm^2)	C_A (μF/cm^2)	R_{BL} (ohm cm^2)	C_{BL} (μF/cm^2)	R_J (ohm cm^2)	R_P (ohm cm^2)	Method	Model used	Comments	References
RUB	12,500	1.8	1,020	8.6	≥100,000	LIS, 130	S	D		Clausen *et al.* (1979)
RDC	460	19	110	8	770	Crypt, 17	R	C		Wills (1984)
TUB	3,500	2.0	3,000	27	—	—	S	L		Warncke and Lindemann (1981)
TUB	—	≃2.0	39,000	≃13	>18,000	—	P	L	Na-gluconate	Lewis *et al.* (1985)
TUB	—	1.97	—	—	—	—	P	L	K-depolarized	Palmer and Lorenzen (1983)
Turtle B.	4,160	3.3	200	8.14	—	—	R	L		Clausen and Dixon (1984)
Frog stomach	150	200	41	99	—	R_A, 24 R_{BL}, 12	S	DD	Resting	Clausen *et al.* (1983)
NGB	—	7	—	18	—	—	S	L		Schifferdecker and Frömter (1978)
	1,220	8	201	26.3	91	—	P	L		Suzuki *et al.* (1982)
	2,310	7	87	32.4	145	LIS, 91	R	D		Lim *et al.* (1984)

[a] Definition of Symbols: RUB, rabbit urinary bladder; RDC, rabbit descending colon; TUB, toad urinary bladder; Turtle B., turtle urinary bladder; NGB, *Necturus* gallbladder; R_A, C_A, apical membrane resistance and capacitance; R_{BL}, C_{BL}, basolateral membrane resistance and capacitance; R_J, tight junction resistance; R_P, resistance along the lateral intercellular space. Method refers to the waveform used: P, square current pulse; S, discrete sinusoids; R, random noise. Model refers to the type of electrical equivalent circuit used: L, lumped model; D, distributed model; DD, both apical and basolateral membrane; C, crypt model; see text for discussion. Normal bathing solutions were used except where noted in Comments column.

model to explain current flow through this structure (Wills, 1984). Finally, in frog gastric mucosa, both apical and basolateral distributed resistances were necessary to account for the impedance data. Thus, apical distributed resistance represents crypts in the colon and tubular vesicles on the apical surface of the parietal oxyntic cells. Basolateral distributed resistance reflects the LIS in the rabbit urinary bladder and *Necturus* gallbladder (Clausen *et al.*, 1983).

In rabbit urinary bladder, which has a reasonably smooth apical membrane, apical capacitance normalized to chamber area is approximately 2 $\mu F/cm^2$. This value can be accounted for by microscopic folds in the membrane. However, if one stretches the tissue, apical capacitance reaches an asymptotic value of about 1 $\mu F/cm^2$ of chamber area (Lewis and Diamond, 1976), which is the value expected for biological membranes (Cole, 1972). Apical capacitance is much greater in epithelia with brush border membranes or crypts, varying from 7 to 200 $\mu F/cm^2$, as would be expected from the area amplification factor observed in electronmicrographs.

Four of the epithelia listed in Table I as having regulated transport systems are also included in Table II along with the conditions under which regulation is observed, the electrolytes or nonelectrolytes that are transported, and any change in capacitance associated with changes in transport rate.

3.1.1. *Gastric Mucosa*

The most impressive increase in apical membrane capacitance occurs in gastric mucosa during histamine-stimulated H^+ secretion. Secretion and area reach a new steady state after 40 min of histamine exposure. Removal of histamine and replacement with cimetidine (a histamine antagonist) reduce H^+ secretion and apical area to control values over a period of 120 min. The magnitude of capacitance changes is in reasonable agreement with morphometric analysis (see Diamond and Machen, 1983, for a review).

3.1.2. *Urinary Bladder*

Both toad and rabbit urinary bladders have Na^+ transport systems that are stimulated by aldosterone and inhibited by amiloride; however, neither of these perturbations alters membrane capacitance. In contrast, antidiuretic hormone (in toad urinary bladder) and osmotic or mechanical perturbations (in rabbit urinary bladder) do result in increased apical membrane capacitance.

Table II. Epithelia That Modulate Membrane Area during Transport

Tissue	Modifier	Ion transported	$\%\Delta C_a$[a]	$\%\Delta C_b$[a]	Comments	References
Frog stomach	Histamine	$H^+ \uparrow$[b]	>200%	0		Clausen *et al.* (1983
	Cimetidine	$H^+ \downarrow$	−69%	0		Clausen *et al.* (1983)
TUB	Amiloride	$Na^+ \downarrow$	0	—		Stetson *et al.* (1982)
	ADH	$Na^+ \uparrow$, $H_2O \uparrow$	30%	33%		Warncke and Lindemann (1981)
		$Na^+ \uparrow$	0	—	H_2O response inhibited with methohexital	Stetson *et al.* (1982)
RUB	Amiloride	$Na^+ \downarrow$	0	0		Clausen *et al.* (1979)
	Mechanical stretch	$Na^+ \uparrow$	>21%	0		Lewis and de Moura (1982)
	Cell swelling	$Na^+ \uparrow$	80%	—	1/2-Osmotic-strength Ringer's	Lewis and de Moura (1982)
Turtle bladder	Acetazolamide	$H^+ \downarrow$	−8%	—		Clausen and Dixon (1984)

[a] $\%\Delta C_x$ = [(exp − control)/control] × 100.
[b] ↑, transport stimulation; ↓, transport inhibition.

3.1.2a. Toad Urinary Bladder. It has been known for a number of years that there is a dramatic increase in Na^+ transport, hydraulic conductivity (water permeability), and urea permeability within a few minutes after addition of ADH to the toad urinary bladder. Concomitant with this stimulation, there is an increase in apical membrane capacitance of approximately 30%. As demonstrated using freeze–fracture techniques, intramembranous particle aggregates appear in the apical membrane during ADH challenge, and the density of these aggregates is directly proportional to the magnitude of water flow across toad bladder in the presence of an applied osmotic gradient. Evidence that these aggregates originate from an intracellular pool was again provided by electron microscopy, which showed tubular vacuoles containing aggregates in the cytoplasm and an inverse relationship between the number in the apical membrane and tubular vesicles. Retrieval of aggregates into tubular vacuoles was recently demonstrated by uptake of horseradish peroxidase into these structures following removal of ADH from the bathing solution (Wade *et al.,* 1981).

Is the capacitance change a result of fusion of tubular vacuoles, or does it result from stimulation of Na^+ transport? Methohexital, a local anesthetic, inhibits H_2O transport and the appearance of apical aggregates without affecting Na^+ transport or the urea permeability response to ADH. Stetson *et al.* (1982) showed that methohexital completely inhibits the increase in capacitance, strongly suggesting that H_2O channels in toad urinary bladder are regulated by insertion/withdrawal whereas Na^+ channels are not. Additional support for this interpretation comes from the observation that colchicine and cytochalasin B (microtubule- and microfilament-disrupting agents, respectively) also reduce the hydroosmotic response and capacitance change but not the Na^+ transport response (Palmer and Lorenzen, 1983). Thus, regulation of water permeability seems to involve a shuttling of membrane between tubular vacuoles and the apical cell surface. Warncke and Lindemann (1981) also measured an increase in basolateral membrane capacitance and a decrease in basolateral membrane resistance during ADH stimulation. However, the resistance change was almost complete before any change in capacitance occurred. This time lag will require further study, but it does suggest that the basolateral membrane may have a role in regulation in H_2O and Na^+ transport.

3.1.2b. Rabbit Urinary Bladder. When the bladder fills with urine there are progressive changes in the ultrastructure of the epithelium. First, macroscopic folds are eliminated, and the microscopic membrane infolding becomes flattened out. To expand further, the bladder must add new membrane to the apical surface, thereby increasing the total surface area. Morphological and electrophysiological data (Minsky and Chlapowski, 1978; Lewis and Diamond, 1976) strongly support the conclusion that cytoplasmic vesicles are moved from the cytoplasm into the apical membrane. More recent work by Lewis and de Moura (1982, 1984) strengthens the idea that vesicle transfer is enhanced during stretch. They found that the most reproducible method for stimulating vesicle insertion was to expose the bladder to half-osmotic-strength NaCl Ringer's solution. After a 15-min lag phase, apical capacitance increased by 80%, and this plateaued after 45 min. A return to solutions having normal osmolality caused capacitance to decrease exponentially with a time constant of about 10 min.

To study the role of the cytoskeleton, tissues were preincubated for 130 min with colchicine, a microtubule-blocking agent. In the absence of an osmotic gradient, this pretreatment caused an 18% increase in apical capacitance. More importantly, subsequent challenge with half-osmotic-strength Ringer's solution caused a more rapid increase in apical capacitance, and this plateaued at a value 100% greater than that following precolchicine treatment. Cytochalasin B, a microfilament-disrupting agent, did not change base-line capacitance values; however, it completely inhibited the apical capacitance increase during osmotic challenge, suggesting that microfilaments are necessary for vesicle translocation or fusion. Other experiments revealed that vesicle retrieval was also microfilament dependent and that the surface area of cytoplasmic vesicles is at least three times that of the apical membrane (Lewis and de Moura, 1984; see also Minsky and Chlapowski, 1978). Interestingly, the Na^+ selectively of channels in vesicular membrane is approximately eight times greater than that in channels in the apical membrane, suggesting that apical Na^+ selectivity is decreased by exposure to urine. This possibility is discussed in Section 3.2.

3.1.2c. Turtle Urinary Bladder. Urinary acidification by the turtle bladder occurs through a proton-translocating ATPase that is situated in the apical membrane (Gluck *et al.*, 1982). Using the fluorescent probe acridine orange, these investigators demonstrated that mitochondria-rich cells contain cytoplasmic vesicles with low intravesicular pH. Proton secretion can be enhanced by elevating serosal CO_2 to 5% from zero, and morphological evidence suggests that there is a parallel loss of vesicles and an increase in apical membrane area during stimulation (Stetson and Steinmetz, 1983). Using fluorescent dextran, Gluck *et al.* (1982) demonstrated endocytosis following inhibition of acid secretion and exocytosis during stimulation. Exocytosis and H^+ transport were both inhibited

Table III. Modifiers of Channel Density and Single-Channel Currents

Condition	Preparation	Channel density[a]	Current[a]	References
Reduced mucosal Na^+	Frog skin	↑	↓	Van Driessche and Lindemann (1979)
	TUB	↑	↓	Palmer *et al.*(1982) Li *et al.* (1982)
	RUB	↑	↓	Lewis *et al.* (1984)
	Hen coprodeum	↑	↓	Christensen and Bindsler (1982)
ADH	TUB	↑	↔	Li *et al.* (1982)
Aldosterone	TUB	↑	↔	Palmer *et al.* (1982)
Ouabain	Frog skin	↓	↔	Van Driessche and Erlij (1983)

[a] ↑, increase; ↓, decrease; ↔, no change.

when tissues were preincubated with colchicine (Gluck *et al.,* 1982; Arruda *et al.,* 1980). Recently, Clausen and Dixon (1984) have shown that inhibition of proton secretion by acetazolamine results in a decline in apical membrane capacitance of 8%. Because the mitochondria-rich cells represent only a small fraction of the total apical membrane area of the turtle bladder, this 8% decrease represents the minimal value for apical membrane endocytosis.

Thus, as in the other three preparations, endocytosis and exocytosis of vesicular membrane containing transport proteins provide a rapid and effective method for the regulation of epithelial transport systems.

3.2. Channel Density

Although originally used to determine whether Na^+ entry across the apical membrane is mediated via a carrier (i.e., a shuttlelike mechanism carrying a maximum of about 10^4 ions/sec) or a pore (i.e., a channel, which may pass more than 10^6 ions/sec), fluctuation analysis also allows one to differentiate between channel insertion and channel modulation. Information concerning changes in channel density (i.e., modulation) and macroscopic current for a variety of epithelia are listed in Table III.

Fuchs *et al.* (1977) reported that apical membrane Na^+ permeability in frog skin decreased when external [Na^+] was increased. This self-inhibition occurred through a reduction in the number of conductive Na^+ channels rather than a modification of single-channel permeability (Van

Driessche and Lindemann, 1979). The same phenomenon has subsequently been reported in toad and rabbit urinary bladder and hen coprodeum. Sodium and amiloride may be competitive or noncompetitive inhibitors, depending on the epithelium.

As mentioned in Section 1, inhibiting Na^+ extrusion across the basolateral membrane with ouabain decreases apical membrane Na^+ permeability; this effect is measurable when intracellular Na^+ (Na_i^+) activity exceeds *ca.* 25 mM (Wills and Lewis, 1980), although it is not known whether intracellular Na^+ acts directly or indirectly. Evidence is accumulating from other tissues that negative feedback is mediated by a rise in intracellular Ca^{2+}. Using fluctuation analysis, Van Driessche and Erlij (1983) showed that the decrease in apical Na^+ permeability under these high-Na_i^+ conditions resulted from a decrease in channel density.

Aldosterone and ADH both stimulate net Na^+ absorption across a number of tight epithelia. Studies of tissue impedance have so far been unable to detect increases in membrane area correlated with Na^+ transport. It is well established that the stimulation of Na^+ transport results from an increase in channel density and not a modification of single-channel properties. The pool from which these channels are recruited cannot be the population of channels that are normally blocked by external Na^+ because self-inhibition is unaffected by ADH and aldosterone. These recruited channels are identical to the prestimulated population in terms of their currents and amiloride-binding kinetics.

Recent evidence suggests that aldosterone activates preexisting quiescent channels in the apical membrane (Garty and Edelmann, 1983; Sariban-Sohraby, *et al.*, 1984a,b), although this does not exclude some vesicle insertion. Furthermore, although impedance measurements cannot detect an increase in capacitance, this does not exclude the possibility that vesicles containing a high density of channels might be inserted and that the capacitance change is simply too small to be measured. Recent experiments by Garty and Edelmann (1983) showed that even though the proteolytic enzyme trypsin was able to hydrolyze both conducting Na^+ channels and quiescent (but aldosterone-stimulatable) channels, it had no effect on quiescent (ADH-stimulated) channels until they were activated by hormone. Either the ADH-stimulated channels are inserted, or they must exist in a trypsin-insensitive state when quiescent. Further pharmacological experiments will be required to differentiate between these possibilities. From fluctuation analysis, it is clear that single-channel properties of aldosterone- and ADH-stimulated channels are similar to those observed in the absence of hormone.

Mammalian urinary bladder is another epithelium in which fluctuation analysis has been used to differentiate between modification and insertion. In the previous section we mentioned that in the mammalian urinary bladder, Na^+-selective permeability is eight times lower in the apical membrane than in the cytoplasmic vesicles and that this difference might be a consequence of some urinary constitutent that alters channel densities or modifies single-channel permeability. Using fluctuation analysis, Lewis *et al.* (1984) found that neither single-channel permeability nor amiloride-binding kinetics was altered after recovery from vesicle fusion, but channel density was decreased by a factor of eight. Further studies revealed that a decrease in channel density could be induced by exposure to urokinase, a urinary enzyme (Lewis and Alles, 1984).

Very recently, fluctuation analysis has been applied to basolateral membrane K^+ channels in mammalian, amphibian, and insect epithelia. However, the question of regulation has not yet been addressed using this technique.

3.3. *Single-Channel Properties*

Table IV lists some of the ion channels that have been identified in epithelia. Most are highly K^+ selective; however, there are reports of less selective cation channels, including one that is somewhat Na^+ selective and amiloride sensitive and a basolateral anion-selective channel. An interesting observation in many of these studies is that channels appear to be clustered on the cell surface. This distribution makes it more difficult to determine even the most basic kinetic properties; for example, estimating the probability of a single channel being in the open state requires knowledge of the number of channels in the patch, and this is sometimes not obvious.

Another somewhat surprising feature in Table IV is that many of the epithelial channels listed are voltage sensitive; for example, open-state probability of basolateral K^+ channels in rabbit urinary bladder increases e-fold for approximately 10 mV of depolarization. This K^+ channel would tend to voltage clamp the basolateral membrane potential during variations in transepithelial transport. Where measured, all of the K^+ channels become more active as membrane potential is depolarized.

Spectral analysis can be used to estimate opening and closing rate constants in multichannel patches, although the values are valid only if the kinetic model is correct (Colquhoun and Hawkes, 1983, p. 164). This approach can also be used to estimate the number of channels in a population. For example, Marty *et al.* (1984) used noise analysis of whole-cell currents (10^3–10^4 channels) to estimate open time constants for a Cl^- channel of small unit conductance (1–2 pS). By the same reasoning, one

Table IV. Ionic Channels Measured Using the Patch Clamp[a]

Tissue	Membrane	Selectivity	γ (pS)	Regulation	Blockers	Kinetics	Density	References
Mouse pancreatic acinar cells	Bl	Cation	35	Ca^{2+}, CCK, ACh; cGMP inhibits	—	T_o = 0.4 sec	1.6 μm^2/ch	Maruyama and Peterson (1982a,b)
Pig pancreatic acinar cells	Bl	K^+	200	Ca^{2+}, CCK, bombesin, ACh, voltage	—	—	4 μm^2/ch	Maruyama *et al.* (1983)
Rabbit C.C.T.	Apical	K^+	90	Ca^{2+}, voltage	Ba^{2+}	—		Hunter *et al.* (1984)
C.C.T.	Apical	K^+	1–4	—	—	—	—	Koeppen *et al.* (1984)
Rabbit proximal tubule	Bl	K^+	30–40	—	—	—	—	Gogeleim and Greger (1984)
	Apical	K^+	Different sizes	—	—	—	—	Gogeleim and Greger (1984)
Rabbit MTAL	Apical	K^+	133	Ca^{2+}, voltage	Ba^{2+}	—	—	Guggino *et al.* (1985)
Chick cultured kidney	Apical	K^+	97	Forskolin, ADH	Ba^{2+}	—	—	Guggino *et al.* (1985)
Rat lacrimal gland	Bl	K^+	250	Ca^{2+}, carbamyl choline, voltage	Ca^{2+}, high conc. TEA	—	5–15 μm^2/ch	Marty *et al.* (1984)
	Bl	Cl^-	1–2	Same as K^+	—	T_o = 221 msec	4–16 μm^2/ch	Marty *et al.* (1984)
	Bl	Cation	25	Ca^{2+}, carbamyl choline	—	T_o = 225 msec	—	Marty *et al.* (1984)
A6 (cultured toad kidney)	Apical	Anion	360	Voltage	SITS	—	—	Nelson *et al.* (1984)
A6	Apical	Na^+	4 and 44	—	Amiloride	—	Lipid bilayers	Sariban-Sohraby *et al.* (1984), Olans *et al.* (1984)
Rabbit urinary bladder	Bl	Anion	64	Voltage	DIDS	1 opened 2 closed	15 μm^2/ch	Hanranhan *et al.* (1984)
	Bl	K^+	200	Voltage	Ba^{2+}	Complex	15 μm^2/ch	Hanranhan *et al.* (1984)

[a] In all cases depolarizing voltages led to increase in time that channel spent open. T_o, open time; Bl, basolateral membrane; C.C.T., cortical collecting tubule; MTAL, medullary thick ascending limb.

may use spectra to test whether there is only a single channel in the patch when P_o is either very large or very small.

Unlike measurements in the time domain, spectral analysis is not limited by the number of channels in the membrane (see Cull-Candy and Parker, 1983; Trautmann and Siegelbaum, 1983). For a simple two-state model, plots of power spectral density yield the following: the corner frequency; the sum of the open and closed rate constants regardless of the number of channels; and the plateau value, which is a function of (1) the rate constants, (2) the single-channel current (which can be directly measured), (3) the number of channels in the path, and (4) the mean current (also directly measured). Because we have more information than we have during ensemble noise analysis, it is possible to calcualte opening and closing rate constants and the number of channels from the spectrum.

It is important to note that when P_o is low, the presence of more than one channel does not drastically affect estimates of the mean open time obtained from duration histograms, but it does cause a serious underestimation of the mean closed time. By combining the mean open time (from a duration histogram) and corner frequency (from spectral analysis), one can calculate the mean closed time even in multichannel patches. Conversely, agreement between the calculated closed time and that derived from a closed-time histogram is good evidence that only one channel is present.

So far we have considered the simplest kinetic model, having one open and one closed state. However, more complex schemes for spontaneously fluctuating channels in epithelia do exist. For instance, Hanrahan *et al.* (1984) reported a Cl^- channel that is open 92% of the time over a voltage range of -80 to -20 mV. Spectral analysis of the records revealed two Lorentzian components having corner frequencies of 30 and 210 Hz, consistent with multiple open and/or closed states. One could propose six kinetic schemes that would give rise to this type of spectrum: the channel could have one open and two closed states arranged in series or cyclically, or one closed and two open states, again in series or in a cyclic pattern (see Dionne, 1981). We can readily exclude the three models having two open states because the open-time histogram could be adequately fit by a single exponential. The cyclical model cannot be characterized without additional information (such as that obtained from first latency histograms). However, these data can be compared with predictions of the linear three-state models CCO and COC. The rate constants calculated using the scheme COC are consistent with both the corner frequencies and the plateau values observed in the spectra, whereas the rate constants calculated using the CCO model predict negative plateau values for the PSD. At present, a sequential scheme with two closed states on either side of the open state best describes the kinetics of the anion channel in rabbit urinary bladder. Thus, a combination of time and frequency domain methods can be used to confirm the number of channels in a patch and can test models that are not distinguishable by either method alone.

So far the patch clamp technique has been used to demonstrate that regulation of ion channels in epithelial cells occurs by activation of quiescent channels by acetylcholine, Ca^{2+}, and cAMP as well as by the modulation of single-channel properties by membrane voltage and intracellular pH.

4. Summary

In this chapter we have attempted to show that studying ion transport in the frequency domain can provide a method for determining the membrane mechanisms involved in transport regulation. All three possibilities have been utilized by different epithelial cells, and in some instances all three mechanisms are used in the same cell.

References

Arruda, J. A. L., Sabatini, S., Mola, R., and Dytko, G., 1980, Inhibition of H^+ secretion in the turtle bladder by colchicine and vinblastine, *J. Lab. Clin. Med.* **96**:450–459.

Christensen, O., and Bindslev, N., 1982, Fluctuation analysis of short-circuit current in a warm-blooded sodium retaining epithelium: Site, current density and interaction with triamterene, *J. Membr. Biol.* **65**:19–30.

Clausen, C., and Dixon, T. E., 1984, Membrane area changes associated with proton secretion in turtle urinary bladder studied using impedance analysis techniques, in: *Current Topics in Membranes and Transport,* Vol. 20 (J. B. Wade and S. A. Lewis, eds.), Academic Press, Orlando, Florida, pp. 47–60.

Clausen, C., and Fernandez, J. M., 1981, A low-cost method for rapid transfer function measurements with direct application to biological impedance analysis, *Pflugers Arch.* **390**:290–295.

Clausen, C., and Wills, N. K., 1981, Impedance analysis in epithelia, in: *Ion Transport by Epithelia.* (S. G. Schultz, ed.), Raven Press, New York, pp. 79–91.

Clausen, C., Lewis, S. A., and Diamond, J. M., 1979, Impedance analysis of a tight epithelium using a distributed resistance model, *Biophys. J.* **26**:291–318.

Clausen, C., Machen, T. E., and Diamond, J. M., 1983, Use of AC impedance analysis to study membrane changes related to acid secretion in amphibian gastric mucosa, *Biophys. J.* **41**:167–178.

Cole, K. S., 1968, *Membranes, Ions, and Impulses,* University of California Press, Berkeley.

Colquhoun, D., and Hawkes, A. G., 1983, The principles of the stochastic interpretation of ion-channel mechanisms, in: *Single-Channel Recording* (B. Sakmann and E. Neher, eds.), Plenum Press, New York, pp. 135–175.

Cull-Candy, S. G., and Parker, I., 1983, Experimental approaches used to examine single glutamate receptor ion channels in locust muscle fibers, in: *Single-Channel Recording* (B. Sakmann and E. Neher, eds.), Plenum Press, New York, pp. 389–400.

Diamond, J. M., and Machen, T. E., 1983, Impedance analysis in epithelia and the problem of gastric acid secretion, *J. Membr. Biol.* **72:**17–41.

Dionne, V. E., 1981, The kinetics of slow muscle acetylcholine-operated channels in the garter snake. *J. Physiol.* (*Lond.*) **310:**159–190.

Fuchs, W., Larsen, E. H., and Lindemann, B., 1977, Current–voltage curve of sodium channels and concentration dependence of sodium-permeability in frog skin, *J. Physiol.* (*Lond.*) **267:**137–166.

Garty, H., and Edelman, I. S., 1983, Amiloride-sensitive trypsinization of apical sodium channels, *J. Gen. Physiol.* **81:**785–803.

Gluck, S., Cannon, C., and Al-Awqati, Q., 1982, Exocytosis regulates urinary acidification in turtle bladder by rapid insertion of H^+ pumps into the luminal membrane, *Proc. Natl. Acad. Sci. U.S.A.* **79:**4327–4331.

Gögelein, H., and Greger, R., 1984, Single channel recordings from basolateral and apical membranes of renal proximal tubules, *Pflugers Arch.* **401:**424–426.

Guggino, S. E., Suarez-Isla, B. A., Guggino, W. B., Green, N., and Sacktor, B., 1985, Ba^{++} sensitive, Ca^{++} activated K^+ channels in cultured rabbit medullary thick ascending limb cells (MTAL) and cultured chick kidney cells (CK), *Kidney Int.* **27:**209.

Hamill, O. P., Marty, A., Neher, E., Sakmann, B., and Sigworth, F. J., 1981, Improved patch-clamp techniques for high-resolution current recording from cells and cell-free membrane patches, *Pflugers Arch.* **391:**85–100.

Handler, J. S., Preston, A. S., and Orloff, J., 1972, Effect of ADH, aldosterone, ouabain and amiloride on toad bladder epithelial cells, *Am. J. Physiol.* **222:**1071–1074.

Hanrahan, J. W., Alles, W. P., and Lewis, S. A., 1984, Basolateral anion and K channels from rabbit urinary bladder epithelium, *J. Gen. Physiol.* **84:**30a.

Hudson, R. L., and Schultz, S. G., 1984, Sodium coupled sugar transport: Effects on intracellular sodium activities and sodium pump activity, *Science* **224:**1237–1239.

Hunter, M., Lopes, A. G., Boulpaep, E. L., and Giebisch, G., 1984, Single channel recordings of calcium-activated potassium channels in the apical membrane of rabbit cortical collecting tubules, *Proc. Natl. Acad. Sci. U.S.A.* **81:**4237–4239.

Koeppen, B. M., Beyenbach, K. W., and Helman, S. I., 1984, Single channel currents in renal tubules, *Am. J. Physiol.* **247:**F380–F384.

Lewis, S. A., and Alles, W. P., 1984, Analysis of ion transport using frequency domain measurements, in: *Current Topics in Membranes and Transport,* Vol. 20 (J. B. Wade and S. A. Lewis, eds.), Academic Press, Orlando, FL, pp. 87–103.

Lewis, S. A., and de Moura, J. L. C., 1982, Incorporation of cytoplasmic vesicles into apical membrane of mammalian urinary bladder, *Nature* **297:**685–688.

Lewis, S. A., and de Moura, J. L. C., 1984, Apical membrane area of rabbit urinay bladder increases by fusion of intracellular vesicles: An electrophysiological study, *J. Membr. Biol.* **82:**123–136.

Lewis, S. A., and Diamond, J. M., 1976, Na^+ transport by rabbit urinary bladder, a tight epithelium, *J. Membr. Biol.* **28:**1–40.

Lewis, S. A., Eaton, D. C., and Diamond, J. M., 1976, The mechanism of Na^+ transport by rabbit urinary bladder, *J. Membr. Biol.* **28:**41–70.

Lewis, S. A., Ifshin, M. S., Loo, D. D. F., and Diamond, J. M., 1984, Studies of sodium channels in rabbit urinary bladder by noise analysis, *J. Membr. Biol.* **80:**135–151.

Lewis, S. A., Butt, A. G., Bowler, M. J., Leader, J. P., and Macknight, A. D. C., 1985, Effects of anions on cellular volume and transepithelial Na^+ transport across toad urinary bladder, *J. Memb. Biol.* **83:**119–137.

Li, J. H. Y., Palmer, L. G., Edelman, I. S., and Lindemann, B., 1982, The role of sodium-channel density in the natriferic response of the toad urinary bladder to a antidiuretic hormone, *J. Membr. Biol.* **64:**77–89.

Lim, J. J., Kottra, G., Kampmann, L., and Frömter, E., 1984, Impedance analysis of *Necturus* gallbladder epithelium using extra- and intracellular microelectrodes, in: *Current Topics in Membranes and Transport,* Vol. 20 (J. B. Wade and S. A. Lewis, eds.), Academic Press, Orlando, FL, pp. 27–46.

Lindemann, B., and Defelice, L. J., 1981, On the use of general network functions in the evaluation of noise spectra obtained from epithelia, in: *Ion Transport by Epithelia* (S. G. Schultz, ed.), Raven Press, New York, pp. 1–13.

Lindemann, B., and Van Driessche, W., 1977, Sodium-specific membrane channels of frog skin are pores: Current fluctuations reveal high turnover, *Science* **195:**292–294.

Marty, A., Tan, Y. P., and Trautmann, A., 1984, Three types of calcium-dependent channels in rat lacrimal glands, *J. Physiol.* (*Lond.*) **357:**293–325.

Maruyama, Y., and Peterson, O. H., 1982a, Single channel currents in isolated patches of plasma membrane from basal surface of pancreatic acini, *Nature* **299:**159–161.

Maruyama, Y., and Peterson, O. H., 1982b, Cholecystokinin activation of single-channel currents is mediated by internal messenger in pancreatic acinar cells, *Nature* **300:**61–63.

Maruyama, Y., Peterson, O. H., Flanagan, P., and Pearson, G. T., 1983, Quantification of Ca^{2+}-activated by K^{+} channels under hormonal control in pig pancreas acinar cells, *Nature* **305:**288–232.

Minsky, B. D., and Chlapowski, F. J., 1978, Morphometric analysis of the translocation of luminal membrane between cytoplasm and cell surface of transitional epithelial cells during the expansion–contraction cycles of mammalian urinary bladder, *J. Cell Biol.* **77:**685–697.

Nelson, D. J., Tang, J. M., and Palmer, L. G., 1984, Single-channel recordings of apical membrane chloride conductance in A6 epithelial cells, *J. Membr. Biol.* **80:**81–89.

Olans, L., Sariban-Sohraby, S., and Benos, D. J., 1984, Saturation behavior of single amiloride-sensitive Na^{+} channels in planar lipid bilayers, *Biophys. J.* **46:**831–835.

Palmer, L. G., and Lorenzen, M., 1983, Antidiuretic hormone dependent membrane capacitance and water permeability in the toad urinary bladder, *Am. J. Physiol.* **244:**F195–F204.

Palmer, L. G., Li, J. H. Y., Lindemann, B., and Edelman, I. S., 1982, Aldosterone control of the density of sodium channels in the toad bladder, *J. Membr. Biol.* **64:**91–102.

Sariban-Sohraby, S., Burg, M., Wiesmann, W. P., Chiang, P. K., and Johnson, J. P., 1984a, Methylation increases sodium transport into A6 apical membrane vesicles: Possible mode of aldosterone action, *Science* **225:**745–746.

Sariban-Sohraby, S., Latorre, R., Burg, M., Olans, L., and Benos, D., 1984b, Amiloride-sensitive epithelial Na^{+} channels reconstituted into planar lipid bilayer membranes, *Nature* **308:**80–82.

Schifferdecker, E., and Frömter, E., 1978, The AC impedance of *Necturus* gallbladder epithelium, *Pflugers Arch.* **377:**125–133.

Stetson, D. L., and Steinmetz, P. R., 1983, Role of membrane fusion in CO_2 stimulation of proton secretion by turtle bladder, *Am. J. Physiol.* **245:**C113–C120.

Stetson, D. L., Lewis, S. A., Alles, W., and Wade, J. B., 1982, Evaluation by capacitance measurements of antidiuretic hormone induced membrane area changes in toad bladder, *Biochim. Biophys. Acta* **689:**267–274.

Suzuki, K., Kottra, G., Kampmann, L., and Frömter, E., 1982, Square wave pulse analysis of cellular and paracellular conductance pathways in *Necturus* gallbladder epithelium, *Pflugers Arch* **394:**302–312.

Taylor, A., Mamelak, M., Reaven, E., and Maffly, R., 1973, Vasopressin: Possible role of microtubules and microfilaments in its action, *Science* **181:**347–350.

Trautmann, A., and Siegelbaum, S. D., 1983, The influence of membrane isolation on single acetylcholine-channel current in rat myotubes, in: *Single-Channel Recording* (B. Sakmann and E. Neher, eds.), Plenum Press, New York, pp. 473–480.

Van Driessche, W., and Erlij, D., 1983, Noise analysis of inward and outward Na^+ currents across the apical border of ouabain-treated frog skin, *Pflugers Arch* **398:**179–188.

Van Driessche, W., and Gogelein, H., 1980, Attenuation of current and voltage noise signals recorded from epithelia, *J. Theor. Biol.* **86:**629–648.

Van Driessche, W., and Gullentops, K., 1982, Conductance fluctuation analysis in epithelia, in: *Techniques in Cellular Physiology,* Vol. 2 (P. F. Baker, ed.), Elsevier, Amsterdam, pp. 1–13.

Van Driessche, W., and Lindemann, B., 1979, Concentration dependence of currents through single sodium-selective pores in frog skin, *Nature* **282:**519–520.

Wade, J. B., Stetson, D. L., and Lewis, S. A., 1981, ADH action: Evidence for a membrane shuttle mechanism, *Ann. N.Y. Acad. Sci.* **372:**106–117.

Warncke, J., and Lindemann, B., 1981, Effect of ADH on the capacitance of apical epithelial membranes. *Adv. Physiol. Sci.* **3:**129–133.

Wills, N. K., 1984, Mechanisms of ion transport by the mammalian colon revealed by frequency domain analysis techniques, in: *Current Topics in Membranes and Transport,* Vol. 20 (J. B. Wade and S. A. Lewis, eds.), Academic Press, Orlando, FL, pp. 61–85.

Wills, N. K., and Lewis, S. A., 1980, Intracellular Na^+ activity as a function of Na^+ transport rate across a tight epithelium, *Biophys. J.* **30:**181–186.

17

The Epithelial Sodium Channel

LAWRENCE G. PALMER

1. Introduction

Epithelial Na^+ channels form an essential component of the Na^+ reabsorptive system in a variety of tissues. They permit entry of Na^+ into the epithelial cell from the outer fluid compartment, e.g., urine, feces, sweat. Furthermore, they regulate the reabsorptive flow of Na^+ across many epithelia and thus are critical to the maintenance of constant salt levels and fluid volumes in the various body compartments of vertebrates.

The basic organization of Na^+-reabsorbing epithelia was described more than 25 years ago by Koefoed-Johnsen and Ussing (1958). The basic idea is that the serosal or blood-facing membrane has many characteristics of a "typical" cell membrane. It is selectively permeable to K^+, poorly permeable to Na^+, and contains an Na^+ pump that uses ATP hydrolysis to keep the intracellular Na^+ activity low and, in the case of the epithelium, maintain active transepithelial Na^+ reabsorption. The outer or mucosal membrane, on the other hand, is selectively permeable to Na^+, an unusual feature for an animal cell membrane. Because of the low concentration of Na^+ within the cell, the ion can diffuse passively into the cell across the outer membrane. Koefoed-Johnsen and Ussing emphasized that this scheme accounted for the vectorial transport of Na^+ across the epithelium and also explained the electrical properties of the tissue, particularly the asymmetric changes of the transepithelial potential in response to changes in the ionic composition of the bathing media: the tissues hyperpolarize with increases in mucosal Na^+ and depolarize with increases in serosal K^+.

This model, originally a description of the frog skin, seems to be valid

LAWRENCE G. PALMER • Department of Physiology, Cornell University Medical College, New York, New York 10021.

Table I. Properties of Na^+ Channels in Toad Bladder

Selectivity	$H^+ > Li^+ > Na^+ \gg K^+, Rb^+, Cs^+$
Single-channel current	0.15 pA with 60 mM Na^+
Single-channel conductance	~3 pS
I–V relationship	Follows constant-field equation
Voltage dependence	Weak
Pharmacology	Blocked by amiloride, $K_I \sim 5 \times 10^{-7}$ M competitive with Na^+
Flux ratio exponent	~1.0

for a number of epithelia including, in mammals, the renal cortical collecting tubule and urinary bladder, the descending colon, and the ducts of the salivary and sweat glands; in amphibian skin, urinary bladder, and colon; and the copradeum of birds (see Table II). In all of these tissues it is likely that the same basic mechanism—the epithelial Na^+ channel—mediates the Na^+ permeability of the apical membrane. This conclusion is based in part on the finding that in all of these tissues Na^+ transport is inhibited by the K^+-sparing diuretic amiloride in micromolar or submicromolar concentrations (see Table II). Amiloride sensitivity has become the pharmacological hallmark of the Na^+ channel.

These epithelia share several other characteristics. They are so-called "tight" epithelia with high electrical resistances (100 to 100,000 ohm·cm^2), except for the salivary gland duct (10–30 ohm·cm^2), and often sustain large transepithelial electrical potential differences. They are situated at the most distal parts of their respective organs and are thus responsible for the final processing of fluids such as urine, feces, sweat, or saliva before excretion. Finally, they all respond to aldosterone, the major mineralocorticoid of vertebrates, by increasing their rates of Na^+ transport.

Thus, the distribution of amiloride-sensitive Na^+ channels correlates with that of high-resistance tight junctions between the epithelial cells and with sensitivity to aldosterone. How these three very different cellular components, one a nuclear/cytoplasmic steroid receptor, another an integral membrane pore protein, and the third an intercellular cement, are coupled during epithelial development remains an important unsolved problem.

In this chapter I describe progress that has been made in the characterization of the Na^+ channel. I focus on our work using the toad urinary bladder as a model (Table I), making comparisons with other tissues in which the channel has been studied.

2. Description of the Na^+ Channel

2.1. Single-Channel Conductance

The identification of the apical Na^+ conductance as a channel-mediated process was made by Lindemann and Van Driessche (1977), who used fluctuation analysis to estimate the turnover rate for Na^+ through individual channels in frog skin to be greater than 10^6/sec, too large for most carrier systems. Using the same technique in the toad bladder, Li *et al.* (1982) and Palmer *et al.* (1982) estimated the single-channel currents to be about 0.15 pA with 60 mM Na^+ in the mucosal medium, an estimated 6 mM Na^+ in the cell, and a membrane voltage of near zero. This translates to a single-channel conductance of about 3 pS. Similar values were reported by Li and Lindemann (1983) in frog skin.

In other tissues in which a transmembrane potential increased the driving force for Na^+ entry, somewhat larger currents have been reported, including frog skin (Helman *et al.*, 1983), rabbit colon (Zeiske *et al.*, 1982), hen copradeum (Christensen and Bindslev, 1982), and rabbit urinary bladder (Lewis *et al.*, 1984) (Table II). The latter three studies were done at a higher temperature (37°C). Recently, using direct measurements of single-channel currents through amiloride-sensitive channels of cultured toad kidney cells with the patch-clamp technique, Hamilton and Eaton (1985) have observed single-channel conductances of 7 to 10 pS.

2.2. Selectivity

Palmer (1982a) found a selectivity sequence for the Na^+ channel in toad bladder of $H^+ > Li^+ > Na^+ \gg K^+, Rb^+, Cs^+$ (Table III). For technical reasons, only the P_{Li}/P_{Na} ratio could be measured directly under biionic conditions. The other permeability ratios were estimated from amiloride-sensitive currents measured under different experimental conditions, using the constant-field equation to derive permeability coefficients. The quantitative sequence arrived at was $P_H:P_{Li}:P_{Na}:P_K = 7:1.3:1:0.0015$. In frog skin, Benos *et al.* (1980) estimated that $P_{Na}:P_K$ was at least 100:1 on the basis of amiloride-sensitive fluxes of $^{22}Na^+$ and $^{42}K^+$. Lewis and Wills (1981) reported $P_{Na}:P_K = 2:1$ in the rabbit urinary bladder. However, when the animals were fed a low-Na^+ diet to increase aldosterone production, the permeability ratio increased to 10:1. Thus, it is possible that the mineralocorticoid-dependent channels in this tissue have selectivities similar to those in the amphibian epithelia.

It is not known whether this high selectivity results from specific

Table II. Distribution of Channels in Tight Epithelia

Epithelium	I_{Na} ($\mu A/cm^2$)[a]	i (pA)[b]	N_0 ($10^6/cm^2$)[c]	Amiloride K_I (μM)	References
Frog skin (depolarized)	16	0.09	200	0.3	Li and Lindemann (1983)
Frog skin (nondepolarized)	17	0.59	40	0.2	Helman *et al.* (1983)
Toad bladder (depolarized)	8	0.15	50	0.2	Palmer *et al.* (1982)
Hen copradeum	180	0.32	500	0.8 to 1.8	Christensen and Bindslev (1982) Bindslev *et al.* (1982)
Rabbit colon	160	0.4	600	0.1	Zeiske *et al.* (1982)
Rabbit urinary bladder	5.9	0.71	5.6	0.37	Lewis *et al.* (1984)

Other Na^+-reabsorbing epithelia sensitive to amiloride
- Rabbit cortical collecting tubule (O'Neil and Boulpaep, 1979)
- Rabbit salivary gland duct (Augustus *et al.*, 1978)
- Human sweat gland duct (Quinton, 1981)
- Toad colon (Bentley, 1968)
- Turtle colon (Thompson and Dawson, 1978)
- Turtle urinary bladder (Nagel *et al.*, 1981)

[a] Average of typical values of amiloride-sensitive short-circuit current at high Na^+. Value depends on hormonal status.
[b] Single-channel current under short-circuit conditions at high Na^+, from noise analysis.
[c] Open-channel density under short-circuit conditions at high Na^+, from noise analysis. Depends on hormonal status.

Table III. Ions That Permeate the Na^+ Channel

Ion	Ionic radii (Å)[a]	Relative permeability[b]
H^+	—	6.7
Li^+	0.68	1.3
Na^+	0.97	1.0
K^+	1.33	0.0015

[a] From *Handbook of Chemistry and Physics*, 47th Edition, Chemical Rubber Company (1966).
[b] From Palmer (1982).

interactions of the cations with a ligand inside the channel (binding site specificity) or if the pore narrows at one point to a size that accepts Na^+ (and smaller) cations but rejects K^+ (and larger) cations (size specificity). There are precedents for ligands that show binding site specificity for Na^+ over K^+ that is quite high. In fact, two of these ligands have been used in the construction of Na^+-selection electrodes. These include Na^+-selective glass such as NAS-11-18, which selects for Na^+ over K^+ by a factor of 300 to 1000 (Eisenmann, 1962), and neutral carriers such as ETH 227 developed by Steiner *et al.* (1979), which has a selectivity ratio of about 100. ETH 227 shows the same selectivity sequence as does the epithelial Na^+ channel, whereas NAS 11-18 selects for Na^+ over Li^+. On the other hand, the lack of permeability of small NH_4-based cations, as discussed below, suggests that the nonhydrated form of the Na^+ ion may be the permeant form. Thus, it is possible that the lumen of the channel is, at one point, just large enough to admit the dehydrated Na^+ or Li^+ ion but too small to admit K^+ readily.

Hille (1971) devised a simple and elegant set of experiments in which he measured the permeability of the excitable Na^+ channel in frog nerve to a series of "organic" cations. From an analysis of which ions would go through the channel and which would not, he deduced an apparent cross section of the pore at its narrowest point of 3×5 Å. This cross section is just large enough to admit a Na^+ ion with a single associated water molecule. When these experiments were repeated in the toad bladder, none of the N-based cations elicited a measurable amiloride-sensitive inward current into the cell from the mucosal medium, even the smallest ions such as NH_3OH^+ and NH_3NH_2, which were nearly as permeant as Na^+ in the excitable Na^+ channel (Hille, 1971). This implies that the narrowest part of the pore in the epithelial channel is considerably smaller than 3 × 5 Å and, since the monohydrated Na^+ ion and the NH_3OH^+ ion have similar sizes, suggests that Na^+ may have to be completely dehydrated to fit into the pore. Thus, it is possible that the pore is only slightly larger, at its narrowest point, than twice the crystal radius of the Na^+ ion of about 1 Å (Hille, 1975). This could account in part for the ability of the pore to distinguish between Na^+ and K^+, whose crystal radius is about 30% larger than that of Na^+. If this picture of the Na^+ channel is correct, it would be virtually impermeable to water. This notion is testable, as it predicts that the Na^+ flux through the channel should be unaffected by the presence of an osmotic gradient, which would produce a solvent drag effect through a larger pore.

2.3. *Current–Voltage Relationship*

The instantaneous current–voltage ($I–V$) relationship of the epithelial Na^+ channel can be described by the constant-field equation. This was first observed in the frog skin by Fuchs *et al.* (1977) and subsequently confirmed in the toad bladder (Palmer *et al.*, 1980), the rabbit colon (Thompson *et al.*, 1982), and the *Necturus* urinary bladder (Thomas *et al.*, 1983). DeLong and Civan (1894) recently showed conformity with the constant-field behavior over an extended voltage range in the frog skin. Henrich and Lindemann (1984) found that the single-channel $I–V$ relationship, measured with noise analysis, also conformed to the constant-field equation. Thus, there seems to be no unusual rectification of Na^+ movement through the channel other than that associated with the ionic gradients. Partial or complete dissipation of the Na^+ gradient across the apical membrane resulted in a linearization of the $I–V$ relationship (Palmer *et al.*, 1980; Palmer, 1985a).

2.4. *The Outer Mouth of the Channel*

Although most monovalent cations do not permeate the Na^+ channel, a large number of them do partially block Na^+ movement through the pore in a manner consistent with the idea that the blocking ions penetrate part of the way into the lumen of the channel. The blocking cation that has been studied most extensively in this regard is amiloride, which is a guanidinium derivative that must be in the form of a monovalent cation to be active (Benos *et al.*, 1979). A clue to the mechanism of amiloride block is the observation that its effectiveness is voltage dependent; voltages tending to drive the ion into the channel increase its ability to block Na^+ movement (Palmer, 1984). This type of phenomenon has been interpreted to indicate that the blocking ion senses part of the potential drop across the pore at its site of interaction (Woodhull, 1973; Latorre and Miller, 1983). Thus, the blocking ion could act as a simple plug of the pore.

A number of observations are consistent with this hypothesis. First, a quantitative assessment of voltage-dependent block indicated that the apparent K_I for amiloride could be expressed as a function of voltage (Fig. 1) by the relationship:

$$K_I(V) = K_I(0) \exp(-\delta FV/RT) \quad (1)$$

This is the form expected of a pore-plugging mechanism, where δ is the fraction of the field sensed at the binding site. Values of δ were independent of amiloride concentration over a range of 0.05 to 0.4 M and independent of the muscosal Na^+ concentration between 29 and 115 mM (Palmer, 1984).

Another important prediction of the pore-plugging hypothesis is that the change in the apparent affinity for amiloride should be instantaneous but that the actual decline in the fractional block of Na^+ currents should depend on the kinetics of the interaction of amiloride with the channel. In voltage-jump experiments, the Na^+ current in the presence of submaximal doses of amiloride relaxed to a new steady state with an exponential time course (Palmer, 1985b). The rate constant for the relaxation was linearly dependent on the amiloride concentration and was predictable from the known kinetics of amiloride block determined with fluctuation analysis (Li *et al.*, 1982). Thus, the change in the apparent affinity for amiloride is much faster than the rate of amiloride binding, which is consistent with the pore-plugging model. Although there are alternative explanations for these data, that of the pore plug is compelling and is the simplest model that can account for the observations.

A number of other organic cations block the Na^+ channel in this voltage-dependent fashion, although most have much lower affinities for their blocking sites than does amiloride (Table IV). These include cations such as NH_4 and its derivatives NH_3OH, CH_3NH_3, and CH_3NH_2OH and guanidinium derivatives such as guanidinium hydroxyguanidinium, methylguanidinium, and formamidinium (Palmer, 1985b). These compounds block Na^+ transport with apparent K_Is ranging from 90 mM (guanidinium) to 1800 mM (CH_3NH_2OH). Values of δ were between 0.12 and 0.24 except for that of CH_3NH_3, which was somewhat lower. Thus, these compounds appear to penetrate the channel to roughly the same extent, although their residence times in the channel appear to vary considerably as judged from their apparent affinities.

All of the cations mentioned above have molecular diameters less

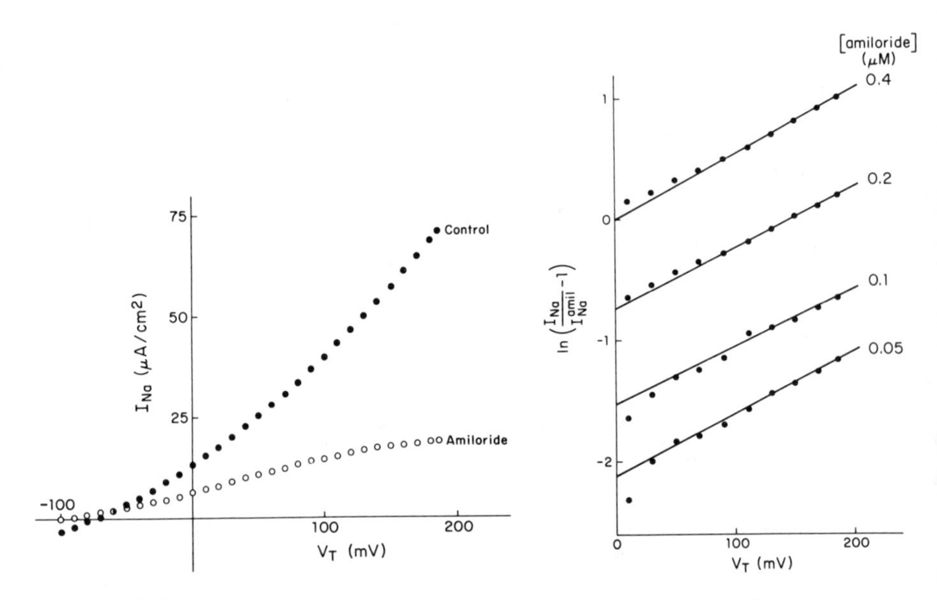

Figure 1. Voltage dependence of amiloride block. The left panel shows I_{Na}–V relationships of a hemibladder under control conditions and with 0.4 μM amiloride in the mucosal solution. Mucosal Na^+ concentration was 115 mM. Transepithelial I–V relationships were measured sequentially with 0, 0.1, 0.2, 0.3, and 0.4 μM amiloride. Currents in the presence of 10 μM amiloride were subtracted from each of the other I–V relationships to generate I_{Na}–V plots. Only data from control (●) and 0.4 μM amiloride (○) traces are shown. On the right, the fractional inhibition of current was calculated and plotted semilogarithmically as a function of voltage as in equation 1. The straight lines show linear least-squares fit of the data between 100 and 200 mV. Values of K and δ for amiloride computed from the fitted lines are:

Amiloride (μM)	K_I (μM)	δ
0.05	0.42	0.13
0.10	0.46	0.12
0.20	0.42	0.13
0.40	0.40	0.14

(From Palmer (1984), reprinted with permission from the *Journal of Membrane Biology*.)

than 5 Å. Ions larger than 5 Å do not appear to block the Na^+ channels at all. These include tetramethylammonium and tetraethylammonium, choline, and N-methyl-*d*-glucamine (Palmer, 1984). These data are consistent with the idea that the blocking cations must enter a portion of the outer mouth of the channel that has a diameter of about 5 Å (Fig. 2). Such a region may serve as a funnel to the narrowest portion of the channel, increasing the effective capture radius for Na^+.

Table IV. Ions That Block Na^+ Channels from the Outside

	Diameter (Å)[c]	K_I(mM)[d]	δ
Organic monovalent[a]			
NH_4^+	3.0	220	0.14
$CH_3NH_3^+$	3.6	660	0.085
NH_3OH^+	3.3	760	0.24
$CH_3NH_2OH^+$	3.6	1800	0.18
Formamidinium	3.6	530	0.14
Guanidinium	4.8	90	0.18
Methylguanidinium	4.8	100	0.16
Dimethylguanidinium	4.8	460	0.12
Inorganic monovalent[b]	Crystal radius (Å)		
Tl^+	1.47	70	—
K^+	1.33	170	0.30–0.35[e]
Rb^+	1.47	250	0.35[e]
Cs^+	1.67	500	0.35[e]
Inorganic divalent[a]			
Mg^{2+}	0.66	360	0.27[e]
Ca^{2+}	0.99	370	0.29[e]
Sr^{2+}	1.12	490	0.32[e]
Ba^{2+}	1.34	940	0.28[e]

[a] Palmer (1985b).
[b] Palmer (1984).
[c] Minimum diameter from CPK models. See Hille (1975).
[d] Concentration required for 50% inhibition at zero membrane voltage.
[e] Value for the inner site of a two-site model assuming the outer site has $\delta = 0$.

Small, relatively impermeant inorganic cations, including the monovalents K^+, Rb^+, and Cs^+ and divalents Mg^{2+}, Ca^{2+}, Sr^{2+}, and Ba^{2+} also block the Na^+ channel in a voltage-dependent manner (Palmer, 1984, 1985b). Although the voltage dependence is qualitatively similar to that of amiloride block, quantitative analysis shows that block by these ions cannot be explained by interaction with a single site with a constant δ value. Instead, as the voltage driving the cations into the channel increases, the apparent δ value increases. This can be explained if the ions interact with two or more sites in series with different δ values, both of which are blocking sites. The first site, which could be near the outside edge of the electric field, would have a higher affinity for the blocker at $V = 0$. As V increased, however, the blocker would be driven into the second site, so that the effective δ value would increase toward a limiting value of that of the inner site.

Making the arbitrary assumption that the δ value of the outer site is zero, one can calculate δ values for the inner site to be in the range of 0.3 for both monovalent and divalent blockers. Note that this implies that

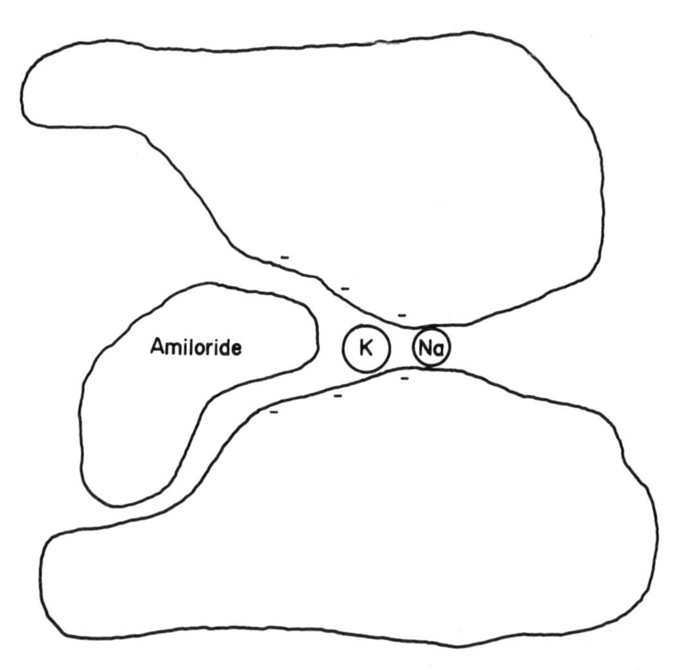

Figure 2. A schematic representation of the epithelial Na^+ channel in the toad bladder. The narrowest part of the pore is too small to allow ions larger than Na^+ to penetrate readily. The outer mouth has an antechamber of about 5 Å diameter and is lined with fixed negative charges.

the voltage dependence for Ca^{2+} block was twice as steep as that for K^+ block, since Ca^{2+} has a valence of +2. This is consistent with the idea that the monovalent and divalent blockers are interacting with the same sites within the outer mouth of the pore.

Although a two-site model can explain the data, there is no particular reason except for mathematical simplicity to prefer such a model over one with three sites or with a long series of sites leading into the narrow portion of the channel.

The affinity of the outer binding site(s) for Na^+ could not be measured directly, as Na^+ goes through the pore. An indirect estimate was made on the basis of the ability of Na^+ and K^+ to competitively inhibit amiloride block (Palmer, 1984). Potassium has been shown to interact competitively with amiloride with an apparent inhibition constant for displacement of amiloride of 170 mM, which is similar to the apparent K_I for block of Na^+ current of 180 mM. Sodium reduced the apparent amiloride affinity competitively with a K_I of 60 to 100 mM. Thus, the selectivity sequence for binding to the site(s) at the outer mouth of the channel is amiloride $\gg$ Na^+ $\approx$ Tl^+ $\approx$ guanidinium $>$ K^+ $>$ NH_4^+ $>$ Rb^+ $>$ Cs^+ $\approx$ NH_3OH.

This inhibitory set of binding sites is much less selective than the selectivity filter that controls translocation. It may serve as a device for raising the local concentration of cations in the mouth of the pore, increasing the translocation rate for Na^+. It is likely that these sites are fixed anions in the mouth of the pore. Raising the mucosal proton concentration reduces the Na^+ current. This block, which is also voltage dependent (Palmer, 1984), is consistent with titration of a group with a pK_a of about 3.0. It is possible that these are carboxyl groups.

2.5. *The Inner Mouth of the Channel*

Relatively little is known about the inner mouth of the channel, since the composition of the cytoplasm is more difficult to alter. One suggestion that has received considerable experimental support, however, is the idea that intracellular Ca^{2+} reduces Na^+ permeability. This notion is based on the finding that Na^+ transport is reduced under conditions in which cytoplasmic Ca^{2+} is presumed to be increased (Grinstein and Erlij, 1978; Taylor and Windhager, 1979; Chase and Al-Awqati, 1982; Garty and Lindemann, 1984) and, more recently, on the observation that Ca^{2+} inhibits amiloride-sensitive Na^+ fluxes in membrane vesicles from the toad bladder (Chase and Al-Awqati, 1983; Garty, 1984).

It is impossible that Ca^{2+} also acts by a pore-plugging mechanism, this time at the inner mouth of the channel. Evidence against this idea, however, is the finding that Ca^{2+} block is not voltage dependent. When apical Na^+ conductance was reduced by addition of Ca^{2+} to the serosal medium under conditions thought to raise cytoplasmic Ca^{2+} (Garty and Lindemann, 1984), the reduction in Na^+ current was independent of voltage (Palmer, 1985a). Thus, if Ca^{2+} does act as a pore plug, it must do so at a point that is outside, or just inside, the electrical field of the membrane. Alternatively, Ca^{2+} could act indirectly on the channel, either by allosteric binding to the channel protein or by promoting a chemical modification.

Acidification of the cytoplasm of the toad bladder by addition of CO_2 was also found to acutely and reversibly depress apical Na^+ permeability (Palmer, 1985a). This effect of intracellular protons was also found to be voltage independent, in contrast to the voltage-dependent block by protons in the mucosal medium (Palmer, 1984).

2.6. Single Filing

Most biological ion-selective channels are thought to be too narrow, at least at their narrowest point, to allow ions or water molecules to pass each other. The movement of ions and water through the pore in single file can lead to unusual behavior, especially if the pore is long and more than one ion can occupy it at one time (Hodgkin and Keynes, 1955). A good test for single-filing behavior is the calculation of the flux ratio exponent n' defined by the equation:

$$M_{12}/M_{21} = [(S_1/S_2)\exp(-ZFV_{12}/RT)]^{n'}$$

where M_{12} and M_{21} are the fluxes in the two directions across the membrane, S_1 and S_2 the concentrations of permeant solute on the two sides, and V_{12} the voltage difference across the membrane (Hodgkin and Keynes, 1955). For simple diffusion through a wide pore $n' = 1$. In addition, for a narrow channel with one ion-binding site, which therefore is occupied by at most one ion, $n' = 1$. However, for a channel that can accommodate n ions, n' can reach values as high as n, depending on the occupancy of the sites (Hille and Schwartz, 1978). For a shuttle or mobile carrier system, in general, $n' \leq 1$.

In the toad bladder, n' was measured as the ratio of the unidirectional Na^+ influx and the Na^+ conductance under conditions of electrochemical equilibrium (Palmer, 1982b). At 10 and 40 mM Na^+ on both sides of the membrane, n' was found to be close to 1. A similar conclusion was reached in the frog skin by Benos *et al.* (1983), who measured undirectional fluxes at various transmembrane potentials and ion gradients and found results consistent with the hypothesis that $n' = 1$. Thus, the Na^+ channel, although it may have multiple ion-binding sites for Na^+, does not appear to exhibit multiple ion occupancy under physiological conditions.

3. Regulation of the Na^+ Channel

3.1. Regulation of External Na^+

Sodium transport across the apical membrane of the toad bladder does not increase linearly with the external Na^+ concentration. Rather, it saturates following roughly the form of the rectangular hyperbola predicted by Michaelis–Menten kinetics, with half-saturation of the process occurring at a Na^+ concentration of around 20 mM (Frazier *et al.*, 1962; Li *et al.*, 1982). This is apparently not because of simple saturation of a binding site within the transport pathway for Na^+. First, the response to a step increase in mucosal Na^+ concentration is an instantaneous overshoot in Na^+ current, which is followed by a slow relaxation to a lower steady-state level over a period of a few seconds (Fuchs *et al.*, 1977), much slower than the kinetics of translocation through the channel (10^6/sec). Second, noise analysis indicates that external Na^+ reduces the number of conducting channels, whereas the current through individual channels increases linearly with the Na^+ concentration (Van Driessche and Lindemann, 1979).

Thus, the saturation of Na^+ transport can be considered as a down-regulation of Na^+ permeability or of the number of conducting channels by mucosal Na^+. This appears to be a direct effect of external Na^+ rather than a secondary rise of intracellular Na^+ or Ca^{2+}. In the toad bladder, increases in intracellular Na^+ over the range of 1 to 12 mM (Palmer *et al.*, 1980) and 10 or 40 mM (Palmer, 1985a) at constant external Na^+ had very little effect on Na^+ permeability.

The mechanism of down-regulation is not clear. Lindemann and Van Driessche suggested a floating receptor model in which Na^+ combines with a regulatory protein in the apical membrane that becomes activated and allosterically effects a closure of the Na^+ channel. Machlup *et al.* (1982) have proposed that the channel can be locked into a long-lived clogged state by the simultaneous occupancy of two Na^+ ions.

3.2. Regulation by Hormones

One of the most interesting aspects of epithelial Na^+ channels is their regulation by hormones. This is a complex subject and is covered here only briefly.

The most important hormonal regulator is aldosterone. Since this hormone was shown to increase the number of conducting channels in the membrane (Palmer *et al.*, 1982) and requires protein synthesis to exert its physiological effects (Edelman, 1978), it is conceivable that the mineralocorticoid directs the *de novo* synthesis of the Na^+ channel. Indirect evidence, however, suggests that the hormone acts on channels that preexist in the apical membrane in closed or inactive form (Palmer and Edelman, 1981; Kipnowski *et al.*, 1983; Garty and Edelman, 1983). Furthermore, recent studies of Na^+ transport in membrane vesicles from A6 cells, a cultured amphibian cell line that expresses amiloride-sensitive Na^+ transport, have suggested that the methylation of a component of the apical membrane may underlie the stimulation by aldosterone (Sariban-Sohraby *et al.*, 1984).

Antidiuretic hormone also stimulates Na^+ transport in the toad bladder, and this agent was also known to increase the number of conducting channels in the apical membrane (Li *et al.*, 1982). This response is known to be mediated by cAMP (Orloff and Handler, 1967). It is not known if cAMP or cAMP-dependent protein kinase affects the channels directly.

Insulin also promotes Na^+ transport and increases the conductance of the toad bladder, suggesting that apical Na^+ permeability is increased (Herrera, 1965; Siegel and Civan, 1976). The mechanism through which insulin acts is not known, but it is of interest that insulin raises intracellular pH in skeletal muscle (Moore *et al.*, 1982). Such an effect would be expected to increase Na^+ permeability in the toad bladder (Palmer, 1984b).

Muscarinic cholinergic agonists such as carbachol depress Na^+ transport in the toad bladder (Weismann *et al.*, 1976; Sahib *et al.* 1978). Sahib *et al.* (1978) suggested the possible involvement of cGMP in this response.

3.3. Other Factors

A number of investigators have proposed a negative feedback system whereby Na^+ permeability would be regulated. An increase in Na^+ influx into the cell as a result of channel activation or increased mucosal Na^+ would lead to a rise in cytoplasmic Ca^{2+} activity as a result of inhibition of Ca^{2+} extrusion from the cell through a Na^+–Ca^{2+} exchanger on the basal–lateral membrane (Grinstein and Erlij, 1978; Taylor and Windhager, 1979; Chase and Al-Awqati, 1982; Garty and Lindemann, 1984). As discussed above, the increase in Ca^{2+} would decrease Na^+ permeability and blunt the rise in Na^+ transport.

The Na^+ channels in the toad bladder are also under metabolic control. This can be observed by measuring Na^+ permeability during metabolic poisoning with agents such as 2-deoxyglucose (Palmer *et al.*, 1980; Garty *et al.*, 1983). Alternatively, cells can be substrate starved overnight, after which Na^+ permeability can be reversibly increased by the addition of a metabolic substrate such as pyruvate or glucose (Palmer *et al.*, 1982; Garty *et al.*, 1983). In the latter case, the stimulation by substrate is dependent on the presence of aldosterone. This regulatory mechanism could protect the cells from flooding with Na^+ under conditions of metabolic stress, when the Na^+ pump is unable to keep up with the Na^+ influx rate.

Cell volume also apparently affects Na^+ transport in the toad bladder. Lipton (1972) showed that hypertonic solutions tended to reduce short-circuit current, whereas hypotonic solutions increased it. Although apical membrane function was not directly assessed, it is likely that the channels were affected, since they are thought to be rate limiting for transport under most circumstances (Macknight *et al.*, 1980).

It should be stressed that it is not known in general which of these responses involve the direct modification of the channels and which involve translocation of channels from intracellular stores. Definitive answers to this question await the development of an irreversible specific marker for the channel.

4. Conclusion

Besides being of vital physiological importance in the regulation of the body fluids, the epithelial Na^+ channel is interesting in its own right as a channel. One feature of particular interest is the high selectivity for Na^+ over K^+. Another is the ability of the channel to respond to a variety of hormones and other factors. The electrophysiological approach that has been described here has been more useful in defining these problems than in solving them. Solutions will require the development of techniques to study the channels in a cell-free system and specific probes to study the synthesis and intracellular processing of the channel protein. This will provide a great challenge for the next decade.

References

Augustus, J., Bijman, J., and Os, C. H., 1978, Electrical resistance of rabbit submaxillary main duct: A tight epithelium with leaky cell membranes, *J. Membr. Biol.* **43**:203–226.

Benos, D. J., Simon, S. A., Mandel, L. J., and Cala, P. M., 1976, Effect of amiloride and some of its analogues on cation transport in isolated frog skin and thin lipid membranes, *J. Gen. Physiol.* **68**:43–63.

Benos, D. J., Mandel, L. J., and Simon, S. A., 1980, Cation selectivity and competition at the sodium entry site in frog skin, *J. Gen. Physiol.* **76**:233–247.

Benos, D. J., Hyde, B. A., and Latorre, R., 1983, Sodium flux ratio through the amiloride-sensitive entry pathway in frog skin, *J. Gen. Physiol.* **81**:667–685.

Bentley, P. J., 1968, Amiloride: A potent inhibitor of sodium transport across the toad bladder, *J. Physiol.* (*Lond.*) **195**:317–330.

Bindslev, N., Cuthbert, A. W., Edwardson, J. M., and Skadhauge, E., 1982, Kinetics of amiloride interaction in the hen copradeum *in vitro, Pflueger's Arch.* **392**:340–346.

Chase, H. S., Jr., and Al-Awqati, Q., 1981, Regulation of the sodium permeability of the luminal border of toad bladder by intracellular sodium and calcium, *J. Gen. Physiol.* **77**:693–712.

Chase, H. S., Jr., and Al-Awqati, Q., 1983, Calcium reduces the sodium permeability of luminal membrane vesicles from toad bladder, *J. Gen. Physiol.* **81**:643–665.

Christensen, O., and Bindslev, N., 1982, Fluctuation analysis of short-circuit current in a warm-blooded, sodium retaining epithelium: Site current, density and interaction with triamterene, *J. Membr. Biol.* **65**:19–30.

DeLong, J., and Civan, M. M., 1984, Apical Na entry in split frog skin: Current voltage relationship, *J. Membr. Biol.* **82**:25–40.

Edelman, I. S., 1978, Candidate mediators in the action of aldosterone on Na+ transport, in: *Membrane Transport Processes*, Vol. 1 (J. H. Hoffman, ed.), Raven Press, New York, pp. 125–140.

Eisenman, G., 1962, Cation selective glass electrodes and their mode of operation, *Biophys. J.* **2**(2, pt. 2):259–323.

Frazier, H. S., Dempsey, E. F., and Leaf, A., 1962, Movement of sodium across the mucosal surface of the isolated toad bladder and its modification by vasopressin, *J. Gen. Physiol.* **45**:529–543.

Fuchs, W., Hviid-Larsen, E., and Lindemann, B., 1977, Current–voltage curve of sodium channels and concentration dependence of sodium permeability in frog skin, *J. Physiol. (Lond.)* **267**:137–166.

Garty, H., 1984, Amiloride blockable sodium fluxes in toad bladder membrane vesicles, *J. Membr. Biol.* **82**:269–280.

Garty, H., and Edelman, I. S., 1983, Amiloride-sensitive trypsinization of apical Na channels: Analysis of hormonal regulation of sodium transport in toad bladder, *J. Gen. Physiol.* **81**:785–803.

Garty, H., and Lindemann, B., 1984, Feedback inhibition of sodium uptake in K^+-depolarized toad urinary bladders, *Biochim. Biophys. Acta* **771**:89–98.

Garty, H., Edelman, I. S., Lindemann, B., 1983, Metabolic regulation of apical sodium permeability in toad urinary bladder in the presence and absence of aldosterone, *J. Membr. Biol.* **74**:15–24.

Grinstein, S., and Erlij, D., 1978, Intracellular Ca^{++} and the regulation of Na^+ transport in the frog skin, *Proc. R. Soc. Lond.* [*Biol.*] **202**:353–360.

Hamilton, K., and Eaton, D. C., 1985, Single channel recordings from the amiloride-sensitive epithelial Na^+ channel, *Am. J. Physiol.* **18**:C200–C207.

Helman, S. I., Cox, T. C., and Van Driessche, W., 1983, Hormonal control of apical membrane Na transport in epithelia: Studies with fluctuation analysis, *J. Gen. Physiol.* **82**:201–220.

Henrich, M., and Lindemann, B., 1984, Voltage dependence of channel currents and channel densities in the apical membrane of the toad urinary bladder, in: *Intestinal Absorption and Secretion* (E. Skadhauge, ed.), MTP Press, Lancaster, pp. 209–220.

Herrera, F. C., 1965, Effect of insulin on short-circuit current and sodium transport across toad urinary bladder, *Am. J. Physiol.* **209**:819–824.

Hille, B., 1971, The permeability of the sodium channel to organic cations in myelinated nerve, *J. Gen. Physiol.* **58**:599–619.

Hille, B., 1975, Ionic selectivity of Na and K channels of nerve membranes, in: *Membranes: A Series of Advances*, Vol. 3: *Lipid Bilayers and Biological Membranes: Dynamic Properties* (G. Eisenman, ed.), Marcel Dekker, New York, pp. 255–323.

Hille, B., and Schwartz, W., 1978, Potassium channels as multi-ion single file pores, *J. Gen. Physiol.* **72**:409–442.

Hodgkin, A. L., and Keynes, R. D., 1955, The potassium permeability of a giant nerve fibre, *J. Physiol. (Lond.)* **128**:61–88.

Kipnowski, J., Park, C. S., and Farnestil, D. D., 1983, Modification of carboxyl of Na^+ channel inhibits aldosterone action on Na^+ transport, *Am. J. Physiol.* **245**:F726–F734.

Koefoed-Johnsen, V., and Ussing, H. H., 1958, The nature of the frog skin potential, *Acta Physiol. Scand.* **42**:298–308.

Latorre, R., and Miller, C., 1983, Conduction and selectivity in potassium channels, *J. Membr. Biol.* **71**:11–30.

Lewis, S. A., and Wills, N. K., 1981, Localization of the aldosterone response in rabbit urinary bladder by electrophysiological techniques, *Ann. N.Y. Acad. Sci.* **372**:56–62.

Lewis, S. A., Ifshin, M. S., Loo, D. D. F., and Diamond, J. M., 1984, Studies of sodium channels in rabbit urinary bladder by noise analysis, *J. Membr. Biol.* **80:**135–151.
Li, J. H.-Y., and Lindemann, B., 1983, Chemical stimulation of Na transport through amiloride-blockable channels of frog skin epithelium, *J. Membr. Biol.* **75:**179–192.
Li, J. H.-Y., Palmer, L. G., Edelman, I. S., and Lindemann, B., 1982, The role of sodium-channel density in the natriferic response of the toad urinary bladder to an antidiuretic hormone, *J. Membr. Biol.* **64:**77–89.
Lindemann, B., and Van Driessche, W., 1977, Sodium specific membrane channels of frog skin are pores: Current fluctuations reveal high turnover, *Science* **195:**292–294.
Lindemann, B., and Van Driessche, W., 1978, The mechanism of Na uptake through Na-selective channels in the epithelium of frog skin, in: *Membrane Transport Processes*, Vol. 1 (J. F. Hoffmann, ed.), Raven Press, New York, pp. 155–178.
Lipton, P., 1972, Effect of changes in osmolarity on sodium transport across isolated toad bladder, *Am. J. Physiol.* **222:**821–828.
Machlup, S., Hoshiko, T., and Frehland, E., 1982, Sodium and amiloride competition in apical membrane channels: A 3-state model for noise, *Biophys. J.* **37:**281a.
Macknight, A. D. C., Di Bona, D. R., and Leaf, A., 1980, Sodium transport across toad urinary bladder: A model "tight" epithelium, *Physiol. Rev.* **60:**615–715.
Moore, R. D., Fidelman, M. L., Hansen, J. C., and Otis, J. N., 1982, The role of intracellular pH in insulin action, in: *Intracellular pH: Its Measurement, Regulation and Utilization in Cellular Functions* (R. Nuccitelli and D. W. Deamer, eds.), Alan R. Liss, New York, pp. 385–416.
Nagel, W., Durham, J. H., and Brodsky, W. A., 1981, Electrical characteristics of the apical and basal–lateral membranes in the turtle bladder epithelial cell layer, *Biochim. Biophys Acta* **646:**77–87.
O'Neil, R. G., and Boulpaep, E. L., 1979, Effect of amiloride on the apical cell membrane cation channels of a sodium-absorbing, potassium-secreting renal epithelium, *J. Membr. Biol.* **50:**365–387.
O'Neil, R. G., and Helman, S. I., 1977, Transport characteristics of renal collecting tubules: Influences of DOCA and diet, *Am. J. Physiol.* **233:**F544–F558.
Orloff, J., and Handler, J., 1967, The role of adenosine 3′-5′ phosphate in the action of antidiuretic hormone, *Am. J. Med.* **42:**757–768.
Palmer, L. G., 1982a, Ion selectivity of the apical membrane Na channel in the toad urinary bladder, *J. Membr. Biol.* **67:**91–98.
Palmer, L. G., 1982b, Na^+ transport and flux ratio through apical Na^+ channels in toad bladder, *Nature* **297:**688–690.
Palmer, L. G., 1984a, Voltage-dependent block by amiloride and other monovalent cations of apical Na channels in the toad urinary bladder, *J. Membr. Biol.* **80:**153–165.
Palmer, L. G., 1985a, Modulation of apical Na permeability of the toad urinary bladder by intracellular Na, Ca and H, *J. Membr. Biol.* **83:**57–69.
Palmer, L. G., 1985b, Interactions of amiloride and other blocking cations with the apical Na channel in the toad urinary bladder, *J. Membr. Biol.* **87:**191–199.
Palmer, L. G., and Edelman, I. S., 1981, Control of sodium permeability in the toad urinary bladder by aldosterone, *Ann. N.Y. Acad. Sci.* **372:**1–14.
Palmer, L. G., Edelman, I. S., and Lindemann, B., 1980, Current–voltage analysis of apical Na transport in the toad urinary bladder: Effects of inhibitors of transport and metabolism, *J. Membr. Biol.* **57:**59–71.
Palmer, L. G., Li, J. H.-Y., Lindemann, B., and Edelman, I. S., 1982, Aldosterone control of the density of sodium channels in the toad urinary bladder, *J. Membr. Biol.* **64:**91–102.

Quinton, P. M., 1981, Effects of some ion transport inhibitors on secretion and reabsorption in intact and perfused single human sweat glands, *Pflüger's Arch.* **391**:309–313.

Sahib, M. K., Schwartz, J. H., and Handler, J. S., 1978, Inhibition of toad urinary bladder sodium transport by carbamylcholine: Possible role of cyclic GMP, *Am. J. Physiol.* **235**:F586–F591.

Sariban-Sohraby, S., Burg, M., Weismann, W. P., Chiang, P. K., and Johnson, J. P., 1984, Methylation increases sodium transport into A6 apical membrane vesicles. Possible mode of aldosterone secretion, *Science* **225**:745–764.

Siegel, B., and Civan, M. M., 1976, Aldosterone and insulin effects on driving force of Na^+ pump in toad bladder, *Am. J. Physiol.* **230**:1603–1608.

Steiner, R. A., Oehme, M., Ammann, D., and Simon, W., 1979, Neutral carrier sodium-ion-selective microelectrode for intracellular studies, *Anal. Chem.* **51**:351–353.

Taylor, A., and Windhager, E. E., 1979, Possible role of cytosolic calcium and Na–Ca exchange in regulation of transepithelial sodium transport, *Am. J. Physiol.* **236**:F505–F512.

Thomas, S. R., Suzuki, Y., Thompson, S. M., and Schultz, S. G., 1983, Electrophysiology of *Necturus* urinary bladder: I. "Instantaneous" current–voltage relations in the presence of varying mucosal sodium concentrations, *J. Membr. Biol.* **73**:157–175.

Thompson, S. M., and Dawson, D. C., 1978, Sodium uptake across the apical border of the isolated turtle colon: Confirmation of the two-barrier model, *J. Membr. Biol.* **42**:357–374.

Thompson, S. M., Suzuki, Y., and Schultz, S. G., 1982, The electrophysiology of rabbit descending colon, I. Instantaneous transepithelial current–voltage relations and the current–voltage relations of the Na-entry mechanism, *J. Membr. Biol.* **66**:41–54.

Van Driessche, W., and Lindemann, B., 1979, Concentration dependence of currents through single sodium-selective pores in frog skin, *Nature* **282**:519–520.

Wiesmann, W., Sinha, S., and Klahr, S., 1976, Effects of acetylcholine (ACh) and carbachol on sodium transport in the toad bladder, *Kidney Int.* **10**:603.

Woodhull, A. M., 1973, Ionic blockage of sodium channels, *J. Gen. Physiol.* **61**:687–708.

Zeiske, W., Wills, N. K., and Van Driessche, W., 1982, Na^+ channels and amiloride-induced noise in the mammalian colon epithelium, *Biochim. Biophys. Acta* **688**:201–210.

VI

ENDOCYTOSIS AS A CELL TRANSPORT PATHWAY

18

Uptake and Intracellular Processing of Cell Surface Receptors

Current Concepts and Prospects

COLIN R. HOPKINS

1. Introduction

Uptake of macromolecules at the cell surface is an essential function that occurs in all eucaryotic cells at some stage in their lifetime. It can be achieved by a variety of processes, but all of them probably involve the invagination and subsequent pinching off of a localized segment of the plasma membrane to form an intracellular vesicle. This form of uptake is readily observed in living cells during phagocytosis and pinocytosis, but these are activities found only in certain specialized cell types. The most widespread form of endocytosis occurs below the resolution limit of the light microscope, where profiles observed by electron microscopy suggest that invaginations in the size range 50–200 nm are responsible. This form of uptake, generally referred to as "micropinocytosis," certainly includes a variety of processes that differ in their selectivity and purpose. Tracer studies suggest that the processes of micropinocytic uptake are responsible for a steady, continuous rate of internalization that, in cultured fibroblasts, for example, internalizes less than 50% of the plasma membrane every hour (Steinman *et al.*, 1976, 1983).

The rapid effect of metabolic inhibitors on such micropinocytotic processes suggests that they impose a major demand on the energy supply of the cell. As yet, however, the molecular mechanisms requiring energy expenditure remain to be defined. Some observations suggest that mech-

COLIN R. HOPKINS • Department of Medical Cell Biology, University of Liverpool, Liverpool L69 3BX, England.

anisms capable of creating a directed flow of membrane may be involved (Bretscher, 1984); others, and especially those made on phagocytosis, implicate actin-based microfilament mechanisms (Stossel *et al.*, 1976; Valerius *et al.*, 1981).

2. Clathrin-Coated Pits

The most extensively studied micropinocytic invaginations of the plasma membrane involved in macromolecular uptake have been those coated on their cytoplasmic surface with a clathrin lattice (Pearse and Bretscher, 1981; Goldstein *et al.*, 1979). The invaginations, the coated pits, are involved in the receptor-mediated uptake of macromolecular ligands, and they have been shown to sequester integral membrane receptor proteins for these ligands to a concentration several thousandfold above those of the surrounding plasma membrane (Bretscher *et al.*, 1980; Kerjaschki *et al.*, 1984; Huet *et al.*, 1980). However, although coated pits are capable of selecting for receptor populations destined for uptake, the finding that receptor populations with very different intracellular destinations are frequently colocalized within them (Carpenter *et al.*, 1979; Via *et al.*, 1982; Geuze *et al.*, 1984) suggests that they are able to discriminate between different internalizing receptor populations.

3. The Role of the Clathrin Lattice

A role for the clathrin-containing lattice in selectively concentrating receptors destined for uptake has long been anticipated, but definitive evidence to show that either clathrin or its associated proteins interact directly with receptor proteins has not thus far been obtained. An additional, or perhaps alternative, role for the clathrin lattice has been looked for in the process of the invagination and pinching off of the coated membrane. Supporting this idea is the capacity of clathrin and its associated proteins to self-assemble into closed geodesic cagelike structures of appropriate dimensions (Pearse, 1982) and the presence of an ATP-dependent enzyme able to dismantle formed lattices (Schlossman *et al.*, 1984). Most of the essentials thus exist for an apparatus that assembles on the inner face of the plasma membrane from a free pool of cytoplasmic components, remains intact during the processes of invagination and vesicle formation, and is then dismantled and returned to the cytoplasmic pool in preparation for a new cycle of assembly–disassembly.

Direct, unequivocal evidence showing that this sequence of events actually takes place in the intact cell is obviously difficult to obtain, and alternative proposals that view peripheral clathrin as a stable component of the plasma membrane, separate from a vesiculation process, have been extensively discussed (Willingham and Pastan, 1984). However, a recent report that most cells do, in fact, contain substantial (14–70%) pools of free clathrin (Gould *et al.*, 1985) argues against the view that lattice components are exclusively membrane associated.

The existence of micropinocytic uptake systems such as those of endothelia, in which inward vesiculation normally occurs in the absence of clathrin lattices (Simionescu, 1982), shows that a lattice is not essential for membrane vesiculation, and there is therefore no *a priori* reason why the clathrin coat should be actively involved in promoting the pinching off of coated vesicles. It is, perhaps, just as likely that the reorganizations of lattice substructure that are seen to accompany the invagination of the plasma membrane (Heuser and Evans, 1980) are simply a means of accommodating for changes in membrane shape while the lattice retains a structured interface with the integral proteins of the membrane.

4. Receptor Recycling

Some receptor populations become sequestered within coated pits without binding ligand. The evidence for this conclusion is largely indirect, but it is derived from a variety of experimental approaches (Anderson *et al.*, 1982; Schwartz *et al.*, 1984; Gonzalez Noriega *et al.*, 1980; Kaplan and Keogh, 1981). The kinds of receptor involved in this form of constitutive internalization are those with continuous, large-capacity uptake, systems such as those for low-density lipoprotein, asialoglycoproteins, and transferrin (Anderson *et al.*, 1982; Schwartz *et al.*, 1984; Watts, 1984). These receptor populations are believed to be part of a recycling pathway that continuously transports surface receptors into the cell and then rapidly, within minutes, returns them to the surface (Brown *et al.*, 1984). Ligand bound during exposure on the cell surface is usually unloaded during the intracellular stage of the cycle, so that a single population of receptors such as those for LDL can internalize concentrations of ligand up to 1500 times their carrying capacity during their lifetime.

Other receptor populations, and especially those for effector molecules such as hormones and growth factors, do not appear to constitutively recycle and normally remain on the cell surface until they bind ligand. Once occupied, these receptor populations recycle ($t_{\frac{1}{2}}$ 2–3 min), and they can be found, en route, within the same coated pits (Carpentier *et al.*, 1982).

One of the best-studied receptor populations for an effector ligand is that of the EGF receptor (Cohen and Carpenter, 1979; Shostak and Carpenter, 1984; Carpenter, 1984). It has been shown that this receptor population is normally resident on the cell surface as a monodisperse, randomly distributed population. On binding ligand, it rapidly redistributes to form ligand–receptor aggregates (Hopkins *et al.*, 1981; Schechter *et al.*, 1979). The precise functional significance of this aggregation is not known (Zidovetzki *et al.*, 1981; Schreiber, 1983), but it is accompanied (or may even be preceded) by the self-phosphorylation of the receptor brought about by the tyrosine-specific kinase activity of its cytoplasmic domain (Hunter and Cooper, 1981; Carpenter, 1984). There is, as yet, no indication that recycling receptors capable of internalizing without ligand aggregate prior to becoming concentrated within coated pits. However, there is some evidence that ligand binding can increase their rate of uptake (Schwartz *et al.*, 1984; Klausner *et al.*, 1984). Interestingly, there is also evidence that occupation by ligand of the transferrin receptor [a receptor population that can internalize without binding ligand but whose internalization can be increased by ligand binding (Klausner *et al.*, 1984; Watts, 1985)] causes the receptor to become phosphorylated (Klausner *et al.*, 1984; Stratford-May *et al.*, 1984).

5. Formation and Longevity of Coated Pits

Estimation of the number of coated pits involved in the uptake of recycling receptor populations such as those for LDL and transferrin and calculation of the rate of ligand internalization suggest that the half-life of a coated pit is about 1 min (Bretscher, 1982; Helenius and Marsh, 1982). In some cultured cells, the distribution of coated pits over the surface is highly ordered, and there is no evidence that their number is not readily altered by changes in the concentration of ligands such as LDL and transferrin. It is therefore probable that the formation and pinching off of pits concerned with this kind of ligand is a constitutive process like the internalization of their receptors. The observation that several different receptor populations can be recruited before a pit pinches off nevertheless suggests that pit formation and longevity may be largely independent of receptor movements. Coated pit/vesicle formation and receptor recycling are therefore probably concomitant but not synchronous.

In the case of receptor populations that bind effector ligands, there appear to be additional factors regulating pit formation. Studies with EGF, NGF, and immunoglobulin, for example, have all shown significant increases in pit number within minutes of ligand binding (Connolly *et al.*,

1981, 1984) Salisbury *et al.*, 1980). In epidermoid carcinoma cells, where we have observed a similar response to EGF, we have found that the size and disposition of the forming pits are defined by the size of the receptor aggregates. These pits are therefore smaller and readily distinguished from the pit populations on these cells containing transferrin receptors (C. R. Hopkins, unpublished observations, 1985).

6. *Uptake from the Fluid Phase*

In addition to their role in the internalization of ligand–receptor complexes, micropinocytic invaginations are also capable of taking up solutes in the bulk phase. Early stereologic studies using a bulk-phase tracer indicated that a fluid phase of less than 10% of the total cell volume could be internalized every hour (Steinman *et al.*, 1983), and in subsequent studies, Helenius and Marsh calculated the displacement of fluid-phase marker by virus to show that in BHK cells a very significant proportion of this bulk-phase uptake occurs via coated pits (Helenius and Marsh, 1982). In keeping with the view that effector molecules induce additional pit formation in their target cells, both EGF and platelet-derived growth factor have been observed to stimulate a burst of bulk-phase uptake within 2–3 min of binding (Haigler *et al.*, 1979; Davies and Ross, 1979).

7. *The Endosome Compartment*

Receptors internalized via coated pits are transferred to a cytoplasmic compartment known as the endosome (Helenius *et al.*, 1983; Hopkins, 1983). Endosome systems characterized by their content of endocytosed tracer have been identified in a wide variety of cell types, but in none of these systems has the morphological organization been fully described. In some cells such as macrophages, the endosome system appears to be predominantly vesicular (Steinman *et al.*, 1976), whereas in others, and especially in epithelial cells, there is a large cisternal component (Wall *et al.*, 1980; Hopkins and Trowbridge, 1983; Willingham *et al.*, 1983).

Kinetic experiments using electron-opaque tracers have shown that ligands and receptors concentrated within coated pits on the cell surface are transferred initially to peripheral endosomal elements and that within a short interval of 10–30 min (depending on the cell type), these tracers reach a deeper juxtanuclear compartment (Wall *et al.*, 1980; Hopkins and Trowbridge, 1983; Willingham and Pastan, 1982; Haigler *et al.*, 1984; Yamashiro *et al.*, 1984). Morphological studies demonstrate that the peripheral elements of the endosome system consist of a network of tubular cisternae interconnecting with vacuolar elements of variable size and form lying beneath the plasma membrane. The deeper, juxtanuclear compartments appear to consist of a less elaborate system of cisternae, which, in some cell types, are connected to multivesicular bodies and arranged in a pericentriolar configuration. The endosome system as a whole displays considerable plasticity of form, and it is difficult to determine the extent to which it exists as a stable continuum. It is not yet clear, for example, whether the movement from the peripheral compartment towards the juxtanuclear area involves the bulk translocation of peripheral elements or the bridging of a discontinuous step by vesicular intermediaries. A third alternative of a transfer between two stable systems by a connecting network of cisternae is also possible.

In the processing of internalized ligand–receptor complexes, integral membrane proteins are transferred from the plasma membrane through the endosome to a variety of membrane boundaries. At least two of these transfers, plasma membrane to endosome and endosome to lysosome (Dunn *et al.*, 1980; Wolkoff *et al.*, 1984), are temperature sensitive and are thus presumably discontinuous, vesicle-mediated transfers. Other discontinuous steps also exist, such as the endosome-to-plasma membrane transfer, which can be selectively inhibited by lysosomotrophic agents (Gonzalez Noriega *et al.*, 1980; Basu *et al.*, 1981; Watts, 1984; Schwartz *et al.*, 1984). In none of these instances have the intermediary vesicles responsible been identified.

Cell fractionation studies have isolated endosome-rich preparations, and although further purification is required before a thorough biochemical analysis of their membranes can be made, it is already apparent that endosomal elements lack the classical marker enzymes of other well-characterized organelles (Quintart, 1983; Dickson *et al.*, 1983; Lamb *et al.*, 1983). Other information available from fractionation studies suggests that in addition to their unusually high content of surface receptor proteins (Debanne *et al.*, 1982; Miskimins and Shimuzu, 1984), endosome fractions also contain relatively high levels of cholesterol (Kielian and Helenius, 1984) and an ATP-dependent proton pump (Galloway *et al.*, 1983; Robbins *et al.*, 1984; Yamashiro *et al.*, 1983).

The proton pump of endosome membranes has been shown to be capable of maintaining a pH of between 5.0 and 5.5 within the endosome lumen (Galloway *et al.*, 1983; Tycko *et al.*, 1983; Merion *et al.*, 1984; Geisow and Evans, 1984). It is in most respects similar to the ATP-dependent pumps identified in coated vesicle fractions and lysosomes (Stone *et al.*, 1983; Harikumar *et al.*, 1983; van Dyke *et al.*, 1984), but it is not yet clear whether these various pumps represent separate populations of integral membrane proteins or whether, like some surface receptors, they are part of a recycling pathway. It is, however, likely that these pumps are unevenly distributed throughout the endosome system, since there is some evidence that in the deeper regions of the endosome system the pH decreases, and there is also an indication that in some later stages of the recycling pathway the lumenal content becomes less acidic (about pH 6.4) (Yamashiro *et al.*, 1984).

8. Selective Routing of Ligands and Receptors

Double-label studies have consistently shown that ligand–receptor complexes and bulk-phase solutes taken up via coated pits are transferred to the same peripheral endosome elements (Abrahamson and Rodewald, 1981; Gonatas *et al.*, 1984). Within the endosome, therefore, the sorting of bulk-phase tracer, free ligand, ligand–receptor complexes, and unoccupied receptors must all take place. The best-documented of these sorting pathways is the route to the lysosome, since this is traversed by the bulk of the nutrient and end-product ligands such as LDL and ASGP (Goldstein *et al.*, 1979; Wall *et al.*, 1980). As shown directly for ASGP, these ligands dissociate from their receptors in the low pH of the endosome (Wolkoff *et al.*, 1984). They are then thought to be transported as luminal content to the lysosome.

Unoccupied receptors and ligand–receptor complexes that remain intact in the endosome environment can be routed out of the endosome; LDL and ASGP receptors, for example, are recycled back to the plasma membrane domain from which they were derived, as are intact complexes such as transferrin–transferrin receptor (Brown *et al.*, 1984; Anderson and Kaplan, 1983; Hopkins, 1983). Alternatively, maternal Ig bound to fc receptors in the neonatal enterocyte (Abrahamson and Rodewald, 1981) and IgA receptors bound to the secretory component receptor in the hepatocyte (Mullock and Hinton, 1981; Geuze *et al.*, 1984) are routed to new plasma membrane domains and so achieve a transcellular transfer; EGF–receptor complexes probably remain intact, but they leave the limiting membrane of the endosome by a process of inward vesiculation, which creates multivesicular bodies (McKanna *et al.*, 1979; Beguinot *et al.*, 1984). These luminal vesicles thus join the luminal content of the endosome and are transferred, along with free, previously dissociated ligands, to the lysosome.

In addition to playing an important role in the sorting of ligands, low pH is also required for the mechanisms involved in transporting macromolecules across the endosome membrane and into the cytoplasm (Helenius and Marsh, 1982; Sandvig and Olsnes, 1982). Viruses gaining access to the endosome, such as influenza, require a low pH in order that their spike proteins can adopt a hydrophobic configuration, which is in some way responsible for the transfer of their nucleocapsids to the cytoplasm (Skehel *et al.*, 1982; White *et al.*, 1983). A low pH is similarly required for the transmembrane transfer of a wide variety of bacterial toxins (Draper and Simon, 1980; Sandvig and Olsnes, 1982; Sandvig *et al.*, 1984a), but it is also probable that translocation mechanisms (such as those for some plant toxins) (Sandvig *et al.*, 1984b) exist for which a neutral or alkaline pH is required. Although the point(s) of departure represented by the cell surface invagination and the cytoplasmic destinations of these receptor-mediated pathways are now quite well defined, there are few insights into the role played by the individual endosomal subcompartments through which they pass. At the molecular level, virtually nothing is known of the mechanisms involved in transmembrane transfer processes.

In future studies of the selective routing of receptor populations it will be important to define the functions of the peripheral endosome components more closely, so that the point at which internalized receptors begin to be recycled back to the plasma membrane can be defined. There are a growing number of reports (Steinman *et al.*, 1983; Besterman *et al.*, 1981) that describe a rapid efflux of either bulk-phase tracer or internalized ligand occurring within the first minutes of uptake, and it is therefore possible that a retrieval of surface receptors may occur from even the most peripheral of endosome elements. Alternatively, this reflux may be primarily concerned with eliminating bulk-phase solute taken in during coated vesicle formation. These solutes presumably represent a substantial (less than 10% of total cell volume/hr) and unwelcome diluent that would be difficult to separate from other endosomal contents once intraluminal ligand dissociation begins (Steinman *et al.*, 1983).

The subcompartmentation and differentiation of the membrane boundaries deeper within the endosome system are being examined in detail. In several systems the sequestration of receptors and ligand and of different receptor populations into separate cisternal elements has been observed (Willingham *et al.*, 1984; Geuze *et al.*, 1984). In many cell types, clathrin-coated microdomains are a major feature of the juxtanuclear area (Bloom *et al.*, 1980), and it has been suggested that within the endosome, as on the cell surface, these specialized segments of the membrane boundary may be responsible for some of the selective transfers that occur (Willingham and Pastan, 1982; Willingham *et al.*, 1983).

Adding to the complexity of the crowded and highly differentiated juxtanuclear area are the biosynthetic pathways en route from the rough endoplasmic reticulum to the cell surface and lysosome. Studies on the trafficking of the mannose-6-phosphate receptor in this area demonstrate that pH-sensitive binding to mobile receptors is a transport mechanism also employed in these pathways (Sly and Fischer, 1982; Brown and Farquhar, 1984). Knowing the extent to which the endosomal and biosynthetic pathways intersect should provide much needed information on the final stages of degradative processing in the endosome–lysosome pathway. Equally important, it should also allow new insights into the options available to receptors being routed from the juxtanuclear area to the different domains of the cell surface. In the biosynthetic pathway, the selection between routes leading to either the apical or basal pole appears to be made in this area (Rindler *et al.*, 1983), and it is likely that similar selections are made in the endosome pathways.

Routes from the endosome to the cell surface need to be described in detail, since in situations in which they provide a transcellular pathway they may have considerable therapeutic potential. For example, pathways such as those followed by cobalamin–intrinsic factor receptors (Levene *et al.*, 1984) and IgA–secretory component (Nagura *et al.*, 1979) could be exploited for drug delivery purposes to provide access routes across the intestinal mucosa to both luminal and abluminal surfaces.

9. *The Internalization of Effector Ligands*

The role of internalization in the processing of effector-molecule–receptor complexes is probably the least-understood functional aspect of receptor-mediated endocytosis (Gorden *et al.*, 1980; Beardmore and Hopkins, 1983). It is apparent that exposure to ligand causes a rapid and often prolonged removal of receptors from target cell surfaces (Gorden *et al.*, 1980; Beguinot *et al.*, 1984; Stoscheck and Carpenter, 1984). The concentrations of ligand required to exert this effect are normally in the physiological range, and it is therefore likely that these processes of "down-regulation" do play a role in modulating target cell sensitivity in some systems. The more important question, however, of whether the internalization of an effector ligand is necessary for its primary physiological response to be provoked cannot be answered unequivocally. The kinetics of stimulus–response coupling in many systems suggest that for some acute effects (such as secretion) the peak of the response precedes internalization. However, it is also evident that many of the effector ligands internalized by receptor-mediated endocytosis have longer-term effects that take place only hours after internalization, such as the initiation of DNA synthesis and an altered patterning of gene expression.

In the case of EGF, there is evidence that all of the information required for mitogenesis is contained within the unoccupied receptor (Schreiber *et al.*, 1983; Schechter *et al.*, 1979), and in all of the experiments thus far described, the initiation of DNA synthesis is always preceded by internalization. There is also a report that the retardation of receptor degradation enhances the mitogenic stimulus (Brown *et al.*, 1980), and a recent experiment demonstrates that the purified receptor is capable of binding and interacting specifically with sites inside the nuclear membrane (Mroczkowski *et al.*, 1984). However, despite these intriguing observations, there is as yet no direct evidence that a transduction mechanism capable of generating a second messenger signal for EGF (or any other internalized effector ligand) resides within the endosome system.

10. Conclusions

In future studies, recent advances in sequencing the genes of surface receptors can be expected to provide powerful new probes with which to study the molecular aspects of selective routing in membranes (Yamamoto *et al.*, 1984; Ullrich *et al.*, 1984; Schneider *et al.*, 1984). It is to be anticipated that in addition to providing new information on the topics raised in this brief review, these studies will open up new insights into related areas of membrane trafficking such as cell motility and the regulation of cell polarity.

References

Abrahamson, D. R., and Rodewald, R., 1981, Evidence for the sorting of endocytic vesicle contents during the receptor-mediated transport of IgG across the newborn rat intestine, *J. Cell. Biol.* **91:**270–280.

Anderson, R. G. W., and Kaplan, J., 1983, Receptor mediated endocycosis, in: *Modern Cell Biology* (B. Satir, ed.), Alan R. Liss, New York, pp. 1–52.

Anderson, R. G. W., Brown, M. S., Beisiegel, U., and Goldstein, J. L., 1982, Surface distribution and recycling of the low density lipoprotein receptor as visualized with antireceptor antibodies, *J. Cell Biol.* **93:**523–531.

Basu, S. K., Goldstein, J. L., Anderson, R. G. W., and Brown, M. S., 1981, Monensin interrupts the recycling of low density lipoprotein receptors in human fibroblasts, *Cell* **24:**493–502.

Beardmore, J., and Hopkins, C. R., 1983, Uptake and intracellular processing of epidermal growth factor receptor complexes, *Biochem. Soc. Trans.* **12:**165–168.

Beguinot, L., Lyall, R. M., Willingham, M. C., and Pastan, I., 1984, Down regulation of the epidermal growth factor receptor in KB cells is due to receptor internalization and subsequent degradation in lysosomes, *Proc. Natl. Acad. Sci. U.S.A.* **81**:2384–2388.

Besterman, J. M., Airhart, R. C., Woodworth, R. C., and Low, R. B., 1981, Exocytosis of pinocytosed fluid in cultured cells: Kinetic evidence for rapid turnover and compartmentation, *J. Cell Biol.* **91**:716–727.

Bloom, W. S., Fields, K. L., Yen, S. H., Haver, K., Schook, W., and Puszkin, S., 1980, Brain clathrin: Immunofluorescent patterns in cultured cells and tissues, *Proc. Natl. Acad. Sci. U.S.A.* **77**:5520–5524.

Bretscher, M. S., 1982, Endocycosis, the sorting problem and cell locomotion in fibroblasts, *Ciba Found. Symp.* **92**:266–281.

Bretscher, M. S., 1984, Endocytosis: Relation to capping and cell locomotion, *Science* **226**:681–685.

Bretscher, M. S., Thomson, J. N., and Pearse, B. M. F., 1980, Coated pits act as molecular filters, *Proc. Natl. Acad. Sci. U.S.A.* **77**:4156–4159.

Brown, K. D., Friedkin, M., and Rozengurt, E., 1980, Colchicine inhibits epidermal growth factor degradation in 3T3 cells, *Proc. Natl. Acad. Sci. U.S.A.* **77**:480–484.

Brown, M. S., Anderson, R. G. W., and Goldstein, J. L., 1983, Recycling receptors: The round trip itinerary of migrant membrane proteins, *Cell* **32**:663–667.

Brown, W. J., and Farquhar, M. G., 1984, Accumulation of coated vesicles bearing mannose-6-phosphate receptors for lysosomal enzymes in the Golgi region of I cell fibroblasts, *Proc. Natl. Acad. Sci. U.S.A.* **81**:5135–5139.

Carpenter, G., 1984, Properties of the receptor for epidermal growth factor, *Cell* **37**:357–358.

Carpenter, G., and Cohen, S., 1979, Epidermal growth factor, *Annu. Rev. Biochem.* **48**:193–216.

Carpentier, J. L., Gorden, P., Anderson, R. G. W., Goldstein, J. L., Brown, M. S., Cohen, S., and Orci, L., 1982, Co-localization of ^{125}I epidermal growth factor and ferritin–low density lipoprotein in coated pits, *J. Cell Biol.* **95**:73–77.

Connolly, J. L., Green, S. A., and Greene, L. A., 1981, Pit formation and rapid changes in surface morphology of sympathetic neurones in response to nerve growth factor, *J. Cell Biol.* **90**:176–180.

Connolly, J. L., Green, S. A., and Greene, L. A., 1984, Comparison of rapid changes in surface morphology and coated pit formation of PC12 cells in response to growth factor, epidermal growth factor and dibutryl cyclic AMP, *J. Cell Biol.* **98**:457–465.

Davies, P. F., and Ross, R., 1978, Mediation of pinocytosis in cultured arterial smooth muscle and endothelial cells by platelet derived growth factor, *J. Cell Biol.* **79**:663–671.

Debanne, M. T., Evans, W. H., Flint, N., and Regoeczi, E., 1982, Receptor-rich intracellular membrane vesicles transporting asialotransferrin and insulin in liver, *Nature* **298**:398–400.

Dickson, R. B., Beguinot, L., Hanover, J. A., Richert, N. D., Willingham, M. C., and Pastan, I., 1983, Isolation and characterization of a highly enriched preparation of receptosomes (endosomes) from a human cell line, *Proc. Natl. Acad. Sci. U.S.A.* **80**:5335–5339.

Draper, R. K., and Simon, M. I., 1980, The entry of diptheria toxin into the mammalian cell cytoplasm: Evidence for lysosomal involvement, *J. Cell Biol.* **87**:849–854.

Dunn, W. A., Hubbard, A. L., and Aronson, N. N., Jr., 1980, Low temperature selectively inhibits fusion between pinocytic vesicles and lysosomes during heterophagy of ^{125}I asialofetuin by perfused rat liver, *J. Biol. Chem.* **255**:5971–5978.

Galloway, C. J., Dean, G. E., Marsh, M., Rudnick, C., and Mellman, I., 1983, Acidification of macrophage and fibroblast endocytic vesicles in vitro, *Proc. Natl. Acad. Sci. U.S.A.* **80**:3334–3338.

Geisow, M. J., and Evans, W. H., 1984, pH in the endosome, *Exp. Cell Res.* **150:**36–46.
Geuze, H. J., Slot, J. W., Strous, G. J. A. M., Lodish, H. F., and Schwartz, A. L., 1983, Intracellular site of asialoglycoprotein receptor ligand uncoupling: Double-label immunoelectron microscopy during receptor mediated endocytosis, *Cell* **32:**277–287.
Geuze, H. J., Slot, J. W., Strous, G. J. A. M., Peppard, J., van Figura, K., Hasilik, A., and Schwartz, A. L., 1984, Intracellular receptor sorting during endocytosis: Comparative immunoelectron microscopy of multiple receptors in rat liver, *Cell* **37:**195–204.
Goldstein, J. L., Anderson, R. G. W., and Brown, M. S., 1979, Coated pits, coated vesicles and receptor-mediated endocytosis, *Nature* **279:**679–685.
Gonatas, N. K., Stieber, A., Hickey, W. F., Herbert, S. H., and Gonatas, J. O., 1984, Endosomes and Golgi vesicles in absorptive and fluid phase endocytosis, *J. Cell Biol.* **99:**1379–1390.
Gonzalez Noriega, A., Grubb, J. H., Talkad, V., and Sly, W. S., 1980, Chloroquine inhibits lysosomal pinocytosis and enhances lysosomal enzyme secretion by impairing receptor recycling, *J. Cell Biol.* **85:**839–852.
Gorden, P., Carpentier, J. L., Freychet, P., and Orci, L., 1980, Internalization of polypeptide hormones: Mechanism, intracellular localization and significance, *Diabetalogia* **18:**263–274.
Gould, B., Huet, C., and Louvard, D., 1985, Assembled and unassembled pools of clathrin: A quantitative study using an enzyme immunoassay, *J. Cell Biol.* (in press).
Haigler, H. T., McKanna, J. A., and Cohen, S., 1979, Rapid stimulation of pinocytosis in human carcinoma cells A-431 by epidermal growth factor, *J. Cell Biol.* **83:**82–90.
Harikumar, P., and Reeves, J. P., 1983, The lysosomal proton pump is electrogenic, *J. Biol. Chem.* **258:**10403–10410.
Helenius, A., and Marsh, M., 1982, Endocytosis of enveloped animal viruses, *Ciba Found. Symp.* **92:**59–76.
Helenius, A., Mellman, I., Wall, D., and Hubbard, A., 1983, Endosomes, *Trends Biochem. Sci.* **8:**245–250.
Heuser, J., and Evans, L., 1980, Three dimensional visualization of coated vesicle formation in fibroblasts, *J. Cell Biol.* **84:**560–583.
Hopkins, C. R., 1983a, The importance of the endosome in intracellular traffic, *Nature* **304:**684–685.
Hopkins, C. R., 1983b, Intracellular routing of transferrin and transferring receptors in epidermoid carcinoma A431 cells, *Cell* **35:**321–330.
Hopkins, C. R., 1985, Appearance of transferrin receptors at the leading edge of epidermoid carcinoma cells, *Cell* (in press).
Hopkins, C. R., and Trowbridge, I. S., 1983, Internalization and processing of transferrin and the transferring receptor in human carcinoma A431 cells, *J. Cell Biol.* **97:**508–521.
Hopkins, C. R., Boothroyd, B., and Gregory, H., 1981, Early events following the binding of epidermal growth factor to surface receptors on ovarian granulosa cells, *Eur. J. Cell Biol.* **24:**259–265.
Huet, C., Ash, C., and Singer, S. J., 1980, The antibody induced clustering and endocytosis of HLA antigens on cultured human fibroblasts, *Cell* **21:**429–438.
Hunter, T., and Cooper, J. A., 1981, Epidermal growth factor induces rapid tyrosine phosphorylation of proteins in A431 human tumor cells, *Cell* **24:**741–752.
Kaplan, J., and Keogh, E. A., 1981, Analysis of the effect of amines on inhibition of receptor-mediated and fluid phase pinocytosis in rabbit alveolar macrophages, *Cell* **24:**925–932.
Kerjaschki, D., Noronha-Blob, L., Sacktor, B., and Farquhar, M. G., 1984, Microdomains of distinctive glycoprotein composition in the kidney proximal tubule brush border, **98:**1505–1513.

Kielian, M. D., and Helenius, A., 1984, Role of cholesterol in fusion of Semliki Forest virus with membranes, *J. Virol.* **52:**281–283.

Klausner, R. D., Harford, J., and van Renswoude, J., 1984, Rapid internalization of the transferrin receptor in K562 cells is triggered by ligand binding or treatment with a phorbol ester, *Proc. Natl. Acad. Sci. U.S.A.* **81:**3005–3009.

Lamb, J. E., Ray, F., Ward, J. H., Kushner, J. P., and Kaplan, J., 1983, Internalization and subcellular localization of transferrin and transferrin receptors in HeLa cells, *J. Biol. Chem.* **258:**8751–8758.

Levine, J. S., Allen, R. H., Alpers, D. H., and Seetharam, B., 1984, Immunocytochemical localization of the intrinsic factor–cobalamin receptor in dog ileum: Distribution of the intracellular receptor during cell maturation, *J. Cell Biol.* **98:**1111–1118.

McKanna, J. A., Haigler, H. T., and Cohen, S., 1979, Hormone receptor topology and dynamics: Morphological analysis using ferritin-labeled epidermal growth factor, *Proc. Natl. Acad. Sci. U.S.A.* **76:**5689–5693.

Merion, M., Schlessinger, P., Brooks, R. M., Moehring, J. M., Moehring, T. J., and Sly, W. S., 1983, Defective acidification of endosomes in Chinese hamster cell mutants "cross resistant" to toxins and viruses, *Proc. Natl. Acad. Sci. U.S.A.* **80:**5315–5319.

Miskimins, W. K., and Shimizu, N., 1984, Uptake of epidermal growth factor into a lysosomal enzyme deficient organelle: Correlation with cell's mitogenic response and evidence for ubiquitous existence in fibroblasts, *J. Cell. Physiol.* **118:**305–316.

Mroczkowski, B., Mosig, G., and Cohen, S., 1984, ATP-stimulated interaction between epidermal growth factor receptor and supercoiled DNA, *Nature* **309:**270–273.

Mullock, B. M., and Hinton, R. H., 1981, Transport of proteins from blood to bile, *Trends Biochem. Sci.* **6:**188–191.

Nagura, H., Nakane, P. K., and Brown, W. R., 1979, Translocation of dimeric IgA through neoplastic colon cells in vitro, *J. Immunol.* **123:**2359–2368.

Pearse, B. M. F., 1982, Structure of coated pits and vesicles, *Ciba Found. Symp.* **92:**246–280.

Pearse, B. M. F., and Bretscher, M. S., 1981, Membrane recycling by coated vesicles, *Annu. Rev. Biochem.* **50:**85–101.

Quintart, J., Courtoy, P. J., and Baudhin, P., 1984, Receptor-mediated endocytosis in rat liver: Purification and enzymic characterization of low density organelles involved in uptake of galactose-exposing proteins, *J. Cell Biol.* **98:**877–884.

Renswoude, J., Bridges, K. R., Harford, J. B., and Klausner, R. D., 1982, Receptor-mediated endocytosis of transferrin and uptake of Fe in K562 cells, *Proc. Natl. Acad. Sci. U.S.A.* **79:**6186–6190.

Rindler, M. J., Ivanov, I. E., Plesben, H., Rodriguez-Boulan, E., and Sabatini, D. D., 1984, Viral glycoproteins destined for apical or basolateral plasma membrane domains traverse the same Golgi apparatus during their intracellular transport in doubly infected Madin–Darby canine kidney cells, *J. Cell Biol.* **98:**1304–1319.

Sly, W. S., and Fischer, H. D., 1982, The phosphomannosyl recognition system for intracellular and intercellular transport of lysosomal enzymes, *J. Biol. Chem.* **18:**67–85.

Steinman, R. M., Brodie, S. E., and Cohn, Z. A., 1976, Membrane flow during pinocytosis, *J. Cell Biol.* **68:**665–687.

Steinman, R. M., Mellman, I. S., Muller, W. A., and Cohn, Z. A., 1983, Endocytosis and recycling of plasma membrane, *J. Cell Biol.* **96:**1–27.

Stone, D. K., Xie, X.-S., and Racker, E., 1983, An ATP-driven proton pump in clathrin coated vesicles, *J. Biol. Chem* **258:**4059–4062.

Stoscheck, C. M., and Carpenter, G., 1984, Down-regulation of epidermal growth factor receptors: Direct demonstration of receptor degradation in human fibroblasts, *J. Cell Biol.* **98:**1048–1053.

Stossel, T. P., and Hartwig, J. H., 1976, Interactions of actin, myosin, and a new actin-binding protein in rabbit pulmonary macrophages II: Role in cytoplasmic movement and phagocytosis, *J. Cell Biol.* **68:**602–619.

Stratford-May, W., Jacobs, S., and Cuatrecasas, P., 1984, Association of phorbol ester-induced hyperphosphorylation and reversible regulation of transferrin membrane receptors in HL60 cells, *Proc. Natl. Acad. Sci. U.S.A.* **81:**2016–2020.

Ullrich, A., Coussens, L., Hayflick, J. S., Dull, T. J., Gray, A., Tam, A. W., Lee, J., Yarden, Y., Libermann, T. A., Schlessinger, J., Downward, J., Mayes, E. L., Whittle, N., Waterfield, M. D., and Seeburg, P. H., 1984, Human epidermal growth factor receptor cDNA sequence and aberrant expression of the amplified gene in A431 epidermoid carcinoma cells, *Nature* **309:**418–425.

Valerius, N. H., Stendahl, O., Hartwig, J. H., and Stossel, T. P., 1981, Distribution of actin binding protein and myosin in polymorphonuclear leukocytes during locomotion and phagocytosis, *Cell* **24:**195–202.

van Dyke, R. W., Steer, C. J., and Scharschmidt, B. F., 1984, Clathrin-coated vesicles from rat liver: Enzymatic profile and characterization of ATP-dependent proton transport, *Proc. Natl. Acad. Sci. U.S.A.* **81:**3108–3112.

Via, D. P., Willingham, M. C., Pastan, I., Gotto, A. M., Jr., and Smith, L. C., 1982, Co-clustering and internalization of low density lipoproteins and $alpha_2$ macroglobulin in human skin fibroblasts, *Exp. Cell Res.* **141:**15–22.

Wall, D. A., Wilson, G., and Hubbard, A. L., 1980, The galactose-specific recognition system of mammalian liver: The route of ligand internalization in rat hepatocytes, *Cell* **21:**79–93.

Watts, C., 1985, Rapid endocytosis of the transferrin receptor in the absence of bound transferrin, *J. Cell Biol.* (in press).

White, J., Kieliman, M., and Helenius, A., 1983, Membrane fusion proteins of animal enveloped viruses, *Q. Rev. Biophys.* **16:**151–195.

Willingham, M. C., and Pastan, I. H., 1982, Transit of epidermal growth factor through coated pits of the Golgi system, *J. Cell Biol.* **94:**207–212.

Willingham, M. C., and Pastan, I. H., 1984, Do coated vesicles exist? *Trends Biochem. Sci.* **9:**93–94.

Willingham, M. C., Haigler, H. T., Fitzgerald, D. J. P., Gallo, M. G., Rutherford, A. V., and Pastan, I. H., 1983, The morphologic pathway of binding and internalization of epidermal growth factor in cultured cells, *Exp. Cell Res.* **146:**163–175.

Willingham, M. C., Hanover, J. A., Dickson, R. B., and Pastan, I. H., 1984, Morphologic characterization of the pathway of transferrin endocytosis and recycling in human KB cells, *Proc. Natl. Acad. Sci. U.S.A.* **81:**175–179.

Wolkoff, A. W., Klausner, R. D., Ashwell, G., and Harford, J., 1984, Intracellular segregation of asialoglycoproteins and their receptor: A prelysosomal event subsequent to dissociation of ligand receptor complex, *J. Cell Biol.* **98:**375–381.

Yamashiro, D. J., Fluss, S. R., and Maxfield, F. R., 1983, Acidification of endocytic vesicles by an ATP-dependent proton pump, *J. Cell Biol.* **97:**929–934.

Yamashiro, D. J., Tycko, B., Fluss, S. R., and Maxfield, F. R., 1984, Segregation of transferrin to a mildly acidic (pH 6.5) para-Golgi compartment in the recycling pathway, *Cell* **37:**789–800.

Zidovetzki, R., Yarden, Y., Schlessinger, J., and Jovin, T. M., 1981, Rotational diffusion of epidermal growth factor complexed to cell surface receptors reflects rapid microaggregation and endocytosis of occupied receptors, *Proc. Natl. Acad. Sci. U.S.A.* **78:**6981–6985.

19

Sorting in the Prelysosomal Compartment (CURL)

Immunoelectron Microscopy of Receptors and Ligands

HANS J. GEUZE, ALAN L. SCHWARTZ, JAN W. SLOT, GER J. STROUS, and JOS E. ZIJDERHAND-BLEEKEMOLEN

1. Introduction

Adsorptive endocytosis provides cells with a means of specifically internalizing exogenous substances (ligands). The specificity for each ligand's delivery is conferred by the nature of the specific receptors on the recipient cell. The initial interaction between ligands and cells is initiated at the extracytoplasmic surface of the plasma membrane. One class of such interactions, receptor-mediated endocytosis (RME), defines a sequence of events that leads to the targeting of the internalized ligands to specific intracellular destinations. It is now well established that during RME clathrin-coated pits at the plasma membrane accumulate receptor–ligand complexes. These specialized structures then pinch off from the surface to form coated vesicles, which have a lifetime of only about 1 min (Petersen and van Deurs, 1983). The subsequent destination of each ligand and receptor then depends on the nature of the receptor system. In many systems ligands are transported to the lysosomes for degradation, whereas

HANS J. GEUZE, JAN W. SLOT, GER J. STROUS, and JOS E. ZIJDERHAND-BLEEKEMOLEN • Laboratory of Cell Biology, Medical Faculty, University of Utrecht, Utrecht, The Netherlands. *ALAN L. SCHWARTZ* • Division of Pediatric Haematology/Oncology, Children's Hospital, Dana Farber Institute and Harvard Medical School, Boston, Massachusetts 02115.

the receptors are spared this fate. These receptors are rerouted back to the plasma membrane via poorly understood mechanisms and pathways for subsequent endocytotic cycles. Examples of such receptor systems are those for asialoglycoproteins (Schwartz, 1984), lysosomal enzymes (Sly and Fischer, 1982), and low-density lypoproteins (Brown *et al.*, 1983).

The obligatory uncoupling of the receptors and ligands has recently been shown to occur in an acidic intracellular compartment that is distinct from the degradative compartment (lysosome). Initial studies by Tycko and Maxfield (1982) demonstrated a pH of about 5 in this so-called "prelysosomal compartment." The discovery that a decrease in pH from 7 to 5 probably effects receptor–ligand uncoupling was consistent with earlier observations that alterations in pH have profound effects on the association of many ligands with their receptors, and that dissociation of ligands from many receptors takes place when the pH decreases to below pH 5.5. Ligands as diverse as asialoglycoproteins, insulin, epidermal growth factor, lysosomal enzymes, and α_2-macroglobulin fall into this category. Low pH (5–5.5) also facilitates the intracellular transmembrane penetration of many viruses and toxins into the cytosol.

Alternatives to the general scheme of receptor–ligand uncoupling, ligand degradation, and receptor recycling are the receptor systems for transferrin, a variety of hormones such as the epidermal growth factor (EGF), and polymeric IgA (pIgA). The transferrin receptor is initially internalized via coated vesicles along with its ligand (ferrotransferrin). However, following loss of the iron moiety in an acidic prelysosomal compartment, the receptor recycles back to the cell surface still associated with its ligand, which is now devoid of iron (apotransferrin) (Ciechanover *et al.*, 1983). Alternatively, the predominant pathway of the EGF receptor appears to be one of internalization followed by receptor degradation, probably in the lysosome (Carpenter, 1984). Thus, exposure of fibroblasts to EGF leads to a dramatic decrease in the number of EGF receptors at the cell surface, a process termed "down-regulation." In liver, however, the EGF receptor has been shown to have some capacity for recycling (Dunn and Hubbard, 1984).

The receptor for pIgA provides an example of yet another receptor pathway distinct from those described above. The IgA receptor mediates the endocytosis and translocation of pIgA across epithelia such as hepatocytes and intestinal adsorptive cells (Kuhn and Kraehenbuhl, 1982). In liver, this transfer is directed from the (basolateral) sinusoidal membrane to the (apical) bile canaliculus. The RME of pIgA is unique because it does not involve recycling of the receptor. Here, the IgA receptor with its covalently bound ligand is transported across the cell in a process called transcytosis. The receptor is ultimately sacrificed (proteolytically cleaved and secreted) with the ligand to the bile.

Clearly, a better understanding of the biological basis of the cell, which underlies these differences in receptor–ligand trafficking, is of enormous importance. What are the precise intracellular routes taken by each ligand–receptor system? Do each of these classes utilize the same transport vesicles and compartments? Where in the cell are these compartments localized, and what are their structural features? Where does receptor–ligand uncoupling occur? Where do recycling receptors segregate from those that undergo degradative "down-regulation" or transcytosis? Does a receptor for endogenous ligands, such as the receptor for newly synthesized lysosomal enzymes, make use of the same compartments as a receptor that directs exogenous ligands such as the asialoglycoprotein receptor?

One approach to addressing these types of questions relies on the *in situ* localization of the receptors and ligands at the subcellular level. Although several methods are available for characterizing ligand localization (e.g., labeling with radioactive or electron-dense tracers), to the best of our knowledge receptor molecules themselves can only be precisely localized by immunocytochemical means. Fortunately, the increasing need for precise subcellular detection has coincided with the recent development of powerful immunoelectron microscope techniques. Of these we have chosen a method that relies on the use of ultrathin cryosections and labeling with colloidal gold particles (Slot and Geuze, 1983). In the following sections we first briefly describe this method. We then describe observations obtained with this method on three receptor systems: the asialoglycoprotein receptor (ASGP-R), the mannose-6-phosphate receptor (MP-R), and the pIgA receptor (IgA-R).

2. *Immunogold Detection of Receptors and Ligands*

The cryosectioning technique developed by Tokuyasu (1978) consists of the mild fixation of cells and tissue with aldehydes, infusion with sucrose as a cryoprotectant, and rapid freezing in liquid nitrogen. Thereafter, ultrathin (150 nm) cryosections can be prepared. In the case of cultured cells, gelatin embedding followed by cross linking of the extracellular gelatinous support is required (Geuze and Slot, 1980). Immunocytochemistry is performed directly on the native cryosections. We prefer an indirect immunolabeling procedure in which the sections are first incubated with a specific antibody and the binding sites are then labeled with protein-A-complexed colloidal gold.

Colloidal gold particles are prepared by reducing chloroauric acid in aqueous solution. The size of the particles depends in part on the reductant used. Some years ago we described a method for the preparation of homodisperse gold particles using white phosphorus and sodium ascorbate as reducing agents. Following gradient centrifugation, homogeneously sized particle fractions of approximately 5 and 10 nm, respectively, were obtained (Slot and Geuze, 1981). Recently we have developed a new method of making colloidal gold sols of any particle size desired within the range of 3 to 17 nm. This technique utilizes the combined reduction capacity of tannic acid and citrate (J. W. Slot and H. J. Geuze, 1985). In studies on receptors and ligands it is essential to compare directly different receptor or ligand localizations. Double or even multiple immunogold labeling is then required. We have shown that this is easily achieved using protein-A–gold preparations of different sizes (Geuze *et al.*, 1981). For double labeling, ultrathin cryosections are successively incubated with antibody 1, small protein-A–gold (e.g., 5 nm), antibody 2, large protein-A–gold (e.g., 10 nm). After labeling, the sections are usually stained with uranyl acetate to enhance membrane contrast. An advantage of stained cryosections is the superb membrane delineation, which is at least as good as that in classical resin-embedded sections.

Colloidal gold particles can also be complexed to ligands directly and may then be administered to living cells in order to trace ligand movement in ultrathin sections of the cells routinely processed for electron microscopy (Geuze *et al.*, 1983b). Apart from the problem of ligand–gold loss from the cells, a major difficulty in this type of study is to make sure that the ligand pathway is not induced by the gold probe rather than by the nature of the receptor system under study. In this regard, it is essential to recognize that in most of our studies described below, receptor and ligand sites have been labeled with gold following immobilization by fixation.

3. Ultrastructural Observations of Different Receptor Systems

3.1. The Asialoglycoprotein Receptor

The ASGP-R is a well-documented receptor system (Ashwell and Harford, 1982; Schwartz, 1984). The receptor is exclusively present on liver parenchymal cells and recognizes galactose-terminal carbohydrates of glycoproteins (asialoglycoproteins). The ASGP-R takes up ASGP ligand from the blood by RME. Hubbard and colleagues have demonstrated that the uptake of ASGP by rat hepatic parenchymal cells in coated vesicles is followed by a rapid transfer to lysosomes through a series of pleiomorphic vesicles (Wall *et al.*, 1980). These vesicles have also been termed endosomes (Helenius *et al.*, 1983) or receptosomes (Pastan and Willingham, 1983) and presumably represent the acidic prelysosomal compartment. In the literature, this compartment has been defined morphologically exclusively on the basis of the presence of ligand visualized by electron-dense labels.

Since no data on the *in situ* localization of ASGP-R or other receptors were available, a few years ago we examined the localization of the ASGP-R in rat liver by immunogold labeling using an affinity-purified antirat ASGP-R antibody (Schwartz *et al.*, 1981). We found ASGP-R exclusively present on liver parenchymal cells. At the cell surface, ASGP-R was not confined to clathrin-coated pits but occurred over the entire plasma membrane surface including that of the bile canaliculus (Geuze *et al.*, 1982). Quantitative immunoelectron microscopy showed that 35% of the total receptors resided at the plasma membrane (85% at the sinusoidal membrane). Of the intracellular receptors, the Golgi complex contained 30% and a tubulovesicular organelle about 45%.

Next we simultaneously examined the localization of the ASGP-R and a ligand. Asialofetuin was selected because it is a foreign protein to the rat and, thus, any immunocytochemical signal would reflect administered ligand. Following administration *in vivo*, the distribution of asialofetuin was compared to that of ASGP-R. These observations led to the identification of this tubulovesicular organelle as the compartment of uncoupling receptors and ligands (CURL). The CURL appeared to contain most of the receptors in the tubules, whereas the ligand was mainly confined to detaching vesicles (Fig. 1). We concluded that the CURL functions as a prelysosomal compartment (Geuze *et al.*, 1983a).

The CURL is composed of units of anastomosing tubules and detaching vesicles. One can distinguish between peripheral CURL, which consists mainly of tubules and is localized in the cytoplasm adjacent to the sinusoidal plasma membrane, and tubulo-vesicles of the trans-Golgi reticulum (TGR) in the Golgi region. The TGR displays more prominent ligand-containing vesicles, some of which have internal vesicles and lipoprotein particles. In our opinion, the vesicular portion of CURL is equivalent to the so-called endosome or receptosome defined in other systems.

We were interested in determining the means of entrance for ligand–receptor complexes into CURL; i.e., do coated vesicles deliver the complexes to the tubules and/or to the CURL vesicles? In order to precisely identify the route of ligand, we have used the human hepatoma cell line Hep G2, which allowed the endocytosis in synchrony of ASGP bound to

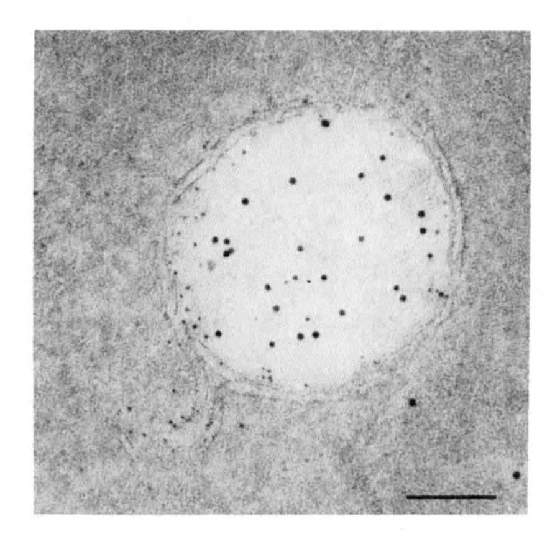

Figure 1. Electron micrograph of ultrathin cryosection from rat liver immuno-double-labeled for the simultaneous demonstration of ASGP-R (small gold particles) and asialofetuin (large particles). Note that the receptor is predominantly located at the pole of a CURL vesicle, where a tubule with heavy labeling of receptor is connected. Most ligand is present free in the vesicle lumen. Bar, 0.1 μm.

colloidal gold. We found that ASGP–gold entered the CURL tubules before delivery to the luminal contents of CURL vesicles (Geuze *et al.*, 1983b) (Fig. 2). A similar coated vesicle–tubule–smooth vesicle route was found by Deschuyteneer *et al.* (1984). As discussed above, in this type of experiment one cannot exclude some influence of the gold particles on the pathway of the ligand.

The Hep G2 cells have also enabled us to experimentally interfere with the presumed acid internal milieu of CURL. "Lysosomotropic" weak bases such as NH_4Cl, chloroquine, and primaquine neutralize the pH of acidic intracellular compartments as a result of protonation of the base. The precise intracellular localization of the acidic compartments (apart from lysosomes) is not known. In order to address this question we have developed monospecific antibodies to primaquine, a chloroquine analogue that contains a free NH_2. This free NH_2 group thus allows both cross linking and fixation. Using immunoelectron microscopy, we found that primaquine accumulated in lysosomes, multivesicular bodies, Golgi cisternae, and CURL vesicles. No significant primaquine labeling was found in CURL tubules (Schwartz *et al.*, 1985). However, since the contents of the narrow CURL tubules (approximate diameter 20 nm) in which primaquine could have been detected is very limited, this may have contributed to these results.

Lysosomotropic agents do not appear to affect the binding or internalization of ASGP ligand in Hep G2 cells but do dramatically block the recycling of internalized ASGP-R back to the cell surface (Schwartz *et al.*, 1984). This is not a result of enhanced lysosomal degradation of receptors, as removal of the agents causes a rapid reappearance of ligand-binding sites at the cell surface (Table I). Thus, following neutralization

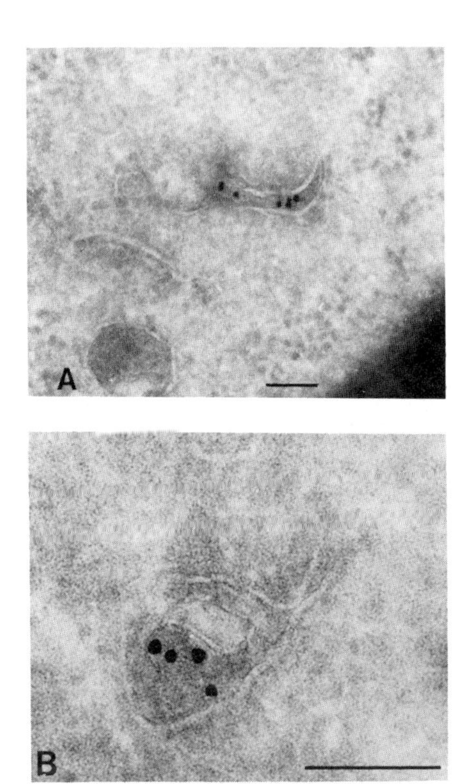

Figure 2. Electron micrographs of ultrathin cryosections from Hep G2 cells that were allowed to endocytose 12-nm colloidal gold particles coated with asialofetuin. The cells were incubated with the gold probe for 10 min (A) and 30 min (B). Bar, 0.1 μm.

of the pH in intracellular compartments, the ASGP-R is sequestered in some unknown site (Fig. 3). Preliminary observations using immunogold labeling have indicated that this reversible block is localized to CURL in the periphery of the cytoplasm as well as in TGR.

In both liver and hepatoma cells, CURL contains an abundant pool of ASGP-R, especially within the tubules. However, the cells differ with respect to receptors in the Golgi complex. In rat liver parenchymal cells, about 20% of the total cellular receptors reside in the membranes of the Golgi cisternae (Geuze *et al.*, 1983a). Although not yet quantified, it is clear that in Hep G2 cells the Golgi complex only contains a few percent of the ASGP-R. It is as yet unclear what this difference in intracellular distribution means in terms of RME and receptor recycling.

Table I. Reversibility of the Effects of Primaquine on Uptake of [^{125}I]-ASOR by Hepatoma Cells[a]

Preincubation drug (mM)	Uptake incubation drug (mM)	2-hr uptake (%)
None	None	100 ± 5
Primaquine (0.2)	Primaquine (0.2)	8 ± 5
Primaquine (0.2)	None	94 ± 5
Primaquine (0.2)	Primaquine (0.2)	101 ± 10
Primaquine (0.2)	Chloroquine (0.2)	29 ± 2

[a] Cells were washed and preincubated for 30 min at 37°C in the absence or presence of 0.2 mM primaquine. After washing, the cells were incubated for 2 hr at 37°C with [^{125}I]-ASOR in the presence of 0.02 or 0.2 mM primaquine or 0.2 mM chloroquine or no drug. Nonspecific uptake was determined, and the results are expressed as percentage of control specific binding ± S.E.M. for three determinations.

3.2. *The Mannose-6-Phosphate Receptor*

A substantial body of evidence now indicates that newly synthesized lysosomal enzymes are selectively targeted to the lysosomes by means of membrane receptors that recognize mannose-6-phosphate residues uniquely present on lysosomal enzymes (Sly and Fischer, 1982). In spite of our present knowledge of the biochemical characteristics of lysosomal enzymes and MP-R, it is still unclear where the lysosomal enzymes are ultimately sorted into lysosomes. Thus, the site of primary lysosome formation has not been elucidated, nor has the precise intracellular localization of the MP-R been unequivocally established. Most data suggest that newly synthesized lysosomal enzymes travel through the Golgi complex. The enzymes involved in the biosynthesis of the mannose-6-phosphate recognition signal are associated with Golgi membrane fractions (Pohlmann *et al.*, 1982). The MP-R itself is a glycoprotein with complex carbohydrates. Finally, lysosomal enzymes also contain the complex or hybrid type of oligosaccharides that are exclusive products of the trans-Golgi complex. Indeed, Willingham *et al.* (1983) were the first to show immunocytochemically the presence of MP-R in the Golgi complexes, in particular the trans-Golgi membranes in Chinese hamster ovary cells. Recently, however, Brown and Farquhar (1984) found in a number of tissues, including rat liver, that MP-R was restricted to one or two cisternae on the cis side of the Golgi complex; MP-R was absent from trans-Golgi elements.

With immunogold multiple-labeling experiments of MP-R together with secretory proteins, lysosomal enzymes, or galactosyl transferase, we have recently investigated this issue in rat liver. We were able to

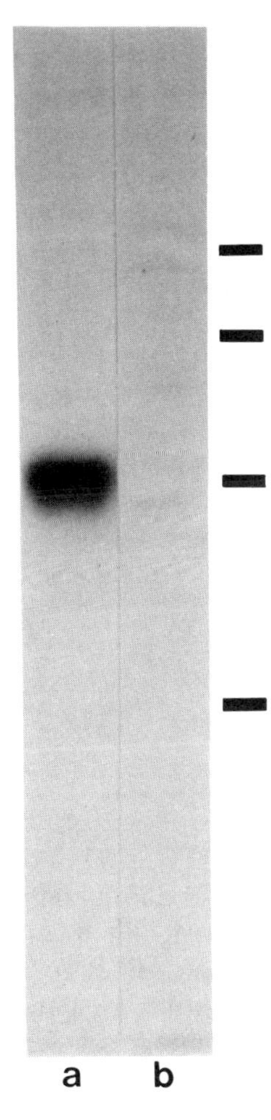

Figure 3. Effect of primaquine on cell surface ASGP-R in Hep G2 cells. The cells were incubated for 30 min at 37°C in media in the absence (lane a) or presence (lane b) of 200 μM primaquine. Following rinsing in PBS at 4°C, the cells were labeled with ^{125}I at 4°C and solubilized. Total cell extract was immunoprecipitated with anti-ASGP-R and analyzed by SDS-PAGE. Molecular mass markers indicate, top to bottom, 93, 68, 46, and 30 kD. Bar, 0.1 μm.

demonstrate MP-R in the plasma membrane, coated pits and vesicles, CURL (Fig. 4), and Golgi complex. The MP-R in the Golgi complex was clearly associated with cathepsin D (Geuze *et al.*, 1984a). These observations, together with the available biochemical data, were suggestive of a cis- to trans-Golgi itinerary of lysosomal enzymes and MP-R. In Hep G2 cells, on the other hand, the Golgi complex contains only little MP-R, similar to that seen with ASGP-R. Double labeling with galactosyl transferase unequivocally demonstrated the MP-R of Hep G2 cells to be present in TGR.

When lysosomotropic agents were administered to Hep G2 cells, MP-R similar to that observed for ASGP-R appeared to accumulate in CURL

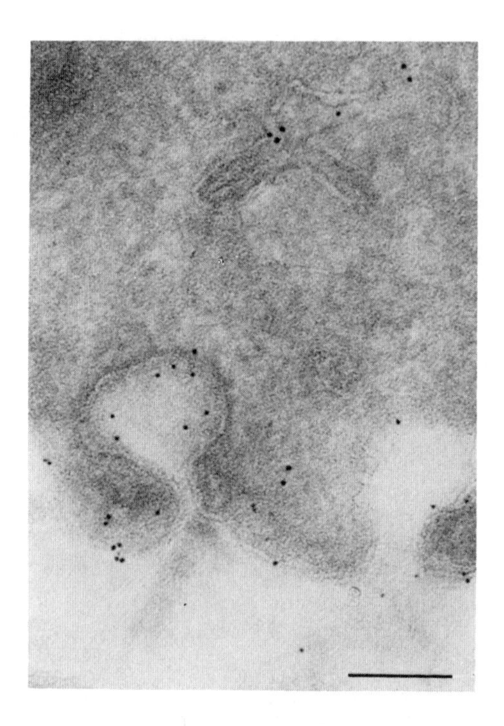

Figure 4. Electron micrograph of ultrathin cryosection from rat liver showing a coated profile of CURL labeled for MPR.

and TGR. This is consistent with the low-pH-dependent uncoupling of MP-R and its ligands (i.e., lysosomal enzymes), which appears to occur in an acidified compartment. Neutralization of the pH prevents the return of the MP-R to the cell surface (Gonzalez-Noriega *et al.*, 1980). It is tempting to speculate that CURL and TGR are involved in the MP-R-mediated transfer of lysosomal enzymes. Nevertheless, CURL tubules contain the bulk of intracellular MP-R.

3.3. The Polymeric IgA Receptor

The IgA-R provides for the directed translocation of IgA from blood and interstitial fluids across epithelial cells. The IgA transfer thus proceeds via RME; however, unlike ASGP-R and MP-R, the IgA-R does not recycle. The major portion of the receptor (the extramembraneous domain) is cleaved from the membrane on or prior to arrival at the cell surface opposite the site of endocytosis. This cleaved portion of IgA-R (the secretory component, SC) together with the covalently linked IgA-ligand is then secreted while the cytoplasmic and transmembrane portions of IgA-R remain cell associated. The intracellular compartments involved in this transcytosis pathway are largely unknown.

Preliminary immunocytochemical observations on the RME of IgA in rat and human intestinal epithelial cells showed that at the cell surface IgA-R was mainly confined to the basal and lateral plasma membrane. The IgA uptake occurred mainly at the lateral cell surface by folds of the plasma membrane rather than by coated pits. Intracellularly, IgA and receptor were present in CURL-like tubules with detaching IgA-transporting vesicles. These probably fuse with the apical plasma membrane for IgA-SC delivery into the intestinal lumen.

Rat liver cells transport pIgA from the basolateral (sinusoidal) cell surface to the apical cell side at the bile canaliculus. In distinction to that seen in the intestinal cells, however, the RME of IgA in hepatocytes involves clathrin-coated pits and vesicles (Geuze *et al.*, 1984b). Following internalization, IgA-R–ligand complexes were next identified in peripheral CURL. Detaching CURL vesicles reminiscent of those in intestinal cells, containing membrane-attached IgA and IgA-R, were found in the subsinusoidal cytoplasm close to the lateral cell side and in the Golgi areas adjacent to the bile canaliculi. Thus, it appears that CURL is involved in the transcytosis of IgA. As in intestinal cells, the Golgi cisternae do not seem not to play an important role in IgA translocation.

3.4. Colocalization of ASGP-R, MP-R, and IgA-R

Using our immunogold double-labeling method, we next directly compared the endocytotic pathways of ASGP-R, MP-R, and IgA-R and their ligands in rat liver parenchymal cells (Hep G2 cells do not possess IgA-R) (Geuze *et al.*, 1984b). All three receptors were found in the Golgi complex, along the entire plasma membrane in coated pits and vesicles, and in CURL. At the cell surface the receptors were randomly distributed, as they were at both smooth and clathrin-coated membrane domains. We also found ASGP ligand and IgA in the same pits, indicating that in liver cells coated pits do not segregate ligands destined for lysosomal degradation from ligands that bypass the lysosomes. Quantitation of label for the three receptors was consistent with nonselective internalization (Table II). We concluded that the smooth-surfaced sinusoidal plasma membrane and the coated membrane are quantitatively similar in respect to the composition of these three receptors. However, coated pits and vesicles may contain greater receptor density, i.e., molecules per unit length of membrane, than the plasma membrane.

In CURL, ASGP-R and MP-R had almost identical distributions. As discussed above, we were unable to demonstrate lysosomal enzymes in association with MP-R in liver coated vesicles or CURL. However, observations on Hep G2 cells have indeed shown a major lysosomal enzyme,

Table II. Percentage of Coated Vesicles Containing Receptor Label in Rat Liver Cells[a]

Receptor	Percentage
MP-R	52
ASGP-R	84
MP-R + ASGP-R	37
No. of vesicles counted	179
IgA-R	45
ASGP-R	79
IgA-R + ASGP-R	35
No. of vesicles counted	251

[a] The table displays the enumerated data of two double-immunolabeling experiments for the simultaneous demonstration of MP-R and ASGP-R and of IgA-R and ASGP-R. Sections from two tissue blocks of each rat were studied. Note that the categories indicated as MP-R, ASGP-R, and IgA-R refer to all vesicles showing label for the appropriate receptor and thus include vesicles containing mixed label.

cathepsin D, present in peripheral CURL as well as in TGR (Geuze *et al.*, 1985).

Although also localized to CURL, IgA-R and its ligand were essentially completely separated from ASGP-R and MP-R. The CURL appeared to contain IgA-R-enriched microdomains. Internalized IgA and IgA-R were found to be transferred from the tubules to detaching vesicles. Thus, CURL is clearly involved in receptor sorting (e.g., IgA-R from ASGP-R and MP-R).

As already mentioned, ASGP-R and MP-R have long lifetimes relative to their ligands and perform multiple rounds of RME (e.g., mean 100–300 cycles), whereas IgA-R has a short intracellular life, ending with the sacrifice of the receptor together with the ligand (one transit). The recycling route of the first two receptors is entirely unknown. As discussed above, all three receptors are present in the Golgi complex. In addition, all three receptors are glycoproteins that pass through the Golgi during biosynthesis and maturation. Thus, in order to address the possible nature of the receptors in Golgi complexes, we had to distinguish between possible recycling receptors and those in transit from their site of synthesis to the plasma membrane. Since inhibition of protein synthesis does not interfere with transport kinetics, we used cycloheximide to deplete newly synthesized receptors from the cells (Geuze *et al.*, 1984c). Using quantitative immunogold double labeling and morphometry in rat liver parenchymal cells, we found that 2 hr and 4 hr after *in vivo* administration of cycloheximide, the densities of ASGP-R and MP-R in the membranes of the Golgi complex were unaltered as compared with untreated liver. Similarly, no effect was found on receptor occurrence in coated pits, coated vesicles, and CURL. As expected, other classes of proteins such as secretory proteins (e.g., albumin), structural cells surface enzymes (e.g., 5′-nucleotidase), and transcytosed ligand (e.g., IgA) and receptor (e.g., IgA-R) were not present in Golgi complexes or other intracellular compartments following treatment with cycloheximide. These observations are consistent with an involvement of the Golgi complex and CURL pools of ASGP-R and MP-R in receptor trafficking and recycling.

Surprisingly, the Golgi complex of Hep G2 cells is probably not involved in receptor recycling to a major degree. The proportion of the Golgi receptor pool in Hep G2 cells (a few percent) is consistent with that expected solely from those receptor molecules in synthetic transit through the Golgi (transit time 1 hr from RER to cell surface) when one considers the slow turnover (lifetime 80 hr) of ASGP-R and MP-R.

4. Concluding Remarks

The CURL, as defined in liver and hepatoma cells, is a tubulovesicular organelle with an acidic internal milieu. The CURL functions as a prelysosomal sorting compartment for receptors for both endogenous and exogenous ligands. We have shown that CURL not only sorts receptors from their ligands via low-pH-induced uncoupling but also segregates recycling receptor species from those transcytosed to another cell surface domain. We believe that CURL functions as a major sorting compartment within the cell, a function fulfilled in the para-Golgi membranes between the Golgi stack and the plasma membrane.

ACKNOWLEDGMENTS. These studies were supported in part by grants from the Koningin Wilhelmina Funds, the Netherlands, the National Foundation, the National Institutes of Health (GM 32477), NATO (818183), and NSF (USA) (INT-8317418).

References

Ashwell, G., and Harford, J., 1982, Carbohydrate-specific receptors of the liver, *Annu. Rev. Biochem.* **51:**531–554.

Brown, M. S., Anderson, R. G. W., and Goldstein, J. L., 1983, Recycling receptors: The round-trip itinerary of migrant membrane proteins, *Cell* **32:**663–667.

Brown, W. J., and Farquhar, M. G., 1984, The mannose-6-phosphate receptor for lysosomal enzymes is concentrated in cis Golgi cisternae, *Cell* **36:**295–307.

Carpenter, G., 1984, Properties of the receptor for epidermal growth factor, *Cell* **37:**357–358.

Ciechanover, A., Schwartz, A. L., and Lodish, H. F., 1983, Sorting and recycling of cell surface receptors and endocytosed ligands: The asialoglycoprotein and transferrin receptors, *J. Cell. Biochem.* **23:**107–130.

Deschuyteneer, M., Prieels, J. P., and Mosselmans, R., 1984, Galactose-specific adsorptive endocytosis: An ultrastructural qualitative and quantitative study in cultured rat hepatocytes, *Biol. Cell* **50:**17–30.

Dunn, W. A., and Hubbard, A. L., 1984, Receptor-mediated endocytosis of epidermal growth factor by hepatocytes in the perfused rat liver: Ligand and receptor dynamics, *J. Cell Biol.* **98:**2148–2159.

Geuze, H. J., and Slot, J. W., 1980, The subcellular localization of immunoglobulin in mouse plasma cells, as studied with immunoferritin cytochemistry on ultrathin frozen sections, *Am. J. Anat.* **158:**161–169.

Geuze, H. J., Slot, J. W., van der Ley, P. A., and Scheffer, R. C. T., 1981, Use of colloidal gold particles in double-labeling immunoelectron microscopy of ultrathin frozen sections, *J. Cell Biol.* **89:**653–655.

Geuze, H. J., Slot, J. W., Strous, G. J. A. M., Lodish, H. F., and Schwartz, A. L., 1982, Immunocytochemical localization of the receptor for asialoglycoprotein in rat liver cells, *J. Cell Biol.* **92:**865–870.

Geuze, H. J., Slot, J. W., Strous, G. J. A. M., Lodish, H. F., and Schwartz, A. L., 1983a, Intra-cellular site of asialoglycoprotein receptor–ligand uncoupling: Double-label immunoelectron microscopy during receptor-mediated endocytosis, *Cell* **32:**277–287.

Geuze, H. J., Slot, J. W., Strous, G. J. A. M., and Schwartz, A. L., 1983b, The pathway of the asialoglycoprotein-ligand during receptor-mediated endocytosis: A morphological study with colloidal gold/ligand in the human hepatoma cell line, Hep G2, *Eur. J. Cell Biol.* **32:**38–44.

Geuze, H. J., Slot, J. W., Strous, G. J. A. M., Hasilik, A., and von Figura, K., 1984a, Ultrastructural localization of the mannose 6-phosphate receptor in rat liver, *J. Cell Biol.* **98:**2047–2054.

Geuze, H. J., Slot, J. W., Strous, G. J. A. M., Peppard, J., von Figura, K., Hasilik, A., and Schwartz, A. L., 1984b, Intracellular receptor sorting during endocytosis: Comparative immunoelectron microscopy of multiple receptors in rat liver, *Cell* **37:**195–204.

Geuze, H. J., Slot, J. W., Strous, G. J., Luzio, J. P., and Schwartz, A. L., 1984c, A cycloheximide-resistant pool of receptors for asialoglycoproteins and mannose 6-phosphate residues in the Golgi complex of hepatocytes, *EMBO J.* **3:**2677–2685.

Geuze, H. J., Slot, J. W., Strous, G. J. A. M., Hasilik, A., and von Figura, K., 1985, Possible pathways for lysosomal enzyme delivery, *J. Cell Biol.* **101:**2253–2262.

Gonzales-Noriega, A., Grubb, J. H., Talkad, V., and Sly, W. S., 1980, Chloroquine inhibits lysosomal enzyme pinocytosis and enhances lysosomal enzyme secretion by impairing receptor recycling, *J. Cell Biol.* **85:**839–852.

Helenius, A., Mellman, I., Wall, D., and Hubbard, A., 1983, Endosomes, *Trends Biochem. Soc.* **8:**245–250.

Kuhn, L. C., and Kraehenbuhl, J.-P., 1982, The sacrificial receptor translocation of polymeric TgA across epithelia, *Trends Biochem. Sci.* **7:**299–302.

Pastan, T., and Willingham, M. C., 1983, Receptor-mediated endocytosis: Coated pits, receptosomes and the Golgi, *Trends Biochem. Sci.* **7:**250–254.

Petersen, O. W., and van Deurs, B., 1983, Serial-section analysis of coated pits and vesicles involved in adsorptive pinocytosis in cultured fibroblasts, *J. Cell Biol.* **96**:277–281.

Pohlmann, R., Waheed, A., Hasilik, A., and von Figura, K., 1982, Synthesis of phosphorylated recognition marker in lysosomal enzymes is located in the cis part of Golgi apparatus, *J. Biol. Chem.* **257**:5323–5325.

Schwartz, A. L., 1984, The hepatic asialoglycoprotein receptor, *CRC Crit. Rev. Biochem.* **16**:207–233.

Schwartz, A. L., Marshak-Rothstein, A., Rup, D., and Lodish, H., 1981, Identification and quantification of the rat hepatocyte asialoglycoprotein receptor, *Proc. Natl. Acad. Sci. U.S.A.* **78**:3348–3352.

Schwartz, A. L., Bolognesi, A., and Fridovich, S. E., 1984, Recycling of the asialoglycoprotein receptor and the effect of lysosomotropic amines in hepatoma cells, *J. Cell Biol.* **98**:732–738.

Schwartz, A. L., Strous, G. J. A. M., Slot, J. W., and Geuze, H. J., 1985, Immunoelectron microscopic localization of acidic intracellular compartments in hepatoma cells, *EMBO J.* **4**:899–904.

Slot, J. W., and Geuze, H. J., 1981, Sizing of protein-A colloidal gold probes for immunoelectron microscopy, *J. Cell Biol.* **90**:533–536.

Slot, J. W., and Geuze, H. J., 1983, The use of protein A–colloidal gold (PAG) complexes as immunolabels in ultra-thin frozen sections, in: *Immunohistochemistry* (A. C. Cuello, ed.), IBRO, John Wiley & Sons, Chichester, New York, pp. 323–346.

Slot, J. W., and Geuze, H. J., 1985, A new method of preparing gold probes for multiple-labeling cytochemistry, *Eur. J. Cell Biol.* **38**:87–93.

Sly, W. S., and Fischer, H. D., 1982, The phosphomannosyl recognition system for intracellular and intercellular transport of lysosomal enzymes, *J. Cell. Biochem.* **18**:67–85.

Tokuyasu, K. T., 1978, A study of positive staining of ultrathin frozen sections, *J. Ultrastruct. Res.* **63**:287–307.

Tycko, B., and Maxfield, F. R., 1982, Rapid acidification of endocytotic vesicles containing alpha-2-macroglobulin, *Cell* **28**:643–651.

Wall, D. A., Wilson, G., and Hubbard, A. L., 1980, The galactose-specific recognition system of mammalian liver: The route of ligand internalization in rat hepatocytes, *Cell* **21**:79–93.

Willingham, M. C., Pastan, I., and Sahagian, G. G., 1983, Ultrastructural immunocytochemical localization of the phosphomannosyl receptor in Chinese hamster ovary (CHO) cells, *J. Histochem. Cytochem.* **31**:1–11.

20

Mechanisms of Molecular Sorting in Endosomes

IRA MELLMAN

1. Introduction

Since the time of Metchnikoff over a century ago, it has been clear that eucaryotic cells interact with their environment by eating some of it via endocytosis. As long as 50 years ago, it was suspected that endocytosis brought about the internalization of enormous amounts of extracellular material and, consequently, of a cell's plasma membrane (Lewis, 1931). Only recently, however, has it been documented that endocytosis is also characterized by a high degree of specificity, both in the initial recognition events leading to the selective uptake of certain extracellular macromolecules and in the subsequent intracellular events that control the fate of internalized membrane, fluid, and solutes.

A wide variety of extracellular macromolecules (including many nutrients, hormones, antibodies, lysosomal enzymes, viruses, toxins, and modified serum glycoproteins) are known to bind to receptors on the mammalian cell plasma membrane (Steinman *et al.*, 1983). In many cases, binding is followed rapidly by endocytosis, which in turn results in the ligand's delivery to and degradation in lysosomes. In contrast, internalized receptors—as well as most other internalized plasma membrane components—usually escape lysosomal degradation and return to the plasma membrane to participate in additional rounds of ligand uptake (Steinman *et al.*, 1983; Brown *et al.*, 1983). The fact that ligands accumulate within the cell while their receptors can continuously recycle between the cell surface and the cytoplasm indicates that endocytosis must involve a series

IRA MELLMAN • Department of Cell Biology, Yale University School of Medicine, New Haven, Connecticut 06510.

of highly regulated events that permit the "sorting" of ligands and receptors. At a minimum, these sorting events must include the dissociation of the receptor–ligand complex, the unidirectional transfer of ligands to lysosomes, and the selective return of free receptors to the plasma membrane. The mechanisms that mediate these processes must be highly efficient, given the magnitude of intracellular membrane traffic that is generated by a cell's ongoing endocytic activity. The quantitative measurements of Steinman *et al.* (1976) have shown that a mammalian cell can continuously internalize as endocytic vesicles an amount of membrane equivalent to twice its entire cell surface area every hour. Individual receptors, such as receptors for low-density lipoproteins (LDL) or asialoglycoproteins, can be internalized and recycled as many as 10–20 times per hour (Steinman *et al.*, 1983; Brown *et al.*, 1983).

The molecular mechanisms that govern the intracellular transport of membrane, ligands, and receptors internalized during endocytosis are only now beginning to be understood and are a major focus of investigation in our laboratory. Nevertheless, a general consensus has been reached within the past 2 to 3 years regarding the basic features of the endocytic pathway (Brown *et al.*, 1983; Hopkins, 1983; Helenius *et al.*, 1983). According to this scheme, receptor–ligand complexes form on the plasma membrane and accumulate at clathrin-containing coated pits. Coated pits are thought by most investigators to pinch off to form coated vesicles in the cytoplasm (for the opposing view, see Pastan and Willingham, 1981). Soon after their formation (i.e., within 15–60 sec), these vesicles lose their coats, an event that may be caused by the activity of an ATP-dependent uncoating enzyme (Braell *et al.*, 1984). Receptors and ligands then appear in a heterogeneous compartment of largely smooth-surfaced vesicles and tubules referred to here collectively as endosomes but also known as CURL, receptosomes, intermediate vesicles, pinosomes, etc. (Helenius *et al.*, 1983; Hopkins, 1983). Endosomes have a slightly acidic internal pH of between 5 and 6.5. (Tycko and Maxfield, 1982; Marsh *et al.*, 1983) and thus provide an environment that favors the dissociation of many ligands from their respective receptors (Helenius *et al.*, 1983; Harford *et al.*, 1983). The newly vacated receptors are then transported directly (via membrane recycling) from endosomes back to the plasma membrane, and the discharged ligands, now free in the endosome lumen, are transferred to lysosomes. Endosomes thus greatly increase the efficiency of receptor-mediated endocytosis by providing an intracellular site for ligand discharge and receptor recycling, thus avoiding the necessity for transport through acidic but hydrolase-rich lysosomes.

Receptor–ligand systems that typify this scheme include those for LDL, asialoglycoproteins, mannose-6-phosphate-containing glycoproteins (lysosomal enzymes), and glycoproteins containing terminal mannose groups (Helenius *et al.*, 1983).

Important variations on this theme also exist. For example, a number of receptors fail to recycle during ligand uptake and are instead transported to lysosomes and degraded after delivery to endosomes. This pathway is characteristic of receptors for the Fc domain of IgG (Fc receptors) (Mellman *et al.*, 1983; Mellman and Plutner, 1984) and for polypeptide hormones such as epidermal growth factor, insulin, and human choriogonadotropin (Kasuga *et al.*, 1981; Beguinot *et al.*, 1984; Ascoli, 1984) and accounts for the irreversible loss of surface receptor activity that occurs following exposure of cells to certain ligands (classically referred to as "down-regulation"). Selective loss of a receptor would render a cell refractory to subsequent exposure to a particular ligand, providing a mechanism by which cells can regulate their response to the environment.

It should be clear from the above discussion that endosomes play an important role in maintaining the orderly traffic of ligands and receptors into and out of the cytoplasm. Consequently, there is considerable interest in the structural and functional properties of endosomes. In the following sections, I review some recent work from my laboratory and from that of my departmental colleague Ari Helenius, which has contributed to our understanding of these important organelles and particularly to the basis of the molecular sorting events they accomplish. I discuss our findings pertaining to the morphology and acidification of endosomes and to endosome isolation and biochemical characterization. Our studies of Fc-receptor-mediated endocytosis in macrophages is also considered, since they provide some insight into the possible mechanisms that control the transport of internalized receptors: the intracellular fate of Fc receptors—either transport from endosomes to lysosomes or from endosomes back to the plasma membrane—can be modulated by the valency of the ligand bound. Finally, I discuss the problem of the structure and biogenesis of the lysosomal membrane, the intracellular site that most membrane proteins manage to avoid by recycling from endosomes.

2. Endosome Ontogeny and Morphology

2.1. The Life History of Endosomes

Although endosomes serve as obligatory intermediates in the transport of ligands (and some receptors) to lysosomes, the way in which endosomes perform this function has yet to be clarified. Endosomes may be transient organelles that arise as a result of the concerted fusion of incoming coated vesicles and have a finite lifetime, during which they participate in membrane recycling. According to this view, an endosome's life ends when its fusion lysosomes deliver endosomal membrane and contents to a hydrolytic compartment. Alternatively, endosomes may be continuously transformed into lysosomes by repeated fusion with Golgi-derived vesicles (i.e., primary lysosomes) containing newly synthesized lysosomal enzymes.

Either of these views is supported by the time-lapse photomicroscopic evidence of Hirsch, Cohn, and others (e.g., Hirsch *et al.*, 1968). These data imply that endosomes form in the peripheral cytoplasm, where they exhibit random saltatory motion for a period of 5–30 min and then migrate rapidly into the perinuclear region of the cell, where they appear to fuse with dense lysosomes. On the other hand, one cannot yet eliminate the possibility that endosomes are stable structures that communicate with the plasma membrane, among themselves, and with lysosomes via specialized transport or "shuttle" vesicles. Such a model is analogous to the situation believed to exist for membrane transport between individual cisternae in the Golgi complex (Helenius *et al.*, 1983).

The problem of endosome ontogeny is of interest in view of its implications for understanding the sorting functions of endosomes. A solution to this problem will require detailed information about endosome morphology and biochemistry.

2.2. Three-Dimensional Structure of Endosomes and Lysosomes

Since endosomes are not yet known to have any unique distinguishing characteristics, they can be identified by electron microscopy or by subcellular fractionation only after being tagged by a suitable endocytic tracer. This is usually accomplished by incubating cells for short times (less than 15 min) at 37°C with a specific ligand or fluid-phase marker (e.g., horseradish peroxidase, HRP). Times of incubation are adjusted to minimize transfer of the internalized marker to hydrolase-containing lysosomes. Alternatively, cells can be exposed to the tracer at an intermediate temperature (e.g., 20°C or less), under which conditions transfer of material from endosomes to lysosomes is often blocked (Dunn *et al.*, 1980).

Little is known about the structure of endosomes. Thus, as mentioned above, they are usually described as a heterogeneous collection of vesicles and tubules (Helenius *et al.*, 1983). Various investigators have proposed that the tubular elements extensively interconnect individual vesicles among themselves (Hopkins and Trowbridge, 1983) or with Golgi cisternae (Pastan and Willingham, 1981; Willingham *et al.*, 1984). We have

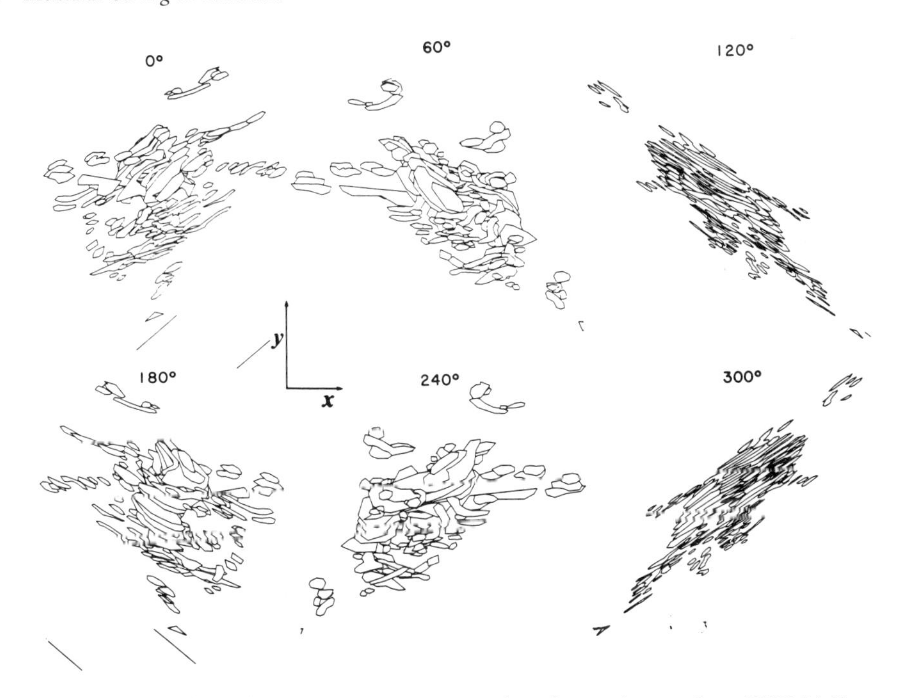

Figure 1. Three-dimensional computer reconstruction of an endosome from BHK-21 fibroblasts. Endosomes visualized in serial thin sections of HRP-labeled BHK-21 cells were entered into the CARTOS graphics system and displayed on an Evans and Sutherland graphic display monitor. Shown here is a single endosome rotated 360° about its *Y*-axis. Both the tubular and vesicular components of the endosome are clearly visible. See text for details.

recently studied the three-dimensional structure of endosomes and their spatial relationships with other organelles, using an advanced computer graphics system (CARTOS) developed by S. Leventhal at Columbia University (Marsh *et al.*, 1986). Serial sections were prepared from baby hamster kidney (BHK-21) fibroblasts whose endosomes had been labeled with HRP. Electron micrographs were taken, and organelles in aligned images were traced into the CARTOS program using a light pencil. Reconstructed images were displayed on an Evans and Sutherland video display monitor, where they could be rotated in three dimensions and quantitated (Fig. 1).

Our results indicate that endosomes in BHK-21 cells exist as discrete entities and consist of a central vesicular portion (0.2–1.0 μm in diameter) from which seven to eight narrow tubules radiate in all directions. Numerous small vesicles often surround the larger structures. No evidence was obtained for direct connections between endosomes. Moreover, although HRP-labeled endosomes could be found in close proximity to Golgi complexes, no connections between these two organelles were observed.

Quantitative measurements indicated that this rather unique arrangement results in a clear segregation of membrane area and internal volume in endosomes. Although up to 70% of an endosome's membrane could be located in the tubular extensions, most (65–70%) of its volume was contained within the vesicular region. Thus, it is tempting to speculate that the tubular elements, having a relatively high surface-to-volume ratio, may be involved in the recycling phase of the endocytic cycle. Such tubules would mediate the preferential transport of membrane as opposed to contents, thus accomplishing by simple geometry one of the major sorting events mentioned above. Previous studies, such as those of Geuze and colleagues (see Chapter 19, this volume), have implicated tubular elements in the recycling of asialoglycoprotein and transferrin receptors (Geuze *et al.*, 1983, 1984).

On the other hand, the endosome's vesicular portion, having a relatively low surface-to-volume ratio, appears geometrically well suited for transport of dissociated ligands and internalized solutes to lysosomes. If endosomes eventually fuse directly with (or are converted into) lysosomes (Helenius *et al.*, 1983), the delivery of endosomal contents could proceed with a minimum sacrifice of endosomal membrane.

The three-dimensional morphology of BHK-21 lysosomes was also studied using the same approach. Interestingly, lysosomes were found to be much more regularly shaped and were often spherical. The lysosomal membrane was not characterized by tubular extensions. Again, one can speculate that the absence of tubules reflects the lysosome's relative noninvolvement in the recycling of internalized membrane (Helenius *et al.*, 1983; Steinman *et al.*, 1983; Besterman *et al.*, 1981; Swanson *et al.*, 1985).

3. Acidification of Endosomes

The only known functional property of endosomes is their capacity for ATP-dependent proton translocation. That endosomes have an acidic internal pH was first suggested by Tycko and Maxfield (1982) and by Marsh *et al.* (1983). Tycko and Maxfield found that various receptor-bound ligands conjugated to the pH-sensitive fluorochrome fluorescein isothiocyanate (FITC) reached a slightly acidic compartment soon after internalization. Insofar as these ligands were shown not to have entered lysosomes, it was concluded that a prelysosomal low-pH compartment must exist. Marsh and co-workers used a substantially different approach, again using intact cells, to arrive at the same result. They relied on the fact that enveloped viruses, such as Semliki Forest virus, penetrate into the cytoplasm after internalization and delivery to a compartment with a pH less than 6.2. Under these conditions, the viral envelope fuses with the endocytic vesicle membrane, allowing the viral nucleocapsid to gain access to the cytosol. By several criteria, this penetration event was found to occur well before the virus was delivered to dense, hydrolase-rich lysosomes. Indeed, efficient virus penetration occurred even under conditions that prevented transfer of endosomal contents to lysosomes (e.g., incubation at 20°C).

3.1. The Endosome H^+ ATPase

We studied the mechanism of endosome acidification *in vitro* using partially purified fractions of endosomes isolated from tissue culture cells and rat liver (Galloway *et al.*, 1983; R. Fuchs and I. Mellman, unpublished results). Endosomes were selectively marked by endocytosis of FITC-conjugated dextran prior to cell homogenization. Our findings indicated that on addition of Mg^{2+}-ATP, isolated endosomes rapidly lowered their internal pH, as indicated by the pH-dependent quenching of the vesicles' FITC fluorescence yield. As expected, the decrease in fluorescence was reversed on addition of proton ionophores (FCCP, nigericin) or lipophilic weak bases (NH_4Cl).

Preliminary characterization of the acidification mechanism of endosomes has suggested that proton translocation is not dependent on the presence of any particular external cation, implying that coupled ion cotransport is not necessary. At least a partial dependence on the presence of a permeant external anion was noted, however. Acidification was not inhibited by classic inhibitors (e.g., vanadate, ouabain) of the class of phosphoenzyme intermediate-type of ion pumping ATPases such as Na^+–K^+ ATPase. This characteristic distinguishes the endosome ATPase from the K^+–H^+ ATPase of the gastric mucosa and the H^+ ATPase of the fungal plasma membrane, both of which are sensitive to inhibition by vanadate (Rudnick, 1985). Similarly, the endosome ATPase can be differentiated from enzymes belonging to the F_1F_0 class of ATPases (e.g., the mitochondrial H^+ ATPase), since endosome acidification was found to be insensitive to potent F_1F_0 inhibitors such as sodium azide, efrapeptin, and oligomycin. The only effective inhibitors of endosome acidification identified thus far are the relatively nonselective alkylating agents N-ethylmaleimide (NEM) and NBD-chloride (Galloway *et al.*, 1983; Robbins *et al.*, 1984). However, these agents are effective at much lower concentrations than are usually required to inhibit Na^+–K^+-ATPase or F_1F_0-ATPase activities. Although the F_1F_0-ATPase inhibitor

DCCD was also found to inhibit endosome acidification, it was effective at concentrations much higher than those required to block the mitochondrial proton pump (Galloway *et al.*, 1983).

3.2. Proton Translocation in Other Endocytic and Secretory Organelles

As discussed in detail elsewhere in this volume (see Chapter 4 by Harikumar and Reeves) we and others have found that essentially the same characteristics also apply to the lysosomal H^+ ATPase (Galloway *et al.*, 1983; Ohkuma *et al.*, 1982). Thus, as might be expected from the functional relationships between endosomes and lysosomes, it is conceivable that the same ATPase is responsible for proton transport into both organelles. More striking, however, is the fact that acidification in virtually all other endocytic and secretory organelles capable of proton translocation also appears to be caused by an endosomelike H^+ ATPase (Rudnick, 1985). Similar ion requirements and inhibitor specificities have been demonstrated for the acidification of coated vesicle fractions from brain and liver (Forgac *et al.*, 1983; Stone *et al.*, 1983), rat liver Golgi and endoplasmic reticulum fractions (Glickman *et al.*, 1983; Zhang and Schneider, 1983), various acidic endocrine secretory granules (e.g., adrenal chromaffin granules, neurosecretory granules, platelet dense granules), and isolated yeast vacuoles (Rudnick, 1985). Such findings suggest the possibility that all endocytic and secretory pathways are acidified by means of a unique class of—if not the same—H^+ ATPase.

Direct comparison of the proton pumps present in different acidic organelles will ultimately require the identification, purification, and characterization of the H^+ ATPases from a variety of endocytic and secretory vesicles. To date, the only published report of a purified, functional H^+ ATPase is from the yeast digestive vacuole, which is believed to consist of at least three polypeptides with molecular weights of 89,000, 64,000, and 19,500 (Uchida *et al.*, 1985). Several laboratories are actively pursuing the purification of proton pumps from relatively accessible mammalian sources such as bovine adrenal chromaffin granules and bovine brain coated vesicles. The isolation of H^+ ATPases from endocytic vesicles, purified fractions of which are far more difficult to prepare, will prove to be a far more difficult task. However, given our recent progress in the purification of endosomal membranes by free-flow electrophoresis (see Section 5.1), we hope to be able to apply isolation procedures developed for the ATPases of other organelles (as they become available) to the isolation of the endosome proton pump.

3.3. Acidification-Defective Mutant Cell Lines

An additional approach to understanding the molecular mechanisms of intravesicular acidification relies on the use of somatic cell genetics. In collaboration with April Robbins at the National Institutes of Health, we have analyzed a series of mutant Chinese hamster ovary (CHO) cell lines that were found to exhibit pleiotropic defects in receptor-mediated endocytosis (Robbins *et al.*, 1983). These cell lines were originally selected for resistance to killing by bacterial toxins (e.g., diphtheria toxin) and by enveloped RNA viruses (e.g., Sindbis virus), both of which agents require acidic pH in endosomes to penetrate into the cytosol to exert their cytotoxic effects. Many of the resistant cell lines were also deficient in the uptake of lysosomal enzymes via the mannose-6-phosphate receptor and were unable to retain newly synthesized lysosomal enzymes intracellularly (Robbins *et al.*, 1983). This phenotype was similar to that exhibited by wild-type CHO cells treated with lipophilic weak bases such as ammonium chloride, suggesting a possible defect in endosome and/or lysosome acidification.

With the FITC-dextran assay, isolated endosomes from several of these mutant cell lines were shown to be either partially or totally defective in ATP-dependent acidification (Robbins *et al.*, 1984). Interestingly, the *in vitro* acidification of lysosomes was apparently not impaired. More recently, we have extended these findings to a series of temperature-sensitive CHO cell lines, which exhibit the wild-type phenotype at the permissive temperature (34°C) and the mutant phenotype at temperatures above 39°C (R. Fuchs, A. Robbins, and I. Mellman, unpublished results). Incubation of these conditional lethals at the nonpermissive temperature for as little as 2–4 hr resulted in the coordinate appearance of the mutant acidification phenotype, toxin resistance, and defects in endocytosis and lysosomal enzyme secretion.

Complementation analysis of the mutant cell lines revealed that endosome acidification in CHO cells is controlled by at least two genes. Thus far, the analysis of somatic cell hybrids has defined two distinct complementation groups, designated *end-1* and *end-2* (Robbins *et al.*, 1984). Complementing hybrids exhibited the wild-type phenotype with respect to ATP-dependent acidification of endosomes *in vitro*, toxin resistance, and lysosomal enzyme uptake and secretion. Importantly, we were also able to isolate revertant cell lines exhibiting the wild-type phenotype, indicating that the pleiotropic defects characterizing the *end-1* and *end-2* mutations were the result of single genetic events (Robbins *et al.*, 1984).

3.4. Endosome Acidification Mutants Also Exhibit Altered Golgi Function

Surprisingly, in addition to their defects in endocytosis and endosome acidification, cells of both the *end-1* and *end-2* classes expressed defects in the biosynthetic pathway. If infected with high multiplicities of virus, all of the acidification-defective mutant cell lines supported the replication of Sindbis virus RNA and the biosynthesis of the viral spike glycoproteins E1 and E2. However, infected cells failed to release virions into the culture medium (Robbins *et al.* 1984). Examination of E1 and E2 produced by the mutant cells revealed that both spike glycoproteins were incorrectly posttranslationally processed. Proteolytic cleavage of the E2 precursor, pE2, was slowed. In addition, the asparagine-linked oligosaccharides on both E1 and E2 were not terminally glycosylated; neither galactose nor sialic residues were added.

Most of the major cell surface glycoproteins of the CHO cell were glycosylated normally in the mutants, indicating that the aberrant processing of E1 and E2 did not result from a generalized glycosylation defect. Several endogenous secreted glycoproteins, however, were found to lack terminal sialic acid.

The fact that terminal glycosylation and, possibly, the proteolytic processing of pE2 occur in the Golgi apparatus suggests that CHO cells that exhibit a defect in endosome acidification also express a defect in Golgi-related functions. The Golgi defect apparently results from the same single mutation as do the defects in the endocytic pathway, since in complementing *end-1* × *end-2* hybrids, in the revertants, and in the temperature-sensitive mutants, the expression of the glycosylation defect correlates with the expression of the endosome acidification defect (Robbins *et al.*, 1984; R. Fuchs, A. R. Robbins, and I. Mellman, unpublished results).

Accordingly, it would appear reasonable to suggest that, like the endosome acidification defect, the Golgi-associated defect is associated with an alteration in ion transport. In this regard, recent findings that rat liver Golgi fractions contain ATP-dependent proton transport activity are of particular interest (Glickman *et al.*, 1983; Zhang and Schneider, 1983). It is also interesting to note that aberrant processing and glycosylation of Sindbid virus E1 and E2 can be produced in wild-type CHO cells treated with low concentrations of monensin, a carboxylic ionophore that dissipates transmembrane proton, sodium, and potassium gradients (Robbins *et al.*, 1984).

Whether either the *end-1* or *end-2* mutation affects the proton ATPase directly or affects some other permeability characteristic of endosomal and/or Golgi membranes is not yet known. However, it is apparent that the Golgi and endosomal membranes share at least one gene product that is necessary for normal transmembrane ion transport or that the endocytic and biosynthetic pathways involve passage through a common organelle whose ion transport characteristics are necessary for normal function.

4. *The Role of Endosome Acidity and Ligand Valency in Fc Receptor Transport*

As discussed above, the major known function of low endosome pH is to permit the efficient discharge of internalized ligands from their receptors and, consequently, to facilitate receptor recycling. The mechanisms that actually direct the targeting of receptors back to the plasma membrane, as opposed to lysosomes or some other organelle, are less well defined. It is also unclear whether acidic pH in endosomes and/or lysosomes plays a role beyond facilitating receptor–ligand dissociation in receptor recycling or ligand transport to lysosomes.

To address these questions, we have studied the endocytosis and transport of immunoglobulin G (IgG) Fc receptors in mouse macrophages. Our findings indicate that acidic pH in endocytic organelles is not necessarily a controlling factor in directing the intracellular traffic of ligands and receptors. Instead, the valency of the bound ligand, at least in the case of Fc receptors, appears to be far more important in deciding the ultimate fate of internalized receptors.

4.1. *The Structure and Function of the Mouse Macrophage Fc Receptor*

Mouse macrophages and macrophage cell lines express at least three distinct receptors for the Fc domain of IgG (Unkeless *et al.*, 1981). The most abundant of these, particularly on the J774 cell line used in our experiments, is a trypsin-resistant receptor IgGl/IgG2b-containing antibody–antigen complex (hereafter referred to simply as FcR). With a specific rat monoclonal antireceptor antibody (designated 2.4G2), this receptor has been purified to homogeneity and characterized in detail (Unkeless, 1979; Mellman and Unkeless, 1980). Additional polyclonal and monoclonal antibodies have been raised against the purified protein (Mellman *et al.*, 1983; Green *et al.*, 1985).

On J774 cells, the mature FcR is a 60-kD intrinsic membrane glycoprotein that exhibits a somewhat heterogeneous electrophoretic mo-

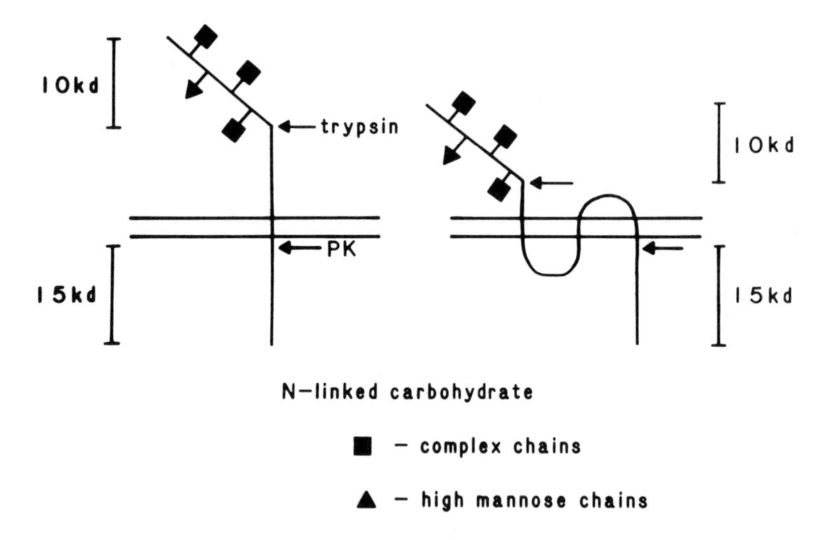

Figure 2. Proposed structures for the mouse macrophage Fc receptor. Based on the biochemical studies of Green *et al.* (1985), two alternatives are proposed. Since at least 15 kD of the receptor is protected from degradation by proteases added from both the cytoplasmic and the extracellular sides of the membrane, the model on the right suggests that the receptor may make more than one pass through the plasma membrane. See text for details.

bility (Mellman and Unkeless, 1980). The receptor can be associated with a second 47-kD band, which appears to be a proteolytic product of the 60-kD species (S.A. Green and I. Mellman, unpublished results). The FcR is synthesized in the rough endoplasmic reticulum as a 53-kD precursor, which is core glycosylated with four N-linked oligosaccharide chains (Green *et al.*, 1985). During processing in the Golgi, at least three of these chains are converted to complex endoglycosidase H-resistant forms. All four oligosaccharide chains appear to be located on a distal 10-kD tryptic peptide of the receptor's ectodomain.

Protease digestions of rough microsomes containing labeled FcR precursor have indicated that the receptor has a large cytoplasmic domain of approximately 15 kD (Green *et al.*, 1985). Consequently, over one-third of the receptor's polypeptide portion (which accounts for 37 kD on SDS gels) may be oriented towards the macrophage cytoplasm. In addition, some 15 kD of the FcR polypeptide is protected from proteolysis from either side of the membrane, suggesting that the receptor may traverse the membrane more than once. Current models for FcR structure based on these biochemical data are illustrated in Fig. 2.

The question of FcR structure is interesting given the receptor's wide array of functional activities. The binding of ligand to the receptor results in the receptor-mediated endocytosis of soluble antibody–antigen complexes via coated pits and coated vesicles (Mellman and Plutner, 1984; Mellman and Ukkonen, 1985; Ukkonen *et al.*, 1985) or the phagocytosis of large 1gG-coated particles (Mellman *et al.*, 1983). In addition, as is discussed below, the pathway of intracellular transport taken by internalized receptors can be modulated as a function of the valency of the bound ligand. A variety of immediate transmembrane events are also signaled by the interaction of FcR with its ligand. These include the localized polymerization of actin-containing microfilaments at the site of particle attachment, the synthesis and release of bioactive lipids (e.g., prostaglandins, leukotrienes), an increase in the rate of oxidative metabolism, and the release of cytotoxic oxygen intermediates and peroxide (Unkeless *et al.*, 1981). The triggering of at least some of these events may be related to recent findings that the receptor exhibits an intrinsic ligand-activated cation channel activity (Young *et al.*, 1983).

4.2. Ligand Valency and Fc Receptor Transport

Several years ago, we found that the endosomal membrane of J774 cells could be radioiodinated *in situ* following the endocytic uptake of lactoperoxidase and glucose oxidase (Mellman *et al.*, 1980; Mellman and Galloway, 1983). Analysis of the major membrane proteins susceptible to labeling demonstrated the presence of FcR among a variety of other internalized proteins. Similar results were obtained for phagosome membrane radioiodinated following the "nonspecific" phagocytosis of lactoperoxidase-conjugated polystyrene latex beads (Muller *et al.*, 1983). Since the receptor was detected on internalized membrane even though the cells had not been exposed to monomeric or complexed IgG, these findings suggest that the Fc receptor was continuously internalized and recycled during the constitutive endocytic activity exhibited by macrophages (and all other cells, for that matter) (Steinman *et al.*, 1983). The continuous nature of this process was supported by the observations that the major iodinatable proteins were found to have turnover rates similar to those of their plasma membrane counterparts and were shown (by electron microscope autoradiography) to be capable of recycling from the phagosome membrane back to the cell surface (Steinman *et al.*, 1983; Muller *et al.*, 1980, 1983; Mellman, 1982).

An entirely different situation occurred in cells exposed to large IgG-coated particles. Following the phagocytosis by macrophages of opsonized sheep erythrocytes or erythrocyte ghosts, a selective and largely irreversible loss of FcR from the plasma membrane was observed (Mellman *et al.*, 1983). The loss was accompanied by a greatly increased rate of FcR turnover, suggesting that receptor-mediated phagocytosis prevented the recycling of the receptor and resulted in its degradation in lysosomes. Thus, it was apparent that the intracellular transport of FcR during endocytosis might be modulated by its ligand.

Conceivably, the failure of FcR to recycle under these conditions resulted from the bypassing during phagocytosis of an organelle, such as the endosome, that is an obligatory intermediate for receptor recycling. Therefore, we also studied the intracellular transport of FcR during the internalization of small soluble ligands taken up by the basic pathway of receptor-mediated endocytosis, as described above. By using monoclonal anti-FcR antibodies, antibody fragments, and soluble IgG–antigen complexes (the receptor's physiological ligand), we have been able to construct FcR-bound ligands of different valencies in order to test a possible role for receptor cross linking in the regulation of receptor recycling. Importantly, each of the ligands used bound to the FcR in a largely pH-insensitive fashion. Accordingly, we were able to dissociate the contributions of ligand valency and endosome acidity in controlling the intracellular transport.

4.2.1. Internalization and Fate of Fc Receptors Bound to Monovalent Probes

Monovalent Fab fragments that recognize the receptor at or near its ligand binding site can be prepared from the rat anti-FcR monoclonal antibody 2.4G2 (Unkeless, 1979; Mellman and Unkeless, 1980). Although the Fab binds with high affinity ($>10^9$ M^{-1}), it can be easily and quantitatively dissociated from surface FcR at 0°C by a brief incubation at pH 4 (Mellman *et al.*, 1984). Little dissociation occurs at pH values that would be encountered in endosomes or lysosomes (i.e., pH 4.8–6.5; Okhuma and Poole, 1978; Tycko and Maxfield, 1982). Thus, the Fab–FcR complex would be expected to remain intact following endocytosis.

Up to 20% of the cell-associated radioactivity was internalized, i.e., became insensitive to removal by incubation of the cells in cold pH 4 buffer following incubation of J774 cells with [^{125}I]-labeled Fab at 37°C (Mellman *et al.*, 1984). However, little Fab degradation was detected, suggesting that the internalized antibody was not delivered to lysosomes. Even after prolonged incubations, most of the intracellular Fab was localized in endosomes. This was indicated by the centrifugation of cell homogenates in Percoll density gradients, which allow the rapid separation of endosomes from the much higher density lysosomes (Merion and Sly, 1983; Galloway *et al.*, 1983; Mellman *et al.*, 1984).

Although internalized Fab–FcR complexes were not transported to lysosomes, they were efficiently returned to the plasma membrane (Mellman *et al.*, 1984). This recycling was demonstrated by incubating cells in labeled Fab at 37°C, removing surface-bound Fab at 0°C by a brief pH 4 treatment, and then warming the acid-stripped cultures to 37°C. At various times after warming, the cells were subjected to a second pH 4 wash, and the amount of [^{125}I]-Fab now accessible to elution was quantitated. With a $t_{1/2}$ of only 1–2 min, most of the intracellular Fab was found to reappear on the cell surface still bound to FcR. Taken together, these observations indicate that monovalent Fab–FcR complexes are internalized, delivered to endosomes, and then rapidly recycled to the plasma membrane.

4.2.2. Pathway of Receptor Bound to Multivalent Ligands

In contrast to the monovalent antireceptor Fab fragments, the FcR-mediated endocytosis of multivalent [^{125}I]-labeled IgG–antigen immune complexes results in the efficient delivery of ligand to lysosomes, as indicated by the release of TCA-soluble ^{125}I, Percoll gradient centrifugation, and electron microscopy (Mellman and Plutner, 1984; Mellman and Ukkonen, 1985; Ukkonen *et al.*, 1985). Delivery to lysosomes and the onset of IgG-complex degradation both occur within 15 min after internalization. More importantly, we have been able to show that the receptor is transported to lysosomes along with its ligand. As in the example of FcR-mediated phagocytosis discussed above (Mellman *et al.*, 1983), the uptake of soluble IgG–antigen complexes is accompanied by the selective loss of FcR from the cell surface as well as by an increased rate of receptor turnover (Mellman and Plutner, 1984). In control cells (or in cells exposed to monovalent Fab fragments), the receptor was found to have a half-life of more than 15 hr, whereas in cells incubated with multivalent antibody–antigen complexes the receptor's half-life decreased to less than 5 hr. In addition, if homogenates of cells exposed to IgG-complexes are centrifuged in Percoll gradients, one can demonstrate a redistribution of immunoprecipitable FcR from the low-density fractions (which contain endosomes and plasma membrane) to the high-density lysosomal fractions (Ukkonen *et al.*, 1985).

Virtually identical results were obtained using preparations of the antireceptor Fab rendered multivalent by adsorption to 5-nm particles of colloidal gold (Ukkonen *et al.*, 1985). Thus, it is apparent that the valency of the bound ligand, as opposed to the presence of Fc domains on intact IgG molecules, is an important determinant in redirecting the transport of internalized FcR to lysosomes.

We have also studied the pathway of FcR transport to lysosomes by electron microscopy, using colloidal gold conjugates of both IgG complexes and the Fab fragment (Mellman and Ukkonen, 1985; Ukkonen *et al.*, 1985). Our data illustrate that FcR-bound ligands are internalized via coated pits and coated vesicles and are then delivered to endosomes. They are initially (less than 5 min) seen as small tubular and vesicular structures in the peripheral cytoplasm and later (5–15 min) as larger, more spherical vesicles often containing multivesicular inclusions. Interestingly, the gold was often associated with the internal membranes as opposed to the limiting membranes of these endosomes. Finally, the gold-labeled ligands appearted in typical electron-dense lysosomes. The presence of ligand in Golgi cisternae was not observed at any stage.

The fact that the endosomal compartment serves as an intermediate in the transport of multivalent ligands to lysosomes as well as in the recycling pathway of monovalent Fab–FcR complexes strongly suggests that endosomes constitute the site from which FcR are sorted (i.e., returned back to the cell surface or transferred to lysosomes). An important unknown, however, is whether the endosomes involved in the FcR recycling pathway are the same as those involved in the pathway to lysosomes.

4.3. The Role of Intravesicular pH in Fc Receptor Transport

The transport to lysosomes of several receptor-bound ligands containing glycoproteins, such as LDL, EGF, and mannose-6-phosphate, was blocked in cells treated with agents that elevate intravesicular pH in endosomes and lysosomes (e.g., Merion and Sly, 1983). Although such data have suggested that endosome–lysosome fusion may depend on transmembrane pH gradients, interpretation is complicated by the fact that in each of these examples the receptor–ligand interaction is itself affected by low pH. Since both IgG complexes and anti-FcR Fab should remain bound to the receptor at endosomal and lysosomal pH, we have been able to study the role of intravesicular pH in intracellular transport without having to consider the role of pH in facilitating receptor–ligand dissociation.

Our results clearly indicate that neither the recycling of monovalent Fab–FcR complexes nor the transport of multivalent IgG complexes to lysosomes is absolutely dependent on low endosome or lysosome pH. Agents such as the carboxylic ionophore monensin and the lysosomotropic weak base NH_4Cl slowed slightly but did not block the normal transport of either ligand (Mellman *et al.*, 1984; Ukkonen *et al.*, 1985). Similar results were obtained for the transport to lysosomes of the fluid-phase endocytic tracer horseradish peroxidase (Ukkonen *et al.*, 1985) and of human choriogonadotropin (Ascoli, 1984), another ligand that does not dissociate from its receptor at endosomal pH. The only effective inhibitor of IgG-complex transport to lysosomes is incubation at temperatures of less than 20°C (Ukkonen *et al.*, 1985); this has been shown for a variety of endocytic tracers (Dunn *et al.*, 1980; Merion and Sly, 1983; Marsh *et al.*, 1983).

4.4. *Mechanisms of Molecular Sorting*

Although the precise mechanisms controlling the transport of ligands and receptors remain to be defined, our studies of FcR-mediated endocytosis have begun to suggest some of the general principles that may be at work. Clearly the most important suggestion comes from the demonstration that the valency of a bound ligand may influence the intracellular pathway of an internalized receptor. FcR bound to nondissociating monovalent ligands are internalized, delivered to endosomes, and recycled back to the plasma membrane. In contrast, when the very same receptors are bound to nondissociating multivalent ligands, they fail to recycle and are transported from endosomes to lysosomes. These pathways are illustrated diagrammatically in Fig. 3.

Why ligand valency should influence receptor transport is less clear. Conceivably, FcR bound to multivalent ligands become aggregated or cross linked in the plane of the membrane. The receptors would presumably remain aggregated on reaching endosomes, since their ligands do not dissociate at acidic pH. The presence of cross-linked receptors in the endosomal membrane may constitute a "positive signal" that triggers selective transport to lysosomes. Alternatively, receptor cross linking may generate a "negative signal," preventing ligated receptors from entering nascent recycling vesicles and increasing the probability that the receptors will be present in an endosome when it ultimately fuses with (or matures into) a lysosome.

Although this paradigm, or even some variation of this paradigm, may not explain the intracellular transport of every receptor, it is interesting to note that in addition to the FcR, a number of examples are known that appear to support the concept that crosslinking by multivalent ligands leads to receptor transport to lysosomes. The LDL and mannose-6-phosphate receptors, for instance, both of which efficiently recycle during the uptake of their pH-sensitive physiological ligands, are prevented from recycling when bound by polyclonal antireceptor antibodies and are transferred to lysosomes (Anderson *et al.*, 1982; von Figura *et al.*, 1984). Similarly, transport of the EGF receptor to lysosomes will occur after the binding of a pentameric antireceptor IgM antibody but not if the monovalent Fab fragment of this antibody is bound (Schreiber *et al.*, 1983).

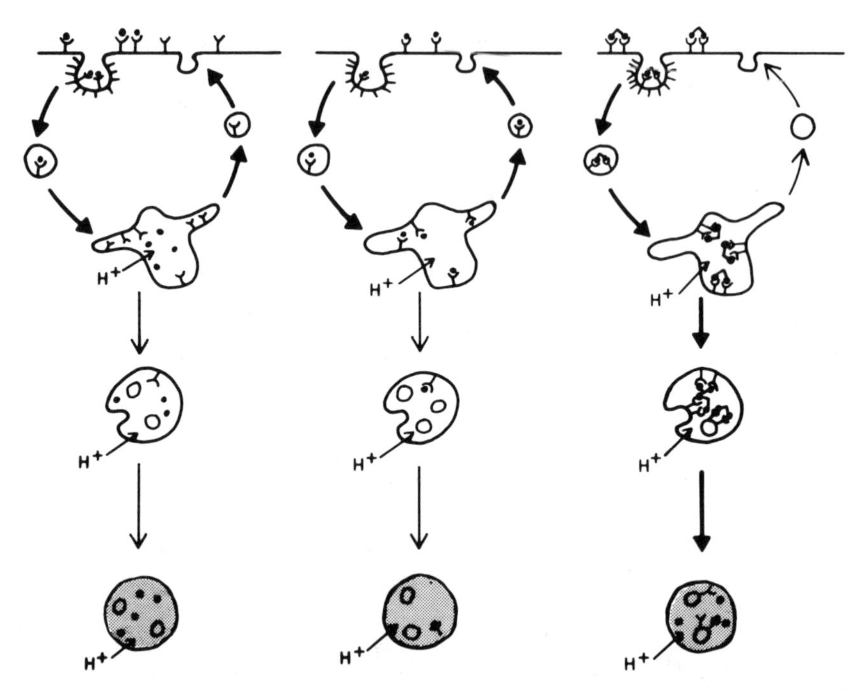

Figure 3. Pathways of receptor-mediated endocytosis. The left panel illustrates the "basic pathway" of endocytosis (Helenius *et al.*, 1983), in which a receptor-bound ligand dissociates at low pH in endosomes, allowing the receptor to recycle back to the plasma membrane while the ligand is transported to lysosomes. In the center panel, the pathway taken by FcR bound to monovalent antireceptor Fab is diagrammed: following internalization, the Fab remains bound to the receptor on delivery to endosomes and returns intact to the cell surface, still bound to the FcR. A similar situation prevails for the uptake and recycling of transferrin. The right panel illustrates the intracellular pathway taken by FcR bound to a nondissociating multivalent ligand. In this situation, it is proposed that receptors become cross linked by the bound ligand, and, as a result, internalized receptors are prevented from recycling and are instead transported from endosomes to lysosomes.

Another interesting example is the transferrin receptor. The endocytosis of transferrin provides a physiologically relevant analogue of the anti-FcR Fab-recycling pathway. Transferrin binds its receptor at the cell surface, where it is internalized and delivered to endosomes. There, the acidic environment facilitates the discharge of transferrin-bound iron but not of transferrin from its receptor. Instead, like the Fab, transferrin recycles along with its receptor directly from endosomes back to the plasma membrane. When transferrin (or many monoclonal anti-transferrin-receptor antibodies) is rendered multivalent by adsorption to colloidal gold, however, both ligand and receptor are transported to lysosomes (Neutra *et al.*, 1985; Hopkins and Trowbridge, 1983; Willingham *et al.*, 1984).

Certain polypeptide hormones such as insulin and EGF presumably bind via monovalent attachment to their receptors and yet can cause receptor transport to lysosomes (Kasuga *et al.*, 1981; Stoscheck and Carpenter, 1984). Although the simple paradigm of receptor cross linking by multivalent ligands cannot apply to these examples, it is conceivable that hormone binding may alter the oligomeric state of their receptors by other means. For example, the receptor phosphorylation that often accompanies hormone binding may induce a conformational change that increases the degree of interaction among adjacent receptors or between the receptors and other membrane proteins. Future work in our laboratory will focus on the possible involvement of receptor oligomerization in controlling receptor recycling and transport to lysosomes.

5. Biochemistry of Endosomes and Lysosomes

The elucidation of the mechanisms controlling receptor transport pathways will require a more complete understanding of the molecular composition of endosomes and lysosomes in addition to further information about receptor structure and assembly. The obvious morphological and functional heterogeneity of endosomes, considered above, must reflect an underlying biochemical heterogeneity. In particular, the tubular endosomes involved in receptor recycling may well have a composition distinct from the endosomes that mediate the last stages of transport to lysosomes (Helenius *et al.*, 1983). Thus, one can expect endosomes to consist of an interacting array of subcompartments, which will ultimately have to be dissected from one another to be understood.

5.1. The Isolation and Characterization of Endosomes and Lysosomes

A first step in the molecular analysis of endocytic organelles must be their physical isolation. Unfortunately, it has thus far proved difficult to prepare highly purified fractions of endosomes. Although easily separated from lysosomes on Percoll gradients, endosomes appear to have density and size characteristics so similar to most other smooth membranes in cell homogenates that separation by conventional means (centrifugation, gel filtration) has not been easy. A variety of different approaches are being tried in a number of laboratories to effect the purification of endosomes. Our approach has relied on the technique of free-flow electrophoresis.

We have adapted the method described by Harms *et al.* (1980), which was originally used to isolate highly purified lysosomal fractions, to obtain an enriched preparation of endosomes (M. Marsh, E. Harms, I. Mellman, and A. Helenius, unpublished data). A crude high-speed membrane pellet is prepared from homogenates of tissue culture cells whose endosomes have been labeled by endocytosis of fluid-phase or receptor-bound tracers. The membranes are lightly trypsinized, a maneuver that seems to release the endosomes from a matrix of cytoskeletal components (C. Hopkins, unpublished results), and injected into a free-flow electrophoresis chamber. Two membrane streams become visible, one of which shifts towards the anode. Most of the cell protein as well as markers for Golgi, ER, mitochondria, plasma membrane, and cytosol are associated with the unshifted stream. Markers for endosomes and lysosomes, however, are almost quantitatively contained in the anodal stream. This combined endosome–lysosome fraction is then combined and centrifuged in a Percoll gradient, which conveniently separates the low-density endosomes from the high-density lysosomes. Based on marker enzyme analysis (ATP-dependent acidification has provided a selective marker for FITC-dextran-containing endosomes), this procedure results in a purification of more than 100-fold.

The biochemical characterization of endosomal and lysosomal membranes is at an early stage. Sodium dodecylsulfate gel electrophoresis of the two fractions using membranes purified from cultured CHO cells has already indicated that the membrane proteins comprising these two organelles are dramatically different from each other as well as from the major membrane proteins found in crude cell homogenates. However, at least some of the major plasma membrane proteins of CHO cells (i.e., those accessible to surface radioiodination using lactoperioxidase) are also present in the endosomal fraction. These proteins are entirely absent from the lysosomal fraction. Instead, lysosomes seem to be specifically enriched in a 120-kD membrane glycoprotein, which we have identified immunologically and designated lgp 120 (lysosomal membrane glycoprotein) (Lewis *et al.*, 1985).

5.2. *Structure of the Lysosomal Membrane*

Monoclonal and polyclonal antibodies generated against purified free-flow electrophoresis fractions from rat liver homogenates have been found to recognize a series of membrane glycoproteins, which appear by immunofluorescence (Fig. 4) and by immunoelectron microscopy to be greatly enriched in the lysosomes of a variety of tissue and cell types. From all immunizations, the antigens identified thus far include lgp 150,

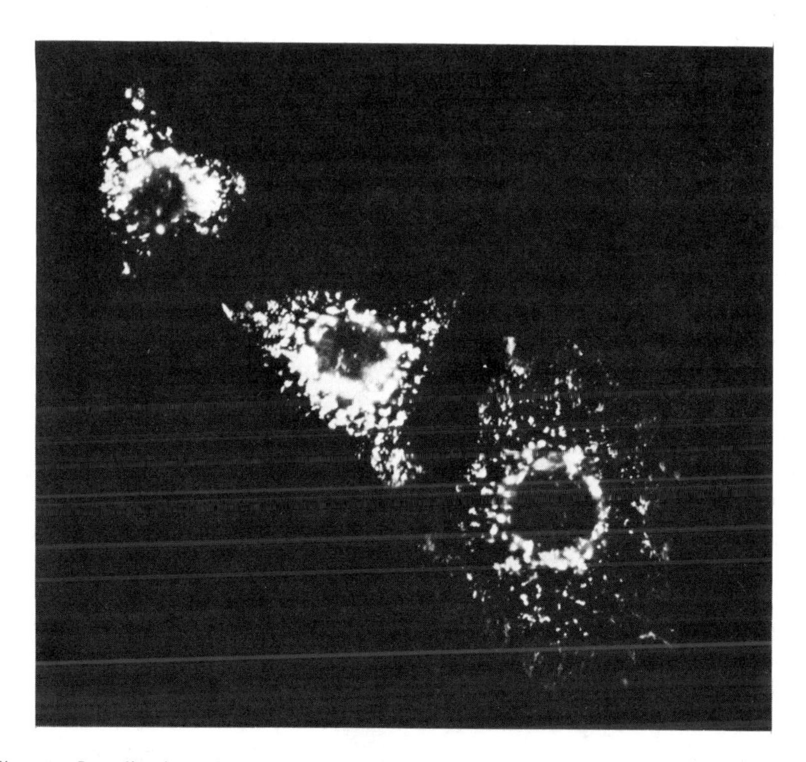

Figure 4. Localization of lysosomal membrane glycoproteins by indirect immunofluorescence. A rabbit antibody specific for two intrinsic membrane glycoproteins of the lysosomal membrane (lgp 100 and lgp 80) was used to stain fixed and permeabilized NRK fibroblasts. Punctate fluorescence indicative of lysosomes is largely concentrated in the perinuclear region of the cell. Identical staining patterns are obtained with monoclonal and polyclonal antibodies to lgp 120 (Lewis *et al.*, 1985).

120, 100, 80, 55, 30, and 20 (Lewis *et al.*, 1985). Of these, the structure of lgp 120 has been the most extensively investigated.

The protein lgp 120 has been classified as an intrinsic protein of the lysosomal membrane (Lewis *et al.*, 1985) based on its ability to bind the nonionic detergent Triton X-114 and its sensitivity to glycosidase digestion. Lgp 120 is heavily glycosylated, containing at least 18 N-linked oligosaccharide chains and possibly an amount of O-linked glycosides. The N-linked sugars are complex (i.e., resistant to digestion by endoglycosidase H) and terminally glycosylated (i.e., they contain sialic acid). The lgp 120 polypeptide is only 42 kD in size. The mature protein is quite acidic and has a pI of less than 4; neuraminidase treatment increases its pI to 8.5.

Although the function of lgp 120 is as yet unknown, its structure provides important information about the biogenesis and properties of the lysosomal membrane. First, the fact that lgp 120 and the other lysosomal membrane proteins we have identified are terminally glycosylated indicates that their biogenesis must involve transport through the trans elements of the Golgi stack. This information is significant because recent data suggest that lysosomal enzymes representative of the content proteins of lysosomes may exit the Golgi from cis cisternae (Brown and Farquhar, 1984). Secondly, although lysosomal enzymes rely on the addition of a mannose-6-phosphate moiety in order to be transported (via the mannose-6-phosphate receptor) to lysosomes, lgp 120 appears not to contain phosphate, indicating that the biogenesis of the lysosomal membrane is significantly different from that of lysosomal enzymes. Finally, the rather extensive degree of glycosylation of lgp 120 (and other lysosomal membrane proteins) suggests the possibility that the lysosomal membrane protects itself from destruction at the hands of its own hydrolases by heavily glycosylating its membrane proteins.

One question of particular interest is the mechanism of targeting of lysosomal membrane proteins. As discussed above, most plasma membrane proteins and receptors that are internalized during endocytosis avoid transit through the lysosomal compartment. Indeed, probably the major function of endosomes is to obviate the need for receptors to be transported to lysosomes in order to deliver their bound ligands. It is also reasonable to assume that the biosynthetic pathway of most secretory and membrane proteins also avoids a lysosomal intermediate. How, then, are membrane proteins such as lgp 120 selectively directed to lysosomes? Although we are currently cDNA cloning and sequencing these proteins, we do not expect that the targeting information will be encoded as an easily recognizable or conserved amino acid sequence. Instead, our experience in studying the control of intracellular FcR transport has led us to believe that targeting to lysosomes in the biosynthetic pathway may also depend partly on the state of oligomeric assembly of a particular membrane protein.

ACKNOWLEDGMENTS. The work described in this chapter has been performed largely by Mark Marsh, Helen Plutner, Cindy Galloway, Victoria Lewis, Sam Green, Renate Fuchs, Christine Howe, Pentti Ukkonen, Philippe Male, Bruce Granger, Terry Koch, and Gary Dean with only slight interference by either the author or his collaborator Ari Helenius on many of these studies. We have been supported by NIH grants (awarded to I. Mellman) GM-29765 and GM-33093.

References

Anderson, R. G. W., Brown, M. S., Beisiegel, U., and Goldstein, J. L., 1982, Surface distribution and recycling of the low density lipoprotein receptor as visualized with anti-receptor antibodies, *J. Cell Biol.* **93:**523–531.

Ascoli, M., 1984, Lysosomal accumulation of the hormone–receptor complex during receptor-mediated endocytosis of human choriogonadotropin, *J. Cell Biol.* **99:**1242–1250.

Beguinot, L., Lyall, R. M., Willingham, M. C., and Pastan, I., 1984, Down regulation of the epidermal growth factor receptor in KB cells is due to receptor internalization and subsequent degradation in lysosomes, *Proc. Natl. Acad. Sci. U.S.A.* **81:**2384–2388.

Besterman, J. M., Airhart, J. A., Woodworth, R. C., and Low, R. B., 1981, Exocytosis of pinocytosed fluid in cultured cells: Kinetic evidence for rapid turnover and compartmentation, *J. Cell Biol.* **91:**716–727.

Braell, W. A., Schlossman, D. M., Schmid, S. L., and Rothman, J. E., 1984, Dissociation of clathrin coats coupled to the hydrolysis of ATP: Role of an uncoating ATPase, *J. Cell Biol.* **99:**734–741.

Brown, W. J., and Farquhar, M. G., 1984, The mannose 6-phosphate receptor for lysosomal enzymes is concentrated in *cis* Golgi cisternae, *Cell* **36:**295–307.

Brown, M. S., Anderson, R. G. W., and Goldstein, J. L., 1983, Recycling receptors: The round trip itinerary of migrant membrane proteins, *Cell* **32:**663–667.

Dunn, W. A., Hubbard, A. L., and Aronson, N. N., Jr., 1980, Low temperature selectively inhibits fusion between pinocytic vesicles and lysosomes during heterophagy of ^{125}I-asialofetuin by the perfused liver, *J. Biol. Chem.* **255:**5971–5978.

Forgac, M., Cantley, L., Wiedenmann, B., Altstiel, L., and Branton, D., 1983, Clathrin-coated vesicles contain an ATP-dependent proton pump, *Proc. Natl. Acad. Sci. U.S.A.* **80:**1300–1303.

Galloway, C. J., Dean, G. E., Marsh, M., Rudnick, G., and Mellman, I., 1983, Acidification of macrophage and fibroblast endocytic vesicles *in vitro*, *Proc. Natl. Acad. Sci. U.S.A.* **80:**3334–3338.

Geuze, H. J., Slot, J. W., Strous, G. J. A. M., Lodish, H. F., and Schwartz, A. L., 1983, Intracellular site of asialoglycoprotein uncoupling: Double label immunoelectron microscopy during receptor-mediated endocytosis, *Cell* **32:**277–287.

Geuze, H. J., Slot, J. W., Strous, G. J. A. M., Peppard, J., von Figura, K., Hasilik, A., and Schwartz, A. L., 1984, Intracellular receptor sorting during endocytosis: Comparative immunoelectron microscopy of multiple receptors in rat liver, *Cell* **37:**195–204.

Glickman, J., Croen, K., Kelly, S., and Al-Awqati, Q., 1983, Golgi membranes contain an electrogenic H^+ pump in parallel to a chloride conductance, *J. Cell Biol.* **97:**1303–1308.

Green, S. A., Plutner, H., and Mellman, I., 1985, Biosynthesis and intracellular transport of the mouse macrophage Fc receptor, *J. Biol. Chem.* **260:**9867–9874.

Harford, J., Bridges, K., Ashwell, G., and Klausner, R. D., 1983, Intracellular dissociation of receptor-bound asialoglycoproteins in cultured hepatocytes: A pH-mediated nonlysosomal event, *J. Biol. Chem.* **258:**3191–3197.

Harms, E., Kern, H., and Schneider, J. A., 1980, Human lysosomes can be purified from diploid skin fibroblasts by free-flow electrophoresis, *Proc. Natl. Acad. Sci. U.S.A.* **77:**6139–6143.

Helenius, A., Mellman, I., Wall, D., and Hubbard, A., 1983, Endosomes, *Trends Biochem. Sci.* **8:**245–250.

Hirsch, J. G., Fedorko, M. E., and Cohn, Z. A., 1968, Vesicle fusion and formation at the surface of pinocytic vesicles in macrophages, *J. Cell Biol.* **38:**619–632.

Hopkins, C. R., 1983, The importance of the endosome in intracellular traffic, *Nature* **304:**684–685.

Hopkins, C. R., and Trowbridge, I. S., 1983, Internalization and processing of transferrin and the transferrin receptor in human carcinoma A431 cells, *J. Cell Biol.* **97:**508–521.

Kasuga, M., Kahn, R., Hedo, J. A., Obberghen, E. V., and Yamada, K. M., 1981, Insulin-induced receptor loss in cultured human lymphocytes is due to accelerated receptor degradation, *Proc. Natl. Acad. Sci. U.S.A.* **78:**6917–6921.

Lewis, V., Green, S. A., Marsh, M., Virkko, P., Helenius, A., and Mellman, I., 1985, Glycoproteins of the lysosomal membrane, *J. Cell Biol.* **100:**1839–1847.

Lewis, W. H., 1931, Pinocytosis, *Bull. Johns Hopkins Hosp.* **49:**17–36.

Marsh, M., Bolzau, E., and Helenius, A., 1983, Penetration of Semliki Forest virus from acidic prelysosomal vesicles, *Cell* **32:**931–940.

Marsh, M., Dean, G., Griffiths, G., Mellman, I., and Helenius, A., 1986, Three-dimensional structure of endosomes in BHK-21 cells, *Proc. Natl. Acad. Sci. U.S.A.*, in press.

Mellman, I., 1982, Endocytosis, membrane recycling, and Fc receptor function, in: *Membrane Recycling, Ciba Foundation Symposium No. 92* (D. Evered, ed.), Pitman, London, pp. 35–58.

Mellman, I., and Galloway, C. J., 1983, Selective labeling and quantitative analysis of internalized plasma membrane, *Methods Enzymol.* **98:**545–555.

Mellman, I., and Plutner, H., 1984, Internalization and degradation of macrophage Fc receptors bound to polyvalent immune complexes, *J. Cell Biol.* **98:**1170–1177.

Mellman, I., and Ukkonen, P., 1985, Internalization and fate of macrophage Fc receptors during receptor-mediated endocytosis, in: *4th Leiden Conference on the Heterogeneity of Mononuclear Phagocytes* (R. van Furth, ed.), A. Nijhoff, Amsterdam, pp. 75–83.

Mellman, I. S., and Unkeless, J. C., 1980, Purification of a functional mouse Fc receptor through the use of a monoclonal antibody, *J. Exp. Med.* **152:**1048–1069.

Mellman, I. S., Steinman, R. M., Unkeless, J. C., and Cohn, Z. A., 1980, Selective iodination and polypeptide composition of pinocytic vesicles, *J. Cell Biol.* **86:**712–722.

Mellman, I. S., Plutner, H., Steinman, R. M., Unkeless, J. C., and Cohn, Z. A., 1983, Internalization and degradation of macrophage Fc receptors during receptor-mediate phagocytosis, *J. Cell Biol.* **96:**887–895.

Mellman, I., Plutner, H., and Ukkonen, P., 1984, Internalization and rapid recycling of macrophage Fc receptors tagged with monovalent antireceptor antibody: Possible role of a prelysosomal compartment, *J. Cell Biol.* **98:**1163–1169.

Merion, M., and Sly, W. S., 1983, The role of intermediate vesicles in the adsorptive endocytosis and transport of ligand to lysosomes by human fibroblasts, *J. Cell Biol.* **96:**644–650.

Muller, W. A., Steinman, R. M., and Cohn, Z. A., 1983, The membrane proteins of the vacuolar system. III. Further studies on the composition and recycling of endocytic vacuole membrane in culture macrophages, *J. Cell Biol.* **96:**29–36.

Muller, W. A., Steinman, R. M., and Cohn, Z. A., 1980, The membrane proteins of the vacuolar system. II. Bidirectional flow between secondary lysosomes and plasma membrane, *J. Cell Biol.* **86:**304–314.

Neutra, M. R., Ciechanover, A., Owen, L. S., and Lodish, H. F., 1985, Intracellular transport of transferrin- and asialoorosomucoid-colloidal gold to lysomes after receptor-mediated endocytosis, *J. Histochem. Cytochem.*, **33:**1134–1144.

Ohkuma, S., and Poole, B., 1978, Fluorescence probe measurement of the intralysosomal pH in living cells and the perturbation of pH by various agents, *Proc. Natl. Acad. Sci. U.S.A.* **75:**3327–3331.

Ohkuma, S., Moriyami, Y., and Takano, T., 1982, Identification and characterization of a proton pump in lysosomes by fluorescein isothiocyanate-dextran fluorescence, *Proc. Natl. Acad. Sci. U.S.A.* **79:**2758–2762.

Pastan, I. H., and Willingham, M. C., 1981, Journey to the center of the cell: Role of the receptosome, *Science* **214:**504–509.

Robbins, A. R., Peng, S. S., and Marshall, J. L., 1983, Mutant Chinese hamster ovary cells pleiotropically defective in receptor-mediated endocytosis, *J. Cell Biol.* **96:**1064–1071.

Robbins, A. R., Oliver, C., Bateman, J. L., Krag, S. S., Galloway, C. J., and Mellman, I., 1984, A single mutation in Chinese hamster ovary cells impairs both Golgi and endosomal functions, *J. Cell Biol.* **99:**1296–1308.

Rudnick, G., 1985, Acidification of intracellular organelles: mechanism and function, in: *Physiology of Membrane Disorders* (T. E. Andreoli, D. D. Fanestil, J. F. Hoffman, and S. G. Schultz, eds.), Plenum Press, New York (in press).

Schreiber, A. B., Liberman, T. A., Laz, I., Yarden, Y., and Schlessinger, J., 1983, Biological role of epidermal growth factor receptor clustering. Investigation with monoclonal anti-receptor antibodies, *J. Biol. Chem.* **258:**846–853.

Steinman, R. M., Brodie, S. E., and Cohn, Z. A., 1976, Membrane flow during pinocytosis. A stereologic analysis, *J. Cell Biol.* **68:**665–687.

Steinman, R. M., Mellman, I. S., Muller, W. A., and Cohn, Z. A., 1983, Endocytosis and the recycling of plasma membrane, *J. Cell Biol.* **96:**1–27.

Stone, D. K., Zie, X.-S., and Racker, E., 1983, An ATP-driven proton pump in clathrin coated vesicles, *J. Biol. Chem.* **258:**4059–4062.

Stoscheck, C. M., and Carpenter, G., 1984, Down regulation of epidermal growth factor receptors: Direct demonstration of receptor degradation in human fibroblasts, *J. Cell Biol.* **98:**1048–1053.

Swanson, J. A., Yirinec, B. D., and Silverstein, S. C., 1985, Phorbol esters and horseradish peroxidase stimulate pinocytosis and redirect flow of pinocytosed fluid in macrophages, *J. Cell Biol.* **100:**851–859.

Tycko, B., and Maxfield, F. M., 1982, Rapid acidification of endocytic vesicles containing alpha-2-macroglobulin, *Cell* **28:**643–651.

Uchida, E., Ohsumi, Y., and Anraku, Y., 1985, Purification and properties of H^+-translocating, Mg^{2+}-adenosine triphosphatase from vacuolar membranes of *Saccharomyces cerevisiae*, *J. Biol. Chem.* **260:**1090–1095.

Ukkonen, P., Lewis, V., Marsh, M., Helenius, A., and Mellman, I., 1985, Transport of macrophage Fc receptors and Fc-receptor-bound ligands to lysosomes, submitted.

Unkeless, J. C., Fleit, H. B., and Mellman, I., 1981, Structural aspects and heterogeneity of immunoglobulin Fc receptors, *Adv. Immunol.* **31:**247–270.

van Renswoude, J., Bridges, K. R., Harford, J. B., and Klausner, R. D., 1982, Receptor-mediated endocytosis of transferrin and the uptake of Fe in K562 cells: Identification of a nonlysosomal acidic compartment, *Proc. Natl. Acad. Sci. U.S.A.* **79:**6186–6190.

von Figura, K., Gieselmann, V., and Hasilik, A., 1984, Antibody to mannose 6-phosphate specific receptor induces receptor deficiency in human fibroblasts, *EMBO J.* **3:**1281–1282.

Willingham, M. C., Hanover, J. A., Dickson, R. B., and Pastan, I., 1984, Morphologic characterization of the pathway of transferrin endocytosis and recycling in human KB cells, *Proc. Natl. Acad. Sci. U.S.A.* **81:**175–179.

Yamashiro, D. J., Tycko, B., Fluss, S. R., and Maxfield, F. R., 1984, Segregation of transferrin to a mildly acidic (pH 6.5) para-Golgi compartment in the recycling pathway, *Cell* **37:**789–800.

Young, J. D.-E., Unkeless, J. C., Young, T. M., Mauro, A., and Cohn, Z. A., 1983, Role for mouse macrophage IgG/Fc receptor as ligand-dependent ion channel, *Nature* **306:**186–189.

Zhang, F., and Schneider, D. L., 1983, The bioenergetics of Golgi apparatus function: Evidence for an ATP-dependent proton pump, *Biophys. Biochem. Res. Commun.* **114:**620–625.

21

Transport of Protein Toxins Across Cell Membranes

SIMON VAN HEYNINGEN

1. Introduction

The protein toxins produced by plants and bacteria are extraordinarily powerful: those of cholera, diphtheria, tetanus, and botulism are responsible for diseases that still kill many thousands of people every year, especially in the Third World. From the point of view of the biologist, their chief interest lies in their high activity: a single molecule of diphtheria toxin will kill a cell. Only hormones have comparable biological activity, and they are products of the organism that they affect; the toxins are produced by alien procaryotes, yet they can profoundly affect eucaryotic life. In this chapter, I give an outline of the properties of some of these proteins, emphasizing the way in which they manage to cross the membrane of the intoxicated cell to arrive at their intracellular target. It must be remembered that they face a double problem: they must first be secreted across the membrane of the cell in which they were synthesized before they can reach another cell. For two useful and recent general reviews, see Eidels *et al.* (1983) and Middlebrook and Dorner (1984).

All the toxins I consider here affect intracellular targets, and all bind to some "receptor" molecule on the cell surface before they enter. The term "receptor" can be misleading, because this binding is not in itself responsible for their activity; it is merely a starting point for their subsequent incorporation into the cell.

Toxins are exotic and interesting molecules, but they are not necessarily typical. Although they have been very widely used as model

SIMON VAN HEYNINGEN • Department of Biochemistry, University of Edinburgh, Edinburgh EH8 9XD, Scotland.

CHAPTER 21

Table I. The Component Structure of Various Toxins[a]

Toxin	Molecular weight of whole toxin	Structure	Active component molecular weight	Target	How joined to binding component	Structure	Binding component molecular weight	Target
Affecting adenylate cyclase								
Cholera	82,000	Subunit A = A1 peptide + A2 peptide	27,000 22,000 5,000	N_s protein of adenylate cyclase	Noncovalent	Five B subunits	11,600 each	Ganglioside GM1
E. coli heat labile	91,000	Subunit A = A1 peptide + A2 peptide	30,000 25,000 5,000	N_s protein of adenylate cyclase	Noncovalent	Five B subunits	11,800 each	Ganglioside GM1 and glycoproteins
Pertussis	117 000	S-1	28,000	N_i protein of adenylate cyclase	Noncovalent	S-2, S-3, two S-4, S-5	23,000 22,000 11,700 9,300 each	Not known
Affecting protein synthesis								
Diphtheria	62,000	A chain	24,000	EF2 (diphthamide)	Disulfide bond	B chain	38,000	Glycoprotein
Pseudomonas aeruginosa	71,000	A fragment	27,000	EF2 (diphthamide)	Not clear	B fragment	45,000	Has no clear function
Ricin (abrin and modeccin rather similar)	62,000	A chain	30.600	60 S ribosomal subunit	Disulfide bond	B chain	31,000	Glycoprotein
Shigella	68,000	A component A1 protein A2 protein	30,000 27,000 3,000	60 S ribosomal subunit	Noncovalent	Six or seven B chains	5,000 each	Glycoprotein
Neurotoxins								
Tetanus	150,000	L chain	50,000	No clear function	Disulfide bond	H chain	100,000	Gangliosides GT1b GD1b
Botulinus	150,000	L chain	50,000	No clear function	Disulfide bond	H chain	100,000	Not known, may be a glycolipid

[a] From van Heyningen (1984), with permission.

systems for studying entry into the cell, this could be misleading. No cell can have evolved a mechanism specifically in order to be killed by diphtheria toxin, and it is therefore almost inevitable that the toxins will take advantage of routes into the cell and receptors that exist for some other purpose. Since the affected cell and the protein will not have evolved together, it seems likely that the design will have to go into the toxin so that it can be "opportunistic" in its approach to the cell. The following sections give examples of these points.

The protein toxins that I consider in this chapter have all evolved a similar overall structure. They are made up of two components: a binding B component that binds to the cell surface and an active A component that subsequently enters the cell and produces the pharmacological effect of the toxin in most, if not all, cases by some enzymatic action (van Heyningen, 1982a). This two-component structure of several such toxins is summarized in Table 1 (from van Heyningen, 1984), from which it is clear that there is no simple pattern to the nature of the different types: some are proteolytic fragments that remain joined to each other by disulfide bonds, whereas some are subunits linked to each other noncovalently. In almost every case, however, the B component by itself can protect cells from the effect of whole toxin by occupying the receptor sites.

The discovery of this two-component structure has turned out to be a very fruitful source of research. If a cytotoxic toxin, or the active component of such a toxin (e.g., the A chain of diphtheria toxin), is linked not to its normal B component but to another protein that will bind to a particular type of cell (e.g., a monoclonal antibody), a protein is produced that should be toxic to that kind of cell alone (see Thorpe and Ross, 1982). The implications for cancer chemotherapy are obvious and have been the subject of much recent published work and some encouraging clinical results (Filipovich *et al.*, 1984). A similar "hydridization" experiment may be possible with cholera toxin (see Section 3.1.2c).

2. *Toxins That Affect Protein Synthesis*

2.1. *Diphtheria Toxin*

2.1.1. *Structure*

Diphtheria toxin is synthesized at the ribosome as a single polypeptide chain, but this can be cleaved proteolytically into the A fragment (M_r 24,000) and the B fragment (M_r 38,000), which remain linked by disulfide bonds. (For general reviews of diphtheria toxin see Collier, 1975; Uchida,

1982.) Clever experiments using cross-reacting mutant forms of the toxin established clearly that the B fragment binds to the cell surface but has no other activity, whereas the A fragment enters the cell, where it folds into an enzymically active conformation that enables it to catalyze a reaction leading to the inhibition of protein synthesis and the death of the cell. This reaction is the ADP-ribosylation of elongating factor 2 (a protein essential in the action of the ribosome) by the reaction:

$$NAD^+ + EF2 = ADP\text{-ribose-}EF2 + \text{nicotinamide} + H^+.$$

The equilibrium of this reaction lies overwhelmingly to the right, and the ADP-ribosylated EF2 is no longer active, so protein synthesis ceases. Since fragment A is acting catalytically, it is understandable that a single molecule is all that is needed to kill a cell.

2.1.2. Mechanism of Entry of Fragment A into the Cell

2.1.2a. Receptor. It is proving remarkably hard to find what the diphtheria toxin receptor is, even though many cell types (e.g., Vero cells) have a very large number of them and the gene for the receptor was mapped some years ago. Proia *et al.* (1981) have evidence that a glycoprotein of M_r about 150,000 is the receptor, but firm proof of this is still lacking and is likely to be very difficult to produce because of the large amount of nonfunctional toxin binding usually found. Recently, Olsnes and his co-workers (1985) have produced some convincing evidence that the toxin binds to the major anion-transport system in Vero cells, an interesting example of a toxin using a receptor there for some other purpose.

2.1.2b. Effects of Acid pH. Early observation showed that cells are protected from diphtheria toxin by ammonium chloride (Kim and Groman, 1965). Later work showed that many other drugs that increase the pH of intracellular vesicles also protect against toxin, although brief previous exposure of the cells to low pH stops this protection (Draper and Simon, 1980; Sandvig and Olsnes, 1981). This suggests that low pH on one side is required for the transport of the toxin across the membrane, and so lysozomes might be involved, since their internal pH is below 5. More recent results have made it appear less likely that entry into the cytoplasm is from the lysozome (which would mean that the toxin molecules would have to survive the hydrolytic enzymes there) but rather from the more recently discovered endosomes (receptosomes) (Willingham and Pastan, 1984). Cells with mutant receptosomes that are less acid are resistant to diphtheria toxin as well as to some viruses that are thought to use the same route (Merion *et al.*, 1983).

Current evidence demonstrates that the toxin can escape directly from these vesicles and that fusion with other organelles is not required (Sandvig *et al.*, 1984; Marnell *et al.*, 1984). There is evidence that a protein factor is required in at least some types of cell (Kaneda *et al.*, 1984). It is not enough for one side of the membrane to be acid; there must also be a gradient of pH. Thus, if the cytosol is also made acid, the toxin can no longer enter (Olsnes *et al.*, 1985).

The best hypothesis seems to be that the toxin binds to some not-clearly-identified receptor, is endocytosed into the endosomes at coated pits, and then passes across the endosome membrane into the cytoplasm. Since so few molecules of toxin need to enter the cell, this entry process does not have to be efficient, and it is possible that specific entry via the coated pit system is not required; uptake might be part of the overall membrane turnover (see Willingham and Pastan, 1984). This mechanism is compatible with the observation (referred to in Section 1.1 above) that artificial hybrid toxin molecules remain able to get into the cell when given non-native binding components, which presumably use a different receptor.

2.1.2c. How the Toxin Crosses the Membrane. Structural and physical chemical work on diphtheria toxin has given information on why a pH gradient is required. Model systems using artificial membranes have shown that a hydrophobic portion of the B chain (Boquet and Pappenheimer, 1976; Falmagne *et al.*, 1985) is exposed at low pH (Sandvig and Olsnes, 1981); this can insert into the membrane and form an ion-permeable channel through which fragment A could pass (Donovan *et al.*, 1981; Kagen *et al.*, 1981; Kayser *et al.*, 1981). Experiments using photoprobes have shown unequivocally that both chains can be inserted into such membranes (Hu and Holmes, 1984). The A chain is a very stable molecule that could perhaps unfold enough to pass through a narrow hole and then refold to an active conformation in the cytoplasm. Thus, the structure of the binding component fits it for its role.

2.2. The Toxin from Pseudomonas aeruginosa

Pseudomonas aeruginosa secretes a protein toxin (Vasil *et al.*, 1977) that may be involved with its ability to cause the infection that frequently follows wounds and burning. The toxin has a complicated subunit structure that is not quite clear, but it kills cells with the same ADP-ribosylation mechanism that diphtheria toxin uses. (Indeed, ADP-ribosylation catalyzed by one toxin can be reversed by the other.) It seems also to enter the cell by the same pathway, using coated pits and endosomes. Low pH is required, at least in some particularly responsive cells such as L-cells and mouse 3T3 cells (FitzGerald *et al.*, 1980). Since the protein is not clearly composed of a binding and an active component, the details connot be exactly the same: an active fragment has been prepared, but not a binding component. Furthermore, the two toxins have very different activities with different types of cells, so different receptors are presumably involved.

2.3. Abrin, Ricin, Viscumin, and Modeccin

2.3.1. Crossing the Membrane

Abrin, ricin, and viscumin are three similar protein toxins isolated from plants (Olsnes and Pihl, 1982). Each has a binding and an active fragment. The active fragment enters the cell and inhibits protein synthesis by some action at the ribosome that has not yet been fully elucidated but does not involve ADP-ribosylation. However, it is clear that these toxins enter cells by a method different from that of the diphtheria or *Pseudomonas* toxins. There is no evidence that low pH is required; if anything, quite the reverse. Reagents such as ammonium chloride or chloroquine that increase intracellular pH increase the sensitivity of cells to these toxins (Sandvig *et al.*, 1979), and the toxins enter most efficiently when the pH of the medium is slightly above neutral (Stirpe *et al.*, 1982). There is, however, some requirement for calcium ions, probably in the movement across the membrane of the A chain, and it appears that entry is not from the cell surface but from an intracellular vesicle (in this case a neutral one).

2.3.2 Role of Each Subunit

It is clear that the binding component of ricin is important to the entry of ricin A chain. Houston (1982) has shown that cells that do not respond to purified A chain will respond if they are first incubated with B chain; in other words, the B chain already on the cell surface must be capable of reacting with A chain *in situ* and then promoting its transport through the membrane. Many artificial hybrids of whole ricin and of the A chain alone have been prepared and shown to kill cells. Experiments in this sort of system have led Herschman (1984) to propose that the A chain by itself can get out of the intracellular vesicles, perhaps through its hydrophobic domain (cf. diphtheria toxin, Section 2.1.2c above). The only role of the B chain is to bind to the outer surface of the cell and allow initial uptake into the vesicles. Photolabeling experiments (Ishida *et al.*, 1983) also show that each of the purified chains, as well as whole ricin, can penetrate into membranes (like those of Newcastle disease virus).

These experiments suggest that penetration by A chain does not need any receptor, unlike penetration by ricin or B chain. Since labeling of individual chains was much greater than of whole toxin, it is possible that they must dissociate from each other before entering the membrane.

2.3.3. *Modeccin*

One other plant toxin, modeccin, should be mentioned. It enters cells from an acidic vesicle as does diphtheria toxin, but the evidence is that the vesicles are different (Sandvig *et al.*, 1984; Draper *et al.*, 1984). This is another example of the fact that even superficially similar toxins can enter cells by different means.

2.4. *Shigella Toxin*

The final toxin to be considered in this category is that of *Shigella*, which is also cytoxic and has an active component linked to a rather complicated component assumed to be involved in binding and itself made up of several small subunits (Olsnes *et al.*, 1981; O'Brien *et al.*, 1983). There are conflicting reports on the effect of low pH on this toxin, and very little is known about its receptor. There is evidence suggesting that endocytotic vesicles are involved and that an intracellular fusion of some kind is required. Here again, different toxins use different mechanisms to enter a cell.

3. *Toxins That Activate Adenylate Cyclase*

3.1. *Cholera Toxin*

3.1.1. *Structure and Receptor*

3.1.1a. The Toxin. Cholera toxin is by far the most studied of those that activate adenylate cyclase (van Heyningen, 1982a,b, 1983a, 1984; Lai, 1980). The heat-labile (protein) toxin of *Escherichia coli* is a very similar protein, and what is true of one is, in general, true of the other.

Cholera is a massive diarrhea caused by *Vibrio cholerae* that grow in the gut and secrete cholera toxin. The toxin alters the ion flow across the gut cells, and this produces a consequent flow of fluid. The detailed mechanism by which the ion transport is altered is not known, but there is little doubt that its root cause is the activation of adenylate cyclase and the consequent increase in the intracellular concentration of cAMP. Cholera toxin activates the adenylate cyclase of essentially all eucaryotic cells, and this is its chief biological interest. Since adenylate cyclase is always intracellular, it follows that either the toxin itself or some signal induced by the toxin must penetrate the membrane.

3.1.1b. The Structure of the Protein. Cholera toxin is composed of two types of subunit. The A subunit (itself two polypeptide chains, A1, M_r 22,000, and A2, M_r 5,000, linked by a disulfide bond) is joined noncovalently to five B subunits (each M_r 11,500), which are probably arranged in a symmetrical ring. The whole protein has been sequenced, and crystals suitable for X-ray diffraction have been prepared. The bacteria secrete a pentamer of subunit B called choleragenoid at the same time as the toxin. Subunit B and choleragenoid bind to the ganglioside receptor, and subunit A has the activity.

3.1.1c. Role of Subunit A. There is no increase in adenylate cyclase activity after toxin has bound to cells until a lag phase of 30 to 90 min (depending on the type of cell) has passed. However, if the cells are lysed and then treated with toxin, there is no lag phase, and neither ganglioside nor isolated subunit B inhibits. Furthermore, isolated subunit A or even A1 peptide is active. These were the original observations that led to the theory that the role of subunit B was to deliver the A1 peptide to the interior of the cell and that the lag phase was the time taken for this process. A simple idea of this scheme is shown in Fig. 1.

Subsequent work has shown that the A1 peptide acts catalytically in very much the same way that diphtheria toxin does (Section 2.1.1 above). It catalyzes the ADP-ribosylation of a GTPase component of the adenylate cyclase complex. This has the effect of keeping the cyclase in a permanently activated conformation.

3.1.1d. The Receptor Ganglioside. The receptor for cholera toxin, ganglioside GM1, is probably the best characterized of all receptors. The gangliosides are a family of complex glycolipids that are particularly suitable to being receptors. They have a hydrophobic portion (composed mostly of sugars and amino sugars) in the aqueous environment. Cells bind large numbers of cholera toxin molecules: more than two million are bound to some mucosal cells. Much of this binding must be nonproductive, since calculation shows that very few molecules of subunit A need to enter the cell for activity.

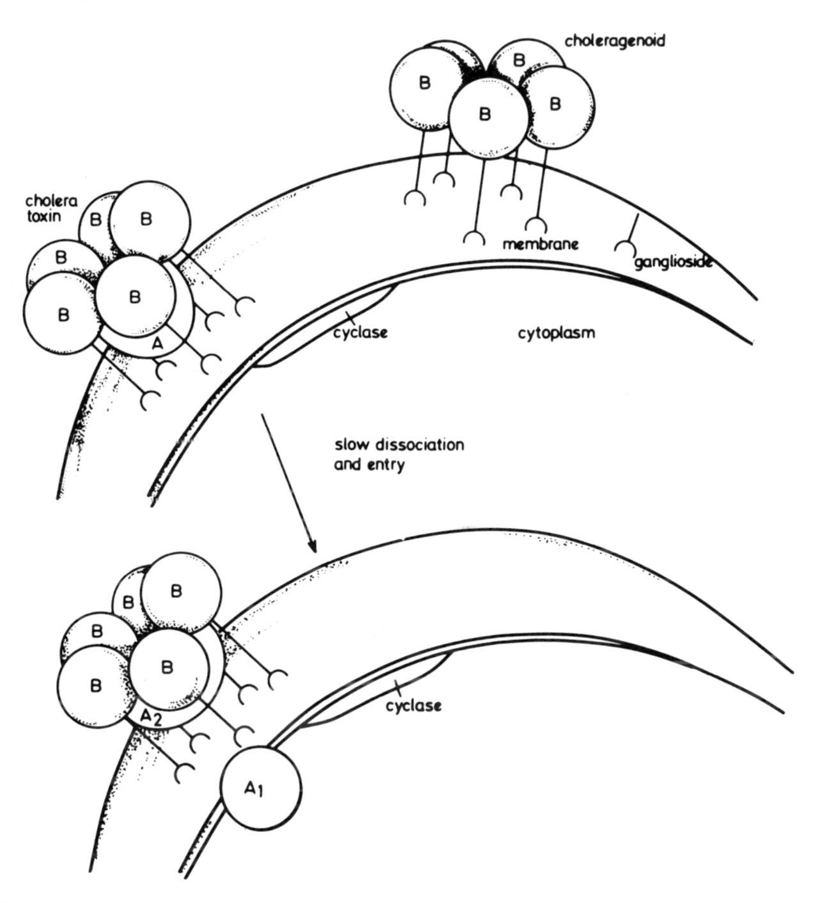

Figure 1. A simplified idea of the role of both subunits of cholera toxin in crossing the cell membrane. (Reproduced, with permission, from van Heyningen, 1984.)

The evidence that GM1 is the receptor is very good and has been reviewed in detail elsewhere (e.g., van Heyningen, 1983a). Gangliosides inhibit the toxin and bind to subunit B. Preincubation of the cells with subunit B protects them from the action of the toxin. Cell lines that lack ganglioside GM1 do not respond to toxin, but if they are given exogenous GM1, they will take it up into their membrane and will then respond to toxin (e.g., Moss *et al.*, 1976). Complexes of toxin with labeled ganglioside have been isolated from cells (Critchley *et al.*, 1979).

The fate of the toxin–ganglioside complex can be followed in the membrane by electron microscopy, which shows that there is considerable lateral movement: the complexes patch and cap in a temperature-dependent way (Revesz and Greaves, 1975; Craig and Cuatrecasas, 1975). Ganglioside GM1 is the only ganglioside that binds the toxin tightly. (There are others that bind much more weakly, and all gangliosides show some, often rather confusing, affinity for all proteins.) The specificity lies in the carbohydrate part of the ganglioside molecule. The binding constant has been measured in several different ways and seems to be of the order of 10^{-10} M.

3.1.1e. Conformational Change on Adding Ganglioside. Experiments using fluorescence (De Wolf *et al.*, 1981), nuclear magnetic resonance (Sillerud *et al.*, 1981), and calorimetry (Dalziel *et al.*, 1984) have all shown a marked conformational change when subunit B binds ganglioside—an almost universal phenomenon when a protein binds a ligand. Less predictably, experiments using fluorescent probes showed that this conformational change was transmitted to subunit A, which also changed conformation when ganglioside bound to subunit B (van Heyningen, 1982c).

3.1.2. Transport of the A1 Peptide across the Membrane

3.1.2a. What Crosses the Membrane? It is clear that A1 peptide must penetrate the membrane, and there is good evidence (e.g., from photoaffinity labeling, Wisnieski and Bramhall, 1981) that it does. However, there is some evidence that both subunits can be found inside the cell (e.g., Tsuru *et al.*, 1984), and although it has been generally assumed that A1 dissociates from the complex before entering the cell, a covalently cross-linked toxin with which this would be impossible was active with pigeon erythrocytes (van Heyningen, 1977). There is much evidence, most of it not very clearly defined, that proteolytic fragments of A1 can be active (e.g., Matuo *et al.*, 1976; van Heyningen and Tait, 1980; Lai *et al.*, 1981); it may be that they penetrate the membrane.

3.1.2b. Is the A1 Peptide Active by Itself? There has been disagreement about the activity of isolated A1 peptide on whole cells: some find activity, although less than with whole toxin (e.g., van Heyningen, 1983b), but others do not. It would seem very difficult for a peptide like A1 that has no particularly hydrophobic areas (Ward *et al.*, 1981) to penetrate the membrane by itself; it is possible that the A2 peptide acts as some kind of leader sequence. On the other hand, the photoaffinity labeling mentioned above (Wisnieski and Bramhall, 1981) does suggest relatively rapid entry of A1 into the membrane, and the lag time with whole toxin is significantly altered by changing membrane fluidity (e.g. Fishman, 1980).

3.1.2c. Function of the Ganglioside. Gangliosides in artificial membranes form clusters that bind toxin tightly and can make conducting channels (see Tosteson *et al.*, 1980). The binding of subunit B, but not subunit A, to ganglioside-containing liposomes frees glucose previously entrapped in them (Moss *et al.*, 1977). Perhaps the gangliosides and the complex formed between them and toxin could by themselves make an entry port for the A1 peptide. However, real membranes are complicated, and there are many other points to consider: patching and capping and a reported requirement for protein synthesis in the membrane transport (Hagmann and Fishman, 1981). Furthermore, a satisfactory theory must explain why each toxin molecule has as many as five binding subunits. There is evidence that the toxin is active only when all five have ganglioside bound (Fishman and Atikkan, 1980). Physicochemical experiments (with model membranes) have provided convincing evidence that a complex is formed with the B subunits binding ganglioside and the A subunit penetrating the bilayer (Dwyer and Bloomfield, 1982). On the other hand, the maximum conformational change in subunit A (Section 3.1.1e above) was achieved with only one ganglioside molecule per whole toxin.

Whatever the function of the ganglioside, it cannot be essential for at least two reasons: the observed intrinsic activity of purified A1 peptide and the activity of artificial hybrid toxins (as mentioned in Section 1.1 above; van Heyningen, 1983b). In experiments, a conjugate of the A1 peptide joined by a disulfide bond to the lectin of *Wisteria floribunda* activated the adenylate cyclase of intact cells that were known to bind the lectin. Figures 2 and 3 show the results of these experiments. Conjugate and toxin had very similar specific activities: the A1 peptide was active, but about two orders of magnitude less so. Furthermore, the characteristic lag phase found with toxin was absent with either peptide or conjugate. In this system, it must have been the lectin–cell interaction that was promoting the entry of the peptide. The absence of the lag phase is evidence that the traditional explanation that it represents the time taken to cross the membrane may be wrong; there is other evidence suggesting that it is the time for the disulfide bond linking the A1 and A2 peptides to be reduced (Kassis *et al.*, 1982).

3.1.2d. Possible Mechanisms. Most of the present evidence suggests that cholera toxin, unlike most of those considered in Section 2 above, enters the cell directly from the plasma membrane. Toxin was taken up endocytotically into the GERL (Golgi/endoplasmic reticulum/lysosome) system of neuroblastoma cells, but it was not active in these experiments, which may have shown nothing but a nonproductive uptake of a mem-

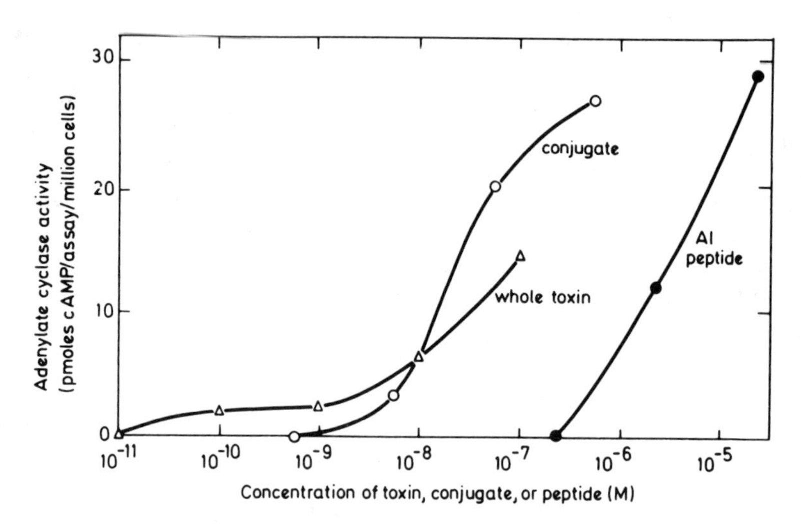

Figure 2. A dose–response curve for the activation of adenylate cyclase in U937 cells by intact cholera toxin, isolated Al peptide, and the artificial conjugate described in Section 3.1.2c. (Reproduced, with permission, from van Heyningen, 1983b.)

brane-bound ligand or the recycling of membrane components (e.g., Joseph *et al.*, 1978). There are reports that some of the reagents that alter the pH of endosomes have an effect on cholera toxin, but these have yet to be followed up (see Houslay and Elliott, 1981).

What does happen at the membrane surface remains mysterious. Original widely published theories that the B subunits formed a ring through which the A1 peptide could pass cannot be true, since the B subunits do not penetrate the membrane. However, there could be some sort of association with a preexisting membrane protein (Gill, 1978).

The function of the toxin–ganglioside binding must be important but not essential, since other ligands can provide the same function. Conversely, since a conjugate of cholera toxin subunit B with diphtheria toxin A chain killed cells (Mannhalter *et al.*, 1980), it appears that the binding to ganglioside can also expedite the transport of other molecules. Whatever the mechanism is, it must be relatively nonspecific.

The simplest theory, first advanced some years ago to explain the observed activity of purified A1 peptide (van Heyningen and King, 1975), is that there is no specific mechanism for entry but that the function of the binding to ganglioside (or other receptor) is to maintain a high concentration of active peptide at the cell surface. This would increase the chance that the peptide could cross the bilayer by some almost random method involving, perhaps, partial dissolution in the membrane (helped

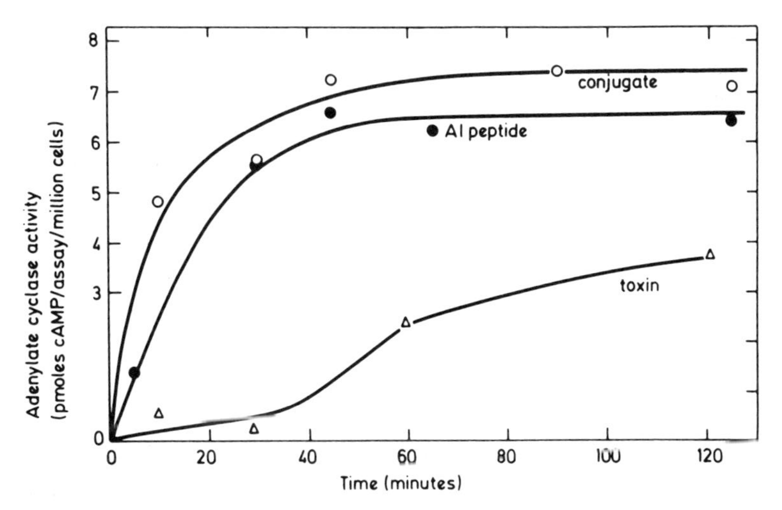

Figure 3. Time course of the activation of adenylate cyclase by intact cholera toxin, isolated Al peptide, and the artificial conjugate in Section 3.1.2c. (Reproduced, with permission, from van Heyningen, 1983b.)

by the hydrophobic nature of the A2 peptide) or even transient openings in the dynamic cell surface.

The binding of five B subunits to five gangliosides may put the A1 peptide in some configuration particularly suitable to this, and the conformational change induced in the A1 peptide and in subunit B by ganglioside binding may enable it to escape from the complex. Once acoss the membrane, the disulfide bond linking A1 and A2 could be easily reduced at the intracellular redox potential, perhaps catalyzed by the thiol:protein disulfide oxidoreductase found associated with the membranes (Moss *et al.*, 1980).

This mechanism is difficult to accept because it is undeniably vague in details. However, this is a good example of the points made in the introduction. Many thousand toxin molecules bind to the cell surface, but only one needs to cross the membrane. Thus, an adequate explanation can invoke what would be a rather rare event, unsuitable for the entry of a molecule vital to the life of the cell (or the whole organism) but enough for an opportunistic toxin. Equally, there is no need to believe that every molecule of A1 has to enter every cell by the same route.

3.2. *Other Toxins That Affect Adenylate Cyclase*

3.2.1. *Pertussis Toxin*

Pertussis toxin (islet-activating protein) is almost certainly part of the mechanism by which *Bordetella pertussis* causes whooping cough (Wardlaw and Parton, 1983). Its active subunit, the S-1 peptide, increases the concentration of cAMP inside the cell in the same way that cholera toxin does but catalyzes the ADP-ribosylation of a different protein in the adenylate cyclase complex.

The binding component is made up of several different polypeptides (see Table I above) and is assumed to be involved in binding and entry chiefly because the S-1 peptide by itself is active only in broken-cell preparations (Tamura *et al.*, 1982). However, very little is known of its action. There is some evidence that the receptor has a sialic acid component, although there is nothing else to suggest a ganglioside. Nothing is known of the mechanism of membrane transport.

3.2.2. Anthrax Toxin

Anthrax toxin is made up of three different proteins: protective antigen (PA), lethal factor (LF), and edema factor (EF). None of these is toxic by itself, but it appears that the role of PA is to combine with an unidentified receptor on the cell surface and bring about the transport of either LF or EF into the cell by an unknown mechanism (Leppla, 1982).

Edema factor increases the intracellular cAMP concentration by a method quite different from that of cholera or pertussis toxin. Although made by a bacterium, it is itself an adenylate cyclase and can be activated by the eucaryotic protein calmodulin.

4. Neurotoxins

4.1. Tetanus Toxin

4.1.1. The Toxin and Its Structure

Following infection of a wound by *Clostridium tetani,* the protein toxin is taken up at neuromuscular junctions and transported by retrograde axonal transport in the α-motoneurons to synapses. There it acts by presynaptic blocking of the release of inhibitory transmitters, especially γ-aminobutyric acid, resulting in the characteristic paralysis of tetanus. Although the toxin has been known and studied for longer than almost any other, comparatively little is known of its action at the molecular level, perhaps because so little is understood of neurochemistry. For general reviews, see van Heyningen (1980), Mellanby and Green (1981).

The toxin is composed of a single polypeptide chain, M_r about 150,000, that can be split by proteolytic enzymes *in vivo* or *in vitro* into an H chain (M_r 100,000) and an L chain (M_r 50,000) that remain linked by disulfide bonds. Neither of these chains has any toxicity by itself.

4.1.2. Binding to Ganglioside

It has been known for more than 20 years that tetanus toxin binds to gangliosides (van Heyningen, 1974). It shows specificity for two gangliosides, GD1b and GT1, both of which have more sialic acid residues than ganglioside GM1 (which binds cholera toxin). There is good evidence that binding is to these gangliosides even at physiological concentrations and *in vivo* (Rodgers and Snyder, 1981). Toxin binding to cultured cells is also affected by ganglioside content (see Yavin and Habig, 1984). Indeed, immunofluorescence experiments with the toxin have been widely used as a marker for neuronal cells.

Purified H chain as well as whole toxin, but not L chain, will bind ganglioside, which shows that H chain can be regarded as a binding component. A proteolytic fragment retaining this binding ability has also been prepared.

Although the binding of toxin to ganglioside is universally accepted, there is less agreement about its significance. It seems hard to accept that it has no biological function, and there is a good correlation between the binding power of toxin derivatives and their toxicity. On the other hand, the effect of toxin on some cells is not abolished by treating them with neuraminidase, which degrades gangliosides GD1b and GT1 so that they will no longer bind toxin (e.g., Habermann, 1981). It is possible that the main role of ganglioside binding is as a mediator of the retrograde axonal transport of the toxin to its site of action rather than in that action itself.

4.1.3. Transport across the Membrane

The observation that only one chain binds the receptor has made it quite irresistible to assume that the L chain will turn out to have some sort of intracellular activity. Unfortunately, there is little real evidence for this idea, and none for any biological activity of isolated L chain.

There is good evidence that the toxin is taken up into cells (Yavin *et al.*, 1982). A pure culture of rat cerebral neurons efficiently internalized [^{125}I]-labeled toxin, mostly in a form in which it retained activity but was contained in some intracellular compartment from which it could not be released by treatments with detergent, neuraminidase, or trypsin. Internalization was inhibited by ganglioside or antitoxin. Strictly speaking, the relevance of these experiments to the action of the toxin has not been proved, but such a transport process does seem likely to be involved.

The actual transport across the membrane may work in a similar way to that of diphtheria toxin. Low pH induces the exposure of a hydrophobic domain in the toxin (Boquet *et al.,* 1984), which can interact with model membranes to form a channel through which the L chain might pass (Boquet and Duflot, 1982). Because so little is known about the toxin, most of these studies inevitably remain rather theoretical.

4.1.4. *Relationship between Receptors for Tetanus Toxin and for Thyrotropin*

The receptor for the hormone thyrotropin in bovine thyroid membranes is related to a ganglioside (Mullin *et al.,* 1976) and shows particularly good correlation with the tetanus toxin receptor. The hormone receptor is not a simple ganglioside but rather a glycoprotein with some similar components. Such molecules may also be able to act as toxin receptors themselves. This could be another example in which a toxin uses a receptor from some other function.

4.2. *Botulinus Toxin*

Botulinus toxin is really a complicated family of neurotoxins with some strong similarities in structure and action to tetanus toxin (Sugiyama, 1980). It is becoming clear that they have a similar two-component structure and that they share similar membrane determinants (Simpson, 1984a); a binding fragment of tetanus toxin inhibits the neuromuscular blocking properties of both intact tetanus toxin and botulinus toxin (Simpson, 1984b). Nothing else is known of their membrane transport.

5. *Conclusion*

This chapter has shown that in spite of the great variety of ways in which protein toxins exert their effects on organisms, there are some surprising generalizations that can be made (van Heyningen, 1982a). All have two components: binding and active. All have parts that can be transported into cells. Many other biologically active compounds have similar properties: for example, the glycoprotein hormones have a common α subunit (which activates adenylate cyclase) and a β subunit that gives the hormones specificity by binding to different types of cells. The main aim of many of those working in this field is to try to make useful generalizations based on what we have observed with toxins. However, we must be careful not to extrapolate too readily from these unusual proteins to molecules important in the normal physiology of the cell.

ACKNOWLEDGMENT. I am grateful to the Medical Research Council of Great Britain for grants towards my own research.

References

Boquet, P., and Duflot, E., 1982, Tetanus toxin fragment forms channels in lipid vesicles at low pH, *Proc. Natl. Acad. Sci. U.S.A.* **79:**7614–7618.

Boquet, P., and Pappenheimer, A. M., 1976, Interaction of diphtheria toxin with mammalian cell membranes, *J. Biol. Chem.* **251:**5770–5778.

Boquet, P., Duflot, E., and Hauttecoeur, B., 1984, Low pH induces a hydrophobic domain in the tetanus toxin molecule, *Eur. J. Biochem.* **144:**339–344.

Collier, R. J., 1975, Diphtheria toxin: Mode of action and structure, *Bacteriol. Rev.* **39:**54–85.

Craig, S. W., and Cuatrecasas, P., 1975, Mobility of cholera toxin receptors on rat lymphocyte membranes, *Proc. Natl. Acad. Sci. U.S.A.* **72:**3844–3848.

Critchley, D. R., Ansell, S., Perkins, R., Dilks, S., and Ingram, J., 1979, Isolation of cholera toxin receptors from a mouse fibroblast and lymphoid cell line by immune precipitation, *J. Supramol. Struct.* **12:**273–291.

Dalziel, A. W., Lipka, G., Chowdry, B. Z., Sturtevant, J. M., and Schafer, D. E., 1984, Effects of ganglioside GM1 on the thermotropic behaviour of cholera toxin B subunit, *Mol. Cell. Biochem.* **63:**83–91.

De Wolf, M. J. S., Fridkin, M., and Kohn, L. D., 1981, Tryptophan residues of cholera toxin and its A and B promoters, *J. Biol. Chem.* **256:**5489–5496.

Donovan, J. J., Simon, M. I., Draper, R. K., and Montal, M., 1981, Diphtheria toxin forms transmembrane channels in planar lipid bilayers, *Proc. Natl. Acad. Sci. U.S.A.* **78:**172–176.

Draper, R. K., and Simon, M. I., 1980, The entry of diphtheria toxin into the mammalian cell cytoplasm: Evidence for lysosomal involvement, *J. Cell Biol.* **87:**849–854.

Draper, R. K., O'Keefe, D. O., Stookey, M., and Graves, J., 1984, Identification of a cold-sensitive step in the mechanism of modeccin action, *J. Biol. Chem.* **259:**4083–4088.

Dwyer, J. D., and Bloomfield, V. A., 1982, Subunit arrangement of cholera toxin in solution and bound to receptor-containing model membranes, *Biochemistry* **21:**3227–3231.

Eidels, L., Proia, R. L., and Hart, D. A., 1983, Membrane receptors for bacterial toxins, *Microbiol. Rev.* **47:**596–620.

Falmagne, P., Capiau, C., Lambotte, P., Zanen, J., Cabiaux, V., and Ruysschaert, J.-M., 1985, The complete amino acid sequence of diphtheria toxin fragment B. Correlation with its lipid-binding properties, *Biochim. Biophys. Acta* **827:**45–50.

Filipovich, A. H., Vallera, D. A., Youle, R. J., Quinones, R. R., Neville, D. M., Jr., and Kersey, J. H., 1984, *Ex-vivo* treatment of donor bone marrow with anti-T-cell immunotoxins for prevention of graft-versus-host disease, *Lancet* **1:**469–472.

Fishman, P. H., 1980, Mechanism of action of cholera toxin: Studies on the lag period, *J. Memb. Biol.* **54:**61–72.

Fishman, P. H., and Atikkan, E. E., 1979, Induction of cholera toxin receptors in cultured cells by butyric acid, *J. Biol. Chem.* **254:**4342–4344.

FitzGerald, D., Morris, R. E., and Saelinger, C. B., 1989, Receptor-mediated internationalization of *Pseudomonas* toxin by mouse fibroblasts, *Cell* **21:**867–873.

Gill, D. M., 1978, Seven toxic peptides that cross cell membranes, in: *Bacterial Toxins and Cell Membranes* (J. Jeljaszewicz and T. Wadstrom, eds.), Academic Press, New York, pp. 291–332.

Habermann, E., 1981, Tetanus toxin and botulinum A neurotoxin inhibit and at higher concentrations enhance noradrenaline outflow from particulate brain cortex in batch, *Naunyn Schmiedebergs Arch. Pharmacol.* **318:**105–111.

Hagmann, J., and Fishman, P. H., 1981, Inhibitors of protein synthesis block action of cholera toxin, *Biochem. Biophys. Res. Commun.* **98:**677–684.

Herschman, H. R., 1984, The role of binding ligand in toxic hybrid proteins: A comparison of EGF-ricin, EGF-ricin A-chain, and ricin, *Biochem. Biophys. Res. Commun.* **124:**551–557.

Houslay, M. D., and Elliott, K. R. F., 1979, Cholera toxin mediated activation of adenylate cyclase in intact rat hepatocytes, *FEBS Lett.* **104:**359–363.

Houston, L. L., 1982, Transport of ricin A chain after prior treatment of mouse leukemia cells with ricin B chain, *J. Biol. Chem.* **257:**1532–1539.

Hu, V. W., and Holmes, R. K., 1984, Evidence for direct insertion of fragments A and B of diphtheria toxin into model membranes, *J. Biol. Chem.* **259:**12226–12233.

Ishida, B., Cawley, D. B., Reue, K., and Wisnieski, B. J., 1983, Lipid–protein interactions during ricin toxin insertion into membranes, *J. Biol. Chem.* **258:**5933–5937.

Joseph, K. C., Kim, S. U., Steiber, A., and Gonatas, N. K., 1978, Endocytosis of cholera toxin into neuronal GERL, *Proc. Natl. Acad. Sci. U.S.A.* **75:**2815–2819.

Kagan, B. L., Finkelstein, A., and Colombini, M., 1981, Diphtheria toxin fragment forms large pores in phospholipid bilayer membranes, *Proc. Natl. Acad. Sci. U.S.A.* **78:**4950–4954.

Kaneda, Y., Uchida, T., Mekada, E., Nakanishi, M., and Okada, Y., 1984, Entry of diphtheria toxin into cells: Possible existence of cellular factor(s) for entry of diphtheria toxin into cells was studied in somatic cell hybrids and hybrid toxins, *J. Cell. Biol.* **98:**466–472.

Kassis, S., Hagmann, J., Fishman, P. E., Chang, P. P., and Moss, J., 1982, Mechanism of action of cholera toxin on intact cells: Generation of A1 peptide and activation of adenylate cyclase, *J. Biol. Chem.* **257:**12148–12152.

Kayser, G., Lambotte, P., Falmagne, P., Capiau, C., Zanen, J., and Ruysschaert, J.-M., 1981, A CNBr peptide located in the middle region of diphtheria toxin fragment B induces conductance change in lipid bilayers, *Biochem. Biophys. Res. Commun.* **99:**358–363.

Kim, K., and Groman, N. B., 1956, Mode of inhibition of diphtheria toxin by ammonium chloride, *J. Bacteriol.* **90:**1557–1562.

Lai, C.-Y., 1980, The chemistry and biology of cholera toxin, *CRC Crit. Rev. Biochem.* **9:**171–206.

Lai, C.-Y., Cancedda, F., and Duffy, L. K., 1981, ADP-ribosyl transferase activity of cholera toxin polypeptide A1 and the effect of limited trypsinolysis, *Biochem. Biophys. Res. Commun.* **102:**1021–1027.

Leppla, S. H., 1982, Anthrax toxin edema factor: A bacterial adenylate cyclase that increases cyclic AMP concentrations in eukaryotic cells, *Proc. Natl. Acad. Sci. U.S.A.* **79:**3162–3166.

Mannhalter, J. W., Gilliland, D. G., and Collier, R. J., 1980, A hybrid toxin containing fragment A from diphtheria toxin linked to the B protomer of cholera toxin, *Biochim. Biophys. Acta* **626:**443–450.

Marnell, M. H., Shia, S.-P., Stookey, M., and Draper, R. D., 1984, Evidence for penetration of diphtheria toxin to the cytosol through a prelysosomal membrane, *Infect. Immun.* **44:**145–150.

Matuo, Y., Wheeler, M. A., and Bitensky, M. W., 1976, Small fragments from the A subunit of cholera toxin capable of activating adenylate cyclase, *Proc. Natl. Acad. Sci. U.S.A.* **73:**2654–2658.

Mellanby, J., and Green, J., 1981, How does tetanus toxin act? *Neuroscience* **6:**281–300.

Merion, M., Schlesinger, P., Brooks, R. M., Moehring, J. M., Moehring, T. J., and Sly, W. S., 1983, Defective acidification of endosomes in Chinese hamster ovary cell mutants "cross-resistant" to toxins and viruses, *Proc. Natl. Acad. Sci. U.S.A.* **80:**5315–5319.

Middlebrook, J. L., and Dorland, R. B., 1984, Bacterial toxins: Cellular mechanisms of action, *Microbiol. Rev.* **48:**199–221.

Moss, J., Fishman, P. H., Manganiello, V. A., Vaughan, M., and Brady, R. O., 1976, Functional incorporation of ganglioside into intact cells: Induction of choleragen responsiveness, *Proc. Natl. Acad. Sci. U.S.A.* **73:**1034–1037.

Moss, J., Richards, R. L., Alving, C. R., and Fishman, P. H., 1977, Effect of the A and B protomers of choleragen on release of trapped glucose from liposomes containing or lacking ganglioside GM1, *J. Biol. Chem.* **252:**797–798.

Moss, J., Stanley, S. J., Morin, J. E., and Dixon, J. E., 1980, Activation of choleragen by thiol:protein disulfide oxidoreductase, *J. Biol. Chem.* **255:**11085–11087.

Mullin, B. R., Fishman, P. H., Lee, G., Aloj, S. M., Ledley, F. D., Winand, R. J., Kohn, L. D., and Brady, R. O., 1976, Thyrotropin-ganglioside interactions and their relationship to the structure and function of thyrotropin receptors, *Proc. Natl. Acad. Sci. U.S.A.* **73:**842–846.

O'Brien, A. D., LaVeck, G. D., Griffin, D. E., and Thompson, M. R., 1980, Characterization of *Shigella dysenteriae* 1 (Shiga) toxin purified by anti-Shiga toxin affinity chromatography, *Infect. Immun.* **30:**170–179.

Olsnes, S., and Pihl, A., 1982, Toxic lectins and related proteins, in *The Molecular Action of Toxins and Viruses* (P. Cohen and S. van Heyningen, eds.), Elsevier, Amsterdam, pp. 51–105.

Olsnes, S., Reisbig, R., and Eiklid, K., 1981, Subunit structure of *Shigella* cytotoxin, *J.Biol. Chem.* **256:**8732–8738.

Olsnes, S., Sandvig, K., Madshus, I. H., and Sundan, A., 1985, Entry mechanisms of protein toxins and picrornaviruses, *Biochem. Soc. Symp.* **50:**171–191.

Proia, R. L., Eidels, L., and Hart, D. A., 1981, Diphtheria toxin: Receptor interaction, *J. Biol. Chem.* **256:**4991–4997.

Revesz, T., and Greaves, M., 1975, Ligand-induced redistribution of lymphocyte membrane ganglioside GM1, *Nature* **257:**103–106.

Rodgers, T. B., and Snyder, S. H., 1981, High affinity binding of tetanus toxin to mammalian brain membranes, *J. Biol. Chem.* **256:**2402–2407.

Sandvig, K., and Olsnes, S., 1981, Rapid entry of nicked diphtheria toxin into cells at low pH. Characterization of the entry process and effect of low pH on the toxin molecule, *J. Biol. Chem.* **256:**9068–9076.

Sandvig, K., Olsnes, S., and Pihl, A., 1979, Inhibitory effect of ammonium chloride and chloroquine on the entry of the toxic lectin modeccin into HeLa cells, *Biochem. Biophys. Res. Commun.* **90:**648–655.

Sandvig, K., Sundan, A., and Olsnes, S., 1984, Evidence that diphtheria toxin and modeccin enter the cytosol from different vesicular compartments, *J. Cell Biol.* **98:**963–970.

Sillerud, L. O., Prestegard, J. H., Yu, R. K., Konigsberg, W. H., and Schafer, D. E., 1981, Observation by ^{13}C NMR of interactions between cholera toxin and the oligosaccharide of ganglioside GM1, *J. Biol. Chem.* **256:**1094–1097.

Simpson, L. L., 1984a, Botulinum toxin and tetanus toxin recognize similar membrane determinants, *Brain Res.* **305:**177–180.

Simpson, L. L., 1984b, The binding fragment from tetanus toxin antagonizes the neuromuscular blocking actions of botulinum toxin, *J. Pharmacol. Exp. Ther.* **229:**182–187.

Stirpe, F., Sandvig, K., Olsnes, S., and Pihl, A., 1982, Action of viscumin, a toxic lectin from mistletoe, on cells in culture, *J. Biol. Chem.* **257:**13271–13277.

Sugiyama, H., 1980, *Clostridium botulinum* neurotoxin, *Microbiol. Rev.* **44:**419–448.

Tamura, M., Nogimori, K., Murai, S., Yajima, M., Ito, K., Katada, T., Ui, M., and Ishii, S., 1982, Subunit structure of islet-activating protein, pertussis toxin, in conformity with the A–B model, *Biochemistry* **21:**5516–5522.

Thorpe, P. E., and Ross, W. C. J., 1982, The preparation and cytotoxic properties of antibody–toxin conjugates, *Immunol. Rev.* **62:**119–158.

Tosteson, M. T., Tosteson, D. C., and Rubnitz, J., 1980, Cholera toxin interactions with lipid bilayers, *Acta Physiol. Scand.* [*Suppl.*] **481:**21–25.

Tsuru, S., Matsuguchi, M., Watanabe, M., Taniguchi, M., and Zinnaka, Y., 1984, Entrance of cholera enterotoxin subunits into thymus cells, *J. Histochem. Cytochem.* **32:**1257–1279.

Uchida, T., 1982, Diphtheria toxin, in: *Molecular Action of Toxins and Viruses* (P. Cohen and S. van Heyningen, eds.), Elsevier, Amsterdam, pp. 1–31.

van Heyningen, S., 1977, Activity of covalently cross-linked cholera toxin with the adenylate cyclase of intact and lysed pigeon erythrocytes, *Biochem. J.* **168:**457–463.

van Heyningen, S., 1980, Tetanus toxin, *Pharmacol. Ther.* **11:**141–157.

van Heyningen, S., 1982a, Similarities in the action of different toxins, in: *Molecular Action of Toxins and Viruses* (P. Cohen and S. van Heyningen, eds.), Elsevier, Amsterdam, pp. 169–190.

van Heyningen, S., 1982b, Cholera toxin, *Biosci. Rep.* **2:**135–146.

van Heyningen, S., 1982c, Conformational changes in subunit A of cholera toxin following the binding of ganglioside to subunit B, *Eur. J. Biochem.* **122:**333–337.

van Heyningen, S., 1938a, The interaction of cholera toxin with gangliosides and the cell membrane, *Current Top. Membr. Transport.* **18:**445–470.

van Heyningen, S., 1983b, A conjugate of the A1 peptide of cholera toxin and the lectin of *Wisteria floribunda* that activates the adenylate cyclase of intact cells, *FEBS Lett.* **164:**132–134.

van Heyningen, S., 1984, Cholera and related toxins, in: *Molecular Medicine,* Vol. I (A. D. B. Malcolm, ed.), IRL Press, Oxford, pp. 1–15.

van Heyningen, S., and King, C. A., 1975, Subunit A from cholera toxin is an activator of adenylate cyclase in pigeon erythrocytes, *Biochem. J.* **146:**269–271.

van Heyningen, S., and Tait, R. M., 1980, Cholera toxin: Structure and function, in: *Hormones and Cell Regulation,* Vol. 4 (J. Dumont and J. Nunez, eds.), Elsevier, Amsterdam, pp. 293–309.

van Heyningen, W. E., 1974, Gangliosides as membrane receptors for tetanus toxin, cholera toxin, and serotonin, *Nature* **249:**415–417.

Vasil, M. L., Kabat, D., and Iglewski, B. H., 1977, Structure–activity relationships of an exotoxin of *Pseudomonas aeruginosa, Infect. Immun.* **16:**353–361.

Ward, W. H. J., Britton, P., and van Heyningen, S., 1981, The hydrophobicities of cholera toxin, tetanus toxin and their components, *Biochem. J.* **199:**457–460.

Wardlaw, A. C., and Parton, R., 1983, *Bordetella pertussis* toxins, *Pharmacol. Ther.* **19:**1–53.

Willingham, M. C., and Pastan, I., 1984, Endocytosis and exocytosis: Current concepts of vesicle traffic in animal cells, *Int. Rev. Cytol.* **92:**51–92.

Wisnieski, B. J., and Bramhall, J. S., 1981, Photolabelling of cholera toxin subunits during membrane penetration, *Nature* **289:**319–321.
Yavin, E., and Habig, W. H., 1984, Binding of tetanus toxin to somatic neural hybrid cells with varying ganglioside composition, *J. Neurochem.* **42:**1313–1320.
Yavin, Z., Yavin, E., and Kohn, L. D., 1982, Sequestration of tetanus toxin in developing neuronal cell cultures, *J. Neurosci. Res.* **7:**267–278.

Index